Soil Biology

Volume 54

Series Editor

Ajit Varma, Amity Institute of Microbial Technology, Amity University, Noida, Uttar Pradesh, India

More information about this series at http://www.springer.com/series/5138

Sushil K. Sharma • Ajit Varma
Editors

Microbial Resource Conservation

Conventional to Modern Approaches

Springer

Editors
Sushil K. Sharma
National Agriculturally Important Microbial Culture Collection (NAIMCC)
ICAR-National Bureau of Agriculturally Important Microorganisms
Maunath Bhanjan, Uttar Pradesh, India

Ajit Varma
Amity Institute of Microbial Technology
Amity University
Noida, Uttar Pradesh, India

ISSN 1613-3382 ISSN 2196-4831 (electronic)
Soil Biology
ISBN 978-3-030-07280-3 ISBN 978-3-319-96971-8 (eBook)
https://doi.org/10.1007/978-3-319-96971-8

This Springer imprint is published by the registered company Springer Nature Switzerland AG
The registered company address is: Gewerbestrasse 11, 6330 Cham, Switzerland

Preface

Microorganisms are the "*silent and unseen majority of life*" on the earth with high magnitude of genetic and metabolic diversity. It is understood that microorganisms are associated with ecosystem processes such as biogeochemical cycle, carbon cycling, and water cycle and thereby maintain functioning and homeostasis of ecosystems. Although the Convention on Biological Diversity (CBD) has not touched upon microorganisms in its documents, scientists like Louis Pasteur, Robert Koch, Martinus Beijerinck, Baas Becking, Joseph Lister, and other contemporaries had long back realized the potential of microorganisms in human life. Today, scientists have isolated an array of microorganisms from the extreme and unique environments such as cold desert, hot arid region, hydrothermal vents, and saline area and having specific traits that are being exploited in agriculture, industry, medicine, and biotechnology. In the early nineteenth century, researchers had realized the unprecedented potential of the natural resources occurring in all kinds of habitats for their long-term conservation so that they can be utilized as and when required in future. Many countries namely, United States of America, Germany, Canada, France, China, and United Kingdom, are earning huge amounts of money utilizing these microbial resources and the technologies generated thereof. In the past, India has focused on the conservation of mainly plant and animal genetic resources but less attention was paid to microbial genetic resources, but after the introduction of CBD in 1992, Government of India also realized the potential of microbes in all spheres of life and thereafter started conservation of these tiny creatures for present and future use for human life. Another reason for conservation is the claim of intellectual property rights on natural resources for their commercial exploitation by specific user under the ambit of access and benefit sharing provision of Nagoya Protocol. Microorganisms are the black box of knowledge, process, and bioactive molecules, and India is home of huge microbial diversity and hence warrants immediate conservation of microbial wealth. Currently, some Microbial Resource Centres (MRCs) across the world are conserving microbial genetic resources, but knowledge and awareness about these practices are known little to the world community.

The aim of this book is to spread knowledge of microbial wealth of the world and acquaint people with strategies and methodologies for conservation of microorganisms. This book covers broad areas like (1) importance of microorganisms in agriculture and allied sectors; (2) introduction to conservation of microorganisms; (3) overview on short-, medium-, and long-term preservation of microorganisms with reference to agricultural importance; (4) culture collections and their role; (5) conventional approach of bacterial, fungal, and cyanobacterial preservation; (6) strategies for preservation of lyo- and cryo-recalcitrant microorganisms; (7) preservation of obligate parasites including VA Fungi through conventional and modern tools and techniques; (8) methods for conservation of microbial community; (9) conservation of ecosystems for conserving associated microorganisms tagged with specific traits; (10) diversity and preservation of truffle, mushrooms, endophytes, phages and microbes of dairy origin; (11) microbes of animal origin, their preservation, and biosafety issues and (12) highlights regulatory mechanisms for access and benefit sharing of microbial resources as per BD Act 2002. These areas are some of the highlights of this book which would provide first hand knowledge about why microorganisms are so important in life and their conservation would be of great use for future generations in the area of agriculture, pharmaceutical, food industry, dairy and biotechnological applications. This book is the first of its kind which is completely devoted to the conservation of microorganisms and worthy for researchers, industry people, and finally society.

The book should be essential reading and reference consortium for microbiologists; mycologists; pathologists; bacterial, cyanobacterial, and fungal taxonomists; agriculture, horticulture, and fisheries scientists; pharmocologistics and forensic scientists; clinicians; veterinarians; and repository managers, who may be interested to unravel the mysteries of microorganisms.

Uttar Pradesh, India Sushil K. Sharma
Uttar Pradesh, India Ajit Varma

Contents

1 **Microbial Genetic Resources: Status, Conservation, and Access and Benefit-Sharing Regulations** . . . 1
Sushil K. Sharma, Sanjay K. Singh, A. Ramesh, Pawan K. Sharma, Ajit Varma, Ees Ahmad, Rajesh Khande, Udai B. Singh, and Anil K. Saxena

2 **Concept of Microbial Preservation: Past, Present and Future** . . . 35
Rohit Sharma, Yogesh Nimonkar, Ashutosh Sharma, Rajesh Singh Rathore, and Om Prakash

3 **Establishment and Management of Culture Collections of Microorganisms (mBRC): An Overview** . . . 55
Duong Van Hop

4 **Specialized Microbial Resource Centers: A Driving Force of the Growing Bioeconomy** . . . 111
Irena B. Ivshina and Maria S. Kuyukina

5 **Agricultural Microbial Genetic Resources: Application and Preservation at Microbial Resource Centers** . . . 141
Tayebeh Soltanighias, Rajesh K. Vaid, and Praveen Rahi

6 **Rhizobia: Culture Collections, Identification, and Methods of Preservation** . . . 175
Manvika Sahgal and Vandana Jaggi

7 **Biobank for Conservation of Arbuscular Mycorrhizal (AM) Fungi** . . . 199
Richa Agnihotri, Hemant S. Maheshwari, Anil K. Sharma, and Mahaveer P. Sharma

8 **Conservation of Fungi: A Review on Conventional Approaches** . . . 223
Sanjay K. Singh, P. N. Singh, S. B. Gaikwad, and D. K. Maurya

9 Veterinary Type Cultures and Their Preservation: Status and Challenges 239
Rajesh K. Vaid, Taruna Anand, K. Shanmugasundaram, and Bhupendra N. Tripathi

10 Biodiversity Conservation of Phages and Microbial Populations . . . 261
Galina Novik, Victoria Savich, and Elena Kiseleva

11 Modern Immunochemical Approaches in Microbiology 303
Elena Kiseleva, Konstantin Mikhailopulo, and Galina Novik

12 Conservation and Application of Microalgae for Biofuel Production 335
Samadhan Yuvraj Bagul, Hillol Chakdar, K. Pandiyan, and Krishnashis Das

13 Desert Truffles in Saudi Arabia: Diversity, Ecology, and Conservation 353
Abeer Hashem, Abdulaziz A. Alqarawi, Manzoor Ahmad Shah, Stephan Wirth, Dilfuza Egamberdieva, B. Tabassum, and Elsayed Fathi Abd_Allah

14 Maintenance, Conservation and Regulation of Microbial Resources for Defense Applications 371
Pritam Sangwan, Kirty Sharma, Mary Celin, and Bala Kiran

15 Diversity, Ecology, and Conservation of Fungal and Bacterial Endophytes 393
Pramod Kumar Pandey, Siddhartha Singh, Mayanglambam Chandrakumar Singh, Amit Kumar Singh, Sudheer Kumar Yadav, Ajai Kumar Pandey, and Punabati Heisnam

16 Archaea: Ecology, Application, and Conservation 431
Dipak T. Nagrale and Shailesh P. Gawande

List of Contributors

Elsayed Fathi Abd_Allah Faculty of Food and Agricultural Sciences, Department of Plant Production, King Saud University, Riyadh, Saudi Arabia

Richa Agnihotri ICAR-Indian Institute of Soybean Research, Indore, Madhya Pradesh, India

Ees Ahmad National Agriculturally Important Microbial Culture Collection (NAIMCC), ICAR-National Bureau of Agriculturally Important Microorganisms, Maunath Bhanjan, Uttar Pradesh, India

Abdulaziz A. Alqarawi Faculty of Food and Agricultural Sciences, Department of Plant Production, King Saud University, Riyadh, Saudi Arabia

Taruna Anand National Centre for Veterinary Type Cultures (NCVTC), ICAR-National Research Centre on Equines, Hissar, Haryana, India

Samadhan Yuvraj Bagul ICAR-National Bureau of Agriculturally Important Microorganisms, Maunath Bhanjan, Uttar Pradesh, India

Mary Celin Centre for Fire, Explosive and Environment Safety, Defence Research and Development Organization, New Delhi, India

Hillol Chakdar ICAR-National Bureau of Agriculturally Important Microorganisms, Maunath Bhanjan, Uttar Pradesh, India

Krishnashis Das ICAR-Indian Institute of Pulses Research, Kanpur, Uttar Pradesh, India

Dilfuza Egamberdieva Institute of Landscape Biogeochemistry, Leibniz Centre for Agricultural Landscape Research (ZALF), Müncheberg, Germany

S. B. Gaikwad National Fungal Culture Collection of India (NFCCI), Biodiversity and Palaeobiology Group, MACS' Agharkar Research Institute, Pune, Maharashtra, India

Shailesh P. Gawande Division of Crop Protection, ICAR-Central Institute for Cotton Research, Nagpur, Maharashtra, India

Abeer Hashem Faculty of Science, Botany and Microbiology Department, King Saud University, Riyadh, Saudi Arabia

Mycology and Plant Disease Survey, Plant Pathology Research Institute, Agriculture Research Center, Giza, Egypt

Punabati Heisnam College of Horticulture & Forestry, Central Agricultural University, Pasighat, Arunachal Pradesh, India

Irena B. Ivshina Institute of Ecology and Genetics of Microorganisms, Ural Branch of the Russian Academy of Sciences, Perm State University, Perm, Russia

Vandana Jaggi Department of Microbiology, G. B. Pant University of Agriculture and Technology, Pantnagar, Uttarakhand, India

Rajesh Khande Department of Biotechnology, Barkatullah University, Bhopal, Madhya Pradesh, India

Bala Kiran Discipline of Biosciences and Biomedical Engineering, Indian Institute of Technology Indore, Simrol, Madhya Pradesh, India

Elena Kiseleva Institute of Bioorganic Chemistry, National Academy of Sciences of Belarus, Minsk, Republic of Belarus

Maria S. Kuyukina Institute of Ecology and Genetics of Microorganisms, Ural Branch of the Russian Academy of Sciences, Perm State University, Perm, Russia

Hemant S. Maheshwari ICAR-Indian Institute of Soybean Research, Indore, Madhya Pradesh, India

D. K. Maurya National Fungal Culture Collection of India (NFCCI), Biodiversity and Palaeobiology Group, MACS' Agharkar Research Institute, Pune, Maharashtra, India

Konstantin Mikhailopulo Institute of Bioorganic Chemistry, National Academy of Sciences of Belarus, Minsk, Republic of Belarus

Dipak T. Nagrale Division of Crop Protection, ICAR-Central Institute for Cotton Research, Nagpur, Maharashtra, India

Yogesh Nimonkar National Centre for Microbial Resource (NCMR), National Centre for Cell Science, Pune, Maharashtra, India

Galina Novik Belarusian Collection of Non-pathogenic Microorganisms, Institute of Microbiology, National Academy of Sciences of Belarus, Minsk, Republic of Belarus

Pramod Kumar Pandey Centre for Biotechnology and Bioinformatics, Dibrugarh University, Dibrugarh, Assam, India

College of Horticulture & Forestry, Central Agricultural University, Pasighat, Arunachal Pradesh, India

Ajai Kumar Pandey College of Horticulture & Forestry, Central Agricultural University, Pasighat, Arunachal Pradesh, India

K. Pandiyan ICAR-National Bureau of Agriculturally Important Microorganisms, Maunath Bhanjan, Uttar Pradesh, India

Om Prakash National Centre for Microbial Resource (NCMR), National Centre for Cell Science, Pune, Maharashtra, India

Praveen Rahi National Centre for Microbial Resource (NCMR), National Centre for Cell Science, Pune, Maharastra, India

A. Ramesh ICAR-Indian Institute of Soybean Research, Indore, Madhya Pradesh, India

Rajesh Singh Rathore National Centre for Microbial Resource (NCMR), National Centre for Cell Science, Pune, Maharashtra, India

Manvika Sahgal Department of Microbiology, G. B. Pant University of Agriculture and Technology, Pantnagar, Uttarakhand, India

Pritam Sangwan Centre for Fire, Explosive and Environment Safety, Defence Research and Development Organization, New Delhi, India

Victoria Savich Belarusian Collection of Non-pathogenic Microorganisms, Institute of Microbiology, National Academy of Sciences of Belarus, Minsk, Republic of Belarus

Anil K. Saxena National Agriculturally Important Microbial Culture Collection (NAIMCC), ICAR-National Bureau of Agriculturally Important Microorganisms, Maunath Bhanjan, Uttar Pradesh, India

Manzoor Ahmad Shah Department of Botany, University of Kashmir, Srinagar, Jammu and Kashmir, India

K. Shanmugasundaram National Centre for Veterinary Type Cultures (NCVTC), ICAR-National Research Centre on Equines, Hissar, Haryana, India

Kirty Sharma Centre for Fire, Explosive and Environment Safety, Defence Research and Development Organization, New Delhi, India

Department of Environmental Science and Engineering, Guru Jambheshwar University of Science and Technology, Hisar, Haryana, India

Sushil K. Sharma National Agriculturally Important Microbial Culture Collection (NAIMCC), ICAR-National Bureau of Agriculturally Important Microorganisms, Maunath Bhanjan, Uttar Pradesh, India

Pawan K. Sharma National Agriculturally Important Microbial Culture Collection (NAIMCC), ICAR-National Bureau of Agriculturally Important Microorganisms, Maunath Bhanjan, Uttar Pradesh, India

Rohit Sharma National Centre for Microbial Resource (NCMR), National Centre for Cell Science, Pune, Maharashtra, India

Ashutosh Sharma National Centre for Microbial Resource (NCMR), National Centre for Cell Science, Pune, Maharashtra, India

Anil K. Sharma Department of Biological Sciences, CBSH, GB Pant University of Agriculture and Technology, Pantnagar, Uttrakhand, India

Mahaveer P. Sharma ICAR-Indian Institute of Soybean Research, Indore, Madhya Pradesh, India

Siddhartha Singh College of Horticulture & Forestry, Central Agricultural University, Pasighat, Arunachal Pradesh, India

Mayanglambam Chandrakumar Singh College of Horticulture & Forestry, Central Agricultural University, Pasighat, Arunachal Pradesh, India

Amit Kumar Singh College of Horticulture & Forestry, Central Agricultural University, Pasighat, Arunachal Pradesh, India

Sanjay K. Singh National Fungal Culture Collection of India (NFCCI), Biodiversity and Palaeobiology Group, MACS' Agharkar Research Institute, Pune, Maharashtra, India

Udai B. Singh National Agriculturally Important Microbial Culture Collection (NAIMCC), ICAR-National Bureau of Agriculturally Important Microorganisms, Maunath Bhanjan, Uttar Pradesh, India

P. N. Singh National Fungal Culture Collection of India (NFCCI), Biodiversity and Palaeobiology Group, MACS' Agharkar Research Institute, Pune, Maharashtra, India

Tayebeh Soltanighias National Centre for Microbial Resource (NCMR), National Centre for Cell Science, Pune, Maharastra, India

B. Tabassum Department of Zoology, Toxicology Laboratory, Government Raza PG College, Rampur, Uttar Pradesh, India

Bhupendra N. Tripathi National Centre for Veterinary Type Cultures (NCVTC), ICAR-National Research Centre on Equines, Hissar, Haryana, India

Rajesh K. Vaid National Centre for Veterinary Type Cultures (NCVTC), ICAR-National Research Centre on Equines, Hissar, Haryana, India

Ajit Varma Amity Institute of Microbial Technology, Amity University, Noida, Uttar Pradesh, India

Duong Van Hop Vietnam Type Culture Collection (VTCC), Institute of Microbiology and Biotechnology, Vietnam National University, Hanoi, Vietnam

Stephan Wirth Institute of Landscape Biogeochemistry, Leibniz Centre for Agricultural Landscape Research (ZALF), Müncheberg, Germany

Sudheer Kumar Yadav Department of Botany, Faculty of Science, Banaras Hindu University, Varanasi, Uttar Pradesh, India

Chapter 1
Microbial Genetic Resources: Status, Conservation, and Access and Benefit-Sharing Regulations

Sushil K. Sharma, Sanjay K. Singh, A. Ramesh, Pawan K. Sharma, Ajit Varma, Ees Ahmad, Rajesh Khande, Udai B. Singh, and Anil K. Saxena

Abstract Majority of the biomass and biodiversity of life on the earth are accounted by microbes, and so far about 10% of the earth's microbial diversity has been characterized. They play a significant role in biogeochemical cycles and extend various ecosystem services. Many microorganisms are rich and serve as untapped reservoirs of metabolic products and, hence, they are potentially important for scientific, industrial and economic purposes. The uninterrupted availability of such microbes for modern scientific security and their ultimate utilization for academia and industry are of paramount importance. Despite countless facts about the role of microorganisms in the biosphere, they have largely been ignored by conservation efforts and never considered part of conservation biology and thus leave number of questions unanswered. Notwithstanding, there are many factors including climate change and habitat destruction affect microbial structure and diversity calls for a consistent environmental ethics for parallel support and protection of microbes and their long-term conservation. It is needless to mention that microbial resources play important roles in developing bio-economy. However, long-term success in conservation strategies requires thorough understanding of basic biology of microorganisms and

S. K. Sharma (✉) · P. K. Sharma · E. Ahmad · U. B. Singh · A. K. Saxena
National Agriculturally Important Microbial Culture Collection (NAIMCC), ICAR-National Bureau of Agriculturally Important Microorganisms, Maunath Bhanjan, Uttar Pradesh, India
e-mail: sushil.sharma1@icar.gov.in; sks_micro@rediffmail.com

S. K. Singh
National Fungal Culture Collection of India (NFCCI), Biodiversity and Palaeobiology Group, MACS' Agharkar Research Institute, Pune, Maharashtra, India

A. Ramesh
ICAR-Indian Institute of Soybean Research, Indore, Madhya Pradesh, India

A. Varma
Amity Institute of Microbial Technology, Amity University, Noida, Uttar Pradesh, India

R. Khande
Department of Biotechnology, Barkatullah University, Bhopal, Madhya Pradesh, India

S. K. Sharma, A. Varma (eds.), *Microbial Resource Conservation*, Soil Biology 54,
https://doi.org/10.1007/978-3-319-96971-8_1

their application through state-of-the-art modern tools leading to longer viability and unaltered genome of microbes. In this regard, specialized training and laboratory infrastructures are required to be extended for significant contributions in protection and successful conservation of microbial gene pool in repositories and natural habitats. Therefore, conservation biologists now are bound to realize that the microbial system on which our livelihoods depend is at a risk of extinction, and this requires serious attention to ensure their sustainability in nature for continuous biogeochemical processes, diversity and abundance. In this way, we are becoming more concerned with the broader aspects of microbial conservation. This review attempts to bring together various aspects with regard to the status of conserved microbes, conservation strategies, methodologies and challenges. Further, this chapter will also appraise about the regulatory mechanisms on sharing of microorganisms at global level under the ambit of Nagoya Protocol of Convention of Biological Diversity (CBD) and Biological Diversity (BD) Act 2002 and Rules 2004 of India.

1.1 Introduction

Diversity of microorganisms is dominant and cosmopolitan on this planet. They are the earth's greatest treasures possessing largest genetic and metabolic diversity that are utilized for improving productivity and sustaining life on earth. Some specific microbes are hostile to their characteristic environments, and it has been assumed that existence of all life forms in the biosphere depends on microorganisms. Microbes belong to all important domains of life like bacteria, archaea and eukarya as well as the viruses. Despite huge importance, $<0.1\%$ of microbial species of the total microorganisms have been characterized, preserved and utilized for various purposes (Alain and Querellou 2009). Every microbial species has its own microbiome, and researchers explore microbial universe and their genetic traits for exploitation for biotechnological purposes in various fields of food industry, medical, pharmaceutical and agriculture (Bull 2004; Challis 2008; Cockell and Jones 2009). These microbes significantly contribute to the global economy through multibillion-dollar biotechnology industry. Apart from its industrial applications, conventional use of microorganisms in taxonomic studies, as standard/reference strains for diagnostic purposes, commercial production of metabolic product, or in biological transformation are also required (Martin 1964). Almost every branch of science uses processes and products of microbial origin. However, it is now well understood that ex situ conservation of microorganisms is essential for ensuring that a source of living cells is readily available for scientific scrutiny of pure and applied nature. Moreover, they need protection from climate change, habitat destruction and various other factors since microorganisms isolated from environmental samples cannot always be recovered. However, even if fresh isolates are obtained, they may lack the desired properties originally expressed. As such importance of microorganisms in biogeochemical processes and industries is well understood by researchers and other professional customers from a scientific and industrial perspective.

However, in order to fully utilize the genetic resources of the uncharacterized microbes, we also need to develop suitable strategy to conserve different microorganisms in their habitats. Anthropogenic interventions and changes in ecosystems are the major threats to existence of many microbes that leads to new challenges of retrieving descriptive data on microorganisms for consideration as part of conservation biology (Sharma et al. 2016) and its impact on ecosystem functioning. With the advent of modern molecular approaches, richness of microbial diversity is increasingly evident and is becoming part of conservation of global genetic resources. The conservation of microbial diversity, their cells or replicable parts (e.g. genomes, plasmids, viruses, cDNAs) (Arora et al. 2005) in the environment has been realized by researchers by understanding and application of rRNA gene barcoding (detection of uncultured and unseen microbes) and use of operational taxonomic units (OTUs) as species representative with the help of next-generation sequencing (NGS) (Sharma et al. 2016).

Recent years have witnessed awareness in conservation of microbial gene pool manyfold, and conservation biologist have realized that microbes are at risk of extinction, and this requires serious attention for their long-term protection (Curtis 2006). There is a widespread concern worldwide for conservation of microorganisms. In this regard, bio-resource centres/culture collections play vital roles in maintaining these microbial gene pools making them available to others for scientific scrutiny and use it for supporting research and development programmes. Basically, resource collections/culture collections are of different types. Some of them are small private collection, while others serve as large service collections, with differing policies and holdings. Some are very small collections holding specific organisms used in individual research activity. Public service collections perform several tasks and serve as custodians of ex situ genetic resources. Functioning of these bio-resource centres is redefined and influenced by several regulations and therefore plays key roles in the conservation strategies (Kirsop and Hawksworth 1994).

World Data Centre for Microorganisms (WDCM) records about 758 microbial culture collections distributed across 76 countries. Most of the bio-resource centres follow operational guidelines of World Federation for Culture Collections (WFCC) and best practices of the Organisation for Economic Development and Cooperation (OECD) (DSTI/STP/BIO (2007)9/REV1). Microbial resource centres (MRCs) collect the microbes and/or microbiomes of high standards for long-term storage and supply of the authentic microbial strains as reference material to researches of academia and industry. They serve as an integral part of a broader activity of protecting and maintaining the ecosystem which helps in stabilizing global environment in order to ensure the availability of earth's biological resources to future generations. With these backgrounds, the present chapter highlights status of conservation of microbial resources and their global sharing following the regulations of access and benefit-sharing mechanisms under the ambit of Nagoya Protocol of CBD with special reference to 'Biological Diversity (BD) Act 2002 and Rules 2004' of India.

1.2 Microbial Resources Centres (MRCs)

Genetic resource can be defined as any material of biological origin containing functional unit of heredity, whereas 'microbial genetic resource' can be defined as any microbial strain that is authenticated, taxonomically defined, physiologically characterized, quality controlled and well documented. Alternatively, microbial genetic materials of actual and potential value are defined as microbial genetic resources. Microbial resource centres (MRCs) play a dynamic role not only in collections and conservation in controlled conditions, but also in sustainable utilization of genetic resources for various applications by academic, research and industry (Table 1.1). They are also necessary for the supply of authentic/reference cultures needed for regulatory compliance to health and trade (Sly 2010). MRCs broadly comply with the framework of Convention on Biological Diversity (CBD) that is implemented to support the conservation and utilization of biodiversity and recognizes the principles of fair and equitable benefit sharing (OECD 2001, 2007; Sly 2010). MRCs ensure long-term maintenance of living microbial material and its replicable parts (genomes, plasmids, viruses, cDNAs) (OECD 2001; Overmann 2015) and their sustainable use and management of related information of molecular and physiological data (Arora et al. 2005; Sigler 2004). They preserve and provide authenticated (technically and legally fit-for-use) genetically stable (consistent quality) microbial and cell cultures. Similarly, MRCs provide access to information on cultures and their characteristics for industrial and scientific research and testing.

Table 1.1 Some important international microbial resource centres/microbial culture collections

Acronym	WDCM no.	Collection name	Country
ATCC	WDCM 1	American Type Culture Collection	USA
JCM	WDCM 567	Japan Collection of Microorganisms	Japan
CBS	WDCM 133	Westerdijk Fungal Biodiversity Institute (Formerly, Centraalbureauvoor Schimmelcultures, Filamentous Fungi and Yeast Collection)	Netherlands
CGMCC	WDCM 550	China General Microbiological Culture Collection Center	China
KCTC	WDCM597	KCTC Korean Collection for Type Cultures	Korea (Rep. of)
CBMAI	WDCM823	Brazilian Collection of Microorganisms from the Environment and Industry	Brazil
IEKC, SSI	WDCM158	The International *Escherichia* and *Klebsiella* Centre (WHO)	Denmark
TISTR	WDCM383	TISTR Culture Collection Bangkok MIRCEN	Thailand
DSMZ	WDCM274	DSMZ-Deutsche Sammlung von Mikroorganismen und Zellkulturen GmbH	Germany
LMG or BCCM/ LMG	WDCM296	Belgian Coordinated Collections of Microorganisms/ LMG Bacteria Collection	Belgium

Source: http://www.wfcc.info/ccinfo

Besides, identification and characterization of species new to science is also undertaken by MRCs. Other important cultures resulting from research and technological perspectives are also accessioned and conserved for supporting basic researches and applications in industry. MRCs also maintain extensive information regarding individual strain and in the form of databases and thus provide access to information on culture characteristics, morphological and physiological characteristics and DNA sequences. The MRCs are the centres of taxonomic expertise and play a vital role in capacity building through hands-on training. Based on requirements, MRCs are of different types such as safe depository to researchers and small and large industries and adhere to the Budapest Treaty on the International Recognition of the Deposit of Microorganisms for the 'Purpose of Patent Procedure' (Winter and Adam 2001). Many MRCs act as service collection to suffice the requirements of basic and applied research, while others serve as private and speciality microbial collections holding specific group of microorganisms. In a nutshell, MRCs meet *the high standards of quality and expertise demanded by the international community of scientists and industry for the delivery of biological information and materials* (OECD 2001). Advancement in information technology (Internet, PC clusters, cloud computing) and sequencing technology (next-generation sequencers, third- and fourth-generation sequencers) helped MRCs to wisely enrich the strain information and resources for delivery of quality research materials to future generation.

1.3 Scenario of Microbial Conservation at Global and Indian Context

1.3.1 Global Scenario

Convention on Biological Diversity (CBD) is a legally binding agreement on biodiversity under the United Nations (UN). As per standard text, negotiations started late in 1987, and the final text was endorsed in Rio de Janeiro, Brazil, in the year 1992 and enforced in December 1993. The World Federation for Culture Collections (WFCC) is a multidisciplinary commission of the International Union of Biological Sciences (IUBS) and a federation within the International Union of Microbiological Societies (IUMS). Its aim is to promote and support the establishment of culture collections and related services, to provide liaison and set up an information network between the collection centres and their users to ensure the long-term perpetuation of important collections in the world. The WFCC pioneered the development of an international database on culture resources worldwide. World Data Center for Microorganisms (WDCM) (http://www.wfcc.info/ccinfo/) maintains the database of culture collections. World directory of culture collections (Sixth Version, 2014) has records of nearly 758 culture collections from 76 countries or regions and holds a total of 2,963,856 microbes. Out of these, 1,221,657 are bacteria, 814,082 are fungi, 38,002 are viruses and 32,125 are cell lines (http://www.wfcc.

info/ccinfo/; accessed on 01 August 2018). The records also contain data on the organization, management, services and scientific interests of these collections. About 402 collections produce catalogue of holdings, and 6406 people work in these culture collections. There are 32 collections in India holding 197,723 cultures. Each of these records is linked to a second record containing the list of species held. The WDCM database forms an important information resource of all microbiological activities and also acts as a nodal agency for managing data activities among WFCC members. The Culture Collections Information Worldwide (CCINFO) is a database management system of culture collections in the world. It includes CCINFO and STRAIN. The CCINFO is a world directory of all registered culture collections, and STRAIN includes list of holdings from registered culture collections (http://www.wfcc.info/ccinfo/). Another important initiative took place to create a robust, reliable and user-friendly WFCC Global Catalogue of Microorganisms (GCM) to help culture collections to manage, disseminate and share the information related to their holdings. It also provides a uniform interface for the scientific and industrial communities to access the comprehensive microbial resource information from any corner of the world. Presently, 120 culture collections from 46 countries and regions have joined the campaign (http://gcm.wfcc.info/overview/; accessed on 01 August 2018).

1.3.2 Indian Scenario

Basic research is considered the backbone to applied discipline. It is a fact that 'taxonomy' as a science has seen some dynamic change and has made significant advances in the past and thereby supported the studies on biodiversity. Considering biodiversity-rich country, proper authentication of the strain is a very important requirement for academic interest and also from the point of protecting intellectual property rights (IPR) on the process and products developed through international patenting. Maintenance of these authentic microbial cultures in a pure state in microbial germplasm bank is a strategic requirement for developing innovative technological processes and products. In addition, awareness of the principles of convention on biological diversity, national regulation governing genetic resources and international patent laws is essential to protect our microbial-based intellectual property rights (IPR) in the global context. As per WDCM updates, 32 microbial culture collections exist in India (Table 1.2). Maharashtra state tops the list with a maximum of 09 numbers of culture collections followed by Uttar Pradesh (04), Delhi (03), Chandigarh (02), Gujarat (02), West Bengal (02), Telangana (02), Tamil Nadu (02), Jammu and Kashmir (01), Haryana (01), Rajasthan (01), Goa (01) and Kerala (01) (Fig. 1.1). Developing sustainable utilization strategies of Indian microbial wealth under the framework of CBD can play important roles in generating bio-economy (http://www.wfcc.info/ccinfo/).

Table 1.2 Major microbial collections in India

Acronym	WDCM no.	Collection name
ABRC	WDCM 912	Anaerobic Bacterial Resource Centre, University of Hyderabad, Hyderabad, Telangana
AYL	WDCM 934	Whylabs Resources Centre for Microorganisms, Hyderabad, Telangana
BAB	WDCM 1058	Bank A Bug, Gujarat Biodiversity Gene Bank, Gandhinagar, Gujarat
BDU	WDCM 976	National Facility for Marine Cyanobacteria, Tiruchirappalli, Tamil Nadu
BT	WDCM 1036	*Bacillus thuringiensis*, M.S. Swaminathan Research Foundation Chennai, Tamil Nadu
CCBB	WDCM 1169	Mushrooms Biodiversity – Western Regional Centre TERI, Belapur, Maharashtra
CCDMBI	WDCM 119	Culture Collection, Department of Microbiology, Bose Institute, Kolkata, West Bengal
CIPDE	WDCM 462	Collection of Insect Pathogens, Department of Entomology, Parbhani, Maharashtra
CM	WDCM 1033	*Chroococcus minor*, Shri Jagdish Prasad Jhabarmal Tibrewala, Jhunjhunu, Rajasthan
DBV	WDCM 173	Division of Standardisation, ICAR-Indian Veterinary Research Institute, Izatnagar, Uttar Pradesh
DMSRDE	WDCM 166	Defence Materials and Stores Research and Development Establishment Culture Collection, DRDO, Kanpur, Uttar Pradesh
DUM	WDCM 40	Mycological Herbarium, Department of Botany, Delhi University, Delhi
EntoPatho	WDCM 1013	Entomopathogen, M.S. University of Baroda, Vadodara, Gujarat
GCC	WDCM 1165	Global Collection of Cyanobacteria, Varanasi, Uttar Pradesh
GFCC	WDCM 946	Goa University Fungus Culture Collection and Research Unit, Goa University, Panaji, Goa
ITCC	WDCM 430	Indian Type Culture Collection, ICAR-Indian Agricultural Research Institute, New Delhi
MCC	WDCM 930	National Centre for Microbial Resources (Formerly Microbial Culture Collection), NCCS, Pune, Maharashtra
MCM	WDCM 561	MACS Collection of Microorganisms, Agharkar Research Institute, Pune, Maharashtra
MPKV	WDCM 448	Biological Nitrogen Fixation Project College of Agriculture, Rahuri, Maharashtra
MRCJ	WDCM 1117	Col. Sir R. N. Chopra, Microbial Resource Center, CSIR-IIIM, Jammu and Kashmir
MTCC	WDCM 773	Microbial Type Culture Collection and Gene Bank, CSIR-IMTECH, Chandigarh, Union Territory
NAIMCC	WDCM 1060	National Agriculturally Important Microbial Culture Collection, ICAR-National Bureau of Agriculturally Important Microorganisms, Maunath Bhanjan, Uttar Pradesh
NCCPF	WDCM 1118	National Culture Collection of Pathogenic Fungi, PGIMER, Chandigarh, Union Territory
NCDC	WDCM 775	National Collection of Dairy Cultures, ICAR-National Dairy Research Institute, Karnal, Haryana

(continued)

Table 1.2 (continued)

Acronym	WDCM no.	Collection name
NCIM	WDCM 3	National Collection of Industrial Microorganisms, CSIR-National Chemical Laboratory, Pune, Maharashtra
NFCCI	WDCM 932	National Fungal Culture Collection of India, ARI, Pune, Maharashtra
NIICC	WDCM 961	NII Microbial Culture Collection, National Institute for Interdisciplinary Science and Technology, Trivandrum, Kerala
NMCC	WDCM 972	North Maharashtra Microbial Culture Collection Centre, North Maharashtra University, Jalgaon, Maharashtra
NTCCI	WDCM 107	Culture Collection, Microbiology and Cell Biology Laboratory, Indian Institute of Science, Bangalore, Karnataka
UMFFTD	WDCM 562	Food and Fermentation Technology Division, University of Mumbai, Mumbai, Maharashtra
VBCCA	WDCM 931	Vishva-Bharati Culture Collection of Algae, Visva Bharati Central University, Santiniketan, West Bengal
VPCI	WDCM 497	Fungal Culture Collection, Vallabhbhai Patel Chest Institute, University of Delhi, Delhi

Sources: www.wfcc.info/ccinfo/collection/coll_by_country/i/91/

1.4 Strategies and Methodologies for Microbial Conservation

MRCs follow various strategies for conserving microbial diversity. Generally, conservation strategies include 'in situ', 'ex situ' and 'in factory' forms of conservation. As a principle, 'in situ' ('on site', 'in place') approach links to conservation in their natural habitats and are considered as the most appropriate way of conserving viable populations of microbes in their natural ecosystem and natural habitats (CBD 1992). The storage and culturing of microorganisms in artificial/culture collections (ex situ) have made enormous contributions to the existing microbial conservation efforts. The systematic link between field conservation efforts and the preservation of important microbial species from such environments would increase the value of culture collections. A system of well-organized MRCs play a distinctive role in preservation and maintenance of distinct wild/isolated/cultivated species\strains and their genetic populations on artificial media. Being taxonomically well described, they provide remarkable research openings on the components of microbial genetic diversity. 'In factory' form of conservation is an intermediate form of conservation, mainly used by the agro-industrial sector (Sharma et al. 2016).

1.4.1 In Situ Conservation Methods

In situ conservation refers to conserving microbial species in their natural habitats by long-term preservation of ecosystems, species and populations under conditions of continuing adaptations. It also protects associated animals together with microbes,

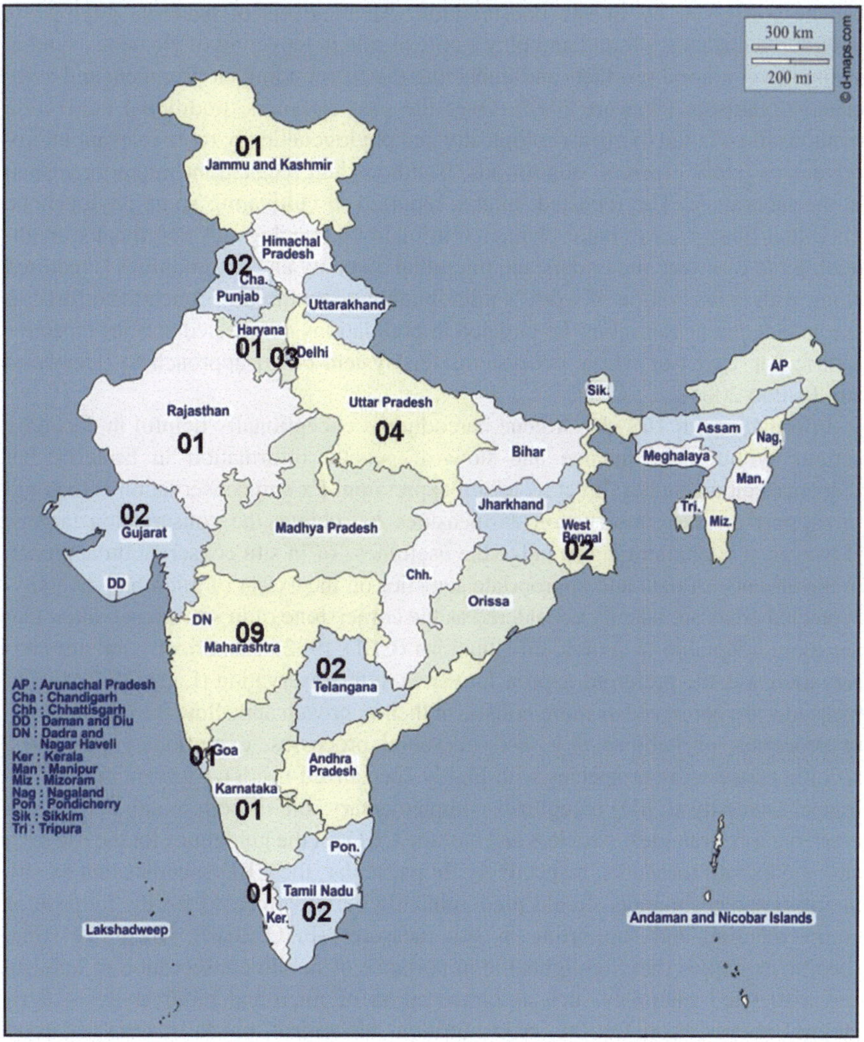

Fig. 1.1 Distribution of microbial culture collections in different states of India

thereby enabling free energy flow. Generally, in situ conservation is considered essential in places where microbiome has not been adequately inventoried. In the new era, creation of wildlife parks, reserved forests, biosphere reserves, eco-parks, etc. where diversity of microbial forms can thrive in various habitats and in natural associations with their plant and animal hosts may lead up to some extent destruction of these natural habitats. There are large numbers of microbial species that are yet to be discovered (Alain and Querellou 2009; Stewart 2012), to determine their complex interactions and critical roles in ecological processes. Bacteria that can be grown in the laboratory are only a small fraction of the total diversity that exists in nature (Stewart 2012). Plethora of microorganisms is known to be non-cultivable and hence can be conserved by in situ conservation. At all levels of bacterial phylogeny, uncultured clades are found and play a critical role in recycling of elements, synthesizing novel natural products and impacting the surrounding environment and associated organisms (Stewart 2012). Over the past 30 years, traditional cultivation methods have failed to grow ecologically and phylogenetically more relevant microorganisms (slow growers, oligotrophs, fastidious and recalcitrant microorganisms) in the laboratory. The repeated failures reported in cultivating some phylogenetic microbial lineages are major challenge in microbial ecology. A coordinated methodology to discover and understand microbial diversity and communities is required along with conservation of world's microbial diversity through concerted efforts. In situ management can either be targeted at populations of selected species (species-centric approach) or whole ecosystems (ecosystem-based approaches) (Heywood and Dulloo 2005).

Information on 16S rRNA gene barcoding is exceptionally helpful in decoding certain evolutionary lineage and flow of genetic information in bacteria, but sequences alone cannot direct its in situ expression. Ex situ conservation is required to support in situ conservation as measures to achieve the conservation targets. However, it is challenging to judge the usefulness of in situ conservation approach in the absence of data and appropriate statistics on the extent of microbial diversity. Protected areas are usually considered as the cornerstone of in situ conservation that are more adaptable to individual situations (CBD 1992). Historically, in situ conservation was the preferred approach over ex situ conservation (Lacy 2010). In situ measures are perceived as more holistic in their approach and allow the conservation of processes or habitats (e.g. soil microbial processes, evolutionary processes, specific ecosystems or species with highly specialized needs). Convention on Biological Diversity (CBD) recognizes complementary role of both in situ and ex situ conservation strategies. Article 8 and 9 of the CBD set the guidelines for the use of in situ and ex situ measures, respectively. In particular, the CBD specifies that ex situ facilities and techniques should predominantly be implemented for the purpose of complementing and supporting in situ measures (FAO 2010). There are some specific examples that highlights the importance of in situ conservation of habitats and associated microbes such as conservation of microbial biodiversity of agricultural interest (soil fertility, crop nutrition, biocontrol, bio-fertilizers) and food security (Tables 1.3, 1.4, 1.5, and 1.6). Extreme ecosystems with wide range of modern and ancient geological environments such as caves, desert soils, fresh and marine water, hot springs, hydrothermal vent, hypersaline areas, cryoconite hole on glacier, unique niches and unusual habitat (digestive tract, rocks, and caves, etc.) do support the occurrence of specific type of microbes that need to be conserved (Sharma et al. 2016) in their habitats. Ecosystem and habitat preservation are in infancy stage require serious attention to start preservation of microorganisms at ecosystem or habitat level to save them from extinction.

Table 1.3 Microbial resources contribute to agroecological and food security

Nitrogen (N_2) fixation: (e.g. *Cyanobacteria, Rhizobium* and *Frankia*)
Plant growth promoting rhizobacteria (PGPR): (e.g. *Bacillus, Pseudomonas, Streptomyces, Azosprillum* sp., and *Bradyrhizobium* sp., as Biofertilizer and production of phytohormones, ACC deaminase, phosphate solubilisation, etc.)
Plant endosymbionts (e.g. *Rhizobium* for biological nitrogen fixation)
Biocontrol agents (e.g. *Bacillus* sp. as pathogens of weeds, fungi, insects through production of antibiotics, siderophores, lytic enzymes)
Inocula to restore soil health and nutrient release (e.g. *Pseudomonas* sp. as phosphate solubilizer)
Source of genes for plant improvement (e.g. insect resistance)
Rhizosphere bacteria for disease suppression
Crop improvement
Sustainable agriculture
Land reclamation
Reference cultures for food safety testing, quarantine and trade
Reference cultures for animal and plant disease testing
Enzymes for food improvement and processing
Cultures for food fermentations and nutritional supplements
Innovative biodiscovery
Locust and other insect pests (*Metarhizium anisopliae, Beauveria bassiana, Isaria fumosorosea,* and *Verticillium lecanii*)
Insect pests belonging to Lepidoptera, Diptera, and Coleoptera (*Bacillus thuringiensis*)

Table 1.4 Microbial resources contribute to industrial and nutritional security

Acetobacter as preservation and fermentation
Arthrobacter as preservation and fermentation
Bacillus as thermo stable enzyme production, fermentation and preservation
Brevibacterium as preservation and fermentation
Clostridium as fermentation and preservation
Enterococcus as fermentation and preservation
Lactobacillus as fermentation and preservation
Lactococcus as fermentation and preservation
Leuconostoc as fermentation and preservation
Listeria as fermentation and preservation
Microbacterium as fermentation and preservation
Micrococci as fermentation and preservation
Pediococcus as fermentation and preservation
Streptococcus as fermentation and preservation

1.4.2 Ex Situ Conservation Methods

There are many approaches for microbial conservation which have been modified from time to time by individual microbial resource centres based on feasibility and availability of resources. The microbial conservation could be achieved either by

Table 1.5 Microbial resources contribute to various pharmaceutical products

Ethanol—*Escherichia coli*
Butanol—*Saccharomyces* sp.
Bioethanol—*Zymononas mobilis*
Battery active material (MnO_2)—*Rhizobium* sp.
Bioinsecticide (Mosquitocidal)—*Bacillus sphaericus*
Phage based biosorbent kit—*Salmonella enteritidis* (with lux genes)
Recombinant polypeptide—*Streptomyces* sp.
D-Mannitol/D-Fructose—*Leuconostoc*
Riboflavin—Engineered *Bacillus subtilis*
Biomining (Uranium, Copper)—*Bacillus* sp., *Pseudomonas* sp.
Deacetoxycephalosporin C—*Acremonium chrysogenum* (genetically engineered)
Evernimicin—*Micromonospora carbonacea*
Penicillin—*Penicillium* spp.
Cephalosporins—*Cephalosporium* spp.
Cyclosporin—*Tolypocladium* spp.
Tetracycline, actinomycine, and adrimycine—*Streptomyces* spp.
Polymyxin B and bacitracin—*Bacillus* spp.

Table 1.6 Microbial resources contribute to various enzymes production for industrial and research application

Cellulase/Hemicellulase—*Volvariella volvacea*
Ligninolytic enzymes—*Coriolus versicolor*
Laundry detergent (alkaline cellulase/protease)—Alkaliphile microorganisms
Cyclomaltodextrin gluanotransferase—Alkaliphile microorganisms (for cyclodextrin production)
Extracellular lipase—*Pseudomonas aeruginosa*
Keratinase—*Bacillus licheniformis*
Alpha-galactosidase—*Aspergillus awamori* and *A. oryzae*
Amylase—*Bacillus circulans*
Extracellular deoxyribonuclease—*Streptomyces thermonitrificans*
Beta-xylosidase (thermostable)—*Thermomonospora curvata*
GTP cyclohydrolase II/3, 4-DH-2-B-4-PS—*Bacillus subtilis*
L-Aspartate beta-decarboxylase—*Alcaligenes faecalis*
Creatininase—*E.coli*
Proteinase—*Tritirachium album*
Serine hydroxymethyltransferase—*E. coli*

detention of the cell vitality to minimum (hypobiosis state) or subjecting the microorganisms to drying, freezing at low temperatures and lyophilization (an anabiotic state) (Uzunova-Doneva and Donev 2005). Several methods are used for ex situ conservation of microorganisms (Gorman and Adley 2004; Berner and Viernstein 2006; Morgan et al. 2006; Smith et al. 2008; Winters and Winn 2010;

Fig. 1.2 Preservation of microorganisms through lyophilization and cryopreservation

Fu and Chen 2011; Garcia 2011; García-García et al. 2014; Mutlu et al. 2015), for example, repeated subcultivation, preservation on agar beads (refrigerator temperatures 5 °C), storage in sterile soil, mineral oils (preventing dehydration and slows down metabolic activity by reducing oxygen availability), silica gel storage, spray-drying, fluidized bed drying, freezing (ultralow temperatures), cryopreservation, lyophilization (freeze-drying), liquid-drying, desiccation, induced anhydrobiosis, vitrification (a non-equilibrium, ultra-rapid cooling technique), sterile distilled water, water-salt solutions, and gelatin discs. Among these methods, cryopreservation and lyophilization (Fig. 1.2) have been considered most valuable, widespread and reliable methods for long-term and stable storage of important microorganisms (Smith and Ryan 2012; Singh and Baghela 2017).

1.4.3 In Factory Conservation

In factory method of microbial conservation means keeping them in normal conditions for practical use. As a principle, two different ways of conservation strategy can be executed: dynamic and static conservation. Dynamic conservation does not impose much restriction on the use of strains, except for the introduction or mixing with cultures of different origin. Static conservation is very restrictive and tries to maintain strains under its original conditions and avoid any kind of changes.

1.5 Preservation of Whole Microbial Community

After successful preservation of pure cultures in culture collections (ex situ) and laboratories, researchers are now trying to preserve whole microbiomes that are involved in certain activities to be useful for biotechnological purposes in agriculture and industry Efforts have been made to preserve selected specific microbial communities. For example, (1) Rothrock et al. (2011) developed a simple preservation protocol for the long-term storage and reactivation of the anammox biomass in which anammox biomass was first frozen in liquid nitrogen (−196 °C) followed by lyophilization in skimmed milk media without glycerol. This is the first report of the successful reactivation of anammox biomass preserved via sub-zero freezing and/or lyophilization. (2) Kerckhof et al. (2014) evaluated the effect of different cryoprotectants on preservation of community structure and functions of several microbiomes. They demonstrated that the use of DMSO, tryptone and soy broth preserved the better functionality than unpreserved cells. (3) Tatangelo et al. (2014) have evaluated the effect of different preservation methods on assessment of bacterial community structure on soil and water samples. (4) Yu et al. (2015) preserved whole microbial communities involved in degradation of switch grass with DMSO and glycerol and showed that samples preserved with DMSO and glycerol experience a consistent shift in community composition though dominant microorganisms were retained in the active community. Despite shifts in the community with storage, the samples were active upon revival under thermophilic and high-solid conditions. (5) Weißbecker et al. (2017) collected soil samples from grasslands of Germany and immediately freeze-dried and stored at 4 °C that allowed preservation of microbial community for 7 days and thus enabled transportation of samples across the continents.

1.6 Preservation of Rust Fungi

Rust fungi are a well-known and economically important group of plant pathogens which are comprised of more than 120 genera and 6000 species. *Puccinia graminis*, the causal agent of stem rust, is well-known for causing devastating epidemics in most of the wheat-grown regions of the world. A new highly virulent strain (Ug99) threatened wheat production worldwide without affecting India as most of the wheat varieties are resistant to Ug99. Similarly, epidemics of poplar leaf rust, *Melampsora* spp., have been a major constraint to the development of bioenergy programmes based on domesticated poplars which resulted in the lack of durable host resistance. Rust fungi are obligate biotrophic parasites with a complex life cycle often including two phylogenetically unrelated hosts. They have evolved specialized structures, haustoria, formed within the host tissue to efficiently acquire nutrients and suppress host defence responses (Duplessis et al. 2011). Preservation of rust fungi was demonstrated by Ryan and Ellison (2003) wherein they reported preservation of *Puccinia spegazzinii*, identified as potential biological control agent to the invasive weed *Mikania micrantha* (Asteraceae). The embedded teliospores and delicate basidiospores of this microcyclic rust are not amenable to direct plunge freezing. In situ cryopreservation technique was the only way out for the long-term cryopreservation of the ten isolates tested. Material from either petiole or stem tissue remained viable after cryopreservation, determined by the ability of the material to produce basidiospores, but infection of the host plant by these basidiospores produced from previously cryopreserved teliospores, embedded in leaf petioles, was not achieved. Garcia et al. (2007) developed a method for long-term preservation of uredospores of *Puccinia melanocephala*, causal organisms of rust in sugarcane. The best results were obtained when uredospores, collected using a vacuum pump from naturally rusted leaves, were dehydrated in silica gel followed by storage at −20 and −80 °C. Testing even after 1 year of storage, uredospores induced rust symptoms on inoculated plants at levels of severity which is adequate to start screening test for sugarcane resistance to rust. Salustiano et al. (2008) evaluated the viability and infectivity of urediniospores of *Puccinia psidii* multiplied on *Syzygium jambos* after preservation in different conditions, like liquid nitrogen (−196 °C), deep freezer (−80 °C), refrigerator (5 °C), BOD incubator (25 °C) and as herbarium specimen (25 °C) for 150 days. The viability and infectivity of urediniospores stored in BOD and as herbarium were not successful, while the urediniospores stored in deep freezer, liquid nitrogen and refrigerator maintained their viability and infectivity for 150 days. Tibolla et al. (2012) reported preservation methods of urediniospores of *Puccinia kuehnii*, in which leaves with symptoms of orange rust were harvested from the sugarcane cultivar SP89 1115 and urediniospores were extracted from the leaves with the aid of a vacuum pump. Dehydration method in silica gel followed by storage at −80 °C was found to allow viability of urediniospores (1.2%) after 180 days.

Table 1.7 Some fungi and bacteria possessing new processes discovered in recent past

Bacteria	Processes	References
Candidatus *Brocadia caroliniensis*	First anammox bacteria identified	Kuenen and Jetten (2001)
Fusarium oxysporum MT-811	Dissimilatory reduction of nitrate to ammonium noticed in this fungus under anaerobic conditions	Zhou et al. (2002)
Saccharomyces cerevisiae CEN.PK 102-5B	Demonstrates an alternative metabolic pathway for butanol and isobutanol production in the yeast *S. cerevisiae*, using glycine as a substrate	Branduardi et al. (2013)
Methylomicrobium alcaliphilum 20Z	Demonstrates that methane assimilation is coupled with a highly efficient pyrophosphate-mediated glycolytic pathway, which is under oxygen limitation participates in a novel form of fermentation-based methanotrophy	Kalyuzhnaya et al. (2013)
Pseudomonas deceptionensis M1T, *Rhizobium leguminosarum* J391 and *Bradyrhizobium diazoefficiens* USDA 110	Microbial catabolism of the marine osmolyte dimethylsulphoniopropionate (DMSP) is thought to be the major biological process generating dimethyl sulphide (DMS). Reporting the discovery and characterization of the first gene, *mddA*, for DMSP-independent DMS production by bacteria. This gene, *mddA*, encodes amethyltransferase that methylatesmethanethiol and generates DMS	Carrion et al. (2014)
Actobacterium woodii	Novel mode of lactate metabolisms involving bifurcating lactate dehydrogenase (LDH) solve low energy substrate lactate in anaerobes without cytochrome, quinines or other membrane-soluble electron carriers and give rationale for the presence of the Rnf complex in the anaerobes	Weghoff et al. (2014)
Candidatus Nitrospira inopinata	Discovery of Comammox: One step nitrification by this single bacterium	Daims et al. (2015) van Kessel et al. (2015) Pinto et al. (2015) Camejo et al. (2017)
Nitrospira moscoviensis	*N. moscoviensis* possesses genes encoding for a urease and cleaves urea to ammonia and CO_2. Ureolysis was not reported earlier in nitrite oxidising bacteria (NOB) but later on discovered ureolytic process in *N. moscoviensis* to supply ammonium to ammonium oxidizing bacteria (AOB) lacking urease	Koch et al. (2015)

(continued)

Table 1.7 (continued)

Bacteria	Processes	References
Clostridium thermocellum DSM1313-derived strains	Demonstrates that *C. thermocellum*, a cellulose degrading bacterium, can fix CO_2 while growing predominantly on cellobiose	Xiong et al. (2016)
Bacteroides thetaiotaomicron	A gut bacterium degrades the most structurally complex glycan, known containing 13 different sugars and 21 distinct glycosidic linkages, utilizing previously undiscovered enzyme families and novel catalytic activities	Ndeh et al. (2017)
Candidatus *Thiolava veneris*	A novel bacterium, discovered from volcanic eruption sites of the Tagoro, Canary Archiplego, Spain, utilizes both organic and inorganic carbon released from volcanic degassing	Danovaro et al. (2017)

1.7 Conservation of Microbes Possessing Novel Processes/Traits

Conservation initiatives have evolved since the late twentieth century from an initial focus on protection of pristine areas and particular (charismatic) species of animals and plants to a more holistic ecosystem-based approach (Salafsky et al. 2002). This planet is home of innumerable microbes with huge metabolic diversity, which operate a number of known and unknown processes and thereby sustaining life on the earth. Present decade has witnessed a number of novel processes discovered in different niches governed by the existing microorganisms or in the newly identified microbes (Table 1.7). Such processes were not known earlier, and were an enigma for the researchers on how the ecosystems are sustained in different niches where human mind cannot predict such processes based on existing knowledge. The microbial resource centres are preserving microbes ex situ following various methods on long-term basis. However, still many microbes require specific methods for long-term preservation and to maintain their viability. Despite the availability of established protocols, only some specific culturable microbes are preserved in some of the resource centres of the world. It is because of the large gap in knowledge of microbial world and methodologies to preserve them. Discovery of new processes/traits in microbes and their ex situ and in situ conservation is of a great challenge to researchers. Hence, comprehensive conservation strategies need focus on conservation of microorganisms in their habitats and ex situ too. In this context, once Tom Curtis (2006) said that '*if the last blue whale choked to death on the last panda, it would be disastrous but not the end of the world. But if we accidentally poisoned the last two species of ammonia oxidizers, that would be another matter. It could be happening now and we wouldn't even know*' emphasizing conservation of niches associated microbes particularly nitrifying bacteria for existence of this world (Fig. 1.3). In recent past, discoveries of many new processes, for example, anammox

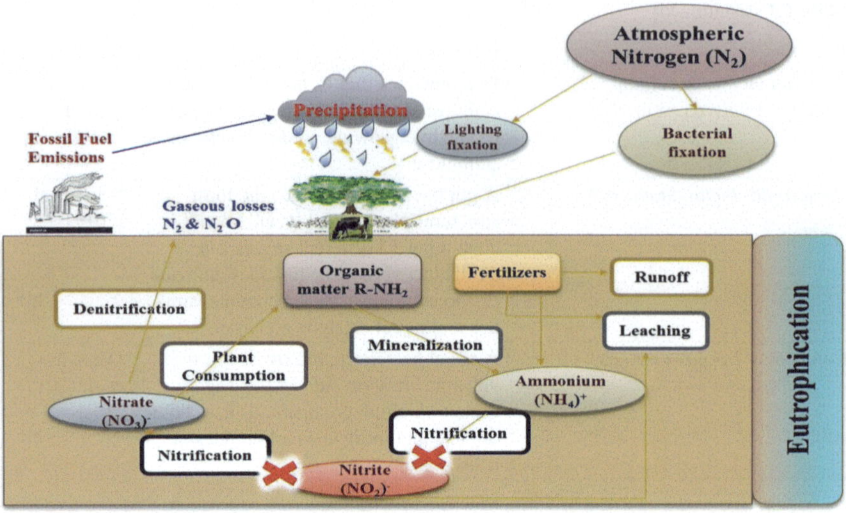

Fig. 1.3 Stopping nitrification jeopardise life on the earth (Source: Curtis 2006)

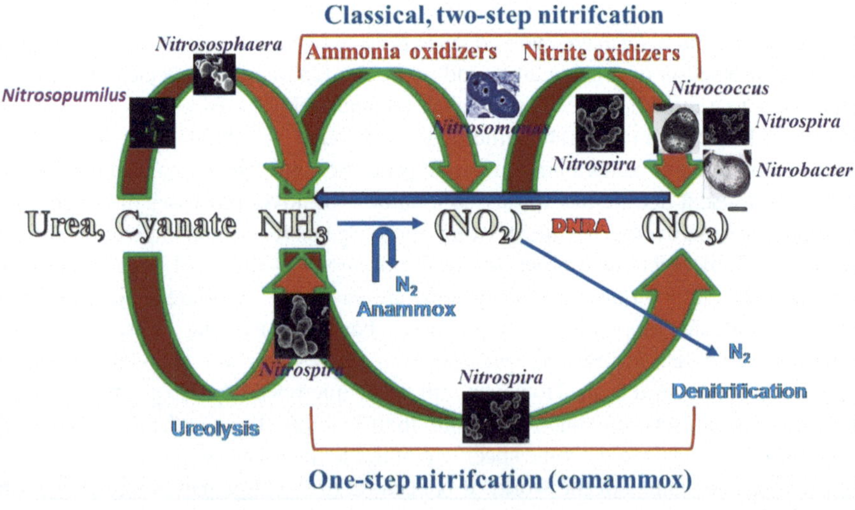

Fig. 1.4 Diagram depicting different processes of nitrogen cycles. *DNRA* Dissimilatory nitrate reduction to ammonium, *Anammox* Anaerobic ammonium oxidation (Sources: Daims et al. 2015; Koch et al. 2015; Lowson and Lucker 2018; Vaishnav et al. 2018)

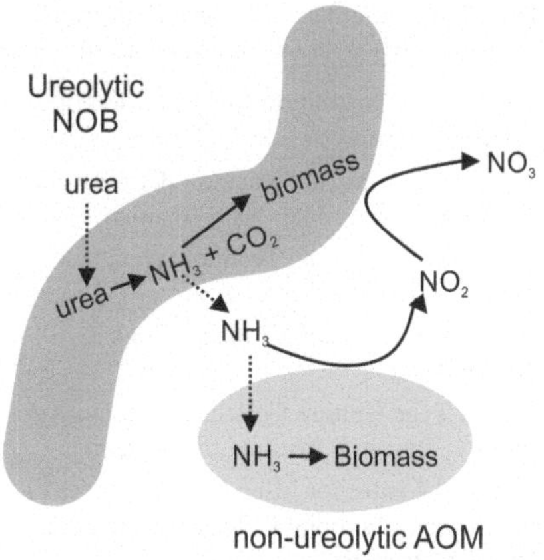

Fig. 1.5 Operation of ureolytic process by nitrite oxidising bacteria (NOB) and utilization of NOB's generated-ammonium by non-ureolytic ammonium oxidizing microorganisms (AOM): an example of reciprocal feeding (Source: Koch et al. 2015)

(anaerobic ammonium oxidation), ureolysis by nitrifying bacteria and comammox (one step aerobic ammonium oxidation by single bacterium; *Nitrospira*), DNRA (dissimilatory nitrate reduction to ammonium) and novel mode of lactate metabolisms in anaerobes, are evidence of continued occurrence of many microbial processes, which were consistently ignored knowingly and unknowingly for years (Figs. 1.4 and 1.5). Thus, ex situ preservation of such microbes is important for studying ecological processes and their uses in teaching, research, biotechnology and agriculture for future generation.

1.8 Biological Diversity Act 2002 and Rules 2004

Before Biological Diversity Act came in to existence, three legislations, namely, Indian Forest Act, 1927; Wildlife Protection Act, 1972; and Environment Protection Act, 1986, were being followed for protecting the environment and forest. The first attempt to bring the biodiversity into the legal framework was made by the introduction of the Biodiversity Bill, 2000, which was passed by both the houses of Parliament in December 2002. The objectives of the Act are (1) conservation of biological diversity, (2) sustainable use of its components and (3) fair and equitable sharing of the benefits arising out of utilization of genetic resources. Apart from these main objectives, the Act has also enforced some of the terms of CBD with inclusion of the following provisions: (1) to set up the National Biodiversity Authority (NBA), State Biodiversity Board (SBB) and Biodiversity Management Committees (BMCs), (2) to respect and protect knowledge of local community traditional knowledge related to biodiversity and (3) to conserve and develop areas of importance from the standpoint of biological diversity by declaring them biological diversity heritage sites.

The Biological Diversity Act, 2002, and Rules, 2004, are implemented by the National Biodiversity Authority (NBA) at the national level, State Biological Board

Table 1.8 Designated National Repositories (NDRs) of India for deposition of safe deposit of type strains, reference strains and samples of microbial resources accessed by foreign citizens

Name of the Institution for deposition of the microbial resources	Category of microbial resources
National Bureau of Agriculturally Important Microorganisms, Mau Nath Bhanjan, UP	Agriculturally important microorganisms (Bacteria, fungi, actinomycetes and cyanobacteria)
Institute of Microbial Technology, Chandigarh	Microorganisms for industrial use (actinobacteria, bacteria, fungi and yeasts)
National Institute of Virology, Pune, Maharashtra	Viruses
Indian Type Culture Collection, Indian Agricultural Research Institute, New Delhi	Fungi
National Centre for Microbial Resources (Formerly, Microbial Culture Collection), Pune, Maharastra	Microbes (bacteria and fungi including yeasts)

Source: National Biodiversity Authority, India; www.nbaindia.org

(SBB) at state level and Biodiversity Management Committees (BMCs) at local levels. Some of the major functions of these authorities are:

- To regulate activities, approve and advice the government of India on matters relating to the conservation of biodiversity, sustainable use of its components and equitable sharing of benefits.
- To grant approval under Sections 3, 4 and 6 of Biological Diversity Act, 2002.
- To notify areas of biodiversity importance as biodiversity heritage sites under this act and perform other functions as may be necessary to carry out the provisions of the Act.
- To take measures to protect biodiversity of the country as well as to oppose the grant of intellectual property rights to any country outside or any biological resources obtained from India.

The National Biodiversity Authority deals with the requests for access to the biological resources as well as transfer of information of traditional knowledge to foreign nationals, institutions and companies. Through this way, piracy of intellectual property rights in and around India is prevented, and it also saves the indigenous people from exploitation.

The NBA has declared Microbial-Designated National Repositories (M-DNRs) under 'Section 39' as an important aspect of infrastructure for biodiversity conservation. These M-DNRs provide various services such as preservation and maintenance of type and reference strains, genomes of organisms and information relating to heredity and function of microbial system (Table 1.8). Most recently, NBA has constituted a 'core expert group' for developing guidelines for identification of repositories for plant, animal and microbe under Section 39 of the BD Act 2002. This is the first ever approach to intensify identification of more repositories for India.

In another effort, NBA has developed Guidelines for Selection and Management of the Biodiversity Heritage Sites (BHS) under 'Section 37' of BD Act 2002 in order to strengthen the biodiversity conservation in traditionally managed areas and to stem the rapid loss of biodiversity in intensively managed areas, which need special attention (www.nbaindia.org/content/106/29/1/bhs.html; accessed on 01 August 2018). Such areas often represent a positive interface between nature, culture, society and technologies so that both conservation and livelihood security can be achieved. The BHS can be defined briefly as 'a well-defined areas that are unique, ecologically fragile ecosystems having rich biodiversity, high endemism, presence of rare and threatened species, keystone species, species of evolutionary significance, wild ancestors of domestic/cultivated species and fossils with or without long history of human association with them'. So far, 11 BHS were identified, (1) Nallur Tamarind Grove, Devanahalli, Bengaluru, Karnataka; (2) Hogrekan, Chikmagalur, Karnataka; (3) UAS, GKVK campus, Bengaluru, Karnataka; (4) Ambaraguda, Shimoga; (5) Glory of Allapalli, Gadchiroli, Maharashtra; (6) Tonglu, Darjeeling, West Bengal; (7) Dhotrey, Darjeeling, West Bengal; (8) Dialong Village, Tamenglong, Manipur; (9) Ammeenpur Lake, Sangareddy, Telangana; (10) Majuli, Majuli, Assam; and (11) Ghariyal Rehabilitation Centre, Lucknow, Uttar Pradesh have been selected to conserve flora and fauna of the respective areas. However, so far there is no area that has been selected which is rich in unique and rare microorganisms that can be utilized for many purposes of human benefit. The main drawback of this guideline is that it does not give any emphasis on selection of sites which are microbially diverse and rich.

1.9 Flaw in BD Act 2002

The major drawback in this act is that it does not give sufficient consideration to conservation; rather it lays more emphasis on preventing profit-sharing from the commercial use of the biological resources. It is true that the foundation of this act was laid to prevent biopiracy by the developed nations. However, the major aim of this act is to protect the biodiversity that required to be considered on priority basis. Therefore, revision in BD Act is highly needed.

1.10 Nagoya Protocol on Access and Benefit Sharing

The CBD was signed by 150 government leaders at the Earth Summit, Rio de Janeiro, in 1992 that came into force in 1993. Article 3 of CBD has recognized that states have sovereign right to exploit their own biological resources and thus provider country has to control access of biological resources by other user countries. Article 15 of the Convention recognizes sovereign rights of provider country over bioresources expect a fair and equitable share of any benefits arising out of their utilization for commercial purposes from users. To facilitate these aims, each member country is required to enact legislation to manage and grant permission to

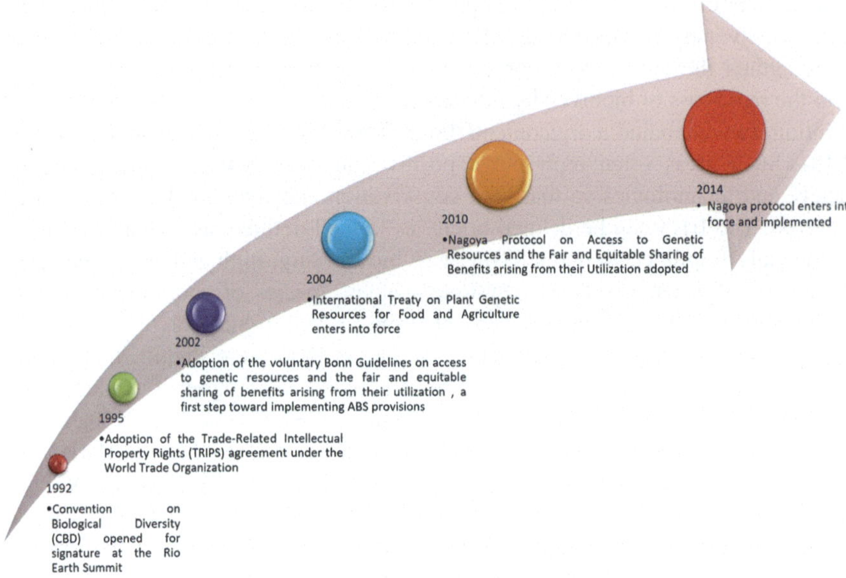

Fig. 1.6 Evolution of Nagoya Protocol on access and benefit sharing of genetic resources (Source: modified from Wynberg and Laird 2018)

the users to access samples from the country where bio-resource is available by seeking and receiving prior informed consent (PIC) and mutually agreed term (MAT) from competent authority of the country. To accomplish this goal, CBD has brought a supplementary global agreement called 'the Nagoya Protocol (NP) on Access to Genetic Resources and the Fair and Equitable Sharing of Benefits Arising from their Utilization' (https://www.cbd.int/convention/). This protocol provides a legal framework for the effective implementation of one of the three objectives of the CBD: the fair and equitable sharing of the benefits arising out of the utilization of genetic resources. The Protocol was adopted on 29 October 2010 in Nagoya (Japan) and come into force on 12 October 2014 (Fig. 1.6). Implementation of the CBD obligates each country to set up their own regulations and restrictions, which in turn resulted in an unmanageable multitude of national regulations. By designing common ways of organizing access to resources, the NP aimed at easing access to countries' genetic resources and ensuring that potential benefits arising from the use of these resources would be shared with provider countries. Importantly, the CBD and the NP are legal binding agreements, and countries that sign them are obligated to implement their provisions nationally. After NP has come into force, new national laws and regulations providing requirements for access to genetic resources are changing traditional views on collection, deposition and distribution of microorganisms (Fig. 1.7).

In order to implement NP globally, CBD had established a platform called 'Access and Benefit-sharing Clearing-House (ABS Clearing-House)' at Montreal, Canada, for exchanging information on access and benefit sharing under Article

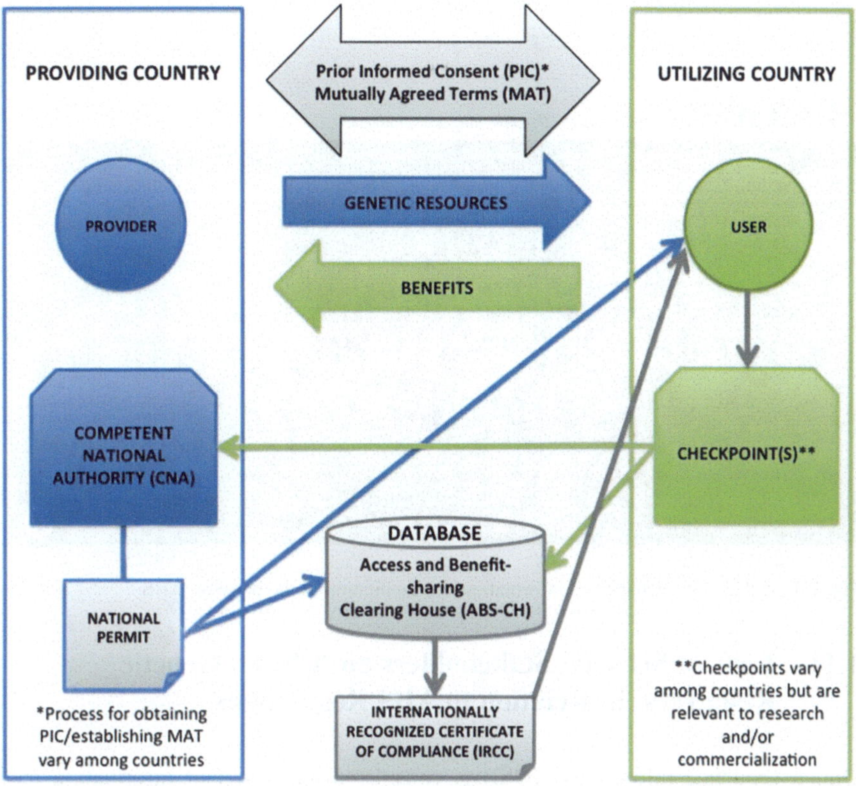

Fig. 1.7 Compliance mechanisms of access and benefit sharing of genetic resources under Nagoya Protocol

14 of the Protocol, as part of the Clearing-House of the Convention established under Article 18, paragraph 3 of the Convention (Fig. 1.8). The ABS Clearing-House is a key tool for enhancing legal certainty and transparency on procedures for access and benefit sharing and for monitoring the utilization of genetic resources along the value chain, involving internationally recognized certificate of compliance. By hosting relevant information regarding ABS, the ABS Clearing-House will offer opportunities for connecting users' and provider's country of genetic resources and associated traditional knowledge. Out of the 198 parties of CBD, 171 have established the National Focal Points (NFP) for ABS, 76 have formed Competent National Authority (CNA); 216 have brought legislative, administrative or policy measures (MSR); and 43 have set up national databases and websites (NDB) (https://absch.cbd.int/; accessed on 01 August 2018).

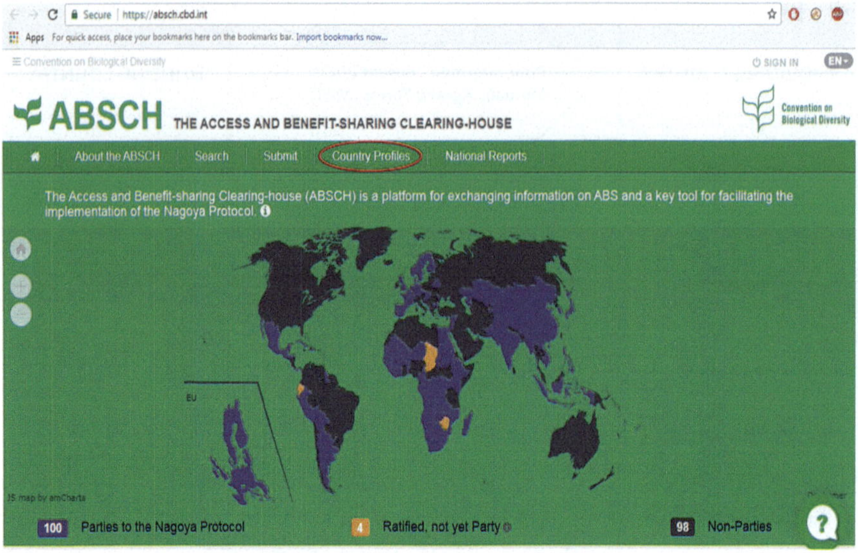

Fig. 1.8 A floor for exchange of information and facilitation to the party on ABS

1.11 View of Selected Stakeholders on 'DSI on Genetic Resources' in Relation to ABS Regulations

Digital sequence information (DSI) is considered an important information that originates by the analysis of the data contained in a digital file with precise order of nucleotides, amino acids or molecular structure of protein. The main function of these sequences is the storage and transmission of genetic information. The issue of digital sequence information on genetic resources emerged as a cross-cutting debatable issue during the 'Conference of the Parties 13 (COP 13) to the Convention on Biological Diversity' in Cancún, Mexico, 4–17 December 2016 (CBD/COP/DEC/XIII/16 dated 16 Dec. 2016). Representatives of 196 nations agreed during the conference that any potential implications of the use of digital sequence information on genetic resources in reference to three objectives of the CBD and the objective of the Nagoya Protocol will be considered at the next meetings of the governing bodies of the Convention and the Nagoya Protocol, in 2018 (COP 14 and COP-MOP 3). Based on the decision during this conference, the fact-finding committee on this issue recently drafted a document called 'The Emergence and Growth of Digital Sequence Information in Research and Development: Implication for the Conservation and Sustainable Use of Biodiversity, and Fair and Equitable Benefit Sharing' (https://www.cbd.int/abs/DSI-peer/Switzerland-FOAG2.pdf) submitted to Secretariat of CBD for comments and suggestions by international community/member party in order to clarify terminology and concept and to assess the extent and the terms and conditions of the use of digital sequence information in the context of CBD and

Nagoya Protocol (NP). In terms of ABS mechanisms, party to NP, researchers, societies and research institutions have their own views. For example:

(a) *India*: As per definition of utilization of genetic resources under NP, utilization of genetic resources is not confined to research and development on the tangible genetic resources, but is extended to activities over the genetic and biochemical composition of resources which is nothing but expression of gene sequence whether intangible (digital) or tangible. Synthesizing DNA from accessed digital information would therefore fall within the scope of utilization under NP. Such utilization of accessed information qualifies for the application of ABS regulatory framework even though there is no physical access of the genetic materials (http://www.cbd.int/abs/DSI-views/India-DSI.pdf).
(b) *Australia*: Digital sequence information (DSI) on genetic resources is not defined under the convention. For the purposes of this submission, Australia defines 'DSI on genetic resources' as electronically held sequence information, which represents biological composition of "genetic material" as defined under the Convention. Australia considers digital sequence information, physical genetic resources and material as distinct entities. To consider DSI on genetic resources, the CBD and NP should require to redefine "genetic material" that must include "functional unity of heredity" or "gene" (http://www.cbd.int/abs/DSI-views/Australia-DSI.pdf).
(c) *Brazil*: Brazilian Law 13.123/2015 defines genetic resources as the genetic information from plant, animal and microbial species or any other species, including substances originating from the metabolisms of these organisms. Therefore, the law of access and benefit sharing recognizes access to dematerialized genetic resources in its framework without the need for access of genetic samples as such (http://www.cbd.int/abs/DSI-views/Brazil-DSI.pdf).
(d) *Japan*: Believe that accumulation, open access and free use of DSI will facilitate development of science and will benefit conservation and sustainable use of biological diversity (http://www.cbd.int/abs/DSI-views/Japan-DSI.pdf).
(e) *USA*: As part of research best practices, DSI and GSD (genetic sequence data) are openly available via international data repositories such as GenBank and the International Nucleotide Sequence Database Collaboration, as well as in journals found in print and online. These repositories and journals further encourage collaboration by providing a free flow of GSD to both researchers and the general public. The open access and collaboration are the key benefits of GSD; regulation of access to and sharing of GSD would likely lead to a significant reduction in data sharing through these and other such mechanisms. GSD regulation could also force changes to procedures for information management within laboratories, with consequent costs and other negative implications for innovation. These dynamics would stifle research, which would then hinder activity to further CBD and Nagoya Protocol objectives (http://www.cbd.int/abs/DSI-views/USA-DSI.pdf).
(f) *Global Genome Biodiversity Network (GGBN)*: Sequence data is neither a genetic material, nor is it a genetic resource. GGBN and its members distinguish

between the genetic materials we hold and make accessible to the scientific community and the data describing this material. We note that it is essential to distinguish between these concepts and definitions. We also challenge the view that simple comparisons of sequence data for non-commercial uses can be considered for utilization; further, we are concerned that including digital sequence data under the scope of the Nagoya Protocol will have a severe limiting effect on essential global science by making data unavailable (http://www.cbd.int/abs/DSI-views/GGBN-DSI.pdf).

1.12 Microbial Commons and ABS Regulation

The concept of microbial commons deals with sharing of microbial resources at institutional levels for addressing the common problems of global concerns for direct benefits to the society, for example, (1) stem rust race 15B and Ug99, (2) microbes with probiotic properties for human health and (3) bioprospecting of microbes from colder regions for their utilization in ice cream for the purpose of regulating freezing points (Dedeurwaerdere 2010). A regulation either separately or under the ambit of ABS needs to be developed in order to facilitate sharing of microbial commons at global level.

1.13 ABS Regulation in India

In pursuance of the Nagoya Protocol on access to genetic resources and the fair and equitable sharing of benefits arising from their utilization to CBD and BD Act 2002, the National Biodiversity Authority (NBA) had made regulations called 'Guidelines on Access to Biological Resources and Associated Knowledge and Benefit Sharing Regulations, 2014' which was enforced on 21 November 2014 (Fig. 1.9). The following sections of BD Act 2002 are dealing with above said purposes.

Section 3 Deals with any foreign person seeking to access any biological resource and associated knowledge including traditional knowledge from India for their commercial utilization (excluding conventional breeding, traditional practices and other practices of agricultural sciences) shall have to take prior approval from the NBA before access to such bio-resources. For this purpose, Form I or Form I with Form A is required to be filled up with prescribed fee.

Section 4 Person who wants to transfer any biological entity or results of research relating to bio-resources to any foreign national has to apply and obtain prior approval from NBA. To this purpose, Form II is required to be filled up.

Section 6 Requires Indians to seek prior approval from NBA for applying intellectual property rights anywhere in the globe. To seek prior approval, an applicant has to fill up Form III.

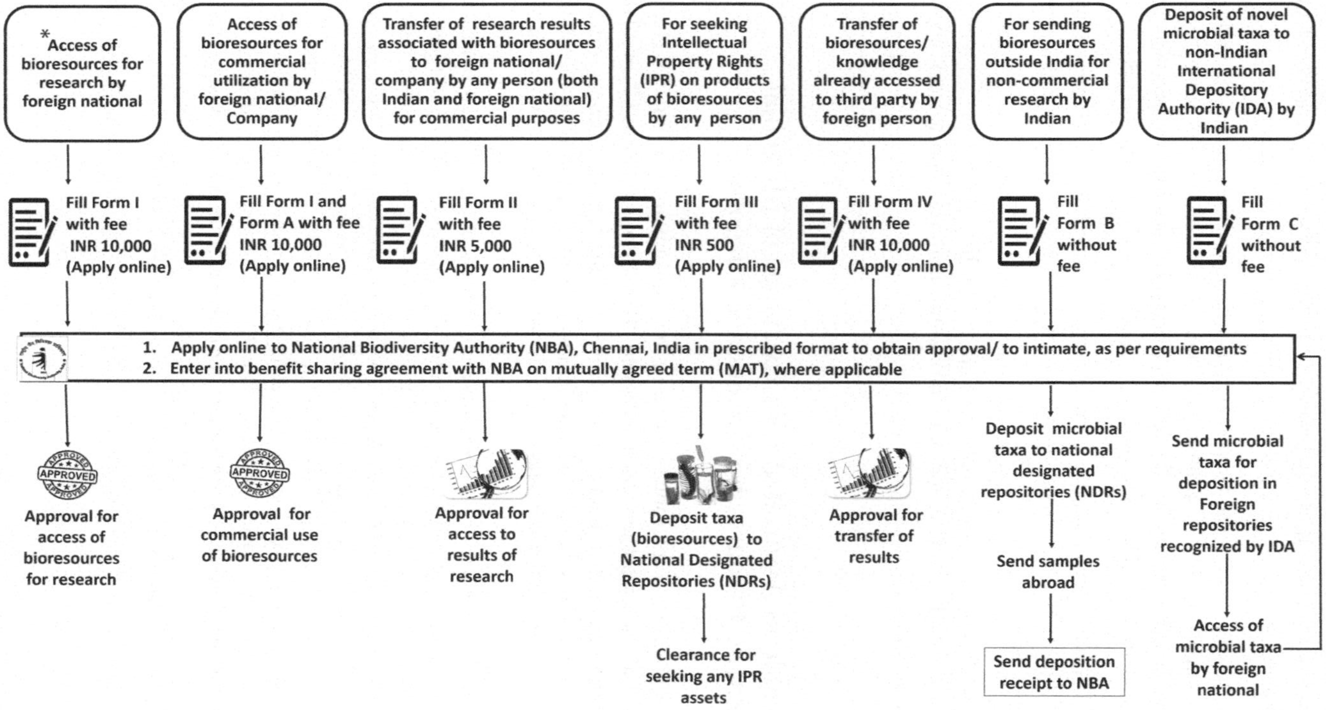

Fig. 1.9 Procedure for access and benefit sharing of bioresources in India (*Indian does not require approval of NBA for access of biological resources for research purpose

Section 7 A person wants to utilize the bio-resource would have to intimate the SBBs from state where the bio-resource is to be accessed. For intimation, an applicant needs to apply through Form IV. Indian person is not covered under above sections.

Section 40 Pursuant to Central Government notifying any item including biological resources normally traded as commodities under this section, the same becomes exempted from the purview of the Act. Currently six mushrooms are also under traded commodities in addition to other commodities of biological origin.

Form B To conduct non-commercial research or research for emergency purpose outside India by Indian researchers/government institutions under memorandum of understanding (MoU).

Form C To deposit the microorganism in non-Indian repository for claim of novel species. The Form C is for prior intimation purposes and to seek approval from NBA.

1.14 Problems Associated with Exchange of Microbial Genetic Resource for Taxonomic Purpose

In case of deposition of microbial culture in the non-Indian repository for claiming novel taxa, Indian researcher has to intimate NBA through 'Form C' in advance before sending cultures outside India. However, foreign researchers still need approval from the NBA before ordering strains of Indian origin from international repositories even for research purposes. Due to this regulatory enforcement by NBA, researchers and culture collections of India are now facing two serious problems: (1) type strains deposited in Indian culture collections are not recognised as valid deposit for publishing in certain journals since they are not available to researchers of foreign countries, and (2) foreign culture collections refuse to accept microbial cultures from Indian researchers to deposit because of the existing clause of seeking 'prior approval' to its further distribution for scientific scrutiny from NBA.

In order to facilitate compliance with the requirements of the NBA and to help researchers performing taxonomical characterization, the concerned Indian culture collections holding the particular strains may be authorized to take the initiative to share cultures worldwide by securing permission from NBA, instead of the researchers, having to seek that permission themselves. The existing regulations must be streamlined in order to facilitate taxonomic work in India.

1.15 Biosafety

Biosafety has become a major concern for several countries creating numerous activities to counter risk assessment, legislation and emergency response. The goal is to implement measures to protect against malicious use of microorganisms, their products, information and technology transfer. The issue to promote biosafety awareness was discussed in the second meeting of the Conference of Parties (CoP) held in November 1995 and established an Open-ended Ad Hoc Working Group on Biosafety. The working group was entrusted to prepare a draft protocol on biosafety focusing on transboundary movement of any living modified organism (LMO) resulting from modern biotechnology considered to have adverse effect on the conservation and sustainable utilization of biological diversity. After several years of negotiations, the Cartagena Protocol on Biosafety was finalized and adopted in Montreal on 29 January 2000 at an extraordinary meeting of the CoPs. Finally, protocol has been considered as important step forward that provides an international regulatory framework to reconcile the respective requirement of trade and environmental protection with respect to a rapidly growing global biotechnology industry. Cartagena Protocol thus creates an enabling environment to the environmentally sound application of biotechnology, making it possible to derive maximum benefit while minimizing the possible risks to the environment and to human health. Different articles (texts and annexes) of protocol states specific requirements to be complied by the party (Cartagena Protocol on Biosafety 2000). In this regard article 8(g) states that Parties should take up at national level, while Article 19, paragraph 3, sets the stage for the development of an international legal binding instrument to address the issue of biosafety.

1.16 Conclusion

Microbes have established their credentials over the years as promising goldmines for various applications. Despite countless facts about the role of microorganisms in the biosphere, they have been largely ignored and never considered in a real sense as part of conservation biology. However, careful examination and concerted efforts are required for conservation of microbial resources on sustained basis. In this regard the uses of conventional as well as modern tools play very important roles in order to minimize the loss of microbial diversity. Similarly, efforts are required to make the awareness about underlying causes of microbial diversity loss and their impact on various fields that include society, environment and agriculture with a view to reduce direct pressures on microbial diversity and to promote sustainable use. It has also been realized to improve the status of microbial diversity by safeguarding ecosystems and genetic diversity and enhance the benefits to all from microbial diversity and ecosystem services. Declining interests in teaching taxonomy and systematics is alarming and drawing serious attention of capacity building to future generation.

Besides, microbial resource centres are likely to have an increasingly important role in taxonomy research training, and these need to be recognized and funded accordingly. A national strategic plan is required to be developed for protecting microbial habitats that will help in meeting large gaps in our knowledge of the distribution and abundance of microbes. Besides, endangered microbial habitats of extreme environments need sound strategies for conservation on high priority basis. Overall, a road map of action plan with long-term financial provision by government and industry is essential for achieving the microbial conservation to an acceptable level. This review emphasizes the current status of conserved microbes, its conservation strategies, methodologies and regulatory framework at global level on conservation and sharing of microorganisms.

Acknowledgements This study was funded by the Indian Council of Agricultural Research (ICAR), New Delhi, India, through fragship programme, 'NAIMCC' and a project entitled 'Microbial diversity analysis of extreme/unique ecological niches'. Dr. S.K. Singh thanks Director, MACS-Agharkar Research Institute, Pune, for extending facilities. We thank Mr. Alok Upadhyay, Technical Assistant, NAIMCC, who helped us in literature search to achieve the task of this book chapter.

References

Alain K, Querellou J (2009) Cultivating the uncultured: limits, advances and future challenges. Extremophiles 13:583–594

Arora DK, Saikia R, Dwievdi R, Smith D (2005) Current status, strategy and future prospects of microbial resource collections. Curr Sci 89:488–495

Berner D, Viernstein H (2006) Effects of protective agents on the viability of *Lactococcus latis* subjected to freeze-thawing and freezing drying. Sci Pharm 74:137–149

Branduardi P, Longo V, Berterame NM, Rossi G, Porro D (2013) A novel pathway to produce butanol and isobutanol in *Saccharomyces cerevisiae*. Biotechnol Biofuels 6:68. https://doi.org/10.1186/1754-6834-6-68

Bull AT (2004) Microbial diversity and bioprospecting. Washington, DC, American Society for Microbiology Press

Camejo PY, Santo Domingo J, McMahon KD, Noguera DR (2017) Genome enabled insights into the ecophysiology of the comammox bacterium "Candidatus *Nitrospira nitrosa*". mSystems 2: e00059-17. https://doi.org/10.1128/mSystems.00059-17

Carrion O, Curson ARJ, Kumaresan D, Fu Y, Lang AS, Mercade E, Todd JD (2014) A novel pathway producing dimethylsulphide in bacteria is widespread in soil environments. Nat Commun 6:6579. https://doi.org/10.1038/ncomms7579

Cartagena Protocol on Biosafety (2000) Convention on biological diversity: text and annexes. Secretariat of the Convention on Biological Diversity, Montreal, p 30

CBD (Convention on Biological Diversity) (1992) Convention on biological diversity Article 2. Use of terms. Secretariat of the Convention on Biological Diversity, Montreal, p 25

Challis GL (2008) Mining microbial genomes for new natural products and biosynthetic pathways. Microbiology 154:1555–1569

Cockell CS, Jones HL (2009) Advancing the case for microbial conservation. Oryx 43:520–526

Curtis T (2006) Microbial ecologists: it's time to 'go large'. Nat Rev. Microbiol 4:488

Daims H, Lebedeva EV, Pjevac P, Han P, Herbold C, Albertsen M, Jehmlich N, Palatinszky M, Vierheilig J, Bulaev A, Kirkegaard RH, von Bergen M, Rattei T, Bendinger B, Nielsen PH,

Wagner M (2015) Complete nitrification by *Nitrospira* bacteria. Nature 582:504–509. https://doi.org/10.1038/nature16461

Danovaro R, Canals M, Tangherlini M, Dell'Anno A, Gambi C, Lastras G, Amblas D, Sanchez-Vidal A, Frigola J, Calafat AM, Pedrosa-Pamies R, Rivera J, Rayo X, Corinaldesi C (2017) A submarine volcanic eruption leads to a novel microbial habitat. Nat Ecol Evol 1:0144. https://doi.org/10.1038/s41559-017-0144

Dedeurwaerdere T (2010) Self-governance and international regulation of the global microbial commons: introduction to the special issue on the microbial commons. Int J Commons 4: 390–403

Duplessis S, Cuomo CA, Lin YC, Aerts A, Tisserant E, Veneault-Fourrey C, Joly DL, Hacquard S, Amselem J, Cantarel BL, Chiu R (2011) Obligate biotrophy features unraveled by the genomic analysis of rust fungi. Proc Natl Acad Sci U S A 108:9166–9171

FAO (2010) The state of in situ management. In: FAO (ed) Second Rep. State world's plant genetic resources for food and agriculture. Food and Agriculture Organisation of the United Nations, Rome, pp. 30–51

Fu N, Chen XD (2011) Towards a maximal cell survival in convective thermal drying processes. Food Res Int 44:1127–1149

Garcia AH (2011) Anhydrobiosis in bacteria: from physiology to applications. J Biosci 36:939–950

Garcia EO, Casagrande MV, Rago AM, Massola NS Jr (2007) Preservation of uredospores of *Puccinia melanocephala*, the causal agent of sugarcane rust. Summa Phytopathol 33 (2):152–156

García-García M, Rocha-Zavaleta L, Valdez-Cruz NA, Trujillo-Roldán MA (2014) Conservation of the mycelia of the medicinal mushroom *Humphreya coffeata* (Berk.) Stey. in sterile distilled water. Methods X 1:19–22

Gorman R, Adley CCN (2004) An evolution of five preservation techniques and conventional freezing temperatures of −20 °C and − 85 °C for long term preservation of *Campylobacter jejuni*. Lett Appl Microbiol 38:306–310

Heywood VH, Dulloo ME (2005) In situ conservation of wild plant species a critical global review of good practices. Food and Agriculture Organisation, Rome

Kalyuzhnaya MG, Yang S, Rozova ON, Smalley NE, Clubb J, Limb A, Nagna Gowda GA, Raftey D, Fu Y, Bringle F, Vuilleumier A, Beck DAC, Trotsenko YA, Khmelenina VN, Lidstrom ME (2013) Highly efficient methane biocatalysis revealed in a methanotrophic bacterium. Nat Commun 4:278. https://doi.org/10.1038/ncomms3785

Kerckhof FM, Courtens EN, Geirnaert A, Hoefman S, Ho A, Vilchez-Vargas R, Vandamme P (2014) Optimized cryopreservation of mixed microbial communities for conserved functionality and diversity. PLoS One 9:e99517

Kirsop B, Hawksworth DL (eds) (1994) The Biodiversity of microorganisms and the role of microbial resource centres. World Federation for Culture Collections

Koch H, Lucker S, Albertsen M, Kitzinger K, Hebold C, Spieck E, Nielsen PH, Wagner M, Diams H (2015) Expanded metabolic versatility of ubiquitous nitrite-oxidising bacteria from the genus *Nitrospira*. Proc Natl Acad Sci 112:11371–11376. https://doi.org/10.1073/pnas.1506533112

Kuenen JG, Jetten MSM (2001) Extraordinary anaerobic ammonium oxidising bacteria. ASM News 67:456–463

Lacy R (2010) Re-thinking ex situ vs in situ species conservation. In: Proceedings of the 65th World Association Zoos Aquariums Gland, Switzerland. WAZA, pp. 25–29

Lowson CE, Lucker S (2018) Complete ammonia oxidation: an important control on nitrification in engineered ecosystems? Curr Opin Biotechnol 50:158–165

Martin SM (1964) Conservation of microorganisms. Annu Rev. Microbiol 18:1–16

Morgan CA, Herman N, White PA, Vesey G (2006) Preservation of microorganisms by drying; a review. J Microbiol Methods 66:183–193

Mutlu BR, Hirschey K, Wackett LP, Aksan A (2015) Long-term preservation of silica gel-encapsulated bacterial biocatalysts by desiccation. J Sol-Gel Sci Technol 74:823–833

Ndeh D, Rogowski A, Cartmell A, Luis AS, Basle A, Gary J, Venditto I, Briggs J, Zhang X, Labourel A, Terrapon N, Buffetto F, Nepogodiev S, Xiao Y, Field RA, Zhu Y, O'Neil MA, Urbanowicz BR, York WS, Davies GJ, Abott DW, Ralet MC, Martens EC, Henrissat B, Gilbert HJ (2017) Complex pectin metabolism by gut bacteria reveals novel catalytic functions. Nature 544(7648):65–70. https://doi.org/10.1038/nature21725

OECD (2001) Biological resource centres: underpinning the future of life sciences and biotechnology. Organisation for Economic Cooperation and Development (OECD), Directorate for Science, Technology and Industry, Committee for Scientific and Technological Policy. http://www.oecd.org/dataoecd/55/48/2487422.pdf

OECD (2007) OECD best practices guidelines for biological resource centers. Organization for Economic Cooperation and Development (OECD), Paris. Available at: http://www.oecd.org/dataoecd/7/13/38777417.pdf

Overmann J (2015) Significance and future role of microbial resource centers. Syst Appl Microbiol 38:258–265

Pinto AJ, Marcus DN, Liaz UZ, Baustista-de-lose Santos QM, Dick GJ, Raskin L (2015) Metagenomic evidence for the presence of comammox *Nitrospira*-like bacteria in a drinking water system. mSphere 1(1):e00054–e00015. https://doi.org/10.1128/mSphere.00054-15

Rothrock MJ Jr, Vanotti MB, Szögi AA, Garcia Gonzalez MC, Fujii T (2011) Long-term preservation of anammox bacteria. Appl Microbiol Biotechnol 92:147–157. https://doi.org/10.1007/s00253-011-3316-1

Ryan MJ, Ellison CA (2003) Development of cryopreservation protocol for the microcyclic rust-fungus *Puccinia spegazzinii*. Cryo Letters 24:43–48

Salafsky N, Margoluis R, Redford KH, Robinson JG (2002) Improving the practice of conservation: a conceptual framework and research agenda for conservation science. Conserv Biol 16: 1469–1479

Salustiano ME, Pozza EA, Ferraz Filho AC, Castro HA (2008) Viability of urediniospores stored in different environment. Trop Plant Pathol 33:313–316

Sharma SK, Gupta AK, Shukla AK, Ahmad E, Sharma MP, Ramesh A (2016) Microbial conservation strategies and methodologies: status and challenges. Indian J Plant Genet Resour 29: 340–342

Sigler L (2004) Culture collections in Canada: perspectives and problems. Can J Plant Pathol 26: 39–47

Singh SK, Baghela A (2017) Cryopreservation of microorganisms. In: Varma A, Sharma AK (eds) Modern tools and techniques to understand microbes. Springer, New York, pp 321–333

Sly LI (2010) Biodiversity and the role of microbial resource centres. In: Presented in biodiversity and world food security: nourishing the planet and its people, conference, Canberra, Australia, pp. 60–67

Smith D, Ryan M (2012) Implementing best practices and validation of cryopreservation techniques for microorganisms. Sci World J 1–2:805659

Smith D, Ryan J, Stackebrandt E (2008) The ex situ conservation of microorganisms: aiming at a certified quality management. In: Doelle HW, EJ DS (eds) Biotechnology. Encyclopedia of life support systems (EOLSS), Developed under the auspices of UNESCO. EOLSS, Oxford

Stewart EJ (2012) Growing unculturable bacteria. J Bacteriol 194:4151–4160

Tatangelo V, Franzetti A, Gandolfi I, Bestetti G, Ambrosini R (2014) Effect of preservation method on the assessment of bacterial community structure in soil and water samples. FEMS Microbiol Lett 356:32–38

Tibolla F, Sumida CH, Peitl DC, Canteri MG, Castro AMC (2012) In vitro preservation methods of *Puccinia kuehnii* urediniospores. Summa Phytopathol 38:198–203

Uzunova-Doneva T, Donev T (2005) Anabiosis and conservation of microorganisms. J Cult Collect 4:17–28

Vaishnav A, Sharma SK, Choudhary DK, Sharma KP, Ahmad E, Sharma MP, Ramesh A, Saxena AK (2018) Nitric oxide as a signalling molecule in plant-bacterial interactions. In:

Egamberdieva D, Ahmad P (eds) Plant microbiome: stress response under series microorganisms for sustainability, vol 5. Springer Nature Singapore, Singapore, pp 183–199

Van Kessel MAHJ, Speth DR, Albertsen M, Nielsen PH, Op den Camp HJM, Kartal B, Jetten MSM, Lücker S (2015) Complete nitrification by a single microorganism. Nature 28:555–559. https://doi.org/10.1038/nature16459

Weghoff MC, Bertsch J, Müller V (2014) A novel mode of lactate metabolism in strictly anaerobic bacteria. Environ Microbiol 17:670–677. https://doi.org/10.1111/1462-2920.12493

Weißbecker C, Buscot F, Wubet T (2017) Preservation of nucleic acids by freeze-drying for next generation sequencing analysis of soil microbial communities. J Plant Ecol 10:81–90. https://doi.org/10.1093/jpe/rtw042

Winter S, Adam G (2001) Pathogen collections, present situations and future challenges. J Plant Pathol 83:83–88

Winters RD, Winn WC Jr (2010) A simple effective method for bacterial culture storage: a brief technical report. J Bacteriol Virol 40:99–101

Wynberg RP, Laird SA (2018) Fast science and sluggish policy: the herculean task of regulating biodiscovery. Trends Biotechnol 36:1–3. https://doi.org/10.1016/j.tibtech.2017.09.002

Xiong W, Lin PP, Magnusson L, Warner L, Liao JC, Maness PC, Chou KJ (2016) CO_2-fixing one-carbon metabolism in a cellulose-degrading bacterium *Clostridium thermocellum*. PNAS USA 113:13180–13185

Yu C, Reddy AP, Simmons CW, Simmons BA, Singer SW, Vander Gheynst JS (2015) Preservation of microbial communities enriched on lignocellulose under thermophilic and high-solid conditions. Biotechnol Biofuels 8:206. https://doi.org/10.1186/s13068-015-0392-y

Zhou Z, Takaya N, Nakamura A, Yamaguchi M, Takeo K, Shoun H (2002) Ammonia fermentation, a novel anoxic metabolism of nitrate by fungi. J Biol Chem 277:1892–1896

Chapter 2
Concept of Microbial Preservation: Past, Present and Future

Rohit Sharma, Yogesh Nimonkar, Ashutosh Sharma, Rajesh Singh Rathore, and Om Prakash

Abstract Microbes are the provider of ecosystem services and act as a key component for environmental sustainability and human health. It is believed that the cultivation and preservation of new taxa with innovative traits is essential for biotechnological growth even in the era of omics-driven science. Cultivation and preservation of entire microbial diversity is impossible even with extra efforts and use of novel high-throughput cultivation approaches. Therefore, development of protocols for community and habitat preservation should be priorities. In current article, we discussed about importance of microbes for biotechnological growth and their preservation for future reference research and application. In addition we also incorporated the loopholes of current preservation research and tried to incorporate the suggestion to fill the existing gaps. We tried to give a brief overview of different preservation methods used for preservation of bacteria and fungi and also incorporated the concept of microbial extinction and need to preserve intact community and microbiome for future culturomics, metagenomics and proteomics.

2.1 Introduction

Recent development in culture-independent metagenomics approach has revealed that the planet earth harbours huge microbial diversity which plays an important role in earth's ecology and environment (Zengler et al. 2002; Giovannoni and Stingl 2007; Alain and Querellou 2009; Epstein 2009; Prakash et al. 2013a, b; Bharti et al. 2017). Now it is clear that microorganisms are the provider of most of the ecosystem services. They are the key components of industrial growth and used in production of vitamins, hormones, novel biomolecules and antibiotics (Heylen et al.

R. Sharma · Y. Nimonkar · A. Sharma · R. S. Rathore · O. Prakash (✉)
National Centre for Microbial Resource (NCMR), National Centre for Cell Science, Pune, Maharashtra, India

Environmental Biotechnology and Genomics Laboratory, School of the Environment, Florida A&M University, Tallahassee, FL, USA
e-mail: omprakash@nccs.res.in

S. K. Sharma, A. Varma (eds.), *Microbial Resource Conservation*, Soil Biology 54,
https://doi.org/10.1007/978-3-319-96971-8_2

2012; Prakash et al. 2013a, b, 2016). Microbes are used as probiotics, biopesticides, biofertilizers and imperative for plant health and growth (Zaidi et al. 2009; Prakash et al. 2016). They are the key player of biogeochemical cycling of materials and essential for sustainable earth and environments (Berg 2009). Microbial activities emit greenhouse gases like methane, nitrous oxide and CO_2 which contributes in global warming and global climate change (Singh et al. 2010; Triantafyllou et al. 2014). They affect global food supply by affecting plant health and crop productivity (Abd El-Salam et al. 2011; Bharti et al. 2017; Berg et al. 2005). Microbes are causative agents of human and animal disease and affect the human health and hygiene (Nicholson et al. 2012). Thus, in "nutshell" if we analyse critically, we find that microbes are the key components for environmental sustainability, planet health and human civilization (Turnbaugh et al. 2007). Therefore, their cultivation and adequate preservation in microbial resource centres is mandatory for reference and teaching purposes and to conduct studies on their role in diverse aspects of life (Prakash et al. 2013a, b). Concept of microbial preservation is as old as the birth of microbiology itself. The increasing number of taxa in culture collections indicates the current interest of scientist in microbial cultivation and preservation, although cultivation and preservation seems a tough job and only few microbiological laboratories are focusing attention on cultivation and preservation of microbial diversity. However, due to use of modern culturomics approaches in microbiology, the rate of cultivation and discovery of novel microbial species have increased tremendously (Lagier et al. 2015, 2016). For instance, recent development in human gut metagenomics revealed that human gut contains more than 1400 different species of diverse bacteria and they are the crucial player of human health and disease, but most of them are still uncultured. Despite all the efforts, substantial part of microbial diversity with immense, medical, agricultural, industrial and environmental potentials are still untapped and need to bring them in culture collections. The objective of current article is to make the researchers well acquainted with different methods of low- and high-cost preservation so that even a simple laboratory with minimum equipment can successfully preserve their microbial diversity. In addition we also tried to expand our vision about recent approaches of preservation including habitat, community and intact microbiome preservation.

2.2 Techniques Used for Preservation of Microorganisms

Considering the widespread importance of microorganism efforts of the microbial preservation for future research, reference, training and application have been made from the very beginning of microbiology. Although strategies of preservation of different groups of microorganisms are almost similar, however they show various responses to different preservation strategies depending on their morphological and physiological traits. A number of different methods such as repeated subculturing; low-temperature [−45 to −70 °C (Winters and Winn 2010)] preservation using agar bits; mineral oil overlay of slants-grown cultures (Nakasone et al. 2004); use of different sterile support materials like silica gel (Streeter 2003; Perez-Garcıa et al. 2006; Smith et al. 2008), distilled water, soil and sand (Bakerspiegel 1953; Liao and

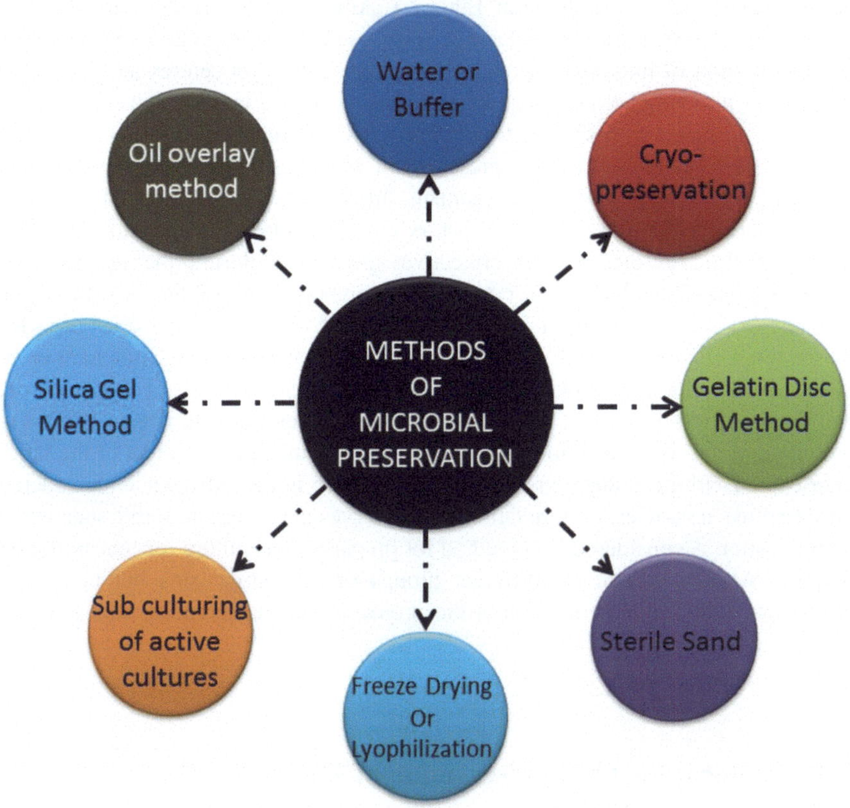

Fig. 2.1 Flow diagram of different methods used for preservation of different groups of microorganisms. Cryopreservation and lyophilisation are the methods of choice for most of the culture collections

Shollenberger 2003); cryopreservation at ultralow temperature in the presence of cryoprotectants (Gorman and Adley 2004; Fuller 2004; Cody et al. 2008; Smith et al. 2008; Chian 2010) and freeze-drying or lyophilization (Israeli et al. 1993; Abadias et al. 2001; Morgan et al. 2006; Berner and Viernstein 2006) have been used for preservation of different groups of microorganism. Cryoprotectants are the essential components for microbial preservation at ultralow temperature (Gomez Zavaglia et al. 2003). They prevent the formation of ice crystals inside cellular cytoplasm and protect the microbial cells from cryogenic effects of ultralow temperature. Based on their mechanism of action, they are classified as penetrating and non-penetrating cryoprotectants. The mechanisms of actions of different cryoprotectants are discussed in detail somewhere else (Prakash et al. 2013a, b). Glycerol, dimethyl sulfoxide (DMSO), sucrose, trehalose (Gomez Zavaglia et al. 2003;Hubalek 2003) and skimmed milk are commonly used cryoprotectants. Different methods of microbial preservation are depicted in Fig. 2.1. A list of well-established culture collections of the world along with their available bacterial stocks and preferred methods of preservation have been given in Table 2.1. Among others, cryopreservation and lyophilization are the most popular and being used on large scale for preservation and distribution of microorganisms in big culture collection centres and industries (Table 2.1), although both the methods have their own pros and cons and response of preservation varies with species to species; even different strains of same species give different response to same preservation method. It has been reported that viability and longevity of microorganisms under preservation depend on some critical factors. The most common factors are (1) nature of suspension medium, (2) types of cryoprotectants or protective agent used during the preservation, (3) rates of cooling during the cryopreservation, (4) stage of the growth of the cultures, (5) cell size of the microorganisms and (6) initial cell densities used for preservation (Palmfeldt et al. 2003; Nagai et al. 2005; Miyamoto-Shinohara et al. 2000). As mentioned above that there are various methods of preservation, employing differs in type of cryoprotectants and preservation strategies, but cryopreservation and lyophilization are widely accepted and used by most of the laboratories and culture collections. Preservation under liquid nitrogen is considered ideal because it induces least mutation and preserves the culture in stable state while lyophilization is considered best method for preparing the culture for transportation purpose. Although there are different groups of microorganisms but in current discussion, we are going to discuss the protocols and preservation strategies for bacteria and fungi only (Table 2.2).

2.3 Protocol for Cryopreservation of Bacteria

Cryopreservation proved to be more satisfactory method for preservation of microorganisms. Lots of work have already been done on this aspect using different groups of microorganisms. The main aim of this section is to refresh the old memories of expert readers of the field as well as to provide an overview and concept of cryopreservation to the fresher. Previous work indicated that almost the same kind of protocols have been used for cryopreservation of different categories of microorganisms with some modifications in concentration and type of cryoprotectants. Unlike lyophilization, in cryopreservation the possibility of desiccation stress is almost zero; therefore less stressed and more active cells from mid or late log phase of the culture are generally preferred in cryopreservation. In brief, first grow the bacteria on solid medium or in liquid broth, and harvest it from mid or late log phase. Suspend the cells in suitable cryoprotectants (generally 10–50% sterilized glycerol or 5–10% DMSO) and subject to snap or controlled freezing (1–5 °C min^{-1} cooling rate was found to be optimum). Alternatively, if controlled cooling device is not available in the laboratory, in that situation keep the suspended culture at −80 °C for 24 h prior to transfer in vapour or liquid phase of nitrogen. After freezing the cells should be stored either in deep freezer or in vapour or liquid phase of nitrogen. For resuscitation, remove the stored cryovials, subject to rapid thawing at 37 °C water bath and suspend or streak on a suitable resuscitation medium for growth (Fig. 2.2).

Table 2.1 List of the major culture collections of the world and their preferred method of culture preservation

S. No.	Culture Collection	Acronym	Preservation methods used	URL	Country
1	Australian Collection of Microorganisms	ACM	CP/FD	http://www.biosci.uq.edu.au/micro/	Australia
2	Agricultural Research Service Culture Collection	NRRL	CP/FD	http://nrrl.ncaur.usda.gov/	USA
3	American Type Culture Collection	ATCC	CP/FD	http://www.atcc.org/	USA
4	Belgian Coordinated Collections of Microorganisms/ LMG Bacteria Collection	LMG	CP/FD	http://bccm.belspo.be/index.php	Belgium
5	Collection du Centre de RechercheenInfectiologie	CCRI	CP/FD	http://www.cri.crchul.ulaval.ca/index.html	Canada
6	Czech Collection of Microorganisms	CCM	CP/FD	http://sci.muni.cz/ccm/index.html	Czech
7	Microbial Type Culture Collection and Gene Bank	MTCC	CP/FD	http://mtcc.imtech.res.in	India
8	National Centre for Microbial Resource (NCMR) previously Microbial Culture Collection	NCMR/ MCC	CP/FD	http://www.nccs.res.in/	India
9	Deutsche Sammlung von Mikroorganismen und Zellkulturen GmbH	DSMZ	CP/FD	http://www.dsmz.de/	Germany
10	Japan Collection of Microorganisms	JCM	CP/FD	http://www.jcm.riken.jp/	Japan
11	KCTC Korean Collection for Type Cultures	KCTC	CP/FD	http://www.brc.re.kr/English/ekctc.aspx	Korea
12	The Netherlands Culture Collection of Bacteria	NCCB	CP/FD	http://www.cbs.knaw.nl/nccb	Netherland
13	Collection de L'Institut Pasteur	CIP	CP/FD	http://www.crbip.pasteur.fr	France
14	All-Russian Collection of Microorganisms	VKM	CP/FD	http://www.vkm.ru/	Russia
15	CABI Genetic Resource Collection	IMI	CP/FD	http://www.cabi.org/	UK

(continued)

Table 2.1 (continued)

S. No.	Culture Collection	Acronym	Preservation methods used	URL	Country
16	National Collections of Industrial Food and Marine Bacteria	NCIMB	CP/FD	http://www.ncimb.co.uk	UK
17	National Collection of Type Cultures	NCTC	CP/FD	UKNCC http://www.ukncc.co.uk/	UK
18	Culture Collection, University of Goteborg	CCUG	CP/FD	http://www.ccug.se/	Sweden

CP Cryopreservation, *FD* Freeze-drying
Data were taken from the Website (http://www.wfcc.info) of "World Federation for Culture Collections (WFCC)"
A total 728 culture collections from 75 countries have been registered under WFCC
Almost all the big collections are using the freeze-drying and liquid N_2 preservation methods for preservation of microorganisms

2.4 Protocol for Lyophilization of Bacteria

Lyophilization gave satisfactory result for the preservation of bacteria yeast and sporulating fungi, but did not give the decent response for non-sporulating fungi (vegetative hyphae), some species of yeast (*Lipomyces, Leucosporidium, Brettanomyces*, *Dekkera*, *Bulleera*, *Sporobolomyces*) and bacteria [*Aquaspirillum serpens*, *Clostridium botulinum* and *Helicobacter pylori* (Smith et al. 2008)]. Almost similar kind of lyophilization protocols are used for preservation of different groups of microorganisms with very little modifications in the nature of suspension medium or lyoprotectants (Leslie et al. 1995; Carvalho et al. 2003; Morgan et al. 2006; Jackson et al. 2006; Berner and Viernstein 2006; Tindall 2007; Patel and Pikal 2011). In lyophilization, selection of right type of suspension medium and lyoprotectants is still a matter of hot research and needs more optimization in the future using different groups of microorganisms. An ideal suspension medium for lyophilization should comprise lyoprotectants (trehalose, sucrose, etc.) as well as matrix materials or excipients (skim milk, BSA, serum and mannitol) (http://www.opsdiagnostics.com/notes/ranpri/rpbacteriafdprotocol.htm). During lyophilization, lyoprotectants protect the organisms from desiccation injury, while excipients give better formulation, shape and stability to the product. Use of stationary phase cultures, borosilicate ampules, 1–2% final moisture contents of lyophilized specimen and storage of freeze-dried cells at 4 °C in the dark are generally recommended for higher cell viability and longer stability in the case of lyophilization (Morgan et al. 2006; Smith et al. 2008). In brief, three basic steps are involved in successful lyophilization. At first, cultivate the microorganisms in desired medium, and harvest the cells from late log or early stationary phase. After that, suspend the cells in suspension medium with lyoprotectant(s), in order to protect the cells from lyo-injury. Finally, freeze the suspended cells, and remove the water under vacuum via sublimation. In the end, seal the vials under vacuum with the help of sealing

Table 2.2 Survival response of different group of microbes to various preservation strategies and cryoprotectants

S. No.	Strains	% Viability	Preservation methods	Cryoprotectants used	Nature of cryoprotectants	References
1	*Microcystis aeruginosa*	70 (30)	Cryopreservation	DMSO (5%)	(PC)	Park (2006)
2	*Microcystis ichthyoblabe*	0 (30)	Cryopreservation	DMSO (5%)	(PC)	Park (2006)
3	*Microcystis aeruginosa*	20 (30)	Cryopreservation	DMSO (5%)	(PC)	Park (2006)
4	*Enterococcus durans, E. faecalis*	ND (8)	Lyophilization	Skim milk + Sodium glutamate	(CUA)	Carvalcho et al. (2003)
5	*Enterococcus* sp., *L.* sp.	ND (8)	Lyophilization	Skim milk + Sorbitol	(CUA)	Carvalcho et al. (2003)
6	*Lactococcus lactis*	39[a]	Lyophilization	Skim milk + Sucrose	(CUA) + (NP)	Berner and Viernstein (2006)
7	*Lactococcus lactis*	62[a]	Lyophilization	Skim milk (10%) + Sucrose	(CUA) + (NP)	Berner and Viernstein (2006)
8	*Lactococcus lactis*	60[a]	Lyophilization	Skim milk (20%) + Sucrose	(CUA) + (NP)	Berner and Viernstein (2006)
9	*Lactococcus lactis*	62[a]	Lyophilization	MRS broth + Skim milk + Sucrose	(CUA) + (NP)	Berner and Viernstein (2006)
10	*Lactococcus lactis*	62[a]	Lyophilization	MRS broth + Mannitol	(NP)	Berner and Viernstein (2006)
11	*Lactococcus lactis*	10[a]	Lyophilization	Skim milk (10%) + Mannitol	(NP)	Berner and Viernstein (2006)
12	*Lactococcus lactis*	62[a]	Lyophilization	Mannitol (5%) + MRS broth	(NP)	Berner and Viernstein (2006)
13	*Escherichia coli*	70[a]	Lyophilization	Trehalose	(NP)	Leslie et al. (1995)
14	*Bacillus thuringiensis*	57[a]	Lyophilization	Trehalose	(NP)	Leslie et al. (1995)
15	*Escherichia coli*	56[a]	Lyophilization	Sucrose	(NP)	Leslie et al. (1995)
16	*Bacillus thuringiensis*	44[a]	Lyophilization	Sucrose	(NP)	Leslie et al. (1995)

(continued)

Table 2.2 (continued)

S. No.	Strains	% Viability	Preservation methods	Cryoprotectants used	Nature of cryoprotectants	References
17	*Saccharomyces cerevisiae*	3.6–7.7 (120)	Lyophilization	Skim milk (10%) + Sodium glutamate (1%)	(CUA)	Miyamoto-Shinohara et al. 2000
18	*Brevibacterium, Corynebacterium*	80 (120)	Lyophilization	Skim milk (10%) + Sodium glutamate (1%)	(CUA)	Miyamoto-Shinohara et al. 2000
19	*Lactococcus lactis*	70[a]	Cryopreservation (−80 °C)	Skim milk (10%) + Mannitol	(NP)	Berner and Viernstein (2006)
20	*Lactococcus lactis*	68[a]	Cryopreservation (−80 °C)	Mannitol (5%) + MRS broth	(NP)	Berner and Viernstein (2006)
21	*Pseudomonas aeruginosa*	91[b]	Cryopreservation (−80 °C)	Skim milk (10%)	(CUA)	Cody et al. (2008)
22	*Escherichia coli*	56[b]	Cryopreservation (−80 °C)	Glycerol (15%)	(PC)	Cody et al. (2008)
23	*Pseudomonas aeruginosa*	60[b]	Cryopreservation (−80 °C)	Glycerol (15%)	(PC)	Cody et al. (2008)
24	*Campylobacter jejuni*	100 (12)	Cryopreservation (−80 °C)	Glycerol (15%)	(PC)	Gorman and Adley (2004)
25	*Erwinia carotovora*	96 (60)	Silica Gel (−20 °C)	Milk	(CUA)	
26	*Pseudomonas solanacearum*	73 (60)	Silica Gel (−20 °C)	Milk	(CUA)	
27	*Pseudomonas glycinea*	52 (60)	Silica Gel (−20 °C)	Milk	(CUA)	
28	*Agrobacterium tumefaciens, Corynebacterium michiganense, C. nebraskense, Erwinia stewartii,*	100 (60)	Silica Gel (−20 °C)	Milk	(CUA)	

	Pseudomonas lachryman, P. phaseolicola, P. syringae, Xanthomonas compastris, X. nigromaculans, X. phaseoli					
29	*A. tumefaciens, C. michiganense, C. nebraskense, E. carotovora, P. phaseolicola, P. solanacearum, P. syringae X. campestris, X. nigromaculan , X. phaseoli*	100 (60)	Silica Gel (−20 °C)	Milk + Glycerol	(CUA) + (PC)	
30	*Erwinia stewartii*	98 (60)	Silica Gel (−20 °C)	Milk + Glycerol	(CUA) + (PC)	
31	*Pseudomonas glycinea*	97 (60)	Silica Gel (−20 °C)	Milk + Glycerol	(CUA) + (PC)	
32	*Pseudomonas lachrymans*	97 (60)	Silica Gel (−20 °C)	Milk + Glycerol	(CUA) + (PC)	

Survival of bacterial strains after 31 days at sustained elevated temperatures of ~30 °C
PC Penetrating compound, *NP* Non- penetrating, *CUA* Complex undefined agents, *ND* No data
[a]Survivability just after preservation (0-d)
[b]One months at 30 °C after power failure

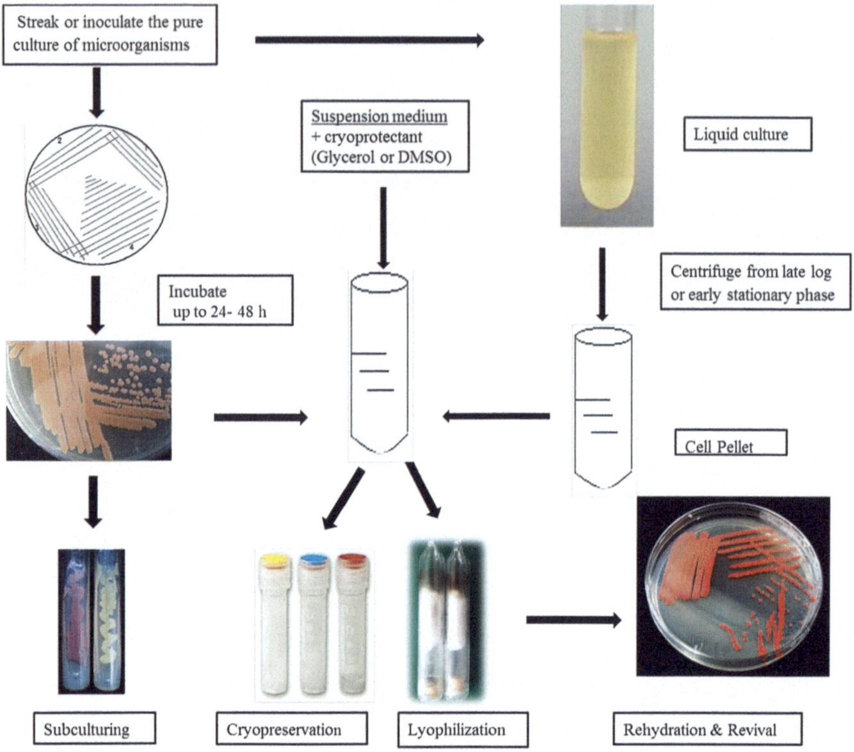

Fig. 2.2 Diagram represents the basic steps of lyophilisation and cryopreservation of bacteria. DMSO (5%) and glycerol (10–15%) are the most common types of cryoprotectants used during cryopreservation, while skimmed milk (10%) frequently used for lyophilization. Late log phase or early stationary phase culture is the most common phase of growth for preservation

torch, check the viability of the cells and store the lyophilized cells at 4 °C in the dark (Fig. 2.2).

2.5 Method of Fungal Preservation

Fungi are eukaryotic organisms and show little difficulty during long-term preservation. Due to immense morphological and physiological diversity, they show differential response and their ability to sustain different preservation methods. Fungi are preserved using an array of different preservation protocols and materials like silica gel, sterile sand, sterile soil, etc. But here we tried to give a brief outline of some common methods used for preservation of different groups of fungi in mycological laboratories across the world.

Subculturing Subculturing is one of the simplest methods of fungal culture preservation or maintenance for routine laboratory work. It involves periodic subculturing or transfer of fungal strains on fresh medium plates or slants. In this method, fungi are generally maintained on appropriate agar medium slants, free from mites and other microbial contamination at 4 °C. Usually common fungal growth media like potato dextrose agar (PDA), potato carrot agar (PCA), Czapek-Dox agar (CDA), etc. are used for maintenance purpose unless the fungus has a specific growth requirement. Subculturing period varies from fungus to fungus which can be from 30 to 90 days depending on the growth characteristic of fungal species. It is mandatory to take care to avoid the dryness of the medium. For subculturing, fungus from old plate should be routinely transferred on fresh medium plates with the help of sterile toothpick or needle and incubated on appropriate temperature for optimum growth. After growth, culture should be again transferred on slants with cotton plugs or screw cap (kept loose) and stored at 4 °C. Generally fungus with spores and sclerotia survive better than vegetative mycelium using this maintenance procedure. It is a low-cost method with easy retrieval and extensively used routine type of cultures but labour intensive and requires continuous monitoring of an expert. Furthermore, repeated subculturing induces mutation; consequently cultures lose their genetic stability with time. Strains may also lose their properties of sporulation and others after several subculturing for long period on artificial medium. Certain fungi like *Phytophthora* sp. MCC 1030 and *Phytophthora* sp. MCC 1045 which do not sustain cryopreservation or freeze-drying are maintained on slants using periodic subculturing.

Storage under mineral oil Covering fungal culture with sterile mineral oil (MO) is another method of preserving fungi. Overlaying of the fungal cultures with mineral oil reduces the penetration of available environmental oxygen which suppresses the metabolic activity and increases the longevity of the strains. In our laboratory, at the National Centre for Microbial Resource (NCMR) Pune, India, fungal cultures are preserved by cutting the disc of actively grown mycelial culture using 3 mm corkborer followed by adding 5–6 discs in 30 ml capacity tubes (HiMedia) containing sterile mineral oil (HiMedia). After inoculation we keep the tubes at 4 °C. It is a good low-cost method, for those fungi not withstanding cryopreservation and freeze-drying. A wide range of fungi are reported to survive under mineral oil for 2–5 years using this method of preservation. Cultures can be revived by inoculating the discs on fresh media plates and incubating them on appropriate temperature in slant position to drain the mineral oil. Fresh revival from mineral oil gives initial slow growth and needs one more round of subculturing to ensure purity and to achieve the normal growth rate. Precaution should be taken to ensure protection from microbial contamination and mite infestation. In addition, the mineral oil layer should not be too deep; otherwise it will prevent the diffusion of oxygen. *Basidiomycete* and non-sporulating fungi, which cannot sustain other methods of low-temperature preservation, can be preserved using this method.

Storage under distilled water For distilled water (DW) preservation, the cultures are grown in optimum conditions and cut using a 3 mm corkborer. Similar to MO preservation, 5–6 discs are inserted in the tube with sterilized distilled water (HiMedia) and the tube kept at 4 °C. It is also a low-cost method and best suited for non-sporulating or low sporulating cultures. Some cultures of *Phytophthora* and *Pythium* have shown deterioration of pathogenicity after preservation in distilled water for 2–3 years as compared to liquid nitrogen. According to our observation at the National Centre for Microbial Resource (NCMR) with cultures isolated from Lonar Lake, Maharashtra, India (hyper saline lake), six fungal isolates of *Cladorrhinum* sp. did not revive from DW after 5 years, whereas five fungal isolates of *Plectosphaerella* are revived only in DW. The other pros and cons are the same as for mineral oil. In a study undertaken at NCMR on revival of endophytic fungi after 2 years, it was found that revival rate of sterile DW was 66% as compared to 34% of glycerol.

Cryopreservation Cryopreservation is a costly but one of the most reliable methods of long-term fungal preservation. Similar to bacteria, in cryopreservation, fungal strains are stored at ultralow temperature in deep freezers (−80 °C) or in liquid nitrogen—LN2 (−196 °C) with cryoprotectants. In brief, at first grow the cultures in optimum conditions of temperature, medium, pH, etc. Once the culture has grown sufficiently, cut the small discs using sterile corkborer. After that, prepare the cryotubes containing 1 ml of 10–15% glycerol or 5–10% DMSO in 2 ml cryotubes and transfer 5–6 discs in each cryotube. These vials are kept at room temperature for 1–2 h which helps in easy penetration of the cryoprotectant. These are internal cryoprotectants which protect cells at low temperature by penetrating inside and replacing water. In this way there is less crystal formation and hence less cell damage/death. Apart from above cryoprotectants, some fungi are also preserved using trehalose, sucrose, etc. as cryoprotectant. These cryovials are then kept in a box with inventory and transferred to deep freezers or LN2 containers. In case of LN2, one vial is kept at vapour phase (−135 to −150 °C) and one at liquid phase (approx. −196 °C), but vapour phase preservation is considered safe.

Freeze-drying or lyophilization Lyophilization is one of the best methods for long-term preservation of the fungi which preserve viability and macroscopic details of the fungi and induces least morphological, physiological and genetic changes (Ryan and Smith 2007). It is first time used by Raper and Alexender (1945) and improved by Heckly (1978) and Jong et al. (1984). In brief for lyophilization, grow the selected fungal strains on appropriate culture medium, and scrap the mycelium and spore in skimmed milk solution. Four to seven days for fast-growing and 10–15 days for slow-growing fungi is considered ideal for lyophilization. After suspending the mycelium and spore, freeze the suspension at −80 °C followed by drying of frozen culture under vacuum (15–25 Pa). Finally seal the tube using the oxygen torch to maintain the vacuum and store at 4 °C (Bunse and Steigleder 1991; Abd El-Salam et al. 2011). Tan et al. (2007) proposed the protocol for lyophilization of freeze-drying of the fungi using shelf freeze-drier and skimmed milk as lyoprotectants. They successfully used the protocol for preservation of 50,000 filamentous fungi at CBS. Ryan and Smith also published the protocol for preservation of fungi by lyophilization in 2007. They mentioned that the generic or basic protocol of lyophilization should be the same, but some modification is essential to get better viability and longevity.

2.6 Extra Attention on Problematic Taxa

Problematic taxa are those groups of organisms which do not respond well for existing preservation methods and not revive successfully during post-preservation recovery. Thus, the way we are focusing on cultivation of rare microbial taxa we should also work to develop the protocol for their successful preservation for future reference, research and application (Prakash et al. 2013a, b; Lagier et al. 2015–2016; Zengler et al. 2002). It has been observed that the bacteria isolated from sea (marine environment) show problem during post-preservation revival. To consider the problem related to preservation of marine microbial diversity, the Department of Biotechnology, Govt. of India, has sanctioned a project for the cultivation and preservation of marine microbial diversity without substantial loss in post-preservation recovery. Research is also needed in the field of preservation of non-sporulating fungi especially human pathogens. These fungi do not sporulate and hence are vulnerable to microbial culture preservation activities. Specific cryoprotectants can be searched for the preservation of problematic fungi. The environmental molecular biologists have different requirement. Unlike whole microbial cells, they work on molecules like DNA or protein; therefore preservation of DNA or protein for future study and comparison is also available option for problematic groups of organisms. For example, Kirk (2012) validly published chytrid genus *Piromyces cryptodigmaticus* for which they submitted sample from the fermenter (from which DNA was isolated) as type material. Preservation of whole sample as *type* material for future reference is an emerging option in case of organisms giving difficult time for cultivation and preservation.

Research in the area of preservation-related issues are very limited, because we think that either it is not a valuable field or there is no any novelty left in the area of preservation, but fact is totally different. We still need more focused research for preservation of problematic cultures, and research funding should be improved for the same to boost the field of culturomics and microbial diversity preservation because growth of biotechnology is mainly related with the cultivation, characterization and preservation of organism with novel traits (Heylen et al. 2012; Hoefman et al. 2012; Prakash et al. 2013a, b). In addition, microbial culture collections or microbial resource centre should expand their interest to preserve different groups of organism from all the three domains of life including bacteria, Achaea and eukaryotes. National Centre for Microbial Resource (NCMR), Pune, India (previously known as Microbial Culture Collection), was established in the year 2009. Holding wise it is preserving largest microbial strains from different niches of India and trying to expand its interest in preservation of different groups of microorganisms like anaerobes, cyanobacteria, photoautotrophic bacteria and fungi.

2.7 Need in the Area of Preservation Research

The main aim of culture collection and microbial resource centres (MRCs) is to preserve the viability and authenticity of their microbial resources and make them available to the scientific community when in demand (Lennon and Jones 2011). It has been found that many cultures lose their property on preservation and subsequent revival. Maintaining the cultures in its wild state after its isolation and purification is the major challenge for culture collections due to mutation and genetic drift during the time of preservation and revival. No culture collection or MRCs can give the guarantee that their method is fool proof and there will be no chance of mutation in culture (Smith and Onions 1994). Stability of a microbial culture is an issue, and many curators think it is not the responsibility of a culture collection. It is partly true because due to less funding and lack of sufficient staff they are unable to track and conduct the extensive research in this area. It is also the responsibility of researcher that after isolation, purification and study of strains they should start to initiate the research on preservation of their valuable traits. Therefore, more work and research are required in the area how to prevent the mutation and prevent strain drift during the preservation of microbial resources. It is advisable to culture collections that industrial requirement of a microbial preservation should also be kept in mind while providing services. Unlike taxonomy and basic research, industries will be more interested in preservation of microbe with valuable industrial traits. Thus, preservation of industrially important microbial properties or activities would be more valuable from industrial point of view. Moreover, the waste management industry and biofertilizer industry mostly work on artificially or naturally isolated microbial consortia. Hence, preservation of consortia is one aspect culture collection should look at. However, with the change in research techniques, requirement of research is also changing.

There is also a current need to study the effect of different cryoprotectants and preservation methods on behaviour, cellular and chemical constituents like cell membrane, cell contents, cytoplasm, cell organelles, proteins, lipids, etc. of microbial cell during the time of preservation. It will assist to understand the behaviour of microbial cells at low temperature. It is proven that cryopreservation has revolutionized the animal breeding programs and human organ transplants; similarly we need to study the benefits of cryopreservation in more detail on microbial cells for its extensive application which is still not completely exploited for microbes. It has been found that some strains lose their valuable properties when the culture is preserved at low temperatures or using other methods. Several researchers have raised their worries related to preservation of microbial activity/property of the microbes. Thus, it is a good area of research on how to conserve the functionality or activity of pure cultures or community during the time of preservation.

2.8 Microbial Extinction and Role of Cultivation and Preservation in Conservation of Microbial Resources

Metagenomics-based studies from diverse habitats have revealed that huge microbial diversity of medicinal, industrial, environmental and biotechnological importance is still untapped. Culturing of novel or uncultured microbes is a difficult task. Despite all the efforts, we are only able to cultivate and preserve tiny fractions of microbes. There are 1.5 million fungi estimated (Hawksworth 1991), and so far we have described only 100,000. Now it is sure that we cannot culture and preserve all the microbial diversity even using the modern high-throughput cultivation methods. Study of the past data on microbial preservation and cultivation indicates that preservation of easily cultivated microorganisms in pure culture has been almost optimized with a few exceptions and almost all culture collections of the world are mainly focusing on ex situ preservation of microorganisms. Initially it was believed that unlike plant and animal there is no need of preservation of microbial diversity because they do not face the problems of extinction and endangerment. But result of recent research has proven the due to climatic variations creates the possibility of microbial extinction (Mackelprang et al. 2015; Prakash et al. 2013a, b). In comparison with plants and animals, the concept of community and habitat preservations is not well established in case of microorganisms. Several factors like lack of kinship, Bas-Baking hypothesis and limited knowledge about microbial ecology play major role for less attention about microbial diversity conservation at community and habitat levels. In addition, high levels of resistance towards changing environmental conditions, quick recovery or resilience after perturbation from external stimuli as well as functional redundancy are also responsible up to some extent for lack of care of microorganisms at community and habitat level (Allison and Martiny 2008; Dai et al. 2012). In addition, initially it was considered that endemism or biogeographical barriers do not exist for microorganism, and concept of everything is everywhere particularly in the case of microorganism was established. But the concept of endemism has been established in microbiology. It has also been proven that microbial extinction is possible due to climatic variations and environmental perturbations. Environmental and climatic perturbation least affects the free-living weed type of microorganism, but it deeply affects the slow grower's rare taxa and host-associated microorganisms. Host extinction definitely leads to the extinction of host-specific microorganism, and we will miss them for their application and ecological services. It has been found that global climate change, habitat destruction, host extinction, excess use of antibiotics, use of disinfectants, sanitation practices, excess consumption of processed food, bottle feeding and caesarean birth cause extinction of valuable microorganisms. Now scientific community started thinking that less attention about microbial cultivation and preservation is a kind of scientific short-sightedness which can lead to loss of valuable and protective gene pool, and valuable ecosystem services eventually result disturbed biogeochemical cycles and ecosystem processes. Data from modelling experiment has shown up to 8% extinction in microbial diversity due to change in environmental parameters. It is also clear now that even the use of high-throughput cultivation method and use of modern cultivation approaches unable to tap all the microbial diversity in short period of time which creates the possibility of extinction or loss of valuable microbial diversity. Therefore community, habitat and intact microbiome preservation is the real solution of this emerging problem. Even extracted DNA and protein can also be preserved from habitat of high ecological significance to study the effect of environmental perturbation and climatic variations on community structure and functions.

Considering the importance of microbial diversity and limitation of cultivation and pure culture preservation now, researchers are trying to work on optimization of intact microbiome or whole microbial community preservation for future culturomics, metagenomics and other kinds of omics studies. Several studies have been recently published in this area. Tatangelo et al. (2014) have evaluated the effect of different preservation methods on assessment of bacterial community structure on soil and water samples. In another study Sorakulova et al. use the biopolymer-based immobilization method for transportation and preservation of bacterial and viral cells to protect them from death and DNA damage. Kerckhof et al. (2014) evaluated the effect of different cryoprotectants on preservation of community structure and functions of several microbiomes (Yu et al. 2015). They demonstrated that use of DMSO, tryptone and soy broth preserved the better functionality than unpreserved cells. In another study Morono et al. (2015) demonstrated the effect of alternating magnetic field on intact environmental sample preservation and gave the term cell alive system (CAS). They also demonstrated that CAS gives better protection and viability to the cells in intact environmental samples and highly useful for intact microbiome preservation. Furthermore Clingenpeel et al. (2014) demonstrate the use of cryopreservation for recovery of cells for single-cell genomics and found that cryopreserved cells give better recovery than ethanol- and paraformaldehyde-based preservation method.

2.9 Perspective

Rapid development in DNA sequencing technology- and culture-independent metagenomics approach revealed the black box of hidden microbial diversity and also showed that despite all the efforts only 1–10% microbial diversity has been retrieved in culture and rest is not yet cultured. Culture-independent metagenomics approach can only give information about structural part of microbial diversity means what they are and who are dominating there but unable to exactly explain the role of microorganism in their respective habitats. Therefore to study the functional role of microorganism in any habitats, its cultivation and preservation is imperative. Furthermore, as discussed earlier that unlike previous belief now it has been proven that microbial extinction is possible due to climatic and environmental perturbation. Due to tough nature of microbial cultivation and lack of right understanding about microbial nutrition and metabolisms, it is impossible to cultivate all the available diversity for industrial application and research purpose. Even after cultivation it is mandatory to study the traits and characteristic of microorganisms for future reference and application but that is also not possible in short period of time. Therefore, cultivation as well as optimization of right preservation methods that protect the phenotypic as well as genotypic stability of the microorganism is very essential for microbial diversity research. Although most of the isolated organisms gave good response with existing preservation method and commonly used cryoprotectants like glycerol, DMSO, etc., there are some fastidious organisms that do not give good response and do not survive with existing preservation method. Thus optimization of cryoprotectants and preservation conditions especially for novel organism isolated first time is mandatory before depositing them in culture collections in order to insure their viability and genetic stability. Another big problem of microbial diversity research is that, due to fear of exploitation of industrial and medicinal traits of organism by other groups, most of the microbiologist involved in diversity research cultivate, purify and maintain the isolated novel culture in their own laboratory without depositing them in other national or international culture collection. And after retirement of the professor or shut down of the laboratory, we lose the valuable diversity and along with associated labour and cost spent on their maintenance and isolation. Thus it is advisable here to deposit the isolated microorganism at least in two culture collections of two different countries and also generate knowledge about intellectual properties right related with use and application of microbial diversity and related benefit sharing in order to get rid of related fear to exploitation by others. In conclusion considering all the pros and cons of cultivation and preservation of microbial diversity-based research, we found the researchers involved in diversity research should focus on how to cultivate the untapped novel microorganism from diverse habitats by designing the novel cultivation method and using the available cutting-edge approaches of culturomics. In addition, after cultivation and purification, scientist must be confident about the proposed preservation protocol that ensures its phenotypic and genotypic stability and should optimize in their own laboratory instead of relying on culture collections or bioresource centre. Furthermore, considering the possibility of extinction of valuable microbial diversity due to habitats destruction and host extinction, scientist must focus to develop the protocols for preservation of intact microbiome or communities for future culturomics, metagenomics and other omics type of research.

Acknowledgements We thank the Department of Biotechnology, New Delhi, for providing funds for the establishment of Microbial Culture Collection (MCC) at NCCS, Pune vide grant letter no. BT/ PR10054/NDB/52/94/2007. We also acknowledge the support from DBT Grant No BT/ PR13969/BCE/8/1142/2015 to complete this work.

References

Abadias M, Teixidó N, Usall J, Benabarre A, Viñas I (2001) Viability, efficacy, and storage stability of freeze-dried biocontrol agent Candida sake using different protective and rehydration media. J Food Protect 64:856–886

Abd El-Salam MH, Hippen AR, El-Shafie K, Assem FM, Abbas H, Abd El-Aziz M, Sharaf O, El-Aassar M (2011) Preparation and properties of probiotic concentrated yoghurt (labneh) fortified with conjugated linoleic acid. Int J Food Sci Technol 46:2103–2110. https://doi.org/10.1111/j.1365-2621.2011.02722.x

Alain K, Querellou J (2009) Cultivating the uncultured: limits, advances and future challenges. Extreomphiles 13:583–594

Allison SD, Martiny JB (2008) Resistance, resilience, and redundancy in microbial communities. Proc Nat Acad Sci 105:11512–11519

Bakerspiegel A (1953) Soil as a storage medium for fungi. Mycologia 45:596–604

Berg G (2009) Plant-microbe interactions promoting plant growth and health: perspectives for controlled use of microorganisms in agriculture. App Microbiol Biotechnol 84:11–18

Berg G, Eberl L, Hartmann A (2005) The rhizosphere as a reservoir for opportunistic human pathogenic bacteria. Environ Microbiol 7:1673–1685

Berner D, Viernstein H (2006) Effects of protective agents on the civability of *Lactococcus latis* subjected to freeze – thawing and freezing drying. Sci Pharm 74:137–149

Bharti N, Sharma SK, Saini S, Verma A, Nimonkar Y, Prakash O (2017) Microbial plant probiotics: problems in application and formulation. In: Probiotics and plant health. Springer, Singapore, pp 317–335

Bunse T, Steigleder GK (1991) The preservation of fungal cultures by lyophilization. Mycoses 34:173–176

Carvalcho AS, Silva J, Ho P, Teixeira P, Malcata FX, Gibbs P (2003) Protective effect of sorbitoland monsodium glutamate during storage of freeze dried lactic acid bacteria. INRA, EDP Sci 83:203–210

Carvalho AS, Silva J, Ho P, Teixeira P, Malcata FX, Gibbs P (2003) Effect of various growth media upon survival during storage of freeze-dried *Enterococcus faecalis* and *Enterococcus durans*. J Appl Microbiol 94:947–952

Chian R-C (2010) Cryobiology: an overview. In: Chian R-C, Quinn R (eds) Fertility cryopreservation. Cambridge University Press, Cambridge, pp 1–9

Clingenpeel S, Schwientek P, Hugenholtz P, Woyke T (2014) Effects of sample treatments on genome recovery via single-cell genomics. ISMEJ 8:2546

Cody WL, Wilson JW, Hendrixson DR, Mclver KS, Wagman KE, Ott CM, Nickerson CA, Schurr MJ (2008) Skim milk enhance the preservation of thawed −80°C bacterial stocks. J Microbiol Methods 75:135–138

Culture collections, the new herbaria for fungal pathogens (PDF Download Available). Available from https://www.researchgate.net/publication/227327530_Culture_collections_the_new_herbaria_for_fungal_pathogens. Accessed 16 Oct 2017

Dai L, Vorselen D, Korolev KS, Gore J (2012) Generic indicators for loss of resilience before a tipping point leading to population collapse. Science 336:1175–1177

Epstein SS (2009) Microbial awakenings. Nature 457:1083–1083

Fuller BJ (2004) Cryoprotectants: the essential antifreezes to protect life in the frozen state. CryoLetters 25:375–388

Giovannoni S, Stingl U (2007) The importance of culturing bacterioplankton in the 'omics' age. Nat Rev 5:820–826

Gomez Zavaglia A, Tymczyszyn EE, De Antoni G, Disalvo EA (2003) Action of trehalose on the preservation of *Lactobacillus delbrueckii* ssp. bulgaricus by heat and osmotic dehydration. J Appl Microbiol 95:1315–1320

Gorman R, Adley CC (2004) An evolution of five preservation techniques and conventional freezing temperatures of −20°C and −85°C for long term preservation of *Campylobacter jejuni*. Lett Appl Microbiol 38:306–310

Hawksworth DL (1991) The fungal dimension of biodiversity: magnitude, significance, and conservation. Mycol Res 95:641–655

Heckly RJ (1978) Preservation of microorganisms. Adv Appl Microbial 24:l–53

Heylen K, Hoefman S, Vekeman B, Peiren J, De Vos P (2012) Safeguarding bacterial resources promotes biotechnological innovation. Appl Microbiol Biotechnol 94:565–574

Hoefman S, Van Hoorde K, Boon N, Vandamme P, De Vos P, Heylen K (2012) Survival or revival: long-term preservation induces a reversible viable but non-culturable state in methane oxidizing bacteria. PLoS One **7**:1–9

Hubalek Z (2003) Protectants used in the cryopreservation of microorganisms. Cryobiology 46:205–209

Israeli E, Shaffer BT, Lighthart B (1993) Protection of freeze-dried *Escherichia coli* by trehalose upon exposure to environmental conditions. Cryobiotechnology 30:519–523

Jackson MA, Erhan S, Poprawski TJ (2006) Influence of formulation additives on the desiccation tolerance and storage stability of blastospores of the entomopathogenic fungus *Paecilomycesfumosoroseus* (Deuteromycotina: Hyphomycetes). Biocontrol Sci Technol 16:61–75

Jong SC, Levy A, Stevenson RE (1984) Life expectancy of freeze-dried fungus cultures stored at 4°C. In: Kocur M, Dasilva E (eds) Proceedings of the fourth international conference on culture collections. World Federation of Culture Collections, London, pp 125–136

Kerckhof FM, Courtens EN, Geirnaert A, Hoefman S, Ho A, Vilchez-Vargas R, Vandamme P (2014) Optimized cryopreservation of mixed microbial communities for conserved functionality and diversity. PLoS One 9:e99517

Kirk PM (2012) Index Fungorum no. 2. Index Fungorum [online]. http://www.indexfungorum.org/Publications/Index%20Fungorum%20no.2.pdf

Lagier JC, Hugon P, Khelaifia S, Fournier PE, La Scola B et al (2015) The rebirth of culture in microbiology through the example of culturomics to study human gut microbiota. Clinic Microbiol Rev 28:237–264

Lagier JC, Khelaifia S, Alou MT, Ndongo S, Dione N, Hugon P et al (2016) Culture of previously uncultured members of the human gut microbiota by culturomics. Nat Microbiol 1:16203

Lennon JT, Jones SE (2011) Microbial seed banks: the ecological and evolutionary implications of dormancy. Nat Rev Microbiol 9:119–130

Leslie SB, Israeli E, Lighthart B, Crowe JH, Crowe LM (1995) Trealose and sucrose protect both membranes and proteins in intact bacteria during Drying. Appl Environ Microbiol 61:3592–3597

Liao CH, Shollenberger LM (2003) Survivability and long term preservation of bacteria in water and phosphate buffered saline. Appl Microbiol 37:45–50

Mackelprang R, Jacobsen CS, Jansson J, Tas N (2015) Permafrost metagenomics and climate change. Annu Rev Earth Planet Sci. 43:439–462

Miyamoto-Shinohara Y, Imaizumi T, Sukenobe J, Murakami Y, Kawamura S, Komatsu Y (2000) Survival rate of microbes after freeze-drying and long-term storage. Cryobiology 41:251–255

Morgan CA, Herman N, White PA, Vesey G (2006) Preservation of micro-organisms by drying: a review. J Microbiol Methods 66:183–193

Morono Y, Terada T, Yamamoto Y, Xiao N, Hirose T, Sugeno M, Inagaki F (2015) Intact preservation of environmental samples by freezing under an alternating magnetic field. Environ Microbiol Rep 7:243–251

Nagai T, Tomioka K, Takeuchi K, Iida M, Kawada M, Sato T (2005) Evaluation of preservation techniques of microorganism resources in the MAFE Genebank. JARQ 39:19–27

Nakasone KK, Peterson SW, Jong SC (2004) Preservation and distribution of fungal cultures. In: Mueller GM, Foster MS (eds) Biodiversity of fungi: inventory and monitoring methods. Elsevier, Amsterdam, pp 37–47

Nicholson JK, Holmes E, Kinross J, Burcelin R, Gibson G, Jia W, Pettersson S (2012) Host-gut microbiota metabolic interactions. Science 336:1262–1267

Palmfeldt J, Radstrom P, Hahn-Hagerdal B (2003) Optimisation of initial cell concentration enhances freeze-drying tolerance of *Pseudomonas chlororaphis*. Cryobiology 47:21–29

Park H-K (2006) Long-term preservation of bloom-forming cyanobacteria by cryopreservation. Algae 21:125–131

Patel SM, Pikal MJ (2011) Emerging freeze-drying process development and scale-up issues. AAPS Pharm Sci Tech 12:372–378

Perez-Garcıa A, Mingorance E, Rivera ME, Del Pino D, Romero D, Tore's JA, De Vicente A (2006) Long-term preservation of *Podosphaera fusca* using silica gel. J Phytopathol 154:190–192

Prakash O, Nimonkar Y, Shouche YS (2013a) Practice and prospects of microbial preservation. FEMS Microbiol Lett 339:1–9. https://doi.org/10.1111/1574-6968.12034

Prakash O, Shouche Y, Jangid K, Kostka JE (2013b) Microbial cultivation and the role of microbial resource centers in the omics era. Appl Microbiol Biotechnol 97:51–62

Prakash O, Sharma R, Singh P, Yadav A (2016) Strategies for taxonomical characterisation of agriculturally important microorganisms. In: Singh D, Singh H, Prabha R (eds) Microbial inoculants in sustainable agricultural productivity. Springer, New Delhi, pp 85–101

Raper KB, Alexender DF (1945) Preservation of molds by the lyophil process. Mycologia 37:499–525

Ryan MJ, Smith D (2007) Cryopreservation and freeze-drying of fungi employing centrifugal and shelf freeze-drying. In: Day JG, Stacey GN (eds) Cryopreservation and freeze-drying protocols. Humana Press, Totowa, pp 127–140

Singh BK, Richard D, Smith BP, Reay DS (2010) Microorganisms and climate change: terrestrial feedbacks and mitigation options. Nat Rev Microbiol 8:779–790

Smith D, Onions AHS (1994) The preservation and maintenance of living fungi, 2nd edn. CAB International, Wallingford. ISBN: 085198-902-0

Smith D, Ryan J, Stackebrandt E (2008) The *ex situ* conservation of microorganisms: aiming at a certified quality management. In: Doelle HW, DaSilva EJ (eds) Biotechnology. EOLSS, Encyclopedia of Life Support Systems (EOLSS), Developed under the Auspices of the UNESCO

Streeter J (2003) Effect of trehalose on survival of *Bradyrhizobium japonicum* during desiccation. J Appl Microbiol 95:484–491

Tan SC, Ingen CW, Stalpers JA (2007) Freeze-drying fungi using a shelf freeze-drier. Cryopreservation and freeze-drying protocols. Methods Mol Biol 368:119–125. https://doi.org/10.1007/978-1-59745-362-2_8

Tatangelo V, Franzetti A, Gandolfi I, Bestetti G, Ambrosini R (2014) Effect of preservation method on the assessment of bacterial community structure in soil and water samples. FEMS Microbiol Lett 356:32–38

Tindall BJ (2007) Vacuum-drying and cryopreservation of Prokaryotes. In: Day JG, Stacey GN (eds) Cryopreservation and freeze-drying protocols. Humana Press, Totowa, pp 73–98

Triantafyllou K, Chang C, Pimentel M (2014) Methanogens, methane and gastrointestinal motility. J Neurogastroenterol Motil 20:31–40

Turnbaugh PJ, Ley RE, Hamady M, Fraser-Liggett C, Knight R, Gordon JI (2007) The human microbiome project: exploring the microbial part of ourselves in a changing world. Nature 449:804–810

Winters RD, Winn WC Jr (2010) A simple effective method for bacterial culture storage: A brief technical report. J Bacteriol Virol 40:99–101

Yu C, Reddy AP, Simmons CW, Simmons BA, Singer SW, Vander Gheynst JS (2015) Preservation of microbial communities enriched on lignocellulose under thermophilic and high-solid conditions. Biotechnol Biofuels 8:206. https://doi.org/10.1186/s13068-015-0392-y

Zaidi A, Khan MS, Ahemad M, Oves M (2009) Plant growth promotion by phosphate solubilizing bacteria. Acta Microbiol Immunol Hung 56:263–284

Zengler K, Toledo C, Rappe M, Elkins J, Mathur EJ, Short JM, Keller M (2002) Cultivating the uncultured. Proc Natl Acad Sci USA 99:15681–15686

Chapter 3
Establishment and Management of Culture Collections of Microorganisms (mBRC): An Overview

Duong Van Hop

Abstract Microbes are the most diverse of all living organisms and play crucial roles in nutrient cycling that support all life forms on Earth. The conservation and preservation of microbes are important for the life sciences, especially the development of sustainable agriculture. This chapter introduces the main requirements for establishing and operating a general public microbial bioresource center, including funding, holdings, facilities, and staff, as well as maintaining, depositing, and distributing microbial cultures. Updated multi-technical aspects of the preservation of bacteria, actinomycetes, fungi, mushroom, microalgae, and microbial communities, in reference to some potential applications in agriculture, are discussed, such as using fungi and mycorrhizae for short-, medium-, and long-term preservation. Additionally, current international collaborations are discussed, including access and benefit sharing of microbial genetic resources under the Convention on Biodiversity and Nagoya Protocol in Asian countries under the Asian Consortium for the Conservation and Sustainable Use of Microbial Resources. A case study on the Nagoya Protocol on access and benefit sharing in a bilateral collaboration between the National Institute of Technology and Evaluation, Japan, and the Institute of Microbiology and Biotechnology, Vietnam National University, Vietnam, is also evaluated.

3.1 Introduction

Microbes are microscopic organisms, including bacteria, actinomycetes, fungi, viruses, and algae. Microbes form most of the biomass on the Earth and play important roles in biogeochemical processes that are important for the survival of all life forms (Sharma et al. 2017a). The discovery of the first microorganisms was made possible by the work of Antonie van Leeuwenhoek in 1677. However,

D. Van Hop (✉)
Vietnam Type Culture Collection (VTCC), Institute of Microbiology and Biotechnology, Vietnam National University, Hanoi, Vietnam
e-mail: hopdv@vnu.edu.vn

S. K. Sharma, A. Varma (eds.), *Microbial Resource Conservation*, Soil Biology 54,
https://doi.org/10.1007/978-3-319-96971-8_3

microbial activities and functions were made evident through the study of lactic acid fermentation by Pasteur (1822–1895) and the study of anthrax by Koch (1876). Rhizobia were isolated and identified by Beijerinck (1888), and microorganisms were recognized as important to humans.

Culture collections were first established in European countries. Frantisek Král (1846–1911) established the first culture collection of bacteria in 1890 at the German University, Prague, and it is the oldest culture collection in the world. In addition, the Culture collection of the Pasteur Institute (1891, France), the Mycothèque de l'Université catholique de Louvain (1982, Belgium), the Centraalbureau voor Schimmelcultures (CBS; 1904, Netherlands), and the Centre for Agriculture and Biosciences International (CABI; 1910, UK) were established. Traditionally, culture collections focus on a single particular group of microbes, such as bacteria, fungi, or entomopathogens.

The Organization for Economic Cooperation and Development (OECD) established the concept of a Biological Resource Center (BRC). BRCs are both service providers and repositories of the living cells, genomes of organisms, and information relating to heredity and the functions of biological systems. BRCs contain collections of cultivable organisms (microorganisms, plants, animals, and human cells), replicable parts of these organisms (genomes, plasmids, viruses, and cDNAs), viable but not yet cultivable organisms' cells and tissues, as well as databases containing molecular, physiological, and structural information relevant to these collections and related bioinformatics (OECD 2007).

As the value of microbial BRCs (mBRCs) has become apparent (Stern 2004), the number of service collections, now greater than 600, has increased. The register of culture collections at the World Data Centre for Microorganisms (WDCM) has data on almost 746 culture collections worldwide that hold over 2.6 million strains of microorganisms, mainly bacteria and fungi. Almost 130 of these are collections affiliated with the World Federation for Culture Collections (WFCC), and as such, they operate under WFCC guidelines (Smith et al. 2014).

3.2 Establishment

3.2.1 Necessity and Impact of mBRCs in the Life Sciences

3.2.1.1 Necessity

Microbes were likely the first forms of life to develop on Earth, approximately 3–4 billion years ago, and were involved in the development of the living world (Schopf 2006). Microbes can live in extreme ecosystems, such as hot spring and very cold areas, as well as in acidic and alkali environments. Microbes likely played a role in the Earth's atmosphere changing from supporting anaerobic to aerobic life and, directly or indirectly, in the evolution of the living world. Microbes are involved in many element (N, C, P, S, and more) cycles. There have been recent revolutions in various fields of science and technology, such as molecular biology, nanobiotechnology, and bioinformatics, and these, along with new materials, provide humans more power to address many problems, such as climate change, food security and safety, and virulent diseases, such as influenza A and influenza Zika. To overcome these problems, maintaining and utilizing biodiversity are crucial. Microbes are much more diverse than plants and animals. The combined number of plant and animal species is estimated as 2–50 million (https://www.factmonster.com/science/animals/estimated-number-animal-and-plant-species-earth), but the figure for microbes is ~1 trillion (http://www.newsweek.com/new-estimate-there-are-1-trillion-species-microbes-earth-454714). At present, most animal and plant species have been identified and characterized, but, for microbes, only an estimated 0.1–10% has been identified and characterized (Kennedy 1999). However, microbes make great contributions to the life sciences. In medical and pharmaceutical fields, the most important application of microbes is in antibiotics. Flemming isolated the first antibiotic from *Penicillium* in 1928 (Tan and Tatsumura 2015). At present, ~16,500 antibiotics have been isolated from microbes, mainly from actinomycetes (~8700) and fungi (~4900) (Bérdy 2005). In addition, more than 4000 bioactive compounds, including amino acids, have various useful applied activities, such as antiviral and anticancer, as well as fungicidal and bactericidal. The global antibiotic market alone reached more than an estimated 1000 billion USD in 2014 (http://www.abpi.org.uk/industry-info/knowledge-hub/global-industry/Pages/industry-market-.aspx#fig 2). Presently, humans are facing problem of antibiotic resistance in microbes. Thus, identifying new antibiotics and using new microbial biotechnologies are important for protecting humans.

In Industrial Applications

Microbes, directly and indirectly, are being used for various applications in industry, such as in brewing and food processing, and in cosmetic and dairy products, such as yogurt and fermented foods. The global market value of microbe-related industries is more than 1000 billion USD. The global beer market alone was valued at approximately 530 billion USD in 2016 and is expected to reach approximately 750 billion USD by 2022 (https://www.zionmarketresearch.com/report/beer-market).

Agricultural and Environmental Applications

According to the US Department of Agriculture (USDA), global rice and grain productions were ~480.3 and 680 million tons in 2016, with a value of more than 3000 billion USD (https://apps.fas.usda.gov/psdonline/circulars/grain-rice.pdf). In addition, some other agricultural products can be used for industry, such as coffee, cacao, rubber, and tea, to satisfy human demands. The biodiversity of microbes has crucial roles in agricultural production, and they are important factors for the bioconversion involved in all important element cycles, such as those of C, N, P, and S. Through their biological activities, many mineral elements and nutrients are provided for agricultural cultivation. Lignocellulose from plants is the most abundant organic carbon source in the biosphere. For biomass production, the plant needs many elements, including C, N, P, K, and S, from the environment. In sunlight, the plant can biosynthesize lignocellulose, but significant amount of soluble organic C and N comes from the soil as microbial biomass and the degradation products of microbial organic materials. Many groups of microbes (such as *Azotobacter* and *Azospirillum*) can fix atmospheric nitrogen in a soluble form that is easily assimilated by plants. Phosphorus, potash, and sulfur exist mostly in insoluble forms in the soil, but some microbes can solubilize and mineralize these elements for plants using various manners, such as organic acid production (acetic, gluconic acid, malic acid, alpha-ketoglutaric acid, and lactic acids) or enzymes (phosphatase and phytase). *Mycorrhizae* improve plant growth by the development of symbiotic relationship with the roots of 95% of plant families and thereby aid in nutrient exchange, increase disease and drought resistance, and ultimately reduce the need for chemical fertilizers. Additionally, they can produce some useful hormones that stimulate plant growth and maintain a good soil structure as carriers for many nutrients and microflora that maintain micro-ecosystems. Presently, humans are facing problem of climate change, and resulting in reduction in agricultural areas in the many parts of the world due to rise and of sea level. The top countries at risk from the negative impacts of climate change belong to Africa and Southeast Asia, including Zimbabwe, Madagascar, Mozambique, Cambodia, Vietnam, and the Philippines as well as other countries, such as Haiti and Bangladesh. The Vietnamese Ministry of Resources and Environment estimates that when the sea level increases by 1 m, the agricultural area in the Cuu Long River Delta of Vietnam will be reduced by 30–40%. The annual rice production will be decreased by 7.6 million tons, and the gross domestic product will be reduced by 10%. In addition, the world's population is dealing with food security and safety issues. With revolutions in the sciences, especially in molecular biology and technology, many new varieties of plants, including genetically modified ones that can grow in extreme conditions, such as saline and sandy soils, have been created, and technology has been applied to agriculture to increase food production. In addition, many kinds of microbial-based fertilizers have been developed to enhance productivity by improving the quality of grain and nutrients available in cultivated soils. The biodiversity and numbers of microbes in soil are strong indicators of its quality and bioproductivity.

Food-borne diseases are risks to humans. Various diseases are related to the accumulation of chemical pesticides and fertilizer residues on food. The use of chemical fertilizers and insecticides in agriculture is decreasing. Various kinds of microbes, including bacteria, yeast, microalgae, and actinomycetes, are now used for microbial fertilizer production. By using microbial fertilizers, the amount of chemical fertilizer used can be decreased by 52% (Gray et al. 2013). The yield of plants and crops is treated for biodisasters caused by mainly insects and pathogenic microbes. Overuse of chemical pesticides is a main reason for human liver and kidney diseases. Many microbes such as *Beauveria*, *Metarhizium*, *Chaetomium*, *Paecilomyces*, *Lecanicillium lecanii*, *Bacillus thuringiensis*, *Bacillus subtilis*, and *Streptomyces* (Usta 2013) have been used successfully as bioinsecticides, fungicides, and bactericides. The trend in green and sustainable agricultural development has gained the public's attention; thus, qualified biofertilizers and pesticides are being developed. The global bioinsecticide market is estimated to reach a value of 3.18 billion USD by

2021. The market is driven by factors such as increasing levels of pest attacks due to climate change and the implementation of biological technologies for pest control (http://www.marketsandmarkets.com/PressReleases/bioinsecticides.asp). The global biofertilizers' market size was estimated at 535.8 million USD in 2014. The agricultural countries like India, China, Vietnam, and the United States introduced national initiative aims at increasing biofertilizer production, distribution, and utilization, which spurred the biofertilizers' market growth. In the United States, the biofertilizer market value was ~140 million USD in 2017, but the figure is expected to reach ~280 million USD in 2022 (http://www.grandviewresearch.com/industry-analysis/biofertilizers-industry).

The world population will reach 7.5 billion in 2017 (http://www.worldometers.info/world-population/), compared with 6.1 billion in 2000. The increase in the human population will result in increasing demands for food, healthcare, energy, and transportation. Many countries are facing urbanization and industrialization, resulting in heavy environmental pollution. Pollution, especially that caused by organic sources, is one reason for the outbreak of various kinds of viral diseases, such as *Vibrio chlorella*, *Escherichia coli* H157, influenza A, and malaria. To address this problem, progress in environmental technology and in the sciences has been made. Microbial products are applied in a variety of these technologies to help treat environmental pollution.

3.2.1.2 Impacts of mBRCs

As with plants and animals, or any organism, microbes are a national resource. The diversity level of microbes has potential for the development of the life sciences. Because mBRCs are part of the important infrastructure for the life sciences and biotechnology, they play several main functions.

Ex Situ Conservatories of Microbes and Custodians of National Microbial Biodiversity

There are three levels to biodiversity, gene, species, and ecosystem. The megadiverse countries are a group of 17 countries, Australia, Brazil, Colombia, the Democratic Republic of the Congo, China, Ecuador, India, Indonesia, Madagas car, Malaysia, Mexico, Papua New Guinea, Peru, the Philippines, South Africa, the United States, and Venezuela, that harbor the majority of the Earth's species and have high numbers of endemic species. Some of these countries are located in, or partially in, tropical or subtropical regions. However, biological diversity has been reduced significantly by certain human activities, mainly industrialization and urbanization. Microorganisms are not only of value for the production of useful substances, they also play unique roles in element cycles with plants and animals. Microorganisms are also the sources of significant gene pools for research and applications in various fields. In addition, a healthy environment requires green and sustainable economic development. Microbial biodiversity is crucial for this goal and should not be lost. Thus, microorganisms can be regarded as part of the cultural heritage and property, and they must be transferred to the next generation in a normal and healthy condition (http://www.iupac.org/symposia/proceedings/phuket97/komagata.html).

Providing Qualified Microbes as Starting Genetic Resources for Scientific, Agricultural, Environmental, Medical, and Pharmaceutical Research and Applications
Maintained microbes are not only preserved but used. Microbes, directly and indirectly, are used in almost all fields related to the life sciences and biotechnology. Scientific investigations and applications must be reproducible. Experiments performed in one laboratory by one set of investigators and facilities replicable in another laboratory. Reliable standardized information on microbes can be very important for particular experiments or applications to be conducted. Microbes are very diverse, but their roles in investigations and their applications are even more diverse. No single scientist can conduct all of the research involving a particular microbe or genetic material, such as a gene or plasmid. Scientific and technological progress, especially in molecular biology, bioinformatics, and nanotechnology, are providing researchers tools and knowledge for microbial investigations. All of these achievements are documented in various forms, such as publications and patents. This knowledge is not only used for conducting more effective and economic research but also for preventing a user from having to "reinvent the wheel" or invest considerable time and money in finding microbial genetic resources and their characterizations. To make mBRCs accessible to the microbiological research community and industrial scientists, dispatching microbial cultures is needed in compliance with national and international regulations.

Repositories of Microbial Genetic Resources and Receiving Safe Deposits for Publication and Intellectual Property Protection
The demand for microbes as starting genetic materials for research and applications is increasing annually. As microbial gene pools, mBRCs must have large quantities of diverse cultures. As custodians of national microbial resources, mBRCs must collect and accept deposits of microbes from internal and external sources. Deposits of microbes for the purposes of patent application and publication should be in compliance with the Budapest Treaty and a Material Transferring Agreement (MTA) with the depositor. In accordance with the Budapest Treaty, International Depositary Authorities keep deposited microbes secret but must furnish the microbes to entitled third parties upon application to the national patent authorities. Some culture collections also provide a special service for the long-term preservation of microbes. The distribution of preserved microbes may be restricted at the discretion of the depositor. Such "safe deposits" of microbes are a way to ensure long-term preservation without loss of ownership.

Conducting Research and Development (R&D) on Microbes
It is required for mBRCs to provide users with qualified microbes and related information. The quality of the microbes includes viability, purity, and identity. The required information mainly is the scientific name, source of isolation, and

conditions for handling and re-culturing. In addition, information on the microbe's biological criteria should be provided to acquaint the user with how to conduct appropriate research and applications. Thus, mBRCs can promote related research and disseminate expertise for preservation. Additionally, they maintain and distribute information related to taxonomy and identification, including the characterization of the most important biological criteria, biosafety, and application potential. The mBRCs maintain many microbes requiring them to balance these activities with other services.

Performing Related Service and Consulting
Based on R&D activities using microbes, mBRCs should be proficient in various techniques and methods and have experience in microbial preservation, taxonomy, identification, and other analyses related to the following activities (Çaktü and Türkoğlu 2011):

1. To provide identification services regarding different kinds of microorganisms in accordance with the expertise of the culture collection.
2. To organize training courses and workshops that are related to the identification and maintenance of microorganisms. Short courses and workshops are essential to train personnel from medical, environmental, industrial, and governmental laboratories that are responsible for isolating and identifying microorganisms, diagnosing disease, quality control, and fermentation, and culture management.
3. To provide general advice in the field of microbiology. This advice is not limited to research and the applications of microbes but includes consulting on internal and external policies and regulations relating to microbial sources.

3.2.2 Requirements for Establishment

The increasing demands on culture collections for authenticated, reliable biological materials and associated information parallel the growth of biotechnology. Recently, worldwide recognition of the necessity to conserve the microbial gene pool for future study and exploitation by mankind has highlighted the need for centers with expertise in cultural isolation, maintenance, identification, and taxonomy. Soon after the Earth Summit in Rio de Janeiro in 1992 approved the Convention on Biological Diversity (CBD) as an organization that promotes the culture collection of microorganisms, the WFCC prepared the first guidelines for the establishment and operation of such activities in 1994 (http://www.wfcc.info/index.php/wfcc_library/guideline/). The guidelines were improved in second (1999) and third (2010) editions to provide a framework for the establishment, operation, and long-term support of microbiological and cell resource centers as fundamental parts of the scientific infrastructure (http://www.wfcc.info/guidelines/). There are several key components that should be taken into account for the establishments, such as (1) organization, (2) the aims or objectives, (3) funding sources, (4) holdings, and (5) human resources.

3.2.2.1 Organization

There is a parental body, under which mBRCs are established. The mBRCs should have the commitment from the organization for the long-term maintenance and implementation of its services. The activities of mBRCs should be appropriate for the organization's strategic plans or objectives. The organization may be classified into one of the two categories, government or nongovernment.

The mBRCs of universities and institutions, which are supported by the government, are the responsible authorities likely house public collections. These consist of a large number of strains and are established for the purpose of public service. The number of cultures varies from collection to collection. For example, the CBS (www.cbs.knaw.nl) in the Netherlands maintains over 18,000 strains of fungi, while there are 38,000 strains in the National Institute of Technology and Evaluation's Biological Resource Center (NBRC), Japan, and 37,000 strains in Thailand Bioresource Research Center of the National Center for Genetic Engineering and Biotechnology, Thailand.

For nongovernment sectors or private organizations, private or in-house culture collections are usually maintained by individuals, laboratories, institutes, hospitals, and commercial firms. The number of holdings may range from small (hundreds of strains) to large (thousands of strains). These culture collections can be used as in-house reference materials or be commercially and publicly accessible. Some collections are comparable to public collections, such as the American Type Culture Collection (ATCC), USA, with over 75,000 cultures. The strains in private culture collections are usually not available to the public.

3.2.2.2 Objectives

The mBRCs should have long-term objectives. These objectives should be related to the scope of the holdings and other envisaged services. In addition, some specific short- and midterm objectives are acceptable and helpful. However, the aims of mBRCs require careful consideration by, and appropriate to, the funding body. The objectives have to be practical and based on funding sources, holding, and available capacities, as well as manpower and facilities. Thus, the service and quality levels at various mBRCs are different. Some mBRCs focus on maintaining particular microbes, such as algae (Plymouth Algal Culture Collection, UK) or fungi (CBS, the Netherlands). The services of some private culture collections only include the preservation and distribution of microbial materials. In general, as standard roles of mBRCs, public mBRCs should provide important high-quality preservation: depository, distribution, taxonomy, and identification services.

3.2.2.3 Funding Sources and Facilities

Funding

Microbial domain Biological Resource Center (mBRC) facilities are ex situ, requiring long-term maintenance and related services. The parental organization must commit to funding the mBRC for the long term to maintain all its activities. Because the funding is crucial for quality and qualified mBRC-related services, the objectives and the activities have to be restricted. To match user's demands on cultures, many kinds of activities and services should be well planned with appropriate funding sources. In general, funding comes from two sources, the parental organization and service-associated fees or both. The parental organization is the government, private sector, or industrial support. Approximately 300 of 746 culture collections are supported by governments (http:/www.wfcc.info/ccinfo/statistics/). For most public mBRCs, the main funding (70–90%) is from the government, while services cover 10–30% of the total running cost. The funding required for developing countries is generally higher than those of developed countries. For many private mBRCs, the running costs come from the parental organization and services, but in some cases, they are partly funded by the government, which can subsidize private mBRCs that belong to quasi-government sectors, such as universities and institutions.

Facilities

Housing, laboratories, and equipment are important for quality control in mBRCs. The differences in holdings and servicing activities in various mBRCs result in differences in facilities. Progress in many fields of science and technology, including molecular biology, bioinformatics, nanobiotechnology, chromatography, and microscopy, has provided excellent tools for important service activities such as taxonomy, identification, and research. The users are in need of qualified microbial materials with well-documented profiles. Many standard mBRCs act as a unique research institutes in microbiology equipped with scanning electronic microscopes, transmission electronic microscopes, next-generation sequencers, matrix-assisted laser desorption/ionization in time-of-flight equipment, and gas chromatography–mass spectrometers. All of the equipment should be regularly maintained to ensure reproducible results.

3.2.2.4 Holdings

Holdings represent the scope and number of microbial genetic resources to be maintained. The holdings are required to match the objectives and the parental organization. Some mBRCs are specific to a particular group of microbes such as fungi, bacteria, algae, or microbes pathogenic to humans, animals, or plants. However, most of the public mBRCs such as the Vietnam Type Culture Collection (VTCC), Vietnam; ATCC, USA; Thailand Bioresource Research Center, Thailand; and NBRC, Japan, hold various kinds of microbes. There is no culture collection that can hold all microbes. To increase the effectiveness of the services provided to the users, the storage capacities, human and financial resources, and other facilities have to be taken into account with careful consideration. To improve services for users, collaborations with other collections in a network at the national (Thailand Network Culture Collection), regional (Asian Consortium Microbial Resources), and global (World Federation for Culture Collections) levels are recommended. One important criterion for culture collections lies in reflecting the diversity of the local and national genetic sources to be maintained. To meet broad user demands related to fundamental and applicable studies, three kinds of cultures, type cultures, reference strains, and patent cultures are preserved. Duplicating cultures from other collections is not recommended. It is economically prudent to maintain complementary rather than duplicate resources; however, some duplication of important reference strains is acceptable. The WFCC recommended that new collections should collectively enrich the available genetic resources rather than duplicate those already existing; however, some authenticated internationally recognized reference strains are acceptable. The strength of a collection is indicated by the quantity and quality of distributed strains and provided services for a broad range of internal and external users. Thus, the diversity of maintained geographical microbes at country and region levels is crucial.

Microorganisms are useful and beneficial for mankind but can be misused against humans. For holdings of microbes pathogenic to man, animals, and plants, or those producing toxic compounds, biosafety and security must be maintained. This is a shared responsibility of government, science, industry, and the community, and they must work together to implement the best practices without impeding scientific development (Smith et al. 2017a). The holdings must be in compliance with all safety and security regulations in force, such as BTWC (Toxin Weapons Convention) or the EC Directive 2000/54/EEC on the control of biological agents. To develop safe practices, it is recommended to follow the guide in the OECD Best Practices on Biosecurity (2007) or World Health Organization (2006) on biorisk management (http://www.who.int/csr/resources/publications/biosafety/WHO_CDS_EPR_2006_6.pdf). National legislation in many countries requires permits or licenses to store, work with, and distribute such microbes. Unfortunately, many countries lack national regulations to manage biorisks (Smith et al. 2017b).

3.2.2.5 Staff

Managing a culture collection is a demanding task. It requires knowledge of not only the organisms themselves but of many other requirements, such as preservation, taxonomy, properties, and potential applications, including other services to meet the customer's requirements. The staff recruited would be expected to have culture collection curation skills.

The number of staff required for full- and part-time positions depends on the holding size and servicing activities, such as accession, preservation, quality control, and other services that the collection is required to provide. Unlike other positions, culture collection management involves repeated activities and requires reproducible results. Thus, particular attention should be paid to the qualifications and experience of the staff in charge of the collection. To maintain the standards and quality of the culture collection and the services provided, retaining sufficient experienced and responsible staff is important. Frequent staff turnover is not recommended. Users of microbial materials are not always experts in microbiology; thus, they often erroneously handle the materials, fail to revive the culture, or contaminate the materials, resulting in claims against the service. In these cases, the curators have to address the claims. Thus, curators must have a wide knowledge of microbiology, not just in taxonomy but also in various related fields, such as physiology, biochemistry, pathology, and molecular biology. Curators should not have other positions or jobs. It is not possible for the culture collection staff to all be specialists in taxonomy, identification, and authentication; however, some basic taxonomic skills are needed for quality control. When in need of a specialists' expertise, such as in taxonomy or identification, it may be preferable to have collaborative arrangements with specialists outside of the organization.

3.3 Management and Operation

The general service collection of microorganisms is part of the essential infrastructure for the life sciences in which microorganisms and their derived materials are employed in both fundamental and applied research activities for various fields, such as biotechnology, industry, agriculture, medicine, pharmacy, and environment. The crucial functions of culture collections are preserving, supplying, and collecting microbial sources.

3.3.1 Preservation

The objective of preserving and storing cultures is to maintain them in viable states without morphological, physiological, or genetic changes until they are required for use. Ideally, microorganisms require special preservation methods to ensure the maintenance of the following crucial indicators:

- Optimal viability/recovery of the preserved culture. Cell death may occur during the preservation process and further losses during storage. To avoid the loss of cells, the preservation process and methods used should minimize the loss of viability over a long period.
- Purity. No contamination should be present in the preserved culture (this does not include any recognized co-culture, such as symbiotic microorganisms or microbial communities, which are not regarded as contaminants). Cultures preserved for both fundamental and applied uses are required to be pure to provide

reproducible results. The preservation method used should minimize the probability of contamination.
- Stability. The preserved culture is valuable and important for various uses in the life sciences and industry. During the preservation process and storage, some mutations may occur or plasmids lost, resulting in the loss or gain of some characteristics. Thus, the preservation method should minimize undesirable changes in morphology, biochemical properties, genetic profiles, immunity levels, and pathogenicity.
- Identity/authenticity. Preserved cultures should be the same as the original culture deposited at the mBRC by the depositor.
- Long-term security. The cultures should be stored using at least two different preservation methods. One should be a long-term preservation method, either lyophilization or quick freezing in liquid nitrogen. It is recommended that a backup collection should be stored at a distant site to protect the cultures from any disaster.

There is a wide variety of available methods for the preservation of microorganisms. The method chosen for preservation depends on some crucial factors, as listed below (Kirsop and Snell 1984):

- *Funding:* Maintaining culture includes the cost of staff, equipment, materials, and facilities, such as laboratories, storage space, and power supply.
- *Holdings*: The number of cultures maintained and the kind of microorganisms may require particular methods, staff skills, equipment, and storage space.
- *Value of the culture:* The consequences of the loss of a culture should be considered when choosing one preservation method.
- *Supplying and transporting cultures:* If the cultures are to be distributed, then replicates are needed. For later distributions, cultures should be prepared in bulk and stored. The convenient distribution of cultures depends on the preservation method used and the number of cultures to be distributed. If cultures are to be supplied by mail, then they must be packaged in a suitable form to survive any possible delay in delivery.
- *Frequency of culture use*: Some cultures are frequently used as assay strains or in industrial production. In these cases, the risk of contamination is high, so the use of stock cultures should be considered.

It may be difficult to choose the most appropriate method for a particular microorganism. For culture of each microorganism, an appropriate method (s) should be chosen by the mBRC, the recommendation of the depositor, or some other reference mBRC, such as the NBRC (Japan), the ATCC (USA), and CABI (England). The available methods range from growing the culture at a reduced metabolism rate to the ideal situation in which the metabolism is suspended. The important available preservation methods may be classified into three categories, sub-culturing, drying, and freezing.

3.3.2 Sub-culturing Preservation Methods

Sub-culturing This is one of the oldest and the most traditional methods of preservation. The microorganism is grown on appropriate medium over a period of days or weeks. Then, the culture is transferred to storage at a low temperature, normally 5 °C for a variety of microorganisms. However, not all microorganisms survive longer at low temperatures. For example, *Neisseria* spp. survives better at 37 °C (Kraus 1979). The process is repeated at an appropriate interval that ensures the preparation of a fresh culture before the old one dies. The period of preservation is different depending on the culture. Many bacteria, yeast, and microalgae strains can survive from some months to years. The advantage is that the method is inexpensive in terms of equipment and little skill is required. Sub-culturing is the simplest and most common method used in small laboratories, especially if cultures are required frequently and quickly. In addition, this method is applicable to a wide range of microorganisms. However, this method has some disadvantages. The most notable disadvantages are:

Contamination: This occurs frequently, especially when large numbers of cultures are involved or the technician is distracted.

Change in characteristics: Sequential transfers often result in high rates of mutation and plasmid loss. The culture's characteristics, such as enzyme activity, immunity, and pathogenicity, are not stable, and undesirable characteristics may arise.

Mislabeling and misplacement: Cultures may be labeled with the wrong name or number. The label may become distorted or illegible. Wrongly inoculated cultures may occur when a large number of cultures are being transferred. Cultures may be inoculated into the wrong tube or several organisms placed in the same tube.

Culture loss: In many cases, the loss is related to equipment failure, such as insufficient stability in incubation and refrigeration temperature.

Storage space: Large numbers of test tube cultures require very large storage and incubation spaces.

3.3.2.1 Filamentous Fungi

1. *On agar medium* Fungi can be grown in various common agar media, such as Malt Agar, Czapek Dox, potato dextrose agar (PDA), yeast extract sucrose, and Sabouraud dextrose agar. However, some fungi require a specific medium, such as *Corticium praticola*, which requires soil-extract agar. The culture is grown in the medium under optimum conditions, including temperature, aeration, pH, and humidity. It is important that the pure culture is in good health prior to use as an inoculum with the full morphological and bio-physiological criteria as the original deposit.

Using spores from a culture as inoculum is optimum, but mycelia of sporeless fungi are acceptable. It is recommended that the culture be undergoing sporulation or growth before storing. In general, the storage temperature is lower, slowing the metabolic rate, which allows more time between sub-culturing. Storage at 4–7 °C or in a cold room can extend the transfer interval to 4–12 months. Cultures growing on agar may be infested by mites. Mites can be detected by the naked eyes at twice weekly intervals. If mites are detected, then the contaminated culture should be destroyed by autoclaving at 121 °C for 15 min (Hunter-Cevera 1996).

To increase the interval transfer, cultures can be stored under oil or water:

2. *For storing under oil* This method is the simplest and most inexpensive for long-term culture preservation, especially for fungi that do not tolerate freeze-drying (Humber 1997). The oil prevents desiccation and diminishes gas exchange, resulting in a reduced fungal metabolism. Cultures kept under mineral oil may remain viable for decades (Mendes da Silva et al. 1994). Some cultures, such as those of *Aspergillus*, *Penicillium*, *Phytophthora*, and *Pythium*, in tubes/bottles with loosened caps can be stored at 15–20 °C for 1–40 years. However, for other organisms such as *Cercospora*, *Arthrobotrys*, *Colletotrichum*, *Conidiobolus*, *Corticium*, *Nodulisporium*, and Basidiomycetes cultures should be transferred at 2-year intervals (Hunter-Cevera 1996). To store fungal cultures under oil, the mature culture is placed on an agar slant or aseptically covered bottle of liquid paraffin [medicinal paraffin, specific gravity of 0.830–0.890 to a depth of 1 cm (Kirsop and Snell 1984)], which underwent autoclaving (121 °C, 15 min) twice, 24 h apart. The culture is covered with a tight cap and stored upright in a rack at 15–20 °C. A sterile loop or needle is used to recover an explant from the culture. Drain the excess oil before placing the sample on fresh appropriate agar petri plate medium. The culture must be monitored for viability and contamination before transferring to a fresh slant agar medium to obtain a pure culture.

3. *Storing under water* The storage of metabolically inactive fungi under sterile distilled water may be the least technologically demanding of any preservation techniques. This method has been used successfully with a wide range of fungi, including human and plant pathogens, and the *Entomophthorales*. The viability levels of some fungi are different from individual cultures. A shelf life of 20 years has been documented with this method (Castellani et al. 1967; Figueiredo and Pimentel 1975). However, it is recommended that cultures be stored for 2–5-year periods (Hunter-Cevera 1996). To recover the culture, the same techniques and attention as described for oil storage are applicable.

3.3.2.2 Mushrooms

A mushroom (or toadstool) is the fleshy, spore-bearing fruiting body of a fungus, typically produced above the ground on soil or its food source. The standard "mushroom" is the cultivated white button mushroom, *Agaricus bisporus*. Thus, the word "mushroom" is most often applied to those fungi (*Basidiomycota* and *Agaricomycetes*) that have a stem (stipe), a cap (pileus), and gills (lamellae) on the underside of the cap. Mushrooms have significant values. Edible mushrooms such as *Agaricus* and *Pleurotus* are useful food sources and with some, such as *Ganoderma*, *Lentinula edodes* (shiitake), and *Grifola frondosa* (maitake), being good sources of functional foods or traditional medicinal materials, although many are toxic to humans and animals, such as *Amanita*, *Cortinarius*, and *Podostroma*.

Mushrooms are invariably stored as mycelial cultures because the spores of heterothallic and secondary homothallic species are produced through asexual processes and have genetic differences, which may eventually result in the failure to fruit. In the absence of hardy structures, like double-walled spores or sclerotic, cultures become susceptible to sudden changes in temperature and pressure.

Sub-culturing is used for current applications of many genera. The techniques are simple and much the same as for filamentous fungi. In our laboratory (VTCC), cultures are sub-cultured onto complete PDA medium slants. The cultures are incubated at 28–30 °C for 7–14 days to achieve full mycelial growth before storing at 5 °C. Some strains prefer to be stored at 10–15 °C. The shelf life is 3–6 months, depending on the individual strain. Using this method, the shelf life of *Lentinula*, *Pleurotus*, and *Agaricus* were up to 6 months (Chang and Miles 2004). Additionally, various mushroom cultures have been stored on sorghum (jowar) grains at low temperatures of 5–8 °C in a refrigerator (Veena and Pandey 2010). The mushroom cultures could be safely stored at low temperatures on sorghum grain free from contamination for longer than 1 year without any growth or morphological changes. The most significant advantage of this method is its ability to conserve milky mushroom (*Calocybe indica*) and some isolates of reishi mushroom (*Ganoderma lucidum*) cultures, which cannot be stored at low temperatures. Recently, Garcia-Garcia et al. (2014) reported a filter paper discs method for storing medicinal mushroom mycelia in sterile distilled water (Fig. 3.1). *Humphreya coffeata* (Berk.) Stey. stored at 4 °C was viable for up to 18 months, with no visible morphological alterations or contamination by bacteria or other fungi. Thus, this method, in addition to being easy and economical, is suitable for the conservation of higher Basidiomycetes. It should be noted that the viability period depends on the fungal species being stored. The method is briefly described below:

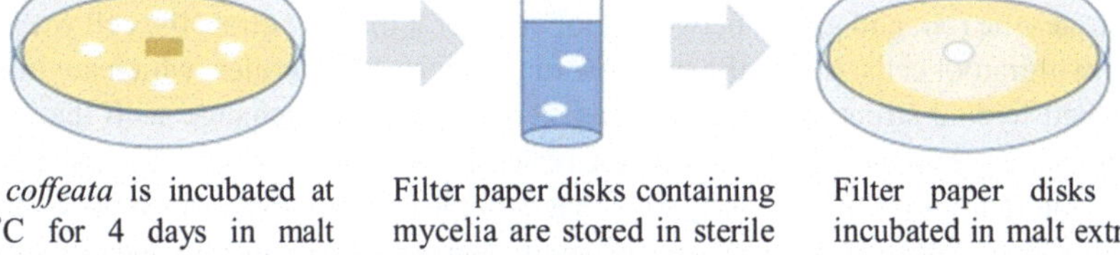

H. coffeata is incubated at 30°C for 4 days in malt extract agar with small disks of filter paper

Filter paper disks containing mycelia are stored in sterile distilled water vials at 4°C for 18 months

Filter paper disks are incubated in malt extract agar after storage of mycelia

Fig. 3.1 Scheme of the filter paper disc method for mushroom preservation

1. Preparing filter paper discs and agar petri medium: Sterile petri dishes (110 × 25 mm) are aseptically filled with 30 ml suitable medium, such as malt extract agar. Filter paper discs with 5 mm diameters (Whatman No. 4) are manually prepared and sterilized.
2. Preparing the mushroom mycelium: The central area of the prepared petri dishes is inoculated with the culture, and 6–8 sterile prepared filter paper discs are then placed around the inoculum point. The petri dishes are incubated at 30 °C for 4–6 days.
3. Storing the mycelial discs: After the culture grows well, mycelia cover the petri dishes and filter paper discs. Using sterile small forceps, the paper discs with mycelia are removed and aseptically placed in vials with sterile screw caps containing 4 ml of sterile distilled water. All of the vials are closed and sealed with Parafilm. Finally, the vials are stored at 4 °C.

   Cultures are recovered by aseptically removing the culture disc and placing it onto a petri dish containing an appropriate medium.

3.3.2.3 Mycorrhizae

Mycorrhizae are symbiotic associations between fungi and the roots of a vascular host plant. Mycorrhizae are commonly divided into *ectomycorrhizae* (ECMs) and *endomycorrhizae* or arbuscular mycorrhizal fungi (AMFs), arbuscular mycorrhizas (AMs), and vesicular-arbuscular mycorrhizae (VAMs). AMFs are a group of obligate biotrophs to the extent that they must develop a close symbiotic association with the roots of a living host plant to grow and complete their life cycle. The hyphae of endomycorrhizal fungi penetrate the cell wall and invaginate the cell membrane. AMFs can symbiotically interact with almost all plants that live on the Earth. They are found in the roots of ~80–90% of plant species (mainly grasses, agricultural crops, and herbs) and exchange benefits with their partners, as is typical of mutual symbiotic relationships. They represent an interface between plants and soil, growing their mycelia both inside and outside of the plant roots. AMFs provide the plant with water, soil mineral nutrients (mainly P and N), and pathogen protection. Spores are produced at the tips of hyphae either within the host root or outside the root in the soil as propagules. These spores germinate, and the germination tube that is produced grows through the soil until it comes into contact with roots. The fungus then penetrates the root and grows between root cells, or it may penetrate the cell wall and grow within root cells. Inside the root, the fungus forms arbuscules, which are highly branched hyphal structures that serve as sites of nutrient exchange with the plant. However, the hyphae of ECMs do not penetrate individual cells within the root. Their dominant form in vitro is hyphae that generally grow slowly in axenic cultures (Molina et al. 2003). Hyphae are usually more sensitive to environmental conditions than spores or conidia (Smith 1993) because of their characteristics (i.e., the absence of conidia and few or no spores produced in vitro).

- ***AMFs***: Maintenance/preservation is mainly through sub-cultivation, either in vitro on a synthetic growth medium, in association with transformed roots, or in vivo in association with (host) plants in the greenhouse. The storage methodologies should preserve a product's high and consistent quality and be simple and inexpensive. AMF viability and efficiency can be maintained for several months at room temperature (20–25 °C), but the inocula must be kept in their packaging and must be partially dried. The main inconvenience during the storage period is that spores can sometimes become dormant, decreasing the germination rates drastically (Oehl et al. 2009). However, a cold-storage period can be used to break dormancy (Juge et al. 2002).
- ***ECMs***: The sub-culturing methods used for mushrooms are applicable. The time between sub-cultures can be extended by reducing the storage temperature. For instance, Tibbett et al. (1999) stored a wide geographical range of *Hebeloma* isolates for 3 years at 2 °C. Additionally, the survival of ECM fungi stored in sterile water at 5 °C was 100% after 1 year but declines for some isolates to 95 and 64% after 2 and 3 years of storage, respectively (Marx and Daniel 1976). Thus, this method is only considered useful for short-term preservation (2–5 years). Maintenance occurs through sub-cultivation at room temperature or preferably under refrigeration. Storing ECM mycelia under water or oil is an easy to apply method. These methods are adequate to maintain some ECM fungi for periods of up to 20 years (Richter 2008).

3.3.2.4 Yeast

Sub-culturing yeast has been using successfully for many years for short-term preservation. The long-term preservation for maintaining yeast is reported by NCYC (National Collection of Yeast Cultures) (Smith et al. 2001). Sub-culturing method can be used for both liquid and solid media. In this method, an inoculum is regularly transferred to a fresh media. All of the yeasts maintained by these methods show high recovery rates (Smith et al. 2001). None of the fermentative strains survive better on agar slants than in broth. Approximately 10% of strains show changes in some characteristics, such as flocculation behavior or other morphological and physiological properties, after long periods, such as 10 years (Kirsop and Snell 1984).

Sub-culturing in Broth Medium

Preparation: Dispense 10 ml medium yeast extract and malt (YM) broth (Difco 071101) into McCartney bottles. Sterilize by autoclaving at 121 °C for 15 min.

Inoculum: Transfer a loop of stock culture aseptically to the prepared McCartney bottle.

Incubation: Incubate the inoculated bottle with shaking at 25 °C for 72 h. Examine visually for growth at 72 h. If the growth is poor, then more incubation time is needed. Examine daily. In some cases, it is necessary to aerate the culture by shaking (for some aerobic genera, such as *Cryptococcus*, *Rhodotorula*, and *Sporobolomyces*).

Storage and shelf life: The common storage temperature is 4 °C. In general, the culture will remain viable under these conditions for 6 months. Fermentative species survive better than non-fermentative species, and some of the latter may need to be sub-cultured more frequently, perhaps at 2-month intervals.

To prepare duplicate cultures, one is for all operations and the other to act as the new stock culture.

Sub-culturing on Agar Slants

In total, 10 ml YM broth (Difco 071102) is dispensed into screw-cap loosened McCartney bottles. Then, the samples are sterilized by autoclaving at 121 °C for 15 min and allowed to sit at an inclined angle to form slants. All of the sub-culturing steps in broth are applicable to this method. The shelf life of many yeast strains will be longer on agar slants than in broth, particularly the non-fermentative genera. However, ascosporogenous strains that sporulate on the agar slants will likely become unstable. If this characteristic is a priority, then these methods should not be used.

Sub-culturing on Agar Slants with an Oil Overlay

In this method, before storing at 4 °C, the culture on the agar slant is overlaid aseptically with sterile oil (B.P. medicinal oil; BDH Chemical Ltd.) to 1 cm above the top of the agar slant. The shelf life is then extended from 2 to 3 years. Care must be taken during sub-culturing not to let the inoculation loop splutter when flaming. Pathogens should not be maintained using this method.

3.3.2.5 Bacteria and Actinomycetes

In general, sub-culturing on agar slants (or petri dishes) and bottles is applicable for bacteria. Because bacteria are very diverse, each organism needs to be grown on a suitable medium under selective conditions. There are hundreds of media for bacteria such as Luria-Bertani, MRS, yeast–glucose agar medium, and sporulation medium for *Streptomyces*. The interval for sub-culture preservation of bacteria on agar media is approximately 1 month, and the period for broth sub-cultures is shorter. Agar petri plates are the preferred media for sub-culture preservation. On plates, the cultures are more easily verified by comparing with a pure culture that shows a type of colony or to see if contaminating colonies arise. In addition, if a plate is contaminated, then it should be discarded. However, in many cases, the target colony can be isolated with a sterile transfer loop and streaked to a fresh agar plate for recovery. For actinomycetes, sub-culturing to agar slants is performed much as it is with fungi rather than with bacteria. Actinomycetes need longer culturing and spore forming periods (10–14 days). For anaerobic bacteria, an agar slope medium can be used, such as MM10 (Caldwell and Bryant 1966) medium. The medium is inoculated by stabbing and then incubating at room temperature in the dark for ~4 weeks. Sub-culturing in broth having a low glucose (BGP or BGPhlf) level is also applicable. The culture broth should be stored at room temperature in the dark with the caps screwed down tightly. The shelf life is ~1 year, and in the special case of the spore-forming anaerobic bacteria, *Clostridia* can survive for several years. The technique is described in item 2.1.2.

3.3.2.6 Algae

When handling many algae strains, practical considerations should be maintained with respect to individual optimum conditions and media. Individual strain shows diverse requirements for media, light, pH, and temperature. Nutrition for algae ranges from obligate phototrophy to heterotrophy. Nutritional requirements vary from strain to strain. Thus, extensive trials are required to determine the best medium for an individual strain. For healthy growth, many strains require extracts from soil, plants, or seawater in the medium.

Sub-culturing on liquid and agar slants is commonly used for short-term preservation. The shelf life is 2–3 months for liquid and 1–6 months for agar slant preservation (Kirsop and Snell 1984). For the latter, the plugs or caps should not be too tight because a gaseous exchange with the external environment is required.

3.3.3 Drying Preservation Methods

The principle of drying methods is the removal of water from the culture, resulting in the suspension of cell metabolism. There are a number of drying methods for culture preservation that are used for various microorganisms, such as on soil/sand, glass beads, silica gels, freeze-drying (lyophilization), and L-drying (liquid drying) (Fig. 3.2). Sodelli's method of preservation is simpler than the lyophilization method that is described by Sharma et al. (2017a). The methods vary from having a high cost and requiring skill, such as for lyophilization and L-drying, to having a low cost and requiring limited skill.

3.3.3.1 Filamentous Fungi

Storing on Silica Gel

Silica gel has been successfully used to maintain diverse species of fungi. It is inexpensive and requires limited technical experience. Fungi can be stored for 7–19 years when covered well and remain morphologically and genetically stable. Briefly, glass bottles are partially (one-third) filled with silica gel and sterilized in an oven (180 °C for 3 h). The bottles are placed in a tray of water, which is placed in a deep freeze (~−20 °C) overnight. Then, the tray is removed from the deep freeze 20–30 min before use. Spore suspension is made in 5 °C sterile 5% (w/v) skimmed milk and added to the silica gel (up to three-quarter of the silica gel's volume). The

A B

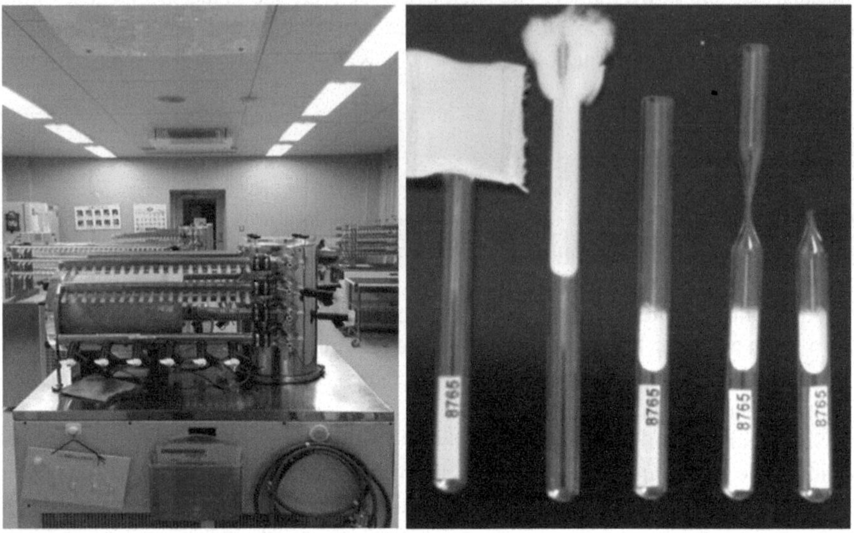

Fig. 3.2 (**a**) Lyophilizer (NBRC, Japan) and (**b**) lyophilized ampoules (Peter Green, National Collection of Industrial Food and Marine Bacteria Ltd.)

bottles are shaken to agitate, coating spores with silica gel. After incubating at 25 °C, when the silica gel is readily separated, the bottles are tightly capped and stored at 4 °C in airtight containers. Recovery is simple. A few silica gel-covered spores are sprinkled on a suitable growth medium and incubated under appropriate conditions (Smith et al. 2001). However, this method is not recommended for fungi with thin-walled spores or spores with appendages or mycelia, for example, the zoospores and oogonia of Oomycetes, such as *Phytophthora*, *Pythium*, *Bartalinia*, and *Cercospora* species (Kirsop and Snell 1984).

Soil/Sand Storage

Preservation in soil/sand is a practical and cost-efficient way to store filamentous sporulating fungi. Preserved cultures remain viable for up to 10 years, and the method reduces the occurrence of mites. Several fungi can be preserved using this method, such as *Fusarium* (Gordon 1952), *Rhizopus*, *Alternaria*, *Aspergillus*, *Circinella*, and *Penicillium* (Atkinson 1953). Soil/sand storage can be used for some fungi that are not stable under oil.

Briefly, a mound of garden soil/sand, with about 20% moisture, is placed into glass bottles at two-thirds capacity. The bottles are sterilized twice, 24 h apart, at 121 °C for 15 min. Then, 1 ml of spore suspension was added in sterile water to the bottles with loose caps and kept at room temperature for 5–10 days to grow and dry. For strains that do not sporulate, a mycelial suspension can be used. Finally, the bottles are stored at 4–7 °C. Recovery is simple. A few grains of soil/sand are sprinkled on a suitable growth medium and incubated under the appropriate conditions.

Lyophilization/Freeze-Drying Method

This method is highly successful for preserving sporulating filamentous fungi. During the freeze-drying process, water is removed directly from frozen materials by sublimation under a vacuum. If carried out correctly, freeze-drying will prevent shrinkage and structural changes and help retain viability. There is a vast array of freeze-drying equipment available, ranging from laboratory bench models to huge industrial installations. This method can be optimized for different organisms or cells.

The freeze-drying method has many advantages over other methods. Generally, preserved cultures have good viability and stability and can be stored for many years. Many genera, including *Almaria*, *Aspergillus*, *Botryotrichum*, *Cephalosporium*, *Cladosporium*, *Cunninghamella*, *Chaetomium*, *Cordyceps*, *Fusarium*, *Gilberella*, *Mucor*, *Neurospora*, *Paecilomyces*, *Penicillium*, *Rhizopus*, and *Trichoderma*, can survive for 14 years (Kirsop and Snell 1984). Storing ampoules requires little space, and they are easily distributed through the mail. However, the method has some disadvantages, such as requiring expensive equipment and having a complex protocol that necessitates a skilled time-consuming technique. Some genera that fail to survive include *Ganoderma*, *Entomophthora*, *Phytophthora*, and *Pythium*. Reduced viability and genetic changes may occur as reported by Ryan et al. (2003).

Some main factors that affect the viability and stability of the culture are as follows:

1. The suspension medium should be convenient to use and protect the spores from freezing- and storage-related damage. The components are skim milk, serum, various kinds of sugars, and peptone.
2. The freezing rate is a very important factor that must be considered to obtain the optimum recovery rate. A slow freezing rate (−1 °C/min) has been successfully applied to many strains. The technique of centrifugal freeze-drying relies on evaporative freezing under reduced pressure, which results in a favorable cooling rate and can be used successfully for many sporulating fungi (Smith 1993).
3. Sealing the ampoules is the most important factor for preventing the deterioration of freeze-dried cultures.
4. The moisture content affects the stability of organisms. To achieve the best recovery, a 5% moisture content should be reached before the sample's temperature rises above −15 °C. Overdrying will kill or cause mutations by damaging DNA; thus, the residual moisture content must not be allowed to fall below 1% (Sneel 1984).
5. Ampoules of freeze-dried samples must be stored out of direct sunlight to avoid shortening the shelf life.

   The centrifugal freeze-drying method (Smith et al. 2001) is briefly described as follows:

The spore suspension is prepared in a mixture of a 10% (w/v) skim milk and 5% (w/v) inositol. The spore suspension aliquots of 0.2 ml are added to each sterile ampoule, covered with a sterile cap. Load the ampoules into a spin freezer, and place it in the dryer. Spin for 30 min, and cool to −40 °C. Evacuate the ampoules in the chamber for 3 h until the moisture content of the sample would be around 5%. Remove the ampoules, and replace the cape by a sterile cottonwood plug; compress the tin at 10 mm depth and 10 mm above the sample. Constrict the plugged ampoules at 10 mm above the plug using a gas torch at about 1 mm. Place the constricted ampoules on the secondary-drying accessory of the freeze dryer, and evacuate over a phosphorous pentoxide desiccant for approximately 17 h. The ampoules are sealed at the constricting point using a crossfire burner under a vacuum. Maintain the ampoules at 4 °C.

To recover a sample from a stored ampoule, score the ampoule at the center of the cottonwood plug using a glass cutter, and then heat with a Bunsen burner until red hot, and apply pressure firmly to the score. Aseptically, open the ampoule, and remove the cottonwood plug, and then add 2–4 drops of sterile distilled water, and leave for 30 min to rehydrate the suspension. Finally, inoculate the suspension onto a suitable growth medium, incubating under appropriate conditions. Check the growth and viability.

3.3.3.2 Mushrooms

The lyophilization method is preferred for the long-term preservation of spore-bearing mushrooms. It is not useful for storing the mycelia of mushrooms. The survival rates of freeze-dried fungal hyphae of a number of ascomycetes have been recorded as being higher than two Basidiomycetes, *Schizophyllum commune* and *Coronus psychromorbidus*. The protocol for storing spore-bearing mushrooms is almost the same as for fungi. However, Singh and co-workers (2004) developed a method for the preservation of mushroom mycelial cultures. Briefly, all of the stock cultures are sub-cultured on pearl millet (*Pennisetum typhoides*) grains in test tubes containing wheat extract agar. All of the inoculated test tubes are then incubated at 25 °C for 14 days, except those of *Volvariella volvacea*, which should be incubated at 32 °C for 10 days. Mycelial cultures multiplied on pearl millet grains are subjected to lyophilization. For lyophilization, glass ampoules are first sterilized in a hot-air oven at 180 °C for 2.5 h and allowed to cool at ambient temperature. The mycelia that are multiplied on pearl millet grains are placed using sterilized forceps into a sterilized ampoule. Each ampoule containing 50–60 grains is plugged with sterilized nonabsorbent cotton. The cotton plugs are pushed inside up to the neck of the ampoules. The constrictions are made from above the cotton region in each ampoule and pre-cooled in a deep freezer to −40 °C. When the freezing chamber reaches −62 °C, the ampoules are attached to the lyophilizer and a vacuum created. The lyophilizer is run overnight. The following day, when the pressure reaches 0.05 mbar, the ampoules are sealed at the point of constriction with the help of a crossfire burner connected to oxygen and gas. The vacuum is tested using a vacuum tester. A purple-colored light inside the sealed ampoule verifies proper sealing. The prepared ampoules are stored at room temperature. For recovery, grains (pearl millet seeds containing mycelia) are placed in pre-sterilized petri plates containing malt glucose agar.

3.3.3.3 Mycorrhizae

For ECMs: Because ECMs lack conidia and produce few or no spores, the freeze-drying method of preservation is not strongly recommended. However, this method is being used for nonspore-bearing cultures too. A *Laccaria fraterna* culture from the late to stationary phase of growth, treated with 10% dimethyl sulfoxide (DMSO) as a cryoprotectant, survived the freeze-drying method. The morphological and physiological characteristics were similar between the freeze-dried and nonfreeze-dried samples (Sundari and Adholeya 1999).

For AMFs: The advantage of AMFs over ECMs is that they are spore-producing fungi. Thus, the drying methods used for filamentous fungi (Smith 1993) can be applied to AMFs. AMFs are mostly cultured in vivo, but a limited number of species are cultured in vitro. Several preservation methods using propagules produced either in vivo or in vitro have been developed. For isolates produced in pots, drying the soil containing AMF propagules is generally achieved at room temperature, for 2–4 days before lyophilization. For isolates produced in vitro, the drying of the carrier (i.e., alginate beads containing AMF propagules) is crucial and should be preceded by incubation in a cryoprotectant (ideally trehalose) at 27 °C for 2 days (Lalaymia et al. 2014). Whether cultured in vivo or in vitro, it is recommended to use propagules from cultures in the late or stationary phase of growth. The preconditioning of AMF propagules for lyopreservation should be conducted in a carrier (i.e., soil or alginate beads) to protect them from preservation stresses and facilitate their manipulation. The carrier should be dried before lyopreservation to prevent water crystallization outside and inside the propagules. For in vitro-cultured AMF propagules, we recommend the use of 0.5 M trehalose as a cryoprotectant, while the in vivo-cultured propagules can be preserved in dried soil without a cryoprotectant.

3.3.3.4 Yeast

Commonly, there are three drying methods for yeast preservation, namely, silica gel, paper replica, and freeze-drying. There is no relationship between survival and taxonomic position, and the factors determining survival are specific to each strain. Thus, a method that is satisfactory for one strain may be unsuitable for others. Whichever method is selected, the most suitable growth conditions for the cells must be established. Both the age of the culture and the oxygen availability during growth can affect the survival rate. Thus, post-logarithmic cells usually survive better than younger cells (Kirsop and Snell 1984).

Silica Gel Storage
Fill McCartney bottles to a depth of 1 cm with grain plain 6–22 mesh non-indicating silica gel, and sterilize in an oven at 180 °C for 90 min. Cool down the bottles, and transfer to an ice tray. Aseptically add a few drops of the yeast suspension prepared in pre-cooled 5% (w/v) skim milk to each gel bottle. Return the inoculated bottles to the ice tray for an additional 30 min. Screw the cap on tightly, and maintain at room temperature for approximately 14 days. The culture is dry when the gel crystals separate. Transfer the culture to an airtight container with a layer of indicator silica gel in the bottom. Store at 4 °C. Check the indicator gel occasionally and redry by heating in an oven at 180 °C for 2 h if necessary. Recovery is the same as for filamentous fungi. Using this method, the shelf life is 2–3 years and up to 5 years as reported by Woods (1976).

Paper Replica Method
Kirsop and Kurtzman (1988) reported that yeast could be dried on filter paper and stored above desiccant silica gel. To prepare the culture for storage, squares of Whatman No. 4 filter paper (10 mm) are used. They are inoculated by an immersion in drops of yeast suspension previously prepared in 5% (w/v) skim milk. The inoculated filter paper squares are dried in a desiccator in the foil packets and stored in an airtight container at 4 °C. The cultures are recovered by removing the filter paper, placing the square on an appropriate agar plate medium, and incubating at a suitable temperature. The shelf life is ~2–3 years.

Freeze-Drying Method
The protocol of centrifugal freeze-drying method (Smith et al. 2001) for yeast is briefly described as follows:

1. Grow the culture without aeration in YM broth (Difco 0711-01) at 25 °C for 72 h.
2. Prepare the yeast suspension by mixing the grown culture and freeze-drying medium. Select one of the following media: (1) 20% skim milk and 10% glutamate; (2) 5% honey, 10% skim milk, and 10% dextran; and (3) 5% honey, 10% skim milk, and 5% glutamate at a culture concentration of at least 10^6 cell/ml.
3. Remove the cotton plugs, and aseptically add 0.2 ml of the cell suspension to each ampoule using a Pasteur pipette, following the steps as described in item 2.1.2.1 previously for filamentous fungi.

The recovery procedure is the same as that for filamentous fungi, except that an aliquot of YM broth from a Bijoux bottle containing 1 ml of sterile broth is added to the materials, and then the suspension is returned to the remaining YM broth in the Bijoux bottle before incubating under appropriate conditions. Check the growth and viability.

3.3.3.5 Bacteria and Actinomycetes

Common drying methods for the storage of bacteria and actinomycetes include gelatin disc, soil, lyophilization, and L-drying.

Gelatin Disc Method

The method is described in detail by Kirsop and Snell (1984):

1. Prepare a bacterial suspension of 0.5 ml in nutrient broth, and add to 3 ml of the gelatin suspension medium (10 g gelatin powder, 2.5 g nutrient broth powder, 5 g meso-inositol, and 100 ml deionized water) that was previously melted at 37 °C, pH 7.2. Place 3 ml aliquot in 6 ml bottles, and sterilize by autoclaving at 121 °C for 15 min.
2. With a pipette, dispense dropwise 0.02 ml of suspension to the base of a plastic petri dish. Approximately, 80 drops can be accommodated in the base.
3. Freeze the drops on the petri base by placing the petri dishes in a freezer (–20 to –40 °C for approximately 2 h).
4. Freeze-dry the drop by transferring the petri dishes quickly to the freeze dryer (which must be loaded with phosphorous pentoxide).
5. Freeze-dry the culture overnight. For a large number of discs in a batch, it may be necessary to replace the phosphorous pentoxide after 2–4 h.
6. Preparing the vials. Into each screw-necked vial, add self-indicating silica gel (to a depth of approximately 10 mm), and pack down tightly with cotton wool, cover with foil, and sterilize at 160 °C for 1 h. The caps are sterilized in an oven at 60–80 °C for 4 h before placing them on the vials.
7. Aseptically transfer the sterilized discs to the vials when the freeze-drying process is finished.
8. Replace the caps of the vials and tighten. Then, store the samples at 5 °C.
9. The shelf life of the method is ~3–4 years. After using this method for 4 years, the NCTC has found little change in phenotypic characteristics of strains used as controls or in identification kits.

   For recovery, select a gelatin disc with sterile fine-nosed forceps, and place it in a test tube containing 1 ml of nutrient broth. Warm the sample at 37 °C until the disc dissolves. A loop of the suspension broth is transferred to a suitable medium agar in a petri plate and streaked for single colonies before incubating at an appropriate temperature.

Soil Storage for Actinomycetes Preservation

The method is described in detail by Dietz and Currie (1996).

A loam soil is the most practical and commonly used for this method. The soil should not have been fertilized or treated with pesticides or herbicides. It is heated at 160 °C for 6 h to destroy spore-forming microorganisms or insects that may be present. The soil is sieved using a 30 mesh sieve. The sieved soil is added to Pyrex tubes (13 × 100 mm) to a depth of 20 mm. The soil tubes are autoclaved separately for 60 min each day for 3 days. It is necessary to check for the contamination of the sterilized tubes by adding a loopful of the soil as seeding to a suitable medium and incubate at 24, 28, and 37 °C. Check for contaminating microorganisms at the appropriate incubation times.

1. Prepare the culture for storage. Make the spore suspension by adding a suitable amount of distilled water to a sporulating culture.

2. Add 1 ml of the spore suspension to each tube.
3. Air-dry the culture tubes at room temperature (24 °C) for a month. Tap the culture tube to loosen the soil particles.
4. Store at 4 °C.
5. The shelf life of a culture stored using this method is 2–3 years. Better results may be achieved if a dried blood preparation (100 g soil, 10 g $CaCO_3$, and 2.5 g dried blood) is used.

For recovery, a sterile loop is moistened in sterile distilled water, and then the soil is aseptically picked up on the wet loop. Place the inoculum on a suitable medium, and incubate at an appropriate temperature.

Lyophilization (Freeze-Drying) Method

Lyophilization is a process in which water vapor is removed directly from frozen materials by sublimation. This method has been used for many years to preserve a wide variety of biological materials and bacterial strains. This is an expensive method but one of the best for the long-term maintenance of microorganisms. The below method is described by Kirsop and Snell (1984).

1. Prepare the culture suspension by adding 2–4 ml of one of the three possible sterile suspensions, (1) 100 ml horse serum, 33 ml Oxoid CM1, and 10 g glucose; (2) 5 g meso-inositol and 100 ml horse serum; or (3) 2.5 g nutrient broth powder No 2 (Oxoid), 5 g meso-inositol, and 100 ml distilled water, to an agar slant of the culture grown to late log or early stationary phase, and mix.
2. Aseptically add 0.2 ml of the cell suspension to each sterile ampoule using a sterile Pasteur pipette.
3. Replace the lint caps, and load the ampoules into the centrifuge plate of the freeze dryer, ensuring the plate is balanced.
4. Place the centrifuge plate in the chamber of the freeze dryer, and evacuate the chamber. After 20 min, switch off the centrifuge, and start the primary drying of at least 3 h.
5. Admit air into the freeze-dryer chamber, and remove the ampoules.
6. Constrict the plugged ampoules at 15 mm above the cotton plug using a constrictor. Alternatively, an air/gas glass blow torch can be used to manually constrict the ampoules.
7. Place the constricted ampoules in the secondary-drying apparatus, which consists of a vacuum system and a phosphorous pentoxide trap. Dry for at least 3 h, and seal each ampoule at the spot of constriction.
8. Test the sealed ampoules with a high-voltage spark tester to ensure the seal is intact by observing a purple to blue illumination inside the ampoule. Maintain the ampoules in the dark at room temperature. Some cultures require storage at 4 °C.
9. The shelf life of most bacteria ranges from 15 to 20 years. A shelf life of 30 years has been reported for members of actinomycetes and enterobacteria and *Staphylococcus* and *Streptococcus*. The method is described in use at the National Collection of Industrial Food and Marine Bacteria. However, few species, such as *Clostridium botulinum*, *Clostridium chauvoei (oedematiens)*, *Peptococcus*

heliotrinreciucans, *Spirillum serpens*, *Bacteroides melaninogenicus*, *Haemophilus canis*, *Leptotrichia buccalis*, *Mycobacterium microti*, and *Neisseria gonorrhoeae*, may have disappointing shelf lives.

For recovery, the stored ampoules are scored at the center of the cottonwood plug using a glass cutter and then heated in a Bunsen burner until red hot, and pressure is applied firmly to the score. The heat should crack the tube at the scored line. Aseptically open the ampoule, and remove the cotton wool plug. Then, add 2–4 drops of sterile distilled water or growth medium, replace the cotton plug, and leave for 30 min to rehydrate the suspension. Finally, inoculate a suitable growth medium with the suspension, and incubate under appropriate conditions. Check the growth and viability.

Lyophilization for Anaerobic Bacteria

A crucial problem for preserving anaerobic bacteria is maintaining viability. Based on the physiological characteristics of various anaerobic bacteria, preservation methods are divided into the following: (1) for obligate anaerobic bacteria, media are prepared and techniques must be used in an oxygen-free atmosphere (Hungate 1950), and (2) for anaerobic facultative bacteria, such as clinical isolates, the media are prepared and inoculated in air and may or may not require oxygen-free headspace (Rosenblatt et al. 1973). There are several strongly recommended media for growing anaerobic bacteria (Kirsop and Snell 1984), such as MM10, VL broth, and BP half broth. Cultures prepared for freeze-drying must be grown rapidly as Hungate cultures in a carbon dioxide headspace (Hungate 1950). It is recommended that for many bacteria, both MM10 and VL broth, supplemented with rumen fluid, fecal extract, liver extract, or lactate, are required. Hungate cultures in a suitable broth medium in screw-capped bottles at the growth stage are harvested by centrifugation (2500 rpm for 15 min). The supernatant is poured off, and 1 ml suspension fluid (7.5% glucose in horse serum sterilized by filtration) is immediately added. Again, the cultures in the screw-capped bottles are centrifuged (2500 rpm for 15 min), the supernatant is poured off, and 1 ml of sterile 16% glucose is added. The suspension should be made just before applying the primary freeze-drying step. The freeze-drying steps are the same as those for freeze-drying bacteria. All of the steps from the culture suspension preparation and freeze-drying process should be performed with minimum exposure to oxygen. The ampoules are stored in the dark at 2–10 °C. The viability varies and depends on the genus, species, and the degree of anaerobiosis required for the particular culture. In general, the viability of anaerobes preserved in this manner remains high for at least 10 years and even up to 22 years.

L-drying Method

L-drying method was first described by Annear (1958). This is a useful alternative method of vacuum drying to preserve bacteria that are sensitive to the freezing step of the common lyophilization method. The intrinsic feature of this process is that liquid cultures are prevented from directly freezing and drying. The method reported by Smith (Smith et al. 2001) is shown below:

1. Prepare the culture suspension by adding 2–4 ml of one of the three sterile suspensions of 100 ml horse serum and 33 ml Oxoid CM1 to an agar slant of the culture grown to the late log or early stationary phase, and mix. In case of medical bacteria, use a suspension of inositol serum (100 ml horse serum and 5 g of meso-inositol).
2. Aseptically add 0.1 ml of the cell suspension to each sterile ampoule using a sterile Pasteur pipette. Cut the sterile cotton wool plugs, and push down the ampoules using a sterile glass rod.
3. Connect the ampoules to the manifold, and clamp above a glass tank containing water at 20 °C, ensuring that the ampoules can be immersed in the water to a depth of 40–50 mm.
4. Connect the manifold through a diaphragm valve and phosphorous pentoxide trap to a rotary pump. Close the valve, and switch the pump on.
5. Open the valve very gradually and degas (i.e., bubble). To prevent violent bubbling in the ampoules, care must be taken to carefully control the rate of degassing by carefully manipulating the valve. When degassing is complete (approximately 5 min), open the valve fully. The contents of the ampoules will appear dry in additional degassing about 30 min.
6. Remove the ampoules from the manifold, and constrict the ampoules following the freeze-drying protocol for bacteria.

The shelf life of most bacteria preserved in this manner is approximately 10 years. Some bacteria such as *Spirilla* and *Azomonas insignis* that are sensitive to freeze-drying have been preserved successfully for 15 years using L-drying. Recovery is the same as for freeze-dried bacteria.

3.3.3.6 Algae

There are two common drying methods for preserving algae:

1. *Air-dried soil*: While not all stages of algal culture can be preserved by drying method, some can. Many algae tolerate adverse conditions in their natural environment, such as during zygotic or cystic stages. Using this method, zygotes of *Hydrodictyon africanum* have been successfully stored for 6 years, and cysts of *Haematococcus pluvialis* have remained viable after 27 years in air-dried soil.
2. *Lyophilization:* Freeze-drying techniques have been successfully employed for the preservation of algae (Holm-Hansen 1964). All of the steps are performed as in the standard protocol for bacteria. The cells are frozen in suspension liquid with an additive such as horse serum for *Bracteacoccus*, *Chlamydomonas*, *Chlorophyta*, *Cyanophyta*, and *Scenedesmus* and skim milk for *Chlorella*, *Phaeodactylum*, and *Stichococcus*. When compared with sub-culture and cryo-preservation methods, the freeze-drying method is not recommended because the average recovery rate is less than 5%. However, the successful lyophilization of the cyanobacterium *Nostocmuscorum* was reported with no observed reduction in viability during 5 years of storage (Holm-Hansen 1973).

Fig. 3.3 Liquid nitrogen tanks

3.3.4 Cryopreservation Methods

The preservation of living organisms by lowering their temperature reduces the metabolism rate until, when all internal water is frozen, no further biochemical reactions occur, and the metabolism is suspended (Franks, 1981). Temperatures of −20, −30, −40, −70, −140, and −196 °C have been used, but in general, temperatures above −30 °C give poor results. Refrigerated storage at −70 °C and storage in liquid nitrogen at −140 °C (nitrogen vapor phase) or −196 °C (nitrogen liquid phase) has been used for a variety of different microorganisms, including bacteria, fungi, yeast, actinomycetes, algae, mycorrhizae, and viruses (Fig. 3.3). Although little metabolic activity takes place at low temperatures, the recrystallization of ice can occur, resulting in damage to the cell structures during storage (Douzou 1982). Thus, adequate care must be taken during freezing and thawing, so the culture does not undergo phenotypic or genotypic changes. Adding cryoprotectants, such as glycerol, DMSO, skim milk, meso-inositol, and animal sera, to the culture suspension helps decrease damage. Cryopreservation is long term and does not require skill or complex techniques, but it does require expensive equipment and electronics, as well as a nitrogen supply. Cross-contamination can occasionally occur.

3.3.4.1 Filamentous Fungi

Storage on Agar Slopes at −20 °C

Agar slopes of fungi cultured at the mature or well-sporulated stage are maintained at −20 °C in deep freezers. For recovery, a small portion of the frozen colony is aseptically placed on a suitable agar medium at room temperature. The frozen stock culture is returned to the freezer as quickly as possible to avoid thawing. The shelf life of these cultures is 6–12 months. In some cases, such as *Aspergillus* and *Penicillium*, cultures can survive for 5 years.

Storage in Cryovials at −80 °C

Cultures of a wide range of fungi are stored successfully in deep freezers at −80 °C. It is a long-term preservation method when liquid nitrogen is not available. The shelf life varies from culture to culture over a large range (4–40 years). The disadvantages of this method are its dependency on electricity and expensive deep freezers and screw-capped cryogenic vials. Preparing the culture is simple. Spore or mycelial suspensions are made in 10% sterile glycerol or DMSO. An aliquot of 0.5 ml of the culture suspension is added aseptically to a sterile cryogenic vial and capped tightly. The vials are frozen at rate of 1 °C min^{-1} until they reach −35 °C (~35–40 min). Then, the vials are placed in a deep freezer at −80 °C. For recovery, each stored vial is used only once by thawing in a 37 °C water bath before inoculating on the suitable medium under appropriate conditions.

Storage in Liquid Nitrogen

Storage in liquid nitrogen (−139 °C in vapor and −196 °C in liquid form) is the most effective long-term preservation method. The shelf life of stored cultures is quite long in comparison with other methods. Under these conditions, the dormant cell does not undergo any phenotypic or genotypic changes. The method can be applied to both sporulating and non-sporulating cultures. At the Commonwealth Mycological Institute (CMI) culture collection, UK, 2900 of the 3286 cultures preserved in this way survived up to 14 years (Kirsop and Snell 1984). There are several storage containers other than glass ampoules that can be used successfully in liquid nitrogen vapor, such as plastic cryogenic vials with screw caps or plastic cryogenic straws. Care must be taken to avoid leaking containers that can result in cross-contamination. The disadvantages of this method include the high costs of nitrogen-controlled containers and a continuous supply of liquid nitrogen. If the supply of nitrogen fails (or the double-jacketed, vacuum-sealed storage vessels corrode and rupture), then the whole collection can be lost. The method applied at CABI (Smith et al. 2001) is as follows:

1. Aseptically prepare fungal suspensions in 10% sterile glycerol, and dispense 0.5 ml aliquots into cryovials (for non-sporulating fungi, plugs of mycelia can be cut and be placed in the cryotubes with 5 ml sterile glycerol). If the culture does not survive in glycerol, then 10% DMSO or a mixture of 5% DMSO and 8% glucose may be used (Smith 1983).

2. Seal cryovials with screw caps. For glass cryotubes, heat and seal using an air–gas torch before checking for leakage.
3. Keep the cultures at 5 °C for 1 h to allow the cells to equilibrate in the glycerol.
4. Cool the prepared cultures at a rate of -1 °C min^{-1} (common for many fungi) from 5 to -50 °C.
5. After the cultures are frozen (-50 °C), transfer them to the liquid nitrogen vessel.

For recovery, unscrew the cap of the cryovial or open the glass cryotube as in the freeze-drying technique before inoculating onto a suitable growth agar medium incubated under the appropriate conditions. Determine the viability and check for contamination.

3.3.4.2 Mushrooms

Freezing is applicable for storing a wide number of mushrooms. The method is preferred for non-sporulation mushrooms. In general, mushrooms can be stored at different freezing temperatures, such as -20 and -70 °C, and in liquid nitrogen at -139 °C (vapor phase) and -196 °C (liquid phase). When storing *Agaricus bisporus* (U-3) and *Pleurotus florida* (PAU-5) at -20 and -196 °C (in liquid nitrogen), for more than 6 months, the latter is the best storage temperature (Kaur et al. 2011). Although there were no differences in the shape, color, or other morphological characteristics of the fruit bodies formed by cultures stored at various temperatures, when the cultures were preserved in glycerol, in liquid nitrogen at -196 °C, in a deep freezer at -20 °C, or at room temperature, declines in endoglucanase enzyme-specific activity of 13.0, 7.6, and 80.2%, respectively, were observed after 20 days of incubation. Additionally, Suman and Jandaik (1991) reported no apparent changes in physiological or morphological characteristics of other mushrooms stored in liquid nitrogen for 10 years. The protocol for storing mushroom mycelia is the same as for filamentous fungi (Sect. 3.3.4.1) with a minor modification in preparing mushroom mycelial samples. The strains are sub-cultured onto complete agar medium plates and allowed to grow for ~10–14 days at 25–28 °C. After the mycelia have fully grown on the plates, they are cut into very small inocula using a hollow stainless-steel tube at 2–3 mm in diameter. The mycelial inoculum is aseptically placed in cryotubes containing 5 ml sterile 10% glycerol. If the culture does not survive in glycerol, then 10% DMSO or a mixture of 5% DMSO and 8% glucose may be used (Smith 1993).

3.3.4.3 Mycorrhizae

Cryopreservation is the most practical and reliable for the long-term storage of mycorrhizae.

- *ECM*: Several cryopreservation methods for storing ECM cultures have been developed recently. Cultures are stored by freezing in cryovials as per the cryovial protocol (Homolka et al., 2003). The isolates are grown on medium supplemented with 5% glycerol (v/v) at 24 °C. Sterile plastic straws open at both ends are used to sample the agar colonized by the fungal hyphae. The straws are transferred into sterile cryovials, sealed, and frozen in a programmable freezer to −70 °C, with a controlled slow cooling rate of −1 °C min^{-1}. They are then plunged into liquid nitrogen. After storage, the cryovials are thawed rapidly at 37 °C. Using this protocol, *E. clandestinum* presented a 100% survival rate, while *S. citrinum* and *S. verrucosum* isolates presented 50% survival rates. These fungi maintained their extracellular laccase production. Freezing the cultures on filter paper discs has been applied successfully to 13 ECM isolates (Stielow et al. 2011). The protocol involves growing the fungal isolates on charcoal filter paper strips (CFSs) placed on the surface of a culture medium for 3–5 weeks. The CFSs are collected in a sterile petri plate, incubated in 10% sterile glycerol (v/v) for 1–2 min and transferred into cryovials by layering the CFSs on top of each other. The closed cryovials were placed for 24 h in the gas phase of a liquid nitrogen tank (i.e., at a cooling rate of approximately −1 to −10 °C/min until reaching between −120 °C and −140 °C), before being directly transferred into liquid nitrogen. The cryovial protocol below (Crahay et al. 2013) efficiently stored 100 ECM fungal isolates and is similar to the method applied to mushrooms in Sect. 3.3.4.2 (Suman and Jandaik 1991):

1. ECM fungal isolates are grown on agar medium in petri plates and incubated under appropriate conditions for 2–4 weeks.
2. A mycelial plug of ~4 mm in diameter is taken from the margin of the growing colony and inoculated into a 2 ml sterile polypropylene cryovial containing 750 μl of sterilized (121 °C for 15 min) suitable agar medium poured at a slope.
3. Cryovials are incubated at 22–23 °C in for 7–9 weeks.
4. At 1–2 h before cryopreservation, 500 μl sterilized (121 °C for 15 min) glycerol cryoprotectant solution (10% v/v) is added to the cryovial.
5. The cultures are cryopreserved following a controlled decrease in temperature (−1 °C min^{-1} from 4 to −50 °C).
6. The cultures are directly transferred into a freezer at −130 °C.
7. For recovery, the ECM fungal isolates are directly thawed in a water bath at 35 °C for 2 min. The culture plugs are transferred to the centers of petri plates containing 30 ml suitable medium and incubated at the appropriate temperature.

Some parameters appear paramount for the long-term preservation of ECM fungi:

1. The preconditioning of the culture prior to preservation: The fungi are cultured on a carrier or in a cryovial to avoid excessive manipulations may damage the hyphae before preservation. Flooding the culture with a cryoprotectant (1–2 h before cryopreservation) can increase the survival rate of the fungi (Lloyd 1994).

2. The age of the culture: Cultures in the late phase or stationary phase of growth are most often better adapted to survive freezing or freeze-drying than young cultures (Lloyd 1994).
3. The cryoprotectant used: Glycerol appears to be the most reliable (and is the most used) cryoprotectant for ECM fungi, as well as for the majority of filamentous fungi (Cavalier-Smith 1998).
4. The cooling rate: The slow cooling of the organism (i.e., ~-1°C min^{-1}) appears to be appropriate for the majority of ECM fungi. Mycelia, which are the predominant form of ECM fungal colonies on synthetic medium, are rich in water. Thus, it is essential to dehydrate, at least partially, the hyphae to avoid injury caused by water crystallization.
5. The thawing rate: Fast thawing by direct immersion in a water bath (i.e., at a temperature between 35 °C and 38 °C) is often reported to prevent the risk of ice recrystallization occurring during a slow thaw.

- *AMFs*: For in vivo cultivation, soils containing isolates of *Glomus*, *Acaulospora*, and *Gigaspora* were first dried at room temperature and then directly stored at -80 °C (Kuszala et al. 2001). The cultures survived for several months. In another case, Kuszala et al. (2001) succeeded in cryopreserving 15 isolates belonging to *Glomus* and *Acaulospora* by immersion in liquid nitrogen.

For in vitro cultivation, Addy et al. (1998) was the first to demonstrate that the hyphae of *Glomus intraradices* were able to survive at temperatures below -12 °C when slowly cooled before freezing (Lalaymia et al. 2014). A method of cryopreservation was then developed (Declerck and Angelo-Van Coppenolle 2000). It was based on the encapsulation of spores in alginate beads, followed by incubation in trehalose (0.5 M) before freezing at -100 °C using a two-step temperature decrease of -1 °C min^{-1} from 20 to -35 °C and -18 °C min^{-1} from -35 to -100 °C. The advantage of in vitro preservation of AMF cultures is the production of pure contamination-free inocula. However, in vitro cultures are only successful for the limited number of species.

The method for the cryopreservation of AMFs developed by Lalaymia and co-workers (2014) should be applicable for both in vivo and in vitro AMF. It is as follows:

1. *Preparing the culture*

   For in vitro: Gelling medium, containing spores and roots of a 5-month-old in vitro culture, is extracted from the petri plates and poured into 100 ml of sterilized (121 °C for 15 min) deionized water and subsequently blended two times for 30 s at the 20,000 rpm in a sterilized (121 °C for 15 min) mixer.

   For in vivo: Pot-cultures that are at least 5-month-old are sampled. Spores are collected by wet sieving and decanting, while roots are collected with forceps and blended in a mixer in 100 ml deionized water for 30 s at 20,000 rpm and then filtered as above. The spores and the supernatant of the blended roots are mixed together.
2. The mixture is filtered using a sterilized (121 °C for 15 min) nylon mesh (40 mm).

3. The supernatant (i.e., spores and mycorrhizal/non-mycorrhizal root pieces) is encapsulated in a 2% (w/v) solution of sodium alginate (50 ± 5 propagules in each bead).
4. The encapsulated propagules are incubated in trehalose (0.5 M) overnight.
5. They are then dried at 27 °C for 48 h (until the bead water content is approximately 8.1 ± 4.6%).
6. The beads are transferred in 2 ml cryovials.
7. The cryovials are cryopreserved in a freezer at −130 °C following a two-step decrease in temperature: a fast decrease (~−12 °C min^{-1}) from room temperature (20 °C) to −110 °C, followed by a slow decrease in temperature (~ −1 °C min^{-1}) from −110 to −130 °C.
8. *Recovery:* Thaw the encapsulated AMF propagules by directly plunging in a water bath at 35 °C.

In vitro The beads are dropped in sterilized (121 °C for 15 min) MSR medium, cooled in a water bath to 40 °C, and then incubated at 27 °C for germination. After 4 weeks, beads containing germinated propagules are associated with a transformed root under in vitro culture conditions to reinitiate the fungal lifecycle.

In vivo The encapsulated propagules are placed directly in contact with plants in pots containing a sterilized (2 × 15 min at 121 °C, with a 12-h interval) substrate. The plants are grown for at least 8 weeks in a growth chamber before the fungal viability is assessed.

From the above cryopreservation protocols, the following six important storage factors were discerned:

1. The culture age is independent of the production mode (i.e., in vitro or in vivo). AMF propagules isolated from cultures in the stationary growth phase are the most suitable for the long-term preservation. At the late growth phase (frequently associated with stress conditions), the fungus accumulates cellular compounds, such as trehalose (Van Laere 1989), polysaccharide, and glycoproteins, to protect the intra- and extracellular mycelial integrity.
2. Drying may reduce ice crystallization during freezing. The natural cryoprotectants produced during drying may reduce ice crystal size or convert the fungal cytoplasm into a glassy state during rapid freezing, thus improving the chances of survival (Tan et al. 2007).
3. *The carrier*: AMF propagules cultured in vivo are usually preserved in the soil substrate in which they are produced. Encapsulation in alginate beads was reported as mandatory for the cryopreservation of in vitro-cultured AMFs (Declerck and Angelo-Van Coppenolle 2000).
4. *The cryoprotectant*: For in vitro-cultured AMF propagules, the use of trehalose (0.5 M) as a cryoprotectant is recommended, while the in vivo-cultured propagules could be preserved in dried soil without a cryoprotectant. Trehalose has the capacity to interact with phospholipids of the cell membrane to maintain their fluidity during freezing and desiccation (Crowe et al. 2001).

5. *Cooling rate*: Fast cooling seems to be preferred by AMFs (Kuszala et al. 2001). Identically, Lalaymia et al. (2014) reported that the most effective preservation of in vitro-produced AMF isolates occurred by direct cooling in the freezer. This may result from AMF propagules (i.e., spores and/or vesicles) being less rich in water as compared with hyphae.
6. *Thawing of AMFs*: Fast thawing by placing cryopreserved AMF propagules cultured either in vitro or in vivo in a water bath at 35 °C is generally preferred to prevent the risk of ice recrystallization after cryopreservation.

3.3.4.4 Yeast

Freezing Cryovials/Ampoules in Liquid Nitrogen
This is a two-stage freezing method in which the cells are initially cooled for dehydrating at −30 °C and then cooled at an uncontrolled rate to −139 °C (vapor form) or −196 °C (immersed in liquid nitrogen). The method has been used for a fairly wide range of yeast, including *Bulera*, *Candida*, *Hansenula*, *Lipomyces*, *Kluyveromyces*, *Pichia*, *Rhodosporidium*, *Rhodotorula*, *Saccharomyces*, *Saccharomycodes*, *Schizosaccharomyces*, and *Zygosaccharomyces*. The shelf life of stored yeasts is expected to qualify as long-term preservation. The National Collection of Yeast Cultures detected no decreases in viability levels in test strains over a period of 26 months, and others recorded good survival rates for up to 4 years. The average survival levels are 65% for *Saccharomyces*, 73% for *Candida*, and 74% for *Brettanomyces*. The method is described as follows:

1. Aseptically prepare the yeast suspension in sterile glycerol: The yeast culture is grown on YM broth (Difco, 0711-01) for 72 h. Prepare each 1 ml sample of the yeast suspension in sterile cryovials/ampoules by mixing 0.5 ml of yeast culture containing 10^6–10^7 cells ml^{-1} with 0.5 ml of filter-sterilized 10% glycerol (0.45 μm pore size). The final concentration of glycerol is ~5%. Care should be taken when screwing the caps to prevent the leakage of liquid nitrogen into the samples.
2. Primary freezing: The cryovials/ampoules are frozen to −30 °C by placing them in a deep freezer. The cooling rate is not critical in this method. Keeping them at −30 °C for 2 h allows the cells to dehydrate.
3. Secondary freezing: All samples are transferred to, and immersed in, liquid nitrogen containers. Care is taken to prevent the samples from thawing.

The recovery protocol is the same as for fungi.

Cryopreservation Straw Method This is an alternative method of freezing yeast cultures. It is used by many culture collections and is recommended by the WFCC.

1. Culture growth: 10 ml YM medium (Difco broth, 700101) for 72 h at 25 °C on a shaker.
2. Equal amounts of the inoculum and 10% sterile glycerol are mixed in a sterile bottle.

3. Remove prepared sterile straws (4 mm × 2.5 cm, sealed at 10 mm from one end; straws are placed in a glass petri dish and sterilized by autoclaving at 121 °C for 15 min.)
4. Add the culture suspension to the straw to approximately two-thirds of its capacity (within 3 mm of the open end).
5. Seal the open end.
6. Test the straw for leaks by holding the sealed straws with forceps and gently squeezing.
7. Place six straws in each 1.8 ml cryotube (mark the cryotubes or straws with the freezing date).
8. Primary freezing: Place the filled cryotube in a methanol bath that has been precooled to −30 °C.
9. Secondary freezing: Transfer the pre-cooled cryotubes to a liquid nitrogen container, and place in a rack. Remember to remove the excess methanol from outside of the cryotubes to prevent it from freezing the tubes to the rack while immersed in the liquid nitrogen.

   For recovery: Remove a single straw, and transfer to a 35 °C water bath. Remove from the water bath, grip one end, and sterilize in 95% alcohol. Suspend the cells by squeezing the straw several times. Cut off the sterilized end with sterile scissors, and transfer the culture suspension to an appropriate growth medium.

3.3.4.5 Bacteria and Actinomycetes

Cryopreservation Using Glass Beads

Storing bacteria at −60 to −86 °C in glass beads is a simple and inexpensive method that avoids the damage that can occur to the cell that is subjected to repeated freezing and thawing. The technique allows individual beads to be removed without thawing the entire tube. The resulting shelf lives are reported to be 7 years (Feltham and Sneath 1982).

1. Prepare the sterile vials containing washed beads. The screw-capped glass 2 ml vials (Creative Beadcraft Ltd.) containing 20–30 prepared 2 mm glass beads (R.W. Jennings Ltd.) are sterilized by autoclaving at 121 °C for 15 min. Slightly loosen the caps prior to, and screw down right after, autoclaving. Prepare the beads. First, the beads are washed in tap water with detergent, followed by dilute 0.1 M HCL to neutralize alkalinity. Then, they are washed again with tap water until reaching the tap water's pH. A final wash in distilled water is performed before the beads are dried at 50 °C in an oven.
2. Growing the bacteria. Bacteria should be grown on the most appropriate medium under optimum conditions. A solid medium plate is recommended to prevent the risk of contamination.
3. Prepare a culture suspension with a cryoprotectant. Aseptically withdraw 1 ml of sterile suspension medium [10 ml bottles of with 15% nutrient broth (Difco

Laboratories) are autoclaved at 121 °C for 15 min], and add to the culture plate. A wire loop is used to emulsify the cell suspension.

4. Distribute the culture suspension to glass bead-containing vials. The culture suspension is aseptically dispensed to glass bead-containing vials using a sterile Pasteur pipette. After the beads are thoroughly moistened, the excess suspension should be removed from the vial. Excess suspension in the vial makes it difficult to remove individual beads after storage.
5. Freezing glass bead vials. The vials are placed in suitably sized trays that are transferred to a commercial deep freezer at 60–80 °C (−70 °C is recommended).

For recovery: A vial is removed from the freezer, and one bead is then removed using a mini-spatula sterilized by flaming in alcohol and then cooling. The bead is rubbed over the surface of a suitable solid plate medium and incubated under the appropriate conditions. Determine the viability, and check for contamination.

Cryopreservation by Freezing Cryovials of Culture in Liquid Nitrogen

This method is an alternative for cultures that cannot survive lyophilization or the L-drying method and for patent deposits, sensitive mutants, genetically manipulated strains, and all bacteriophages, including other cultures with phage particles. It is recommended to add cryoprotective agents using the following criteria: (1) nontoxic to the cell, (2) penetrates the cell membrane easily, and (3) binds either an electrolyte that increases in concentration during freezing or water molecules to delay freezing. Common cryoprotectants include glycerol and DMSO. The shelf lives for a wide range of bacteria are over 10 years. The basic steps are simple.

1. Cells are grown on a suitable broth medium to mid or late phase.
2. The cell count is adjusted to $2–6 \times 10^6$ cells/ml. Sterilized glycerol or DMSO is added and mixed to a final 5% concentration.
3. The cell suspension is aseptically dispensed in 0.5 ml aliquots into 2 ml sterile cryovials or cryotubes. These are maintained at 30 °C for 30 min to allow for cell dehydration. The samples are checked for leaks.
4. Primary freezing: The temperature is cooled to −30 °C at a rate of 1–3 °C/min.
5. Secondary freezing: The samples are transferred to a liquid nitrogen container. It is recommended to use the vapor phase.

For recovery: The cryovials (or cryotubes) are removed and immediately placed in a 37 °C water bath to thaw the cultures as rapidly as possible. A sterile wire loop is used to streak the culture on a suitable agar plate medium under the appropriate conditions.

3.3.4.6 Algae

Compared with other groups of microorganisms, relatively little research has been carried out on the development of long-term preservation methods for algae and cyanobacteria. Cryopreservation has been successfully employed to maintain algae (Morris 1981). Most protocols use a simple two-step system with controlled/semicontrolled cooling from room temperature to a subzero holding temperature before immersion in liquid nitrogen. There are several different cryoprotectants in use, but DMSO and methanol have been reported to protect many algae better than glycerol (Morris 1981). Care should be taken to remove DMSO when recovering aliquots, because even trace amounts can inhibit photosynthesis. However, the majority of protocols in current use have been developed with a pre-freezing culture regime, cryoprotectant choice (DMSO or methanol), cooling rate, and thawing regime to minimize the damage to algal cells and maximize the viability level.

The following method for algal preservation is in use at the Culture Collection of Algae and Protozoa, UK:

1. Grow the culture in the appropriate medium under controlled conditions. The 50 ml flasks, containing 30 ml medium, should be incubated at 15 °C under suitable light (dark) conditions. Generally, 30 days are a standard interval for a culture to reach the late log or early stationary phase.
2. Aseptically transfer 15 ml of culture to sterile centrifuge tubes, and harvest the algae by centrifugation at 500 rpm for 10 min at an ambient temperature. Collect the algae, and resuspend in 15 ml fresh sterile medium.
3. Add 10 ml of sterile 10% DMSO to 10 ml of dense algae, and gently mix to a final 5% DMSO culture suspension.
4. Aseptically dispense 1 ml of the culture into sterile 2 ml cryovials/cryotubes. Screw caps down, or seal the cryovials, and maintain at room temperature for 5 min.
5. Primary freezing occurs by transferring the culture vials to a pre-cooled freezer (−30 °C) and incubating for 15 min.
6. Secondary freezing occurs by transferring the culture to a liquid nitrogen container and immersing it in liquid nitrogen.

For recovery: Thaw the vial by removing and placing it in a water bath at 37 °C. Wipe the vial with 70% (v/v) ethanol to sterilize the outside, transfer the whole content of the culture to a bottle containing 30 ml suitable sterile medium, and incubate under appropriate conditions.

3.3.4.7 Preservation Methods for Microbial Communities

In the environment, microorganisms are extremely important to the recycling of elements and nutrients, to balancing trophic chains, and to vital physiological activities in plants and animals, as well as to the conservation of natural habitats. It is crucial to store cultures in mBRCs, but they do not conserve the full diversity of microorganisms found in nature. Functional diversity is very important in the ecological assessment of microorganisms within the ecosystem, mainly because little is known about the relationship between the structural and functional diversity levels of these microorganisms (Yamanaka et al. 2003). Because of advances in technology in metagenomics, metaproteomics, and metatranscriptomics as well as molecular biology, including bioinformatics, much progress has been made in understanding and applying microbial exosystems, such as methanotrophic co-cultures, oxygen-limited autotrophic nitrification/denitrification biofilms and anaerobic AOBs, human intestinal microbial ecosystems, and rumen microbial ecosystem, from various microbial communities (microbiomes) in environmental, industrial, and healthcare-related fields. The microbiomes are likely mixed cultures of both uncultivable and cultivable isolates with many differences in physiological and biochemical characteristics. The assessment of stored microbiomes is not the same as for pure isolates in mBRCs. A metagenomic analysis is the most common method, compared with metatranscriptomic and metaproteomic analyses for assessing the quality of stored microbiomes. The same stored microbiomes may not give the same results when different analysis methods are used. In recent decades, developing methods for storing microbiomes has received more attention.

Yu et al. (2015) stored switchgrass-degrading microbial communities for 56 days at −80 °C with glycerol and DMSO as the cryoprotectants. The relative abundances of *Firmicutes* and *Actinobacteria* were also greater after being stored with DMSO. The storage conditions were not favorable for *Anaerolinea*, *Roseiflexales*, and *Steroidobacter* but did favor *Micromonosporaceae*, *Chelatococcus*, and *Thermobacillus*. Samples preserved with DMSO and glycerol did experience a consistent shift in community composition, although dominant microorganisms were retained in the active communities. In another study, microbial communities in marine sponges were stored by both lyophilization and freezing with or without RNAlater solution. Total DNA and RNA from stored samples were extracted by various methods for a denaturing gradient gel electrophoresis analysis of the community's 16s rDNA genes. It is likely that lyophilization and freezing in liquid nitrogen provide better results (Simister et al. 2011). When preserving microbiomes found in infant fecal samples for sequencing 16s rDNA and oligosaccharides, Lewis et al. (2016) reported that lyophilization was an acceptable method of sample preservation for the purposes of studying microbial communities. The optimization of the cryopreservation of three microbiomes, methanotrophic co-culture, oxygen-limited autotrophic nitrification/denitrification, and human fecal material, was studied based on analyzing operational taxonomy units before and after cryopreservation (Kerckhof et al. 2014). The cryopreservation protocols that used a cryoprotectant (DMSO, trehalose, or tryptic soy broth) succeeded in preserving both community structure and functionality. In addition, Gray and co-workers attempted to compare DNA preservation methods for mixed environmental bacterial samples using DNA, RNAlater, DMSO-EDTA-salt, FTA card, and FTA Elute cards (Gray et al. 2013). The liquid-based preservatives (DNAgard, RNAlater, and DMSO-EDTA-salt) outperformed the card-based methods. Generally, long-term preservation methods, such as lyophilization and cryopreservation in liquid nitrogen or freezing (−80 °C), are preferred for microbiomes. Glycerol, DMSO, or trehalose should be used as a cryoprotectant.

3.3.5 Providing Qualified Microbes as Starting Genetic Resources for Scientific, Agricultural, Environmental, Medical, and Pharmaceutical Research and Applications

3.3.5.1 Culture Quality

An important service of mBRCs in supporting life science and biotechnology development is distributing preserved cultures to users. The user expects the culture to have good qualities: (1) reference cultures come from a reliable source, (2) standardized cultures produced from expertly preserved authenticated seed stocks with appropriate traceability, (3) cultures are as described with valuable information documented, and (4) an "authentic" strain. Before being supplied, the culture should undergo the following chain of events: (1) received from depositor; (2) viability, purity, identification, and identity (authenticity) confirmed; and (3) preserved as a stock until supplied. The quality of a culture depends on continual working practices, equipment, media, reagents, personal skills, and experiences from deposition to distribution. Thus, all of the mBRCs need to address key functions: authentication, best practices in preservation and supply, and confirming the validity of the associated information provided (Smith et al. 2014). The best industrial practices based on ISO-9001:2000 are recommended. For quality assurance, confirming the identity of the culture (the culture being distributed should be the same as deposited to mBRCs) is not easy. This is very important for the reproducible performance of the culture. The identity needs to be confirmed before and after preservation, as well as during storage, by experienced experts and taxonomists. In some cases, the depositor is required, if available. Supporting techniques, such as sequencing, fingerprinting, and analyzing protein profiles by MALDI-TOF, are applicable and useful. To reduce the variation in a stored culture, it is advisable to minimize the numbers of transfers or generations from the original culture, where this is appropriate. The mBRC must use master (or seed) and distribution stocks. The mBRC must produce the master stock from the original culture. This master stock must be used to generate the distribution stock. The mBRC must use the distribution stock for distributions to users.

3.3.5.2 Supplies

User Requirements

In general, a user (individual or organization) is requested to fill and sign the Order and Accession forms from the mBRCs. The form contents should have some clear main points: (1) user's official contact information (contact name, delivery address, and telephone/fax number); (2) the form of culture (living or freeze-dried ampoules); (3) written and signed documentation proving that the user has appropriate containment facilities and authorization to import and handle such a culture, in particular, safety and security requirements needed for handling plant, human, and animal pathogens; and (4) in compliance with local, national, and international legislation, such as the CBD and Nagoya Protocol on access and benefit sharing (ABS) for bioresources. In some cases, an MTA is added to the forms.

When ordering, users are advised not to quote accession numbers unless they are sure that a particular culture is most suitable for their purpose. It is frequently more satisfactory to state the purpose for which the culture is required and to leave the choice of strain to the collection staff.

Available Cultures

It is understood by users that mBRCs should have the capacity to provide the timely and reliable delivery of catalogued strains. However, the number of catalogued strains and the total number of holdings are not the same. The available cultures are documented and available for distribution. For many stored cultures, the long-term preservation by lyophilization or cryopreservation is time-consuming and labor-intensive in checking viability, purity, and identity. If the situation changes, the user should be informed.

Pricing

The monetary value of the cultures needs to be set. The price depends on the level of difficulty in handling the culture, its preservation method, characterizing properties, and confirming its identity and associated information. In some cases, the cost is as high as €3000 (Smith et al. 2014). A discount can be applied for teaching and scientific activities and bulk purchases. Normally, a value-added tax and transportation fees are not included in the price. Gratis exchange agreements are in use in many mBRCs. However, the memorandum of understanding is applicable for related organizations and individual users.

Restrictions

All culture parcels are required to be opened in an appropriate laboratory.

If the strain is a nonindigenous pathogen, the customer has to obtain the relevant license before the strain can be delivered. For cultures belonging to hazard groups (2 or 3), the appropriate containment facilities are needed for safety and security. The user should be informed of any special restrictions and the relevant paperwork that is required from them, prior to providing the requested pathogen cultures.

Delivery and Packing

The mBRC must provide at least the following information to the user: (1) biological material identifier, accession number, and batch number; (2) an estimate of the shelf life, storage conditions, storage instructions and, if appropriate, growth conditions; (3) instructions for opening ampoules or vials (when appropriate and in all cases where materials are being provided to new users); (4) a safety data sheet, including the containment level required for handling the biological material, disposal measures, and measures to take in case of spillage; and (5) an MTA, an essential requirement to protect the IPR and mandatory where they are required. They are used to relay the depositor's and/or country of origin's requirements on the use of the culture. The packing of biological materials and its transport by postal and other transport services is controlled by international and regional agreements and national laws. For noninfectious microorganisms of Risk Group 1, the cultures can be sent by airmail and other means of transport. For infectious microorganism, mBRCs should follow the WHO Guideline on International Regulation for the Packing and Transport of Infectious Substances. However, pathogenic microorganisms that qualify as dangerous goods cannot be sent by air in UN-certified packages if the postal services of the countries through which it passes do not allow the microorganisms in the postal system. Currently, there are few private carriers that transport dangerous goods internationally, with the additional costs of packing and documentation needed at the airport. The carrier is responsible for the shipment, correct packing, documentation, marking, and labeling (Smith et al. 2001).

3.3.5.3 Repositories of Microbial Genetic Resources and Receiving Safe Deposits for Publication and the Protection of Intellectual Property Rights

Servicing deposits of cultures from depositors (individual and organizer) is a useful activity of mBRCs. There is no culture collection that has enough diverse microbes to meet the increasing demand for life science-related activities. Deposited cultures are a source of enrichment for mBRCs. For the depositor, before publishing and applying for the patent, the culture is required to be deposited at an mBRC. In addition, there are other advantages to the depositor: (1) safe preservation, (2) worldwide accession, (3) removal of the burden of distribution, and (4) exchanging cultures. For safe deposits, to ensure long-term viability, cultures need to be stored in freeze-dried ampoules or cryopreserved in liquid nitrogen. The depositor is required to fill the deposit form. The culture to be deposited should be provided with the following information: (1) name, number, or other identifier; (2) depositor's name and address; (3) original source from which the culture was isolated, as well as the place and date of isolation; (4) growth media and conditions and preservation conditions; (5) hazard information and biosafety level; (6) conditions and any restrictions on the distribution of the deposited culture; and (7) the rights of the depositor and mBRC (in compliance with international and national regulations). The price and termination conditions need to be set, including maintenance services and interval quality checks. Not all deposits receive an accession number and are maintained. Sometimes viability and identification checks are not successful. The failure rate is more than 10% annually in some mBRCs. The problems could relate to contamination, identification, or reculture.

3.3.5.4 Conducting R&D on Microbes

Based on the techniques, skills, and knowledge required for checking the viability, purity, and identities of cultures, mBRCs should be more involved in conducting R&D activities on stored microorganisms. The output from these activities would make the cultures more valuable by adding information and increase the value of the mBRC as a microbiology research unit.

Enrichment of Biodiversity and Isolation Techniques
To meet the needs of different users, mBRCs need a large number of diverse cultures, especially reference cultures, such as type cultures. The collection should be independent from internally and externally deposited and exchanged cultures. The curators should be aware of the requirements for cultures from customers, especially national users. Strains of cultures from indigenous sources are crucial for enriching the size and diversity of the collection. Discovering novel cultures is a good enrichment activity. The same sample taken for isolation, testing using different techniques, and growing in media may give rise to various kinds of cultures and increase the biodiversity level. In our laboratory, novel species of filamentous fungi are isolated by directly selecting single spores with a Skerman's micromanipulator. Novel isolation techniques should be developed at the mBRCs.

Taxonomy and Identification
Isolates require a taxonomic analysis and identification to determine whether the isolate represents new taxa or known one. For new taxa, the isolate may have novel properties and applications. For known taxa, the available knowledge and experiences relating to the taxa should be useful to direct further studies on the isolate. Performing taxonomic analyses and identification experiments requires manpower, expertise, and expensive equipment. In fact, not all culture collections (especially private and in-house collections) can implement these activities. However, the trend of many public mBRCs, such as the Thailand Bioresource Research Center, Thailand; VTCC, Vietnam; and Leibniz-Institute DSMZ–German Collection of Microorganisms and Cell Cultures, Germany, includes performing R&D. In total, 330 culture collections provide identification services (Smith et al. 2014). Based on progress in supporting fields, such as molecular biology and bioinformatics, research activities are not limited to individual pure cultures but have expanded to include the study of the biodiversity of microbial communities. The mBRCs need to update any changes in the nomenclature of the preserved cultures. The Leibniz-Institute DSMZ–German Collection of Microorganisms and Cell Cultures website (www.dsm.de) has a section "nomenclature up to date" that can be visited.

Characterization of Microorganisms
Many mBRCs have large collections of microorganism that include cultures that have not been characterized yet. While mBRCs cannot conduct intensive studies on all of the preserved cultures, some important information on the microorganisms that support its practical application are useful to users.

Microorganisms are useful in various fields of the life sciences. Their uses include (1) type strains (for bacteria) and ex-type strains (for fungi) that are used as reference strains for taxonomic studies, (2) enzyme producers, (3) secondary metabolite producers (vitamins and bioactive compounds, such as bacteriocins, fungicides, and antiviruses), (4) antibiotic producers (penicillin, tetracycline, and vancomycin), (5) food and fermented food production, (6) biocontrol agents (such as *Bacillus thuringinesis* and *Bauveria bassiana*), (7) environmental microbes, (8) biorefinery and bioremediation, and (9) pathogens. To identify such properties, many screening methods are conducted, such as (1) screening enzymes, (2) screening antibiotics, (3) screening bioactive compounds, and (4) fermentation studies. The data obtained is added to each culture's profile and stored in a database.

3.3.5.5 Documentation and Databases

For public services, mBRCs need to make the preserved cultures and associated information accessible. Documentation and database are needed for all culture collections. All of the data needs to be maintained for all strains in written or electronic records. Depositors are responsible for assuring the quality of the data associated with deposited cultures. The data will be updated and maintained with the culture. The mBRCs can store data and produce catalogues (written or electronic form) based on authenticated and validated information. Data should also be retained for traceability in compliance with relevant national laws, regulations, and policies. For security, mBRCs should adhere to national data protection regulations. If mBRC data are made available online or on disc, nonconfidential information should be presented. The data for searching and retrieving information from the catalogue and the database must be identical. It is strongly recommended to use standard terminology and formats for data exchanges and protocols for data transmission. For more practical use of the data, a standardized uniform format should be used. Recently, WDCM minimum dataset (MDS) and recommended dataset (RDS) were developed by the Global Catalogue of Microorganisms project (Wu et al. 2017).

Many culture collections use common software programs (Excel or Access), but some are using software specific for mBRCs, such as iCollect software (TBRC: Thailand Bioresources Center) and BioloMICS software (CBS).

3.3.5.6 Performing Related Services and Consulting

Training

To efficiently operate and manage an mBRC, extensive knowledge, experience, and expertise, such as in general microbiology, taxonomy, molecular biology, and bioinformatics, are required. In addition, much progress in these fields has been made, resulting in conventional methods, techniques, and knowledge becoming outdated and replaced. Organizing training courses conducted by experienced specialists will provide the existing staff with new knowledge and techniques and also provide manpower for the sustainable development of the mBRC. However, conducting qualified technical training course is costly in manpower and facilities. Only ~280 of 700 WFCC members can offer training services (Smith et al. 2014). International training courses conducted by scientists from reputed mBRCs are practical. Conducting training courses for local young scientists improves the mBRC's reputation and can be supported by the government.

Consultancy

Technical Advice and Consultancy

It is notable that many users are not microbiologists. Thus, they may have some difficulties in handling and performing a study using microorganisms. The support from experienced staff should improve their experience. The contents of training courses vary. Many mBRCs conduct courses on some common topics with the available staff and facilities, such as (1) collection management, (2) isolation of microorganism, (3) preservation methods, (4) taxonomy and identification, (5) food safety, and (6) special techniques (Smith et al. 2001).

Consultancy on Related Policies and Regulations

Because of the fast progress of global science and technology, fostering internal and external collaborations is important for any country to exchange and update knowledge. A number of internal and external collaborations are established based on microorganisms. Policies and regulations for collaboration based on plant and animal use are more common than on microorganisms. In fact, policies and regulations should protect the sovereign and intellectual property rights, benefit from local biodiversity, and support internal and external collaborations in compliance with international and national laws and regulations under the light of the CBD and ABS. However, not all regulations that are applicable to animals and plants are suitable for microorganisms. Many plants and animals have a long history of traditional knowledge, while a good number of the microorganisms annually deposited represent novel species. It was reported that ~900 new fungal species and ~700 new bacterial species were described in a year (Blackwell 2011); however, they do not all have a ready application. Before use, those microorganisms may require much study before they have an effective use. This process may be risky, and it is not easy to evaluate the benefits of microorganisms. This issue should be addressed by policymakers for practical guidance.

3.4 International Collaboration in Asian mBRCs

3.4.1 International Collaboration, Access and Benefit Sharing Under the CBD and Nagoya Protocols on ABS

3.4.1.1 International Collaboration

In the long history of culture collection development, there is no culture collection that has a large enough holding and expertise to satisfy the wide range of user demands. The collaboration among mBRCs in the exchange of knowledge, expertise, experiences, and cultures is important for culture collection management. The mBRC collaborations have been established for the following different geographical areas: (1) national level, Thailand (Thailand Network on Culture Collection) and the Philippines (the Philippine Network of Microbial Culture Collection); (2) regional level, Europe (European Culture Collection Organization), Asia (the Asian

Consortium for the Conservation and Sustainable Use of Microbial Resources), and Latin American (the Latin American Association for Microbiology); and (3) global level, WFCC, Microbial Resource Centers, and WDCM. The worldwide development of organizations and the management of culture collections are strong indicators of the increasing demand for microbial resources in the life sciences. The WDCM website on 31 December 2016 listed 713 culture collections from 74 countries or regions holding 1,051,947 bacteria, 788,397 fungi, and a total of more than 2.5 million microorganisms. Among the culture collections, 232 from Europe, 250 from Asia, 181 from the Americas, 41 from Oceania, and 11 from Africa are registered with WDCM (Sharma et al. 2017b). Thus, the roles of international organizations are crucial. The WFCC Executive Board has recently updated the WFCC Guidelines for the Establishment and Operation of Collections of Cultures of Microorganisms. The dissemination of the WFCC guidelines at the local level reinforces awareness (http://www.wfcc.info/). In recognition of genetic resources for world development, the Organization for Economic Cooperation and Development in collaboration with the WFCC issued several guidelines for the management of mBRCs, including *guidance for the operation of Biological Resource Centers (BRCs)-2004*, *OECD best practice guidance for BRCs-2007*, and *certification and quality criteria for Biological Resource Centers* (www.oecd.org). To ensure the quality of provided cultures, many mBRCs are certified and accredited by ISO standards ISO9001:2000, ISO 9001-2001, and ISO 9001-2008 (Smith et al. 2014).

3.4.1.2 ABS Under the CBD and Nagoya Protocol

Because culture collection collaborations involve living cultures, they differ from other collaborations. A living culture is not limited in commercial research activities, and it may be the starting genetic source for developing a commercial product or applying for patent. The Budapest Treaty on the international recognition of the deposit of microorganisms for the purpose of patent procedures was the first international attempt to support culture exchanges and to protect the intellectual property rights of the depositor.

The CBD in 1993 was a milestone marking the worldwide recognition of protecting the Earth through the conservation of biodiversity for bio-economic development. The main content of the convention is found in the following three objectives: (1) conservation of biodiversity, (2) sustainable use of its components, and (3) fair and equitable sharing of benefits arising from the utilization of genetic resources. Having an agreement between the user and provider on benefit sharing is crucial for collaborations. The CBD was a great achievement for international legislation on collaborations based on genetic resources. It provided an open platform for collaboration with key concepts and guidelines in Articles 15, 16, and 17. However, the procedure for establishing collaboration was not described in the convention. The Bonn Guidelines were adopted in 2002 to support collaboration negotiations. The prior informed consent procedure (PIC) and mutual agreed terms (MAT) are key elements for ABS. For the first time, the benefits in terms of

Fig. 3.4 NP meeting, Nagoya, Japan, 2010.Source: Katsuhiko Ando (Personal communication)

monetary and non-monetary payment were mentioned in the guidelines. The Nagoya Protocol on 12 October 2014 represented internationally significant progress on ABS (Fig. 3.4). In light of the CBD, the Nagoya Protocol provides an international mechanism developed to elaborate and negotiate based on ABS. The Nagoya Protocol consists a total of 36 articles and includes 2 annexes. To encourage technology transfers through access to genetic resources, the Nagoya Protocol draws and highlights the following important aspects of the negotiation: (1) ABS procedures are initiated through a clearing house from each party as a national focal point (Articles 13 and 14); (2) ABS must be in compliance with domestic legislation and regulatory requirements (Articles 15 and 16); (3) the appropriate capacities in terms of manpower, facility, and policies are required for negotiation and technology transfer (Article 22); (4) technology transfer is encouraged (Article 17); and (5) ABS content and concept of benefit in terms of monetary and non-monetary payments are clarified (Article 5 and Annex I).

After the Nagoya Protocol on ABS was adopted in October 2014, working with genetic resources and associated data has required greater attention. Biologists must ensure that they have a legal understanding of how they can and cannot use the genetic resources in their research (Smith et al. 2017a). However, at the moment, only 100 countries have adopted the NP, with 98 countries abstaining (https://absch.cbd.int/).

3.4.2 ACM: The Asian Consortium for the Conservation and Sustainable Use of Microbial Resources as a Case Study of Asian Collaboration in Culture Collection in Light of the Nagoya Protocol on ABS

3.4.2.1 ACM Establishment and Activities

Asia covers an area of 44,579,000 square kilometers and is notable for its overall large size and population but also its dense and large settlements. It also has vast barely populated regions, within the continent of 4.4 billion people, which represents 59% of the world's population. Asia contains temperate to tropical climates. It has a diversity of fermented foods from megadiverse countries, such as China, India, Indonesia, Malaysia, and the Philippines (https://en.wikipedia.org/wiki/Megadiverse_countries). Asian countries have a high rate of economic development, with a 5% average growth rate in the annual gross domestic product compared with 2.5% for the world. The protection, conservation, and sustainable utilization of biodiversity have been gained much attention in Asian countries. Collaborations among mBRCs were established in 2004 by representatives of 12 Asian countries (Cambodia, China, Indonesia, Japan, Korea, Laos, Malaysia, Mongolia, Myanmar, the Philippines, Thailand, and Vietnam). To facilitate international collaborations among mBRCs and encourage biotechnology utilizing microbial resources in Asia, the ACM aims are as follows: (1) establishment of a framework for international cooperation to encourage microbiological research, (2) construction of mechanisms for academia and industry to utilize microbial resources, (3) establishment of the Microbial Resource Centre network, (4) establishment of international standards for biological material transfer and benefit sharing, and (5) improvement and sharing of standardized techniques. The ACM focused on three task forces: (1) Asian BRC Network, (2) Human Resource Development, and (3) Management of Material Transfer. Through task force activities, much progress has been made to increase the capacities of ACM members in mBRC management, manpower development, and biotechnology. To date, about 100 international workshops and technical training courses have been conducted by experienced international scientists inside and outside of Asia to update the knowledge and techniques of thousands of young Asian scientists in microbiology, culture collection management, taxonomy and identification, and bioinformatics. Under the ACM umbrella, a number of international bilateral and internal collaborations have been established for technology transfer among ACM member countries (Fig. 3.5). There is, however, a large gap between the advanced mBRCs in Japan, China, and Korea and those of some developing countries, such as Vietnam, Laos, Cambodia, Myanmar, and Mongolia. To encourage technology transfers, many scientists are involved in exchange programs or specific research projects. In accordance with the CBD and Nagoya Protocol, the task force team on the Management of Material Transfer has developed the Network of International Exchange of Microbes under ACM, a mechanism for microbial

Fig. 3.5 ACM 1 meeting, October 2004, Tsukuba, Japan. Source: Katsuhiko Ando (Personal communication)

strain transfer and distribution to end users for non-commercial research use (Ando et al. 2014).

3.4.2.2 A Case Study on Bilateral Collaboration Between the National Institute of Technology and Evaluation (NITE), Japan, and the Institute of Microbiology and Biotechnology (IMBT), Vietnam National University, Vietnam

Biotechnology is a priority field in Vietnam policies on science and technology. The conservation of microbial diversity is recognized to be important for the various fields of the life sciences, such as agriculture, food safety, healthcare, and environment. The Vietnam Ministry of Science and Technology has been implementing the national program on the conservation of plants, animals, and microorganisms since 1997. The government ratified both the CBD (1994) and Nagoya Protocol (2014), but the manpower and policy levels need much improvement for internal and external collaboration. In 2004, NITE and IMBT implemented a collaborative project on taxonomic and ecological studies of microorganisms in Vietnam and their utilization. The project was conducted based on the following agreement on common interests of each party: (1) IMBT's need to establish a Vietnam Type Culture Collection (VTCC) and strengthen its ability to study microbial diversity and management and (2) NITE's interest in culture collection enrichment, studying tropical microbial diversity in Vietnam and performing Network of International Exchange of Microbes under ACM mechanisms for microbial strain exchanges, transfers, and distributions to end users for non-commercial research use. After a 10-year implementation period, the project was fruitful for both parties.

Technical training course at IMBT, Hanoi

IMBT scientist studying at NITE, Japan

Report on taxonomic and ecological studies of microorganisms in Vietnam and their utilization

Fig. 3.6 The activities of the collaborative project

The benefits to the IMBT from the collaboration included a great improvement in the capacity to study and manage microbial resources (Fig. 3.6). This is illustrated by the following points:

1. The VTCC is a national mBRC, holding ~10,000 microbial cultures, and 3600 strains are available from an online catalogue (http://vtcc.imbt.vnu.edu.vn/). The VTCC became a member of WFCC in 2008 and registered 2082 cultures in the Global Catalogue of Microorganism in 2013 (http://gcm/wfcc.info). The final report of the project is the first report on the biodiversity of microbes in Vietnam.
2. Technology transfer and the training of young Vietnamese scientists. During the project period (2004–2015), ~20 young IMBT scientists and 30 experienced Japanese scientists were involved in exchange programs in Japan and Vietnam. Through those activities, IMBT staff learned updated techniques, methods, and knowledge in studying microbial diversity and culture collection management; in the isolation, preservation, identification, and taxonomic studies of common microbes (filamentous fungi, actinomycetes, bacteria, yeast, and microalgae); and bioinformatics. In addition, both NITE and IMBT co-conducted 18 workshops and technical training courses in Hanoi that were attended by 150 participants from many institutes and universities in Vietnam.
3. Improved capacity of the VTCC. By implementing the project, the capacity of the VTCC is much improved. Now, the VTCC can extend its activities beyond preservation to the identification and taxonomic study of microbial isolates. At the current time, the VTCC is conducting some R&D activities on screening useful strains for the establishment of a Vietnamese database of cultures with valuable information on stored microbes. Implementing the project was a good case study and allowed in the cataloguing of microbial resources under the CBD in Vietnam. The practical experience gained was useful for amending current Vietnam policies and regulations under the CBD and Nagoya Protocol for worldwide collaboration.

The benefits to NITE from the collaboration were as follows:

1. NITE scientists gained a better understanding of the diversity of tropical microorganisms and their roles in diverse fermented food processes in Vietnam. In total, ten manuscripts written by co-authors from NITE and IMBT were published internationally, and six manuscripts are in progress.
2. By implementing the Network of International Exchange of Microbes under ACM mechanisms, NITE enriched its stored isolates with tropical strains from Vietnam, which are potential genetic sources for R&D.
3. The collaborative project was a good case study for NITE to better understand user demands and related existing policies and regulations. It may be useful for further long-term collaboration in biotechnology with developing countries.

The CBD and Nagoya Protocol identified the sovereign rights of the original country over genetic resources and procedures and established principles for ABS in terms of monetary and non-monetary benefits as practical platforms for collaboration between users and providers. However, successful negotiations based on common interests for collaboration regarding accession and material transfers of microbial resources are not so easy. The concept of fair equitable benefit sharing is quite sensitive. The NITE–IMBT collaborative project was successful and fruitful for both parties. The achievements were not limited to increasing the VTCC's capacity but impacted related legislation in Vietnam to better negotiate collaborations. The key element for this collaboration was that it required being "translated" into the detailed and clear interests of each party in compliance with international and national policies and fruitful negotiations based on a practical understanding.

3.5 Conclusion

The demand for biotechnology and bioindustry development needs more public attempts on studying microbial diversity and its roles in life science. The number of some thousands of new species of microbes found annually is a strong indicator for the attention from scientist's community. The useful roles of microbes in life science development has been discovered.

For any life science activities related to microbes, the quality of the cultures as the starting genetic material is the most important. It is strongly recommended to a new established mBRCs to follow crucial guidelines from WFCC and to update knowledge, techniques, methods, and experiences appropriate to the maintained cultures in culture collection management and operation from long history of experienced mBRCs such as ATCC (USA), CABI, (UK), DSMZ (Germany), and NBRC(Japan).

Promoting international collaboration among mBRCs is in need for life science development. The collaboration is not only limited in information, techniques, methodology, and education but also extended to the exchange in the microbial genetic source. Still there are some differences in regulations in national management of genetic resources from country to country based on particular common interest. However, CBD and NP are excellent platforms for developing bilateral and multilateral collaborations between and among mBRCs.

Acknowledgments We thank Lesley Benyon, PhD from Edanz Group (www.edanzediting.com/ac), for editing a draft of this manuscript.

References

Addy HD, Boswell EP, Koide RT (1998) Low temperature acclimation and freezing resistance of extraradical VA mycorrhizal hyphae. Mycol Res 102:582–586

Ando K, Jin TE, Funabiki R (2014) Network of international exchange of microbes under the ACM (NIEMA) – a transfer and exchange system of microbes for microbial resource centres for non-commercial purposes according to the CBD and the Nagoya Protocol. Microbiol Cult Coll 30:85–96

Annear DI (1958) Observations on drying bacteria from the frozen and from the liquid state. Austral J Exp Biol 36:211–221

Atkinson RG (1953) Survival and pathogenicity of *Alternaria raphani* after five years in dried soil cultures. Can J Bot 31:542–547

Beijerinck MW (1888) Die bakterien der Papilionaceenknollchen. Bot Ztg 46:725–804

Bérdy J (2005) Bioactive microbial metabolites. J Antibiot 58:1–26

Blackwell M (2011) The fungi: 1, 2, 3 ... 5.1 million species? Am J Bot 98:426–438

Çaktü K, Türkoğlu EA (2011) Microbial culture collections: the essential resources for life. Gazi Uni J Sci 24:175–180

Caldwell DR, Bryant MP (1966) Medium without rumen fluid for nonselective enumeration and isolation of rumen bacteria. Appl Microbiol 14:794–801

Castellani O, Ribeiro LV, Fernandes JF (1967) Differentiation of *Trypanosoma cruzi* in culture. J Protozool 14:447–451

Cavalier-Smith T (1998) A revised six-kingdom system of life. Biol Rev 73:203–266

Chang ST, Miles PG (2004) Mushrooms: cultivation, nutritional value, medicinal effect, and environmental impact. CRC, Boca Raton, FL

Crahay C, Wevers J, Munaut F, Colpaert JV, Declerck S (2013) Cryopreservation of ectomycorrhizal fungi has minor effects on root colonization of *Pinus sylvestris* plantlets and their subsequent nutrient uptake capacity. Mycorrhiza 23:463–471

Crowe JH, Crowe LM, Oliver AE, Tsvetkova N, Wolkers W, Tablin F (2001) The trehalose myth revisited: introduction to a symposium on stabilization of cells in the dry state. Cryobiol 43:89–105

Declerck S, Angelo-Van Coppenolle MG (2000) Cryopreservation of entrapped monoxenically produced spores of an arbuscular mycorrhizal fungus. New Phytol 148:169–176

Dietz A, Currie SA (1996) Actinomycetes. In: Hunter-Cevera JC, Belt A (eds) Maintaining cultures for biotechnology and industry. Academic, San Diego, pp 85–99

Douzou P (1982) Effects of low temperatures on biological membranes. Trends Biochem Sci 7:344

Feltham RK, Sneath PH (1982) Construction of matrices for computer-assisted identification of aerobic Gram-positive cocci. J Gen Microbiol 128:713–720

Figueiredo MB, Pimentel CPV (1975) Metodos utilizados para conservacao de fungos na micoteca da secao de micologia fitopatologica do Instituto Biologico. Summa Phytopathol 1:299–302

Franks F (1981) Biophysics and Biochemistry of low temperature and freezing. In: Morris GJ, Clarke A (eds) Effects of low temperature of biological membranes. Academic, London, pp 3–19

Garcia-Garcia M, Rocha-Zavaleta L, Valdez-Cruz NA, Trujillo-Roldan MA (2014) Conservation of the mycelia of the medicinal mushroom *Humphreya coffeata* (Berk.) Stey. in sterile distilled water. MethodsX 1:19–22

Gordon WL (1952) The occurrence of *Fusarium* species in Canada: II. prevalence and taxonomy of *Fusarium* species in cereals. Can J Bot 30:209–251

Gray MA, Pratte ZA, Kellogg CA (2013) Comparison of DNA preservation methods for environmental bacterial community samples. FEMS Microbiol Ecol 83:468–477

Holm-Hansen O (1964) Isolation and culture of terrestrial and fresh-water algae of Antarctica. Phycologia 4:43–51

Holm-Hansen O (1973) The use of ATP determinations in ecological studies. Bull Ecol Res Comm 17:215–222

Homolka L, Lisa L, Nerud F (2003) Viability of basidiomycete strains after cryopreservation: comparison of two different freezing protocols. Folia Microbiol (Praha) 48:219–226

Humber RA (1997) Fungi: preservation of cultures. In: Lacey LA (ed) Manual of techniques in insect pathology. Academic, London, pp 269–279

Hungate RE (1950) The anaerobic mesophilic cellulolytic bacteria. Bacteriol Rev 14:1–49

Hunter-Cevera JC (1996) The importance of culture collections in industrial microbiology and biotechnology. In: Samson RA, Stalpers JA, DVD M (eds) Culture collections to improve the quality of life. Centraalbureau voor Schimmelcultures, Baarn, The Netherlands, pp 158–162

Juge C, Samson J, Bastien C, Vierheilig H, Coughlan A, Piche Y (2002) Breaking dormancy is spores of the arbuscular mycorrhizal fungus Glomus intraradices: a critical cold-storage period. Mycorrhiza 12:37–42

Kaur L, Dhanda S, Sodhi HS, Kapoor S, Khanna PK (2011) Storage and preservation of temperate mushroom cultures, *Agaricus bisporus* and *Pleurotus florida*. Indian J Microbiol 51:234–238

Kennedy AC (1999) Bacterial diversity in agroecosystems. Agric Ecosyst Environ 74:65–76

Kerckhof FM, Courtens ENP, Geirnaert A, Hoefman S, Ho A, Vilchez-Vargas R, Pieper DH, Jauregui R, Vlaeminck SE, deWiele TV, Vandamme P, Heylen K, Boon N (2014) Optimized cryopreservation of mixed microbial communities for conserved functionality and diversity. PLOS One 9(6):e99517

Kirsop BE, Kurtzman CP (1988) Yeasts and their uses living resources for biotechnology yeasts. Cambridge University Press, Cambridge

Kirsop BE, Snell JJS (1984) Maintenance of microorganims. Academic, London

Koch R (1876) Untersuchungen ueber Bakterien V. Die Aetiologie der Milzbrand-Krankheit, begruendent auf die Entwicklungsgeschichte des *Bacillus anthracis*. Cohns Beiträge zur Biologie der Pflanzen,Bd. II, Heft 2, p. 277. Breslau 1870, J. U. Kerns

Kraus SJ (1979) Culture methods for Neisseria gonorrhoeae. Arch Androl 3:343–349

Kuszala C, Gianinazzi S, Gianinazzi-Pearson V (2001) Storage conditions for the long-term survival of AM fungal propagules in wet sieved soil fractions. Symbiosis 30:287–299

Lalaymia I, Cranenbrouck S, Declerck S (2014) Maintenance and preservation of ectomycorrhizal and arbuscular mycorrhizal fungi. Mycorrhiza 24:323–337

Lewis ZT, Davis JC, Smilowitz JT, German JB, Lebrilla CB, Mills DA (2016) The impact of freeze-drying infant fecal samples on measures of their bacterial community profiles and milk-derived oligosaccharide content. Peer J 4:e1612

Lloyd DH (1994) The preservation and maintenance of fungi. 2nd edn. Smith D, Onions AHS. Veterinary dermatology, 5: 215-215.

Marx DH, Daniel WJ (1976) Maintaining cultures of ectomycorrhizal and plant pathogenic fungi in sterile water cold storage. Can J Microbiol 22:338–341

Mendes da Silva AM, Borba CM, Oliveira PC (1994) Viability and morphological alterations of *Paracoccidioides brasiliensis* strains preserved under mineral oil for long periods of time. Mycoses 37:165–169

Molina R, Caldwell BA, Castellano MA, Horton T, Smith JE (2003) Mycorrhizae: ectomycorrhizal fungi. In: Encyclopedia of environmental microbiology. Wiley, New York

Morris GJ (1981) Cryopreservation. An introduction of cryopreservation in culture collections. Institute of Terrestrial Ecology, Cambridge

Oehl F, Sieverding E, Ineichen K, Mäder P, Wiemken A, Boller T (2009) Distinct sporulation dynamics of arbuscular mycorrhizal fungal communities from different agroecosystems in long-term microcosms. Agric Ecosyst Environ 134:257–268

Richter DL (2008) Revival of saprotrophic and mycorrhizal basidiomycete cultures after 20 years in cold storage in sterile water. Can J Microbiol 54:595–599

Rosenblatt JE, Fallon A, Finegold SM (1973) Comparison of methods for isolation of anaerobic bacteria from clinical specimens. Appl Microbiol 25:77–85

Ryan MJ, Smith D, Bridge PD, Jeffries P (2003) The relationship between fungal preservation method and secondary metabolite production in *Metarhizium anisopliae* and *Fusarium oxysporum*. World J Microbiol Biotechnol 19:839–844

Schopf JW (2006) Fossil evidence of Archaean life. Philos Trans R Soc Lond B Biol Sci 361:869–885

Sharma SK, Kumar R, Vaishnav A, Sharma PK, Singh UB, Sharma AK (2017a) Microbial cultures: maintenance, preservation and registration. In: Varma A, Sharma AK (eds) Modern tools and techniques to understand microbes. Springer, Cham, pp 335–367

Sharma SK, Saini S, Verma A et al (2017b) National agriculturally important microbial culture collection in the global context of microbial culture collection centres. Proc Natl Acad Sci India Sect B Biol Sci:1–14

Simister RL, Schmitt S, Taylor MW (2011) Evaluating methods for the preservation and extraction of DNA and RNA for analysis of microbial communities in marine sponges. J Exp Marine Biol Ecol 397:38–43

Singh SK, Upadhyay RC, Kamal S, Tiwary M (2004) Mushroom cryopreservation and its effect on survival, yield and genetic stability. Cryoletters 25:23–32

Smith D (1983) A two stage centrifugal freeze drying method for the preservation of fungi. Trans Br Mycol Soc 80:333–337

Smith D (1993) Tolerance to freezing and thawing. In: Jennings DH (ed) Tolerance of fungi. Marcel Dekker, New York, pp 145–171

Smith D, Ryan MJ, Day JG (2001) The UKNCC biological resource: properties, maintenance, and management. UKNCC Secretariat, Egham, Surrey, UK

Smith D, Mccluskey K, Stackebrandt E (2014) Investment into the future of microbial resources: culture collection funding models and BRC business plans for biological resource centres. SpringerPlus 3:81

Smith D, Da Silva M, Jackson J, Lyal C (2017a) Explanation of the Nagoya Protocol on access and benefit sharing and its implication for microbiology. Microbiology 163:289–296

Smith D, Martin D, Novossiolova T (2017b) Microorganisms: good or evil, MIRRI provides biosecurity awareness. Curr Microbiol 74:299–308

Sneel JS (1984) Maintaining cultures for biotechnology and industry. Academic, San Diego

Stern S (2004) Biological resource centers: knowledge hubs for the life sciences. Brookings Institution Press, Washington, DC

Stielow B, Bratek Z, Orczán AKI, Rudnoy S, Hensel G et al (2011) Species delimitation in taxonomically difficult fungi: the case of Hymenogaster. PLOS One 6(1):e15614

Suman BC, Jandaik CL (1991) Preservation of culture of *Agaricus bisporus* (Lange)Singh.in liquid nitrogen and its effect on yield and characters of fruiting bodies. Ind J Mycol Pl Pathol 21:34–37

Sundari SK, Adholeya A (1999) Freeze-drying vegetative mycelium of Laccaria fraterna and its subsequent regeneration. Biotechnol Tech 13:491–495

Tan SY, Tatsumura Y (2015) Alexander Fleming (1881–1955): discoverer of penicillin. Singapore Med J 56:366–367

Tan CS-H, Van Ingen CW, Stalpers JA (2007) Freeze-drying fungi using a shelf-freeze drier. In: Day JG, Stacey GN (eds) Cryopreservation and freeze-drying protocols. Humana, Totowa, NJ, pp 119–125

Tibbett M, Sanders FE, Cairney JWG (1999) Long-term storage of ectomycorrhizal basidiomycetes (*Hebeloma* spp.) at low temperature. J Basic Microbiol 39:381–384

Usta C (2013) Microorganisms in biological pest control—a review (Bacterial toxin application and effect of environmental factors). In: Silva-Opps M (ed) Curr Prog in Biol Res. InTech, Rijeka

Van Laere A (1989) Trehalose, reserve and/or stress metabolite? FEMS Microbiol Lett 63:201–209

Veena SS, Pandey M (2010) A simple method for culture conservation of some commercial mushrooms. Mycosphere 1:191–194

Woods R (1976) UK Federation for culture collection. Newsletter 2, 5

Wu L, Sun Q, Desmeth P, Sugawara H, Xu Z, McCluskey K, Smith D, Alexander V, Lima N, Ohkuma M, Robert V, Zhou Y, Li J, Fan G, Ingsriswang S, Ozerskaya S, Ma J (2017) World data centre for microorganisms: an information infrastructure to explore and utilize preserved microbial strains worldwide. Nucleic Acids Res 45:D611–D618

Yamanaka T, Helgeland L, Farstad IN, Fukushima Midtvedt HT, Brandtzaeg P (2003) Microbial colonization drives lymphocyte accumulation and differentiation in the follicle-associated epithelium of Peyer's patches. J Immunol 170:816–822

Yu C, Reddy AP, Simmons CW, Simmons BA, Singer SW, Vander Gheynst JS (2015) Preservation of microbial communities enriched on lignocellulose under thermophilic and high-solid conditions. Biotechnol Biofuels 8:206

Chapter 4
Specialized Microbial Resource Centers: A Driving Force of the Growing Bioeconomy

Irena B. Ivshina and Maria S. Kuyukina

Abstract The growth of the world economy is accompanied by a decrease in the quality of the environment. The unsustainable resource use, hence the depletion of the resource base, and as a consequence the catastrophic biodiversity loss and the climatic change are indeed alarming signs of our time. It becomes obvious not only for professionals that the way out of the challenging ecological situation is to reduce the unfavorable impacts of economy through development and implementation of advanced technologies, primarily biotechnologies. The advancement of biotechnology largely depends on the state of biological collections. For many years, the authors have been studying the biodiversity of microorganisms with regard to environmental stresses and disturbances and are well acquainted with the collection activities. In this review we attempt to emphasize the pivotal role of specialized microbial resource centers in the period of building the knowledge-based bioeconomy. The advantages of their integration into international networks are shown. The biotechnology significant collections supported in Russia are presented. The problems of the collection evolution and human resources are discussed.

4.1 Introduction

However, paradoxical it may be, the science today, having created powerful means of natural resource management, begins to look for ways to protect from the consequences of its own inventions. Successful it would seem, exploration of natural resources and advancement of modern industries have led to a decreased quality of natural environment and an increased number of habitats in which organisms are in critical conditions of environmental contamination with technogenic pollutants.

Concerns about the quality of the environment and its remediation from ecotoxicants determine the need for thorough study of microorganisms that are

I. B. Ivshina (✉) · M. S. Kuyukina
Institute of Ecology and Genetics of Microorganisms, Ural Branch of the Russian Academy of Sciences, Perm State University, Perm, Russia
e-mail: ivshina@iegm.ru

S. K. Sharma, A. Varma (eds.), *Microbial Resource Conservation*, Soil Biology 54,
https://doi.org/10.1007/978-3-319-96971-8_4

able to respond to adverse changes in the habitat and trigger the appropriate adaptive reactions. Therefore, it is no accident that in the focus of broad international initiatives are microorganisms associated with human activities and participating in remediation of affected ecosystems (DIVERSITAS 1996; CBD 2011; OECD 2007; MOSAICC 2011). Today, microbial genetic resources become an intellectual property item and a resource base for the growth of bioindustry, with their significance being not inferior to the mineral resources.

One of the effective approaches to study and preserve microorganisms is to maintain them ex situ in laboratory repositories. Microbial collections as ideal sources to discover new processes, biodegraders of xenobiotics, and producers of novel bioactive compounds and as providers of valuable biological information are, in fact, investments in the future that can change the world. Against the background of the global development of applied microbiology, which rapidly grows into a highly profitable bioeconomy sector with billion turnovers, the depositories of microorganisms and collection activities become increasingly important and acquire the real value (Bull 1996; Constanza et al. 1997; Kirsop and Canhos 1998; ten Kate 1999; Miyazaki 2006; OECD 2009; Smith 2012; Overmann 2015).

Many countries worldwide aiming to develop their competitive economies take measures to strengthen and replenish traditional pre-genomics microbial culture collections, as well as to create their databanks. During the past decade, collections have laid the basis for the intensive formation of microbiological resource centers (microbial domain Biological Resource Centers sensu OECD, mBRCs), modern depositories, and providers of biological material and relevant professional expertise (OECD 2001, 2004, 2007; WFCC 2010). This shift (from the term culture collections to biological resource centers and "microbial BioBanks") reflects the evolution of microbial repositories and their increasingly larger social and economic roles following the scientific and technical progress, as well as the pressure of socioeconomic, legal, and political sectors (Desmeth 2017).

At present, microbial collections develop (proactively or reactively) from providers of microbial materials for the scientific sector to resource providers for the society at large (Desmeth and Bosschaerts 2008). The formation of national collections proceeds dynamically not only to meet the needs of research and industries but also to guarantee the biosafety of their industrial uses, to protect intellectual property rights in biotechnology valuable strains, and to control their distribution.

At the same time, preference is given to specialized nonmedical collections (institutional, academic). They cooperate according to the uniform rules, meet the needs of qualified users, and are intended to study and preserve microorganisms of specific taxonomic groups isolated from natural ecosystems and possessing potentially valuable properties. Such collections are specific in that they are not only centers for collecting and storing microbial cultures and accumulating comprehensive information on their properties and prospective uses, including legal advice for users on the intellectual property rights, as well as high-quality services for research and development in life sciences.

Large multipurpose collections cannot essentially provide all these services for a number of reasons. The increased flow of valuable cultures (subject to obligate deposition) due to improved methods of selective isolation of microorganisms from natural communities and modern identification techniques, together with a large-scale detection and cultivation of previously unknown microorganisms, contributes to the enlargement of the collection holdings and related information. The abovementioned impose certain restrictions on the effective operation of collections. It is no coincidence that the formation of universal multi-profile "giant" collections has stalled worldwide, whereas a network of specialized microbial culture collections is formed on a mass scale.

Today, such collections ensuring the completely characterized strains are conceived as next-generation collections and being reformed into modern multifunctional mBRCs. According to the needs of the time, mBRCs assume the functions of service providers not only in conservation and distribution of properly stored microbial cultures of biotechnology significance but also useful scientific information on issues, reflecting the current situation in microbial diversity and its potential use in various biotechnology sectors. These include prospecting of fossil fuels, biodegradation of emergent pollutants, development of environmentally friendly and economically sound technologies for environmental protection, and production of new compounds and advanced materials. Adaptation of mBRCs to the new socioeconomic environments is not only fundamental (academic) but also economic issues, as it is ultimately associated with biotechnology.

Such recognized centers with modern bio-depositories, relevant research facilities, advanced data resources, qualified personnel, and an adequate government and public support guarantee an appropriate safety of holdings and have an unchallenged academic reputation. They operate in a number of countries based on the recognized microbial collections, in particular the Belgian Coordinated Collections of Microorganisms (BCCM), the Brazilian Collection of Microorganisms from the Environment and Industry (CBMAI), the China General Microbiological Culture Collection Center (CGMCC), the Culture Collection of Beijing Agriculture University (CCBAU, China), the Centre de Ressources Biologiques de l'Institut Pasteur (CRBIP, France), the German Collection of Microorganisms and Cell Cultures (DSMZ), the Japan Collection of Microorganisms (JCM), the Korean Culture Center of Microorganisms (KCM), the Centraalbureau voor Schimmelcultures (CBS, The Netherlands), the All-Russian Collection of Microorganisms (VKM), the Russian National Collection of Industrial Microorganisms (VKPM), and the Coleccion Espanola de Cultivos Tipo (CECT, Spain). OECD considers collections of microorganisms and their modern counterparts—mBRCs interconnected into networks as key elements of the research infrastructure in life sciences and bioeconomy (OECD 2007).

Many microbial culture collections established worldwide join the World Federation for Culture Collections (http://www.wfcc.info) which successfully coordinates cooperation of collections from different countries (Fig. 4.1). To date, the WFCC-MIRCEN World Data Center for Microorganisms (WDCM, http://wcdm.org and http://wfcc.info/wdcm.db) registered and included in the world catalog of collections WFCC-CCINFO (http://www.wfcc.info/ccinfo/home/) 726 collections (service, specialized, research) belonging to different parent organizations (most often to scientific institutions) from 75 countries (WDCM 2017). The total number of strains

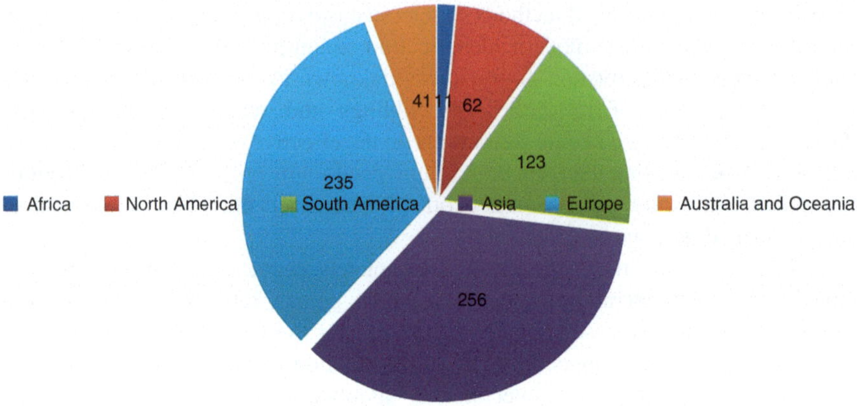

Fig 4.1 Regional distribution of world microbial collections (as for 25.07.2017)

maintained exceeds 2.5 million, of which 41% are bacteria, 31% are fungi, and 28% are others, including viruses and cell lines (WFCC 2017). About 40% of these depositories receive permanent government support. According to data on the use of cultured microorganisms available at the Analyzer of Bio-resource Citations (ABC, http://abc.wfcc.info/), 142,758 papers in reference to 76,807 strains stored in 131 WDCM member culture collections from 50 countries have been published in 50,307 scientific journals since January 1953. It is worth noting that only a low proportion (about 13%) of member collections have produced online catalogues publicly available and meeting the OECD requirements (OECD 2007). This might indicate that the multifaceted potential of such collections is currently in a latent form. Catalogs help to respond to rapid changes in the taxonomy and nomenclature of microorganisms in the context of the improved classification system, to make microbial cultures more accessible for users, and to guarantee the prompt cooperation of collections with the experts' community and users of cultures and related data (Vasilenko et al. 2011). The current status, characteristics, history, and evolution of microbial collections in the world are given in a recent review (Sharma et al. 2017).

The purpose of numerous microbial collections is to maintain viable authentic microorganisms, to identify and preserve their original biological properties, and to ensure the continuity in transferring collection strains to users in strict compliance with current international regulations, including (1) the Convention on Biological Diversity (CBD, http://www.cbd.int), (2) Best Practice Guidelines for Biological Resource Centers (OECD 2007), (3) the Nagoya Protocol (NP) on Access and Benefit Sharing (ABS, http://www.cbd.int/abs/), and (4) national and regional laws implementing this Protocol, such as the European Union (EU) Regulation 511/2014. According to the NP, access to biological resources is made available provided that a fair and equal portion of benefits derived from their use is returned to those who contribute to the conservation and deposit of these resources. This Regulation forms a registry of collections, having integrated numerous legal constraints and, thereby, providing the social and economic players with microorganisms secured by the legal environment for their uses. The appropriate legal arrangement of the activities of microbial collections should make the circulation of microbiological genetic resources transparent and extremely clear for all participants in the process.

This review focuses on the role of specialized microbial collections as a key determinant in the innovative development of new biotechnology and bioindustry. The purpose of this work is to draw the attention of the international community to the importance of protecting and preserving "our microbial heritage" and to the relevance and need of establishing and developing national mBRCs and centralized networks in the period of building knowledge-based bioeconomy and biopolicy.

4.2 Specialized Collections of Microorganisms in the Time of Bioeconomy Building

Until recently, the understanding that "a factory may be set up, a patient cured or a crop saved" (Kirsop 1992) as a result of taxonomic research and the in-depth characterization of a microorganism did not go beyond a narrow community of professionals; and preservation of collections of microorganisms was solely on the shoulders of its scientific and technical personnel.

Everything suggests that the situation is now changing for the better. Not only professional biologists but also policy makers, businessmen, and ordinary contemporaries, regardless of their occupations, begin not only to realize the events taking place in the world of microbes and our crucial dependence on it but also to adopt the strategies they evolved. The society becomes aware of and concerned that ignoring microbial diversity leads to negative consequences in medicine, environmental protection, and biotechnology. Against the background of pathogenization of saprotrophs in the technogenically polluted environment, a range of potentially dangerous microorganisms with "nonprofessional" parasitism expands, a certain "blurring" of clear boundaries between pathogens and non-pathogens occurs, and the idea of the versatility of pathogenicity factors of microorganisms is increasingly asserted. At the same time, new opportunities are revealed for positive human interactions with the microbial world (Lederberg 2000). Of particular relevance are nowadays the words of Kalakoutskii (2016) that "if until recently the image of the 'enemy' (less often, of a 'friend') dominated in the relationship between a human and the world of microbes, now it becomes obvious that it is necessary to build bridges for 'peaceful coexistence' with this huge world."

The communities of culture collections promote the search for optimal ways of constructive interactions of humans with the microbial world which opens up prospects to forecast and manage the environmental situation and to develop and implement advanced technologies. In addition, the study of microorganisms and their involvement in human activities becomes increasingly possible based on the microorganisms' description in silico, in rapidly growing relevant gene and protein databases (GenBank, EMBDatabank, wwPDB) (Cole et al. 2009; Hindrê et al. 2012).

During the last decade, due to the improved techniques of a biological experiment, the range of study objects (from cultured, monstrous, forms to uncultured, phantom, and cryptic ones) is expanding enormously, providing investigation of new biological functions. At the same time, there is a return of attention to microbial strains deposited in the collections, the biotechnological potential of which until recently seemed to be exhausted. All these are mainly due to technological advances in phylogenomics and comparative genomics, ecological metagenomics and proteomics, and systemic and structural biology. Additionally, there is a steadily growing fundamental interest in the in-depth study of genomes and metabolisms of previously not studied microorganisms, including those not cultured on traditional nutrient media. Relying on the present-day discoveries of functional metagenomics and mass spectrometry, there are studies on the microorganisms' survival mechanisms under the growing technogenous pollution of the environment. Moreover, an active search for stable and highly active enzymes, both universal and unique, produced by microorganisms and involved in pollutant biodegradation and intended for potential applications in ecobiotechnology is carried out (de Vasconcellos et al. 2010; Dellagnezze et al. 2014; Sierra-García et al. 2014; Ufarté et al. 2015). These studies also aim to develop effective biocatalysts for industrial production.

At this moment, however, these technologies are available not for every microbial collection. Many culture collections are insufficiently staffed and equipped. Therefore, the taxonomic and functional diversity of microorganisms presented ex situ is still not fully described. In practical terms, the thorough revision of the collection cultures in accordance with the improving system of microorganisms' classification has addressed still not all microbial collections. This situation is largely due to the methodical and technical difficulties encountered by many collections, including the updating of characteristics, correct identification and reidentification of microorganisms, description of novel taxa in compliance with international standards, and identification of the metabolic potential to the full. To add to the above difficulties is the lack of experience and also high-skilled taxonomists that, in turn, is due to the lack of modern equipment and, finally, insufficient financing.

Poorly characterized collections, not only in terms of taxonomy but also in physiology, conservation, and integration of genotypic, phenotypic, and environmental data, make it difficult to select strains for targeted screening of novel bioactive metabolites (Bull et al. 2000). There is an urgent need to update and improve methods for quality control of bio-resources stored. Addressing these issues is ultimately important for assessing the potential of microbial cultures maintained in the collections in regards to their biotechnological exploitation.

To implement a unified strategy for studying and preserving microbiological resources in order to obtain socioeconomic benefits from their use, there is an urgent need for more effective cooperation between ex situ collections at the global level (Smith et al. 2008; Stackebrandt 2010). Incidentally, there is no a single structure today able to provide an appropriate coverage of microorganisms and related information (Shimura 2003).

Perfect examples of consolidated efforts of microbial collections and the community of microbiologists are national, regional (ECCO), and world (WFCC) public associations and powerful international networks of living culture collections accessible for data exchange. More examples include their integration under international mega-projects (WFCC-MIRCEN WDCM, CCINFO; Strain Info, http://www.straininfo.net/; WFCC Global Catalogue of Microorganisms, GCM, http://gcm.wfcc.info; Common Access to Biological Resources and Information, CABRI, http://www.cabri.org/; Microbial Resource Research Infrastructure, MIRRI, http://www.mirri.org; the European Consortium of Microbial Resources Centres, EMbaRC, http://www.embarc.eu; the Global Biological Resource Centre Network, GBRCN, http://www.gbrcn.org; Transparent User-friendly System of Transfer, TRUST, http://bccm.belspo.be/projects/trust) and the creation of national consortia of collections operating in different organizations and cities in one country (e.g., BCCM, Belgium). All these initiatives are relevant to the international agreements, such as the Cartagena Protocol on Biosafety (CBD 2001) and the Nagoya Protocol on the Regulation of Access to Genetic Resources and the Fair and Equitable Sharing of Benefits Arising from Their Utilization (CBD 2011) to the Convention on Biological Diversity (CBD 1992).

A recent example of coordinated international efforts to provide access to bioresources suitable for use in biotechnology might be the Pan-European Rhizosphere resources Network (PERN, http://www.PERN-BRIO.eu), the European–Russian initiative on microbiological resources of the rhizosphere created as part of the FP7 BRIO project. PERN aims at enhancing the interactions between the linked collections and between collections and biotechnology companies (Declerck et al. 2015).

It is generally accepted that the rhizosphere microbiome, sometimes called the second genome of a plant (Berendsen et al. 2012), is significant for plant growth and health. Currently, numerous data on the beneficial properties of plant-associated microorganisms have been accumulated. They stimulate the plant's protective mechanisms (Whipps 2001; Lugtenberg and Kamilova 2009; Mendes et al. 2013) and resistance to various environmental stresses, for example, salinity (Cohen et al. 2009), heavy metal-contaminated soils (Belimov et al. 2009), temperature shock (Jones 2013), etc.

PERN was created as a result of consolidated efforts of specialists and collections from Western Europe (BCCM/MUCL, http://bccm.belspo.be/about-us/bccm-mucl and BCCM/LMG, http://bccm.belspo.be/about-us/bccm-lmg, Belgium; Mycotheca Universitatis Taurinensis, MUT, http://www.mut.unito.it/en, Italy; SAF Swiss Collection of Arbuscular Mycorrhizal Fungi, http://www.agroscope.admin.ch, Switzerland) and Russia (VKM, http://vkm.ru/; Regional Specialized Collection of Alkanotrophic Microorganisms, IEGM, http://www.iegm.ru/iegmcol/; and Collection of Rhizosphere Microroganisms, IBPPM, http://ibppm.ru/), which maintain rhizosphere microorganisms. It is a virtual pool of agroecological species of microorganisms and related data for professionals in various fields of science and bioindustry.

Cooperation under the BRIO project, centrally coordinated in Brussels by BCCM (BCCM, http://bccm.belspo.be/), initiated a joint research (carried out by each of the consortium participant at the premises of their own institutions, but according to joint experimental protocols), including exchange of cultures and data between PERN member collections and use of biotechnological potential of microbial

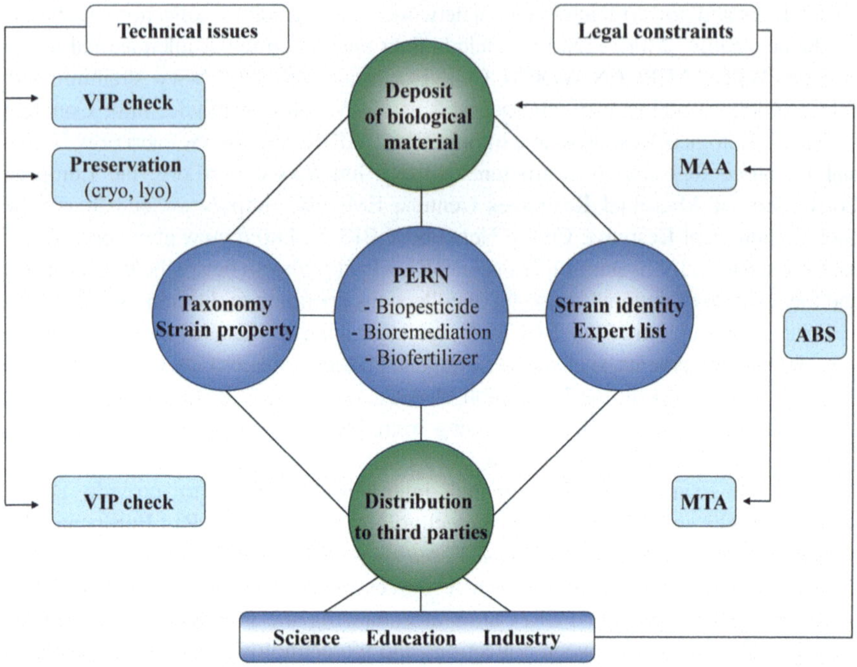

Fig 4.2 Technical issues and legal constraints in the Pan-European Rhizosphere resources Network (PERN). *ABS* access and benefit sharing, *MAA* Material Accession Agreement, *MTA* Material Transfer Agreement, *VIP* viability, identity and purity. Modified from Declerck et al. (2015)

funds maintained in these collections. This collaboration contributed to the expansion of the phylogenetic and functional diversity of rhizosphere microorganisms and the enhanced efficiency in all areas of the collection activities in compliance with regulatory documents and administrative procedures on quality standards, distribution, and deposition of biological material (Fig. 4.2).

Totally, the diversity of these collections exceeds 63,000 strains (importantly, isolated from food, fodder, and agro- and industrial environments). Of these, 1100 are rhizosphere microorganisms with well-documented properties accessible through PERN in compliance with the international rules. Selection of these strains relied upon strict criteria to discover their potential as biofertilizers (770 strains), biopesticides (81 strains), and bioremediation (291 strains). This number is growing constantly (Declerck et al. 2015).

PERN supports the operation of the virtual collection and facilitates access to microorganisms suitable for use in agriculture and/or environmental remediation. It allows users to make online searches by the "taxon browser" and the "strain property browser." A comprehensive set of data, including the description of cultures and the use of procedural protocols for isolation, conservation and long-term storage, identification and characterization of the microorganisms stored, and a list of external experts in applied microbiology and biotechnology (in particular, biofertilizers, biopesticides, bioremediation etc.), is available at the website. Additionally, also accessible is a complete administrative documentation on exchange and distribution of information and biological material technically and legally fit for use in research and commercial applications, taking into account new national quality standards for mBRCs at the international level.

The created cooperative, dynamic network of collections interacting in their interests, goals, and needs and able to compete, to some extent, against each other is a good means of sharing useful information, evaluating reasonably the results, and comparing the methodical level of their research and also a window of opportunity for rapid data dissemination and cost cutting.

A common representative database and a coordination center provide standardized approaches to the storage of rhizosphere microorganisms and make it possible to work out standard protocols for their detection and identification procedures, to establish new scientific links between collections and with industrial companies, and to ensure conditions for material transfer beneficial to all interacting parties. Finally, the opportunity for cooperative research, training, technology transfer, and culture exchange can be seen as a nonmonetary benefit to PERN member collections.

Such cooperation is of strategic importance for the survival of microbial collections. In this connection, it is possible to cite the obvious thesis that the decisive role in the activity of ecosystems is played by a consortium formed as a protective system rather than an individual (Zavarzin 1995). This is as obvious as the fact that "one is not a warrior in the field."

4.3 A Practice-Verified Development Strategy of Russian Specialized Microbial Resource Centers

The importance of microbial collections increases as the bioeconomy penetrates deeper into everyday life, striking in its scale and extraordinary growth rate in the modern world. Bioeconomy addresses a number of issues on the improved quality of life, with health and environmental conditions as priorities. It is closely connected with the actual development of biotechnology (OECD 2009). Bioeconomy (or the "green" economy of reduced environmental damages) is, in fact, the most high-tech sector of the economy based on the large-scale applications of post-genomic technologies using renewable bioresources and efficient bioprocesses to produce valuable products and energy (The Lisbon European Council 2000; European Commission Conference Report 2005).

The most significant outcomes of applied biotechnology are biofuel, specifically bioethanol, biodiesel, and biogas, and local progress in environmental protection, biodetection of anthropogenic changes, and also biological remediation of contaminated areas using site-specific biodegraders. Many biotechnology aspects today are still of only theoretical relevance and associated with, for example, continuous replenishment of the database on sequenced genomes and functional genes of microorganisms.

It is obvious that since different countries worldwide are at different levels of development, particular directions and sectors of bioeconomy in each country are of different priority. As concerns Russia with its great natural resources, priority should be given to waste treatment from extractive industries. In such a case, the development and use of environmental technologies will become economically viable. Moreover, in order to maintain the leading position as a supplier of raw materials and derived products in the "post-oil world," Russia needs to develop bioeconomy sectors utilizing renewable resources and based on beneficial nonpathogenic strains of microorganisms stored in microbiological collections.

In Russia, there are currently about 100 collections of nonpathogenic microorganisms (bacteria, archaea, filamentous fungi, and yeasts), of which only 25 collections are represented in WDCM (Table 4.1). Russian collections of microorganisms, housed mainly by research institutes and universities, are not legal entities. The profiles of collections are as a rule determined by the interests of parent organizations, with the latter providing the financing of collections.

All-Russian Collection of Microorganisms (VKM, WDCM 342, http://www.vkm.ru/VKM-rus.htm) is the largest microbial collection in terms of diversity and one of the largest in terms of its total holding in Russia. It encompasses about 20,000 strains (bacteria, including *Actinobacteria*, *Archaea*, filamentous fungi, and yeasts). The collection maintains over 2500 type strains, as well as patent strains and other biotechnology promising strains. VKM is recognized as an International Depositary Authority (IDA) for biological material for patent purposes under the Budapest Treaty. The scientific research is related to the taxonomy of microorganisms and microbial diversity in various ecosystems. The collection is not only a prompt provider of relevant data on maintained cultures but also is integrated into global networks, such as StrainInfo, GenBank, WDCM, MycoBank (Ozerskaya et al. 2012; Vasilenko and Ozerskaya 2012), MIRRI (Stackebrandt et al. 2015), and PERN (Declerck et al. 2015). In recent years, the VKM research team works to create a world-class mBRC based on its collection, including the development of information resources.

In 2014, the Russian National Collection of Industrial Microorganisms, VKPM (WDCM 588), acquired the status of the national Bioresource Center (BRC VKPM, http://www.genetika.ru/vkpm). Currently, the collection holding amounts to more than 20,000 industrial and genetically engineered strains of microorganisms of biotechnological application. VKPM is recognized as an IDA for biological material for patent purposes under the Budapest Treaty. The development program of this collection, in addition to performing infrastructure functions in the field of microbial bioresources, includes improvement of the instrumental, technological, and methodical base to carry out researches for priority areas of biotechnology, as well as high-tech service activities.

Smaller specialized collections are distributed unevenly across the European and Asian parts of Russia (80% and 20%, respectively), reflecting the historically uneven distribution of population density, as well as high-tech and labor-intensive industries

Table 4.1 List of Russian microbial culture collections registered with WFCC

Acronym	WDCM number	Collection name	Institution and region
ACCS	WDCM 936	Algae Culture Collection of Siberia	Siberian Federal University, Krasnoyarsk, http://www.sfu-kras.ru
ACSSI	WDCM 1132	Algal Collection of Soil Science Institute	Institute of Physicochemical and Biological Problems in Soil Science, Russian Academy of Sciences, Pushchino, Moscow region, http://acssi.org/
BCAC	WDCM 1023	Bashkortostan Collection of Algae and Cyanobacteria	M. Akmullah Bashkir State Pedagogical University, Ufa, https://bspu.ru
BOROK	WDCM 602	The Collection of Algae	Institute for Biology of Inland Waters, Russian Academy of Sciences, Yaroslavl, http://ibiw.ru/index.php?p=project/algo/index
CALU	WDCM 461	Collection of Algae St. Petersburg (Leningrad) State University	Centre for Culture Collection of Microorganisms, St. Petersburg State University, St. Petersburg, http://microbio.museums.spbu.ru/
CBMW	WDCM 986	Collection of Bacteriophages and Microorganisms MicroWorld	Research-and-Production Center MicroWorld Ltd, Moscow, http://micro-world.ru
CCCS	WDCM 1024	Culture Collection of Ciliates and their Symbionts	St. Petersburg State University, St. Petersburg, http://www.bio.spbu.ru
CCIBSO	WDCM 836	Culture Collection IBSO	Institute of Biophysics, Siberian Branch, Russian Academy of Sciences, Krasnoyarsk, http://bl.ibp.ru/
CIAM	WDCM 890	Collection of Nonpathogenic Microorganisms for Agriculture	All-Russia Research Institute for Agricultural Microbiology, Russian Academy of Sciences, St. Petersburg—Pushkin, http://www.arriam.spb.ru/
CSMS	WDCM 1121	Collection of Soil Microorganisms "Symbiont"	Institute of Biochemistry and Genetics, Russian Academy of Sciences, Ufa, http://ibg.anrb.ru
EMTC	WDCM 974	Collection for Extremophile Microorganisms and Type Cultures	Institute of Chemical Biology and Fundamental Medicine, Siberian Branch, Russian Academy of Sciences, Novosibirsk, http://www.niboch.nsc.ru/
IBPPM	WDCM 1021	Collection of Rhizosphere Microorganisms	Institute of Biochemistry and Physiology of Plants and Microorganisms, Russian Academy of Sciences, Saratov, http://collection.ibppm.ru
ICES	WDCM 1035	International Collection of Epidemic Strains	I.I. Mechnikov North-Western State Medical University, St. Petersburg, http://szgmu.ru/

(continued)

Table 4.1 (continued)

Acronym	WDCM number	Collection name	Institution and region
IEGM	WDCM 768	Regional Specialized Collection of Alkanotrophic Microorganisms	Institute of Ecology and Genetics of Microorganisms, Ural Branch, Russian Academy of Sciences, Perm, http://www.iegmcol.ru/
IPPAS	WDCM 596	Culture Collection of Microalgae IPPAS	Institute of Plant Physiology, Russian Academy of Sciences, Moscow, http://www.ippras.ru/
KMM	WDCM 644	Collection of Marine Microorganisms	B. Elyakov Pacific Institute of Bioorganic Chemistry, Far-Eastern Branch, Russian Academy of Sciences, Vladivostok, Primorsky Kray, http://www.piboc.dvo.ru/
LE-BIN	WDCM 1015	Komarov Botanical Institute Basidiomycetes Culture Collection	Komarov Botanical Institute, Russian Academy of Sciences, St. Petersburg, http://www.binran.ru/en/structure/podrazdelenia/laboratoriya-biokhimii-gribov/
LTI	WDCM 554	Cryobank of Microorganisms-Destructors	Department of Molecular Biotechnology, St. Petersburg State Technological Institute, St. Petersburg, http://technolog.edu.ru/en/fakultety-2/2-fakultet/kafedry-2-fakulteta/molecular-biotechnology.html
PGC	WDCM 641	Peterhof Genetic Collection of Microalgae	Biological Research Institute of St. Petersburg State University, St. Petersburg, http://www.bio.spbu.ru/faculty/collections/genetics.php?print=Y
RCAM	WDCM 966	Russian Collection of Agricultural Microorganisms	All-Russia Research Institute for Agricultural Microbiology, Russian Academy of Sciences, St. Petersburg, http://www.arriam.spb.ru/
RIA (NRCA)	WDCM 337	Microorganisms—Producers of Antibiotics Culture Collection "GosNIIsintezbelok" JSC	"Scientific Research Institute of Protein Biosynthesis" Joint Stock Company, http://www.sintezbelok.ru/
SYKOA	WDCM 1125	Strain Collection of Microalgae and Cyanobacteria from Northern and Arctic Regions	Institute of Biology, Ural Branch, Russian Academy of Sciences, Syktyvkar, Komi Republic, http://ib.komisc.ru/sykoa/eng/collection/
VIZR	WDCM 760	Collection for Plant Protection	All-Russian Institute of Plant Protection, Russian Academy of Sciences, St. Petersburg—Pushkin, http://www.cl.spb.ru/vizrspb/

(continued)

Table 4.1 (continued)

Acronym	WDCM number	Collection name	Institution and region
VKM	WDCM 342	All-Russian Collection of Microorganisms	G.K. Skryabin Institute of Biochemistry and Physiology of Microorganisms, Russian Academy of Sciences, Pushchino, Moscow region, http://www.vkm.ru/
VKPM	WDCM 588	Russian National Collection of Industrial Microorganisms	National Bioresource Center—Russian National Collection of Industrial Microorganisms (BRC VKPM), State Research Institute of Genetics and Selection of Industrial Microorganisms (GosNII Genetika), Moscow, http://www.genetika.ru/vkpm/

across the country. Below is a brief description of Russian microbiological collections most relevant for biotechnology, with their localization from west to east. These collections are different in terms of microorganisms they maintain with different physiological properties (luminous, hydrocarbon-oxidizing, producers of antibiotics and bioactive compounds, plant symbionts) and environmental affinity (marine, agricultural, extremophilic). These collections could underpin the formation of appropriate biotechnological clusters that determine the socioeconomic development of the country and strengthen its food base and biosafety. In our opinion, the bioresource collections operating in the Urals, Siberia, and the Far East are of particular importance for the regional development of the country.

- Russian Collection of Agricultural Microorganisms at the All-Russia Research Institute for Agricultural Microbiology (RCAM, WDCM 966, www.arriam.spb.ru) is located in St. Petersburg, Russian Northwest. The RCAM collection contains about 5000 cultures of bacteria, filamentous fungi, and yeasts, including root-nodule bacteria—microsymbionts of over 30 leguminous genera. There are strains that degrade pesticides, accumulate heavy metals and radionuclides, produce vitamins and phytohormones, stimulate plant growth, and generate anti-stress agents for drought and pollution conditions. Collection holdings are useful in developing microbe-based sustainable agriculture in Russia to improve crop productivity and quality and to replace hazardous agrochemicals widely used in the former Soviet Union countries with harmless microbial preparations (Tikhonovich and Provorov 2011).
- Culture Collection of Microscopic Algae and other Microorganisms at Saint Petersburg State University (CALU, WDCM 461, www.eng.spbu.ru/tandr/rc/microorganisms/) is celebrating its 60th year of continuous operation in 2018. The CALU collection contains over 1500 strains of freshwater cyanobacteria, microalgae, and algae parasites. A range of biotechnological products, such as pigments, antibiotics, enzyme inhibitors, toxins and other valuable metabolites

was discovered based on the collection holdings (Pinevich et al. 2004). The novel application of microalgal and cyanobacterial cultures is the production of biofuels, a developing technology designed to compensate the depletion of fossil fuels (Larkum et al. 2012).

- Collection of Unique and Extremophilic Microorganisms at the Research Centre of Biotechnology, Russian Academy of Sciences, Moscow, Central Russia (UNIQEM, http://www.fbras.ru/en/services/ckp/tskp-kollektsiya-uniqem) contains more than 2500 strains of microorganisms. It includes representatives of bacteria and archaea of various physiological groups (including thermo-, acido-, alkali-, halophiles, etc.) isolated from saline and alkaline soil and water, hot springs, and anthropogenic habitats. There are microorganisms of novel taxonomic status difficult to cultivate and are not duplicated in other collections. UNIQEM contains type (reference) strains, intellectual property items, and microorganisms with unique properties and biotechnological potential.
- Regional Specialized Collection of Alkanotrophic Microorganisms at the Institute of Ecology and Genetics of Microorganisms, Ural Branch of the Russian Academy of Sciences (IEGM, WDCM 768, http://www.iegm.ru/iegmcol/index.html; http://www.iegmcol.ru) is located in Perm, the Urals. The IEGM collection contains over 2000 pure, identified, nonpathogenic thoroughly characterized hydrocarbon-oxidizing actinobacterial strains isolated in the course of years of expeditions and field research. The maintained strains were isolated from rhizosphere, soil, surface and stratal water, snow, air, and core samples taken in contrasting eco-geographical regions. The profile of the collection is determined by the fact that the Perm Krai is one of the oldest oil- and gas-producing regions in the European part of Russia confronting chronic environmental problems, particularly oil pollution. The collection includes many extremotolerant polytropic bacteria (psychroactive strains and halo-, baro-, osmo-, xero-, acido-, and alkalotolerants); producers of amino acids, enzymes, and biosurfactants; and degraders of organic pollutants, including crude oil and oil products (Ivshina 2012). Potential applications include biosynthesis of valuable products from hydrocarbons, bioremediation of hydrocarbon-contaminated environments (Kuyukina and Ivshina 2010), and development of biologically active precursors for pharmaceuticals. IEGM collection is a member of WFCC and ECCO (http://www.eccosite.org), a MIRRI collaborating party and a BRIO partner (http://www.pern-brio.eu/about.php).
- Collection of Biotechnology Relevant Microorganisms (http://microbes.biores.cytogen.ru/microbe/) at the Institute of Cytology and Genetics, Siberian Branch of the Russian Academy of Sciences, Novosibirsk, Western Siberia, contains more than 1600 strains (cultures, DNA samples) of microorganisms (fungi, bacteria, archaea) representing various physiological groups, including anaerobes and extremophiles. The collection was created in order to discover novel microorganisms promising for biotechnology and bioengineering and to study their genetics and metabolism. Most of the collection strains are isolated by the Institute's staff from previously unexplored unique extreme ecosystems.

- Specialized Collection of Luminous Bacteria at the Institute of Biophysics, Siberian Branch of the Russian Academy of Sciences (CCIBSO, WDCM 836, www.ibp.ru/collection/default.php) is located in Krasnoyarsk, Eastern Siberia. The collection maintains over 700 strains of luminous bacteria, valuable sources of lux genes, and biologically active compounds. Producers of luciferases were used for toxicity testing of heavy metals, phenols, drugs, toxins, and pesticides. A microbiosensor based on luminous bacteria was applied to monitor water quality in rivers and industrial waste waters. The collection strains and enzymatic bioluminescence kits could be used in medicine, eco-toxicological monitoring, and toxicity control of newly synthesized chemicals (Medvedeva et al. 2005).
- Collection of Marine Microorganisms at the Pacific Institute of Bioorganic Chemistry, Far-Eastern Branch of the Russian Academy of Sciences (KMM, WDCM 644, www.piboc.dvo.ru) is located in Vladivostok, Far East. The KMM collection contains over 6000 strains of marine bacteria and micromycetes isolated from sea water, sea-bottom sediments, and marine invertebrates. Research activities focus on screening for new microbial producers of bioactive secondary metabolites and/or commercially attractive enzymes. Several strains synthesizing antitumor and antiviral compounds have been discovered along with novel producers of antibiotics, cytostatics, and biosurfactants (Stonik et al. 2007). The KMM collection could contribute significantly to the marine biotechnology sector in Russia.

Below, an exclusive specialized collection in the Urals, where the contributors of this chapter work, is presented in more detail to illustrate the evolution of a microbiological collection into an mBRC. The IEGM collection is biotechnology-oriented and focuses on maintaining microorganisms that oxidize natural and anthropogenic hydrocarbons and thus participate in the biogeochemical processes of the biosphere (Ivshina and Kuyukina 2013).

Hydrocarbon pollution will remain a global environmental concern for quite some time due to the unprecedented large-scale development of natural resources, especially petroleum hydrocarbons. Today, the scale of environmental pollution with oil and oil products significantly exceeds the remediation efforts applied. Hence, there is an ever-increasing fundamental attention to studies of hydrocarbon-oxidizing microorganisms (their genomes, metabolism, and adaptive reactions) and the search for key bio-oxidants, new biodegraders, and their biotechnological applications, in particular for pollution detection and oil bioremediation (Ivshina et al. 2017).

In the late 1970s and early 1980s, at the setup of the collection, the biology of alkanotrophs was poorly studied, and identification of strains isolated from the environment was an extremely hard task. We initiated the studies on their distribution and interaction patterns with the habitat and accompanying microflora and developed the isolation methods for this difficult-to-isolate group. The program of the year-long research included various activities—the adequate sampling of natural material, isolation of dominant species, and identification of bacterial adaptation mechanisms to extreme environments. All these efforts resulted in the reference collection of biodegraders of various hydrocarbons used to form bacterial preparations with hydrocarbon-oxidizing capabilities and to develop environmental technologies most closely approximate to natural processes.

The main outcome of the performed research was the unique work experience with alkanotrophic cultures and the creation of an actually operating collection of hydrocarbon-oxidizing microorganisms isolated from various habitats covering the contrasting ecological and climatic zones of the former Soviet Union.

The first issue of the *Catalogue of Strains* of the Regional Specialized Collection of Alkanotrophic Microorganisms (1994) was published in Russian and English; the electronic version of the catalogue is regularly updated (Fig. 4.3). The catalogue information is included in the Consolidated Catalogue of Microbial Cultures Held in Russian Non-medical Collections and uploaded to the WFCC-MIRCEN WDCM website (http://www.wfcc.info/datacenter.html). The collection database contains information about the properties of strains, sources of isolation, medium formulations, and bibliographic references. It also allows installing new scientific information like microphotographs, sequences, strain description, and graphic information. The collection has been recognized as a unique research facility (http://www.ckp-rf.ru/usu/73559) on a competitive basis. The collection potential is exploited in fundamental and applied research in the field of microbiology and biotechnology.

The core of the IEGM collection is the genus *Rhodococcus* actinobacteria prominent among other microorganisms by a great variety of degraded pollutants. Their biological features, with the main ability to degrade hydrocarbons, indicate the exceptional genome versatility and fully correspond to the "doctrine of catabolic failure-free microorganisms." Today, this is one of the most exploited bacterial groups in biotechnology (Larkin et al. 2006; Martínková et al. 2009; Ivshina et al. 2017). The collection is of significance because many *Rhodococcus* species are represented not by single strains (often only type strains) but by numerous natural isolates from various habitats. This allows the targeted screening for active producers of valuable substances and biodegraders of organic pollutants.

Rhodococcus strains isolated from natural samples at sites with high anthropogenic loads are characterized by emulsifying and biodegrading abilities toward individual hydrocarbons and petroleum products, increased resistance to heavy metals (Cd, Cr, Cu, Mo, Ni, Pb, V, Zn), and maintain their activity under extreme acidity (pH 2.0–6.0) and salinity (2–6% NaCl). The selected nonpathogenic strains of rhodococci with active oxygenase complexes are suitable objects for screening of new producers of valuable substances, degraders, and transformers of complex organic compounds and also for constructing new strains and designing new effective technologies (Table 4.2).

At present, the IEGM collection serves the basis for the establishing *Rhodococcus* Center at Perm State University. Despite the obvious technological advantages of this group of actinobacteria as promising biocatalysts in degradation and transformation of complex organic compounds, their collection holdings around the world are limited. Given the high scientific and potential commercial values of rhodococci and the importance of proper storage of microbial cultures, it seems appropriate to

REGIONAL SPECIALIZED COLLECTION OF ALKANOTROPHIC MICROORGANISMS

***Rhodococcus ruber*[1] (Kruse 1896) Goodfellow and Alderson 1977[2]**

IEGM[3] 231[4]

Coidentity: VKPM AC-1899

<- I.B. Ivshina, OEGM 29-1B-1[5]. Isolated from: water, spring[6], oil-extracting enterprise, Perm region, Russia[7]. Taxonomy/description: (55, 95, 97, 245, 343)[9]. Shows positive result with *Rhodococcus ruber* primers in species-specific PCR (245)[9]; analysis of 16S rRNA gene sequence (GenBank accession number KF155234); whole genome was sequenced and deposited at DDBJ/EMBL/GenBank under accession numbers CCSD01000001 to CCSD01000115 (343)[9].Properties: uses propane and *n*-butane as sole carbon source (95)[9]; produces biosurfactants when growing on *n*-alkanes (C_{12}-C_{17}) (74, 248, 254, 327, 347, 348)[9]; degrades high-porous ceramic materials (237)[9]; degrades paracetamol (265)[9]; forms cholesterol oxidase; resistant to Cd^2+, Mo^{6+}, Ni^{2+}, Pb^{2+}, VO^{2+}, VO_3^-, VO_4^{3-}, accumulates molybdenum and nickel (286, 329)[9]; adheres to liquid hydrocarbons (*n*-hexadecane) (317, 320)[9]; bioremediation agent for oil-contaminated soil (324); uses *n*-hexadecane as sole carbon source (327)[9]; resistant to 1-butanol, ethanol (341, 349)[9]; produces glycolipid biosurfactant with immunomodulating properties (344)[9]. (Medium 5 or 8, 11, 28°C)[8].

Medium: 5, 8, 11

Reference(s): 55, 74, 95, 97, 237, 245, 248, 252, 253, 254, 257, 258, 261, 263, 264, 265, 267, 268, 269, 271, 275, 277, 286, 294, 317, 320, 324, 327, 329, 341, 343,344, 347, 348, 349

http://www.iegmcol.ru/strains/rhodoc/ruber/r_ruber231.html

Fig. 4.3 Sample catalogue information on *Rhodococcus* strains maintained at the Regional Specialized Collection of Alkanotrophic Microorganisms. [1]Valid genus/species name of bacteria; [2]Authors who described and re-described the species, year of validation; [3]Collection acronym; [4]Strain accession number in the collection; [5]An individual or organization from where the strain was acquired; strain accession number upon acquisition; [6]Isolation substrate; [7]Geographical location of the strain isolation site; [8]Nutrient medium, cultivation temperature, conservation and storage methods; [9]References to publications where this strain was used

suggest a possible redistribution of *Rhodococcus* strains stored in different collections and to centralize them in a single collection.

Strains are selected to the collection by taxonomic and functional criteria considering their prospects for biotechnology and environmental protection. It seems possible to use the IEGM collection holding to solve problems, such as the production of fodder from unconventional sources (propane, *n*-butane), enzymatic

Table 4.2 Biotechnologically relevant *Rhodococcus* spp. strains from the IEGM Collection

Strain	Geographical origin and isolation substrate	Biotechnologically relevant traits	References
R. erythropolis IEGM 270	Oil-polluted soil, oil-extracting enterprise, Perm region, Russia	Accumulates caesium ions; resistant to Cr^{6+}, Zn^{2+}	Ivshina et al. (2002)
R. erythropolis IEGM 275	Oil-polluted soil, oil-extracting enterprise, Perm region, Russia	Produces psychro-active biosurfactants when growing with *n*-alkanes (C_{12}–C_{17}); bioremediation agent for oil-contaminated soil	Kuyukina et al. (2013), RU Patent 2525943
R. erythropolis IEGM 487	Water, the Baykal lake, Irkutsk region, Russia	Forms cholesterol oxidase; transforms β-sitosterol	Nogovitsina et al. (2011), RU Patent 2472857
R. erythropolis IEGM 708	Oil-shale from settling pit, oil-extracting enterprise, Perm region, Russia	Resistant to Cd^{2+}; bioremediation agent for oil-contaminated soil	Ivshina et al. (2013), RU Patent 2180276, 2193464
R. erythropolis IEGM 767	Oil contaminated sludge, Perm region, Russia	Degrades paracetamol; resistant to toluene, drotaverine hydrochloride and organic solvents	Ivshina et al. (2006, 2012), Korshunova et al. (2016), RU Patent 2475542
R. opacus IEGM 249	Soil, lavsan (polyether fibre) production plant, Belarus	Transforms thioanisole; degrades hydrocarbons; used as a gel-immobilized biocatalyst; resistant to ethanol	Korshunova et al. (2016), RU Patent 2525943
R. rhodochrous IEGM 66	Soil, UK	Transforms methyl phenyl sulfide (thioanisole) into optically active (R)-sulfoxide; transforms triterpenoid betulin into betulone	Elkin et al. (2013), Grishko et al. (2013), Tarasova et al. (2017), RU Patent 2477316, 2529365
R. rhodochrous IEGM 608	Water, Perm region, Russia	Degrades drotaverine hydrochloride; drotaverine hydrochloride-resistant; adheres to liquid hydrocarbons	Ivshina et al. (2012), RU Patent 2496866
R. ruber IEGM 219	Water, the Upper Ilitch river, Russia	Uses propane, *n*-butane and liquid *n*-alkanes (C_5–C_7, C_{11}–C_{16}) as sole carbon sources; produces psychro-active biosurfactants when growing with *n*-alkanes (C_{10}–C_{17}); exhibits antilysozyme activity; adheres to liquid hydrocarbons; resistant to 1-butanol and ethanol	Rubtsova et al. (2012), Korshunova et al. (2016), RU Patent 2216525

(continued)

Table 4.2 (continued)

Strain	Geographical origin and isolation substrate	Biotechnologically relevant traits	References
R. ruber IEGM 231	Water, spring, oil-extracting enterprise, Perm region, Russia	Produces glycolipid biosurfactant with immunomodulating and antiadhesive activities; modifies high-porous ceramic materials; degrades paracetamol; forms cholesterol oxidase; resistant to Cd^{2+}, Mo^{6+}, Ni^{2+}, Pb^{2+}, VO_2^{+}, VO_3^{-}, VO_4^{3-}, accumulates molybdenum and nickel; bioremediation agent for oil-contaminated soil; resistant to 1-butanol and ethanol	Gein et al. (2011), Kuyukina et al. (2015, 2016), Ivshina et al. (2014, 2016), RU Patent 2298033
R. ruber IEGM 327	Turf soil, oil-gas field, Perm region, Russia	Produces biosurfactants when growing with *n*-alkanes (C_{10}–C_{16}); resistant to Ni^{2+}, accumulates nickel; adheres to solid surfaces	Ivshina et al. (2001), RU Patent 2193464, 2180276, 2475542
R. ruber IEGM 896	Soil, *Poa pratensis* rhizosphere, fluoropolymer production plant, Perm region, Russia	Degrades aromatic hyarocarbons (benzene, phenol, toluene, naphthalene, biphenyl, chlorobiphenyls) and acids (benzoic, chlorobenzoic, gentisic, salicylic), degrades polychlorobiphenyls	Plotnikova et al. (2012), Egorova et al. (2013)

transformation of carbon compounds, oil and gas prospecting, optimization of secondary oil recovery, control and clean-up of hydrocarbon contamination of air and water, and bioremediation of oil-contaminated soils (Fig. 4.4).

As part of the structural research unit of the Institute of Ecology and Genetics of Microorganisms (IEGM), the collection performs multi-complex tasks. The information in Table 4.2 and Fig. 4.3 indicates that fundamental researches using bioresources of the IEGM collection are closely related to applications ranging from complete genome sequencing and analyses of functional genes of the most active biodegraders of organic pollutants and producers of valuable bioactive compounds to the production of effective biocatalysts for hydrophobic compound transformation (oil hydrocarbons, organic sulfides, phytosterols, a polycyclic

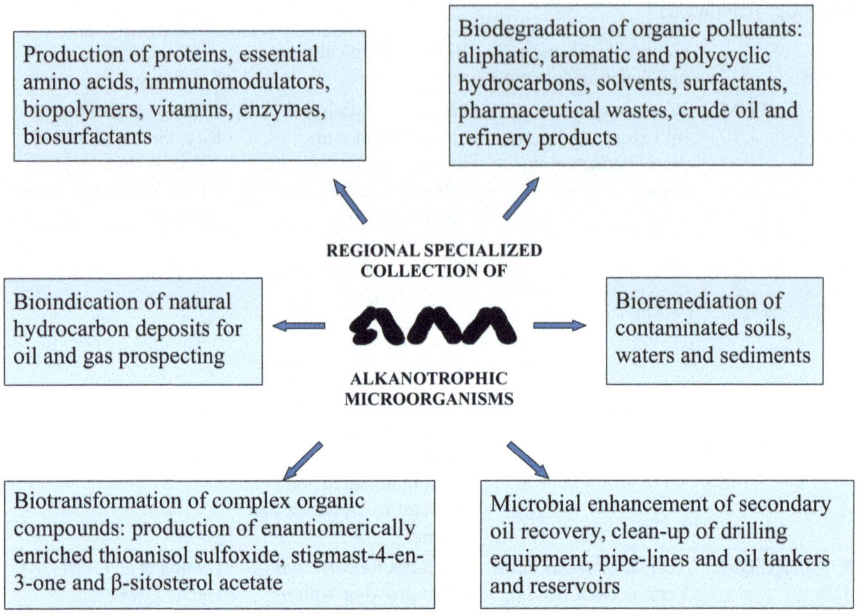

Fig. 4.4 Potential applications of IEGM collection strains

triterpenoid betulin, and pharma pollutants) and development of bioremediation technologies.

Realization of the biotechnological potential of IEGM collection (Fig. 4.5) requires a cooperative interaction of a broad range of specialists. In particular, when developing bacterial catalysts with guaranteed functional activity or bioremediation agents for oil-contaminated soils in cold climates, many problems were solved in the framework of interdisciplinary projects in cooperation with colleagues from various universities and academic institutions (physicists, chemists, pharmacists, mathematicians), as well as industrial companies. The results are summarized in joint publications and patents (see Table 4.2) and are a step forward in solving the problems of sustainable use of biological resources. Over the last years, the collection resources have been used by regional ecobiotechnology companies, such as OOO Priroda-Perm (www.priroda-perm.ru) and OOO Vostok (www.vostokbiohim.ru). One of the purposes of this publication is to facilitate the potential users' access to the IEGM collection.

The collection provides training courses on isolation and identification of hydrocarbon-oxidizing actinobacteria. The qualified staff of the collection conducts lecture courses for PhD students and undergraduates from Perm State University. Additionally, the collection runs workshops on microbiology for students of the Microbiology and Immunology Department of the above University, and students are involved in scientific projects. The collection conducts extensive outreach activities, closely cooperates with schools, biological and ecological colleges. The cooperation includes excursions and a set of popular science lectures. Many of the pupils go to the Biology faculty of Perm State University. The most gifted graduates

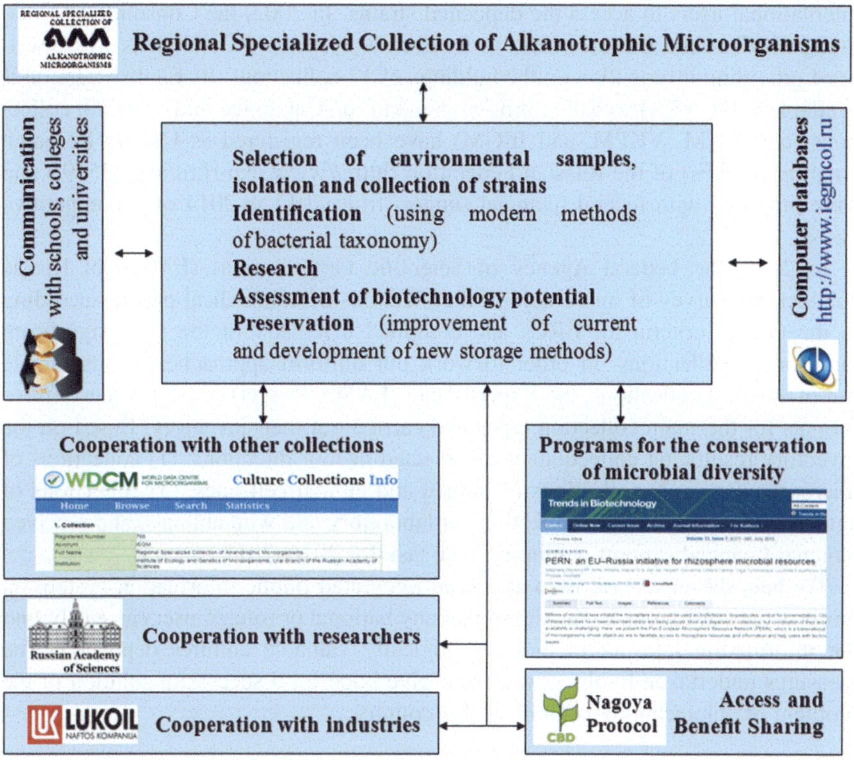

Fig. 4.5 A flow diagram of activities at the IEGM collection

continue their education taking the PhD position at IEGM. Owing to this educational "chain," the collection has opportunities to select people who are committed to scientific research and collection activities per se.

In recent years, the government authorities in Russia have been supporting the bioresource collections. The adopted State Coordination Program for the Development of Biotechnology in the Russian Federation until 2020 (2012) envisages an infrastructure that includes large national specialized bioresource centers; in fact, these are state centers of biotechnology. In order to ensure the safety functioning of existing biological collections and the evolving BRCs, the Program provides for the provision of long-term financing and tax incentives and the improved customs legislation relating to the transfer/exchange of biomaterials. In order to fulfill the functions of BRCs, Russian collections need to expand international cooperation in terms of strain exchange and joint consideration of organizational and methodological issues and to create a national infrastructure for its consistent integration into the global information network of microbial genetic resources.

Currently, only five Russian collections have online catalogues in English (most catalogues have Russian versions), which makes it extremely difficult for international users to access the deposited strains. In 2002, the Consolidated Catalogue of Microbial Cultures Held in Russian Non-medical Collections was developed providing information on the holdings of 17 collections. Its English version is available at the VKM website (http://www.vkm.ru/eCatalogue.htm). Several collections (e.g. VKM, VKPM, and IEGM) have been registered as Unique Research Facilities (URFs) of the Russian Federation (http://www.ckp-rf.ru/usu/73559/) and were provided with federal financial support from 2011 to 2014 on a competitive basis.

In 2016, the Federal Agency of Scientific Organizations (FASO) of Russia conducted a survey of microbiological collections of nonmedical profile according to the OECD criteria for BRCs and compiled a registry of the most significant bioresource collections. In order to work out uniform approaches to sustainable use of existing collections, the Expert Panel (FASO, Russia) developed descriptive formats for the main collection types and carried out their inventory. Based on the inventory results, 64 collections were selected in four directions, (1) collections of microorganisms, (2) collections of human and animal cell lines, (3) collections of agricultural plants, and (4) collections of laboratory and wild animals, and received targeted financial support within the state task for 2016–2017.

To date, the main task is to create an integrated public information system on bioresource collections in Russia, so that any national or foreign user can easily find the information relevant to biotechnologically valuable cultures deposited. The measures undertaken by the government give hope for a successful solution of the problem of "biotechnologization" of this country.

4.4 Conclusion

Professionally managed collections of microorganisms and their modern counterparts, mBRCs (or "microbial BioBanks" that *provide the highest rates on deposits and are guaranteed against bankruptcy*), are in "the pivotal position between the oil-based economy and knowledge-based bioeconomy" (Desmeth 2016). In fact, microorganisms stored in biological collections play an indispensable role in shaping the sustainable future (OECD 2004). Microbial collections are the richest resources for biotechnology. Their potential is huge. They are used for a wide range of purposes, including healthcare, bioremediation, environmental protection, biological control, and production of enzymes and added value chemicals.

To realize their potential, culture collection require significant material and human resources. Keeping in the category "healthy and safe" (Yamasato 1992) is not easy, as to operate a collection is not only labor-intensive but also expensive. According to calculations, a very approximate average cost of enrichment, screening, isolation, characterization, preservation, and long-term maintenance of a single microbial strain at a high-quality level to be further used in basic and applied research makes up over 10,000 € (Overmann 2015). This figure could be even higher, because the calculations do not take into account the travel costs and time spent on sampling. This confirms the need for financial support for microbial collections.

The microorganisms' applications in various biotechnology sectors require reliable methods of ensured long-term preservation of cultures maintaining their original properties. Today, many methods of preserving microbial cultures have been developed, but none of them is universal. The collection-related experience shows that each particular taxonomic group of microorganisms requires an individual selection of effective storage methods; and this involves high material costs (Ivshina et al. 1994; Suzuki 2017). Successful implementation of specific scientific tasks and exploitation of biotechnological potential of microbial collections depend on the correct choice and method of culture storage.

In the current situation, many collections struggling for their survival and effective operation are forced to spend a lot of effort in searching for sources of funding. Often, some of them experience methodological difficulties when evaluating the qualitative profiles of their holdings. Overcoming the difficulties is facilitated by the transformation of existing collections, with significantly improved quality, into mBRCs, accompanied by agglomeration based on national and international regulatory acts. The trajectory of microbial resources from mBRCs to the user (access—application) should be extremely transparent to all parties involved in the exchange of these resources (Desmeth et al. 2011). Against the backdrop of the rapid growth of bioeconomy accompanied by the ever-increasing volume of biological resources in demand, there is a growing need for state-of-the-art information systems and adequate handling of cultures in compliance with international standards (Wu et al. 2017). However, today, the community of highly qualified microbiologists worldwide, who know how to build up such information arrangements, is extremely narrow.

Pursuant to their concept, microbial collections should not be only a scientific-research center but also a kind of scientific and educational center. Applying the interdisciplinary approach to collection work (in the context of a combination of traditional and innovative activities ranging from information support and networking with international information resources, biosafety and biosecurity, intellectual property rights to international law) requires the systems approach to training and retraining of collection staff on the basis of lead mBRCs. The intentional education and training of a specialist is a costly affair. Nevertheless, today we should begin to negotiate the ways of solving this general problem. The labor market demand for qualified taxonomy and bioinformatics specialists dictates a new approach to their training through undergraduate and graduate programs. This should be implemented by a general (international) curriculum, to be developed, for special courses in biodiversity and systematics of microorganisms (Trchounian and Netrusov 2016).

Microbial collections, providing scientific and information services, need continuous targeted governmental and public support, for the available microbial genetic resources are unique and, in case of loss, their restoration through re-isolation from natural populations is difficult or impossible at all (Overmann 2015; Overmann and Scholz 2017). There are although many researchers who prefer to go to "natural one-stop shops."

To fit in the modern scientific process, the collections need to retain qualified personnel to improve the equipment status, including culture conservation technologies, to provide regular upgrading of databases and search services, and an accelerated culture exchange through the compliance with the rules for handling of microorganisms. These rules determine the legitimate exchange of microbial resources, complying with national and international regulations. The collection should ultimately be included in the list of state priorities. The above mentioned is one of the prerequisites for ensuring the effective growth of bioeconomy sectors, including a raise in energy efficiency through renewable bioresources, efficient waste management and the greening of the industrial sector, increased agriculture sustainability, and novel food products and health technologies.

Acknowledgments The work was supported by the Federal Agency for Scientific Organizations program for support the bioresource collections and partly supported by the state tasks 01201353247 and 6.3330.2017/4.6.

References

Belimov AA, Dodd IC, Hontzeas N, Theobald JC, Safronova VI, Davies WJ (2009) Rhizosphere bacteria containing 1-aminocyclopropane-1-carboxylate deaminase increase yield of plants grown in drying soil via both local and systemic hormone signaling. New Phytol 181:413–423

Berendsen RL, Pieterse CMJ, Bakker PAHM (2012) The rhizosphere microbiome and plant health. Trends Plant Sci 17:478–486

Bull AT (1996) Biotechnology for environmental quality: closing the circles. Biodivers Conserv 5:1–25

Bull AT, Ward AC, Goodfellow M (2000) Search and discovery strategies for biotechnology: the paradigm shift. Microbiol Mol Biol Rev 64:573–606

Catalogue of Strains of Regional Specialized Collection of Alkanotrophic Microorganisms (1994) Moscow, NAUKA, 163 pp. http://www.iegmcol.ru/strains/index.html

CBD (1992) Convention on biological diversity. http://www.cbd.int/convention/text/default.shtml

CBD (2001) Cartagena protocol on biosafety to the convention on biological diversity. Secretariat of the Convention on Biological Diversity, Montreal, 30 pp

CBD (2011) Nagoya protocol on access to genetic resources and the fair and equitable sharing of benefits arising from their utilization to the convention on biological diversity. Secretariat of the Convention on Biological Diversity, Montreal, p 25. https://www.cbd.int/abs/text/default.shtml

Cohen AC, Travaglia CN, Bottini R, Piccoli PN (2009) Abscisic acid produced by endophytic *Azospirillum* alleviates drought effects in maize. Botany 87:455–462

Cole JR, Wang Q, Cardenas E, Fish J, Chai B, Farris RJ, Kulam-Syed-Mohideen AS, McGarrell DM, Marsh T, Garrity GM, Tiedje JM (2009) The ribosomal database project: improved alignments and new tools for rRNA analysis. Nucleic Acids Res 37:141−145. http://ec.europa.eu/research/conferences/2005/kbb/pdf/kbbe_conferencereport.pdf

Constanza R, d'Arge R, de Groot R, Farber S, Grasso M, Hannon B, Limburg K, Naeem S, O'Neill RV, Paruelo J, Raskin RG, Sutton P, van den Belt M (1997) The value of the world's ecosystem services and natural capital. Nature 387:253–260

De Vasconcellos SP, Angolini CFF, García INS, Dellagnezze BM, da Silva CC, Marsaioli AJ, dos Santos Neto EV, de Oliveira VM (2010) Screening for hydrocarbon biodegraders in a metagenomic clone library derived from Brazilian petroleum reservoirs. Org Geochem 41:675–681

Declerck S, Willems A, van der Heijden MGA, Varese GC, Turkovskaya O, Evtushenko L, Ivshina I, Desmeth P (2015) PERN: an EU–Russia initiative for rhizosphere microbial resources. Trends Biotechnol 33:377–380

Dellagnezze BM, de Sousa GV, Martins LL, Domingos DF, Limache EEG, de Vasconcellos SP, da Cruz GF, de Oliveira VM (2014) Bioremediation potential of microorganisms derived from petroleum reservoirs. Marine Poll Bull 89:191–200

Desmeth P (2016) Culture collections in a pivotal position between oil-based economy and knowledge-based bio-economy. In: Ivshina IB, Kuyukina MS, Kamenskikh TN, Alfimova LA (eds) Microbial diversity: resource potential: proceedings of the IV International conference. Institute of Ecology and Genetics of Microorganisms, Ural Branch, Russian Academy of Sciences; Perm State University. Moscow, Perm, Russia, pp 108–109

Desmeth P (2017) Nagoya protocol applied to microbial genetic resources. In: Kurtböke I (ed) Microbial resources: from functional existence in nature to applications. Elsevier, New York, pp 205–217

Desmeth P, Bosschaerts M (2008) The necessary adaptation of culture collections to the new socio-economic environment at global level. In: Ivshina IB, Kuyukina MS, Kamenskikh TN, Alfimova LA (eds) Microbial diversity: current situation, conservation strategy and biotechnological potential: Proceedings of 3rd international conference. Perm, pp 144–146

Desmeth P, Kurtböke I, Smith D (2011) Tools to implement the Nagoya protocol on access and benefit sharing in microbiology: ABS, an intrinsic preoccupation of the World Federation for Culture Collections (WFCC). Information document for ICNP1. http://www.cbd.int/abs/doc/protocol/icnp-1/wfcc-en.pdf

Diversitas (1996) Diversitas: an international programme of biodiversity science (Operational plan), Diversitas, Paris. http://unesdoc.unesco.org/images/0016/001632/163285eo.pdf

Egorova DO, Demakov VA, Plotnikova EG (2013) Bioaugmentation of a polychlorobiphenyl contaminated soil with two aerobic bacterial strains. J Hazard Mater 261:378–386

Elkin AA, Kylosova TI, Grishko VV, Ivshina IB (2013) Enantioselective oxidation of sulfides to sulfoxides by *Gordonia terrae* IEGM 136 and *Rhodococcus rhodochrous* IEGM 66. J Mol Cat B: Enzym 89:82–85

European Commission Conference Report (2005) New perspectives on the knowledge-based bio-economy. European Commission, Brussels

Gein SV, Kuyukina MS, Ivshina IB, Baeva TA, Chereshnev VA (2011) *In vitro* cytokine stimulation assay for glycolipid biosurfactant from *Rhodococcus ruber*: role of monocyte adhesion. Cytotechnology 63:559–566

Grishko VV, Tarasova EV, Ivshina IB (2013) Biotransformation of betulin to betulone by growing and resting cells of *Rhodococcus rhodochrous* IEGM 66. Process Biochem 48:1640–1644

Hindrê T, Knibbe C, Beston G, Schneider D (2012) New insights into bacterial adaptation through *in vivo* and *in silico* experimental evolution. Nat Rev Microbiol 10:352–365

Ivshina IB (2012) Current situation and challenges of specialized microbial resource centres in Russia. Microbiology 81:509–516

Ivshina IB, Kuyukina MS (2013) Turning Russian specialized microbial culture collections into resource centers for biotechnology. Trends Biotechnol 31:609–611

Ivshina IB, Kamenskikh TN, Kuyukina MS, Richkova MI, Shadrin OA, Rybalka LV, Zvereva LV, Chumakov OB (1994) Methods for preservation of alkane-utilizing *Rhodococcus* spp. expedient for culture conservation practice. Microbiology 63:69–74

Ivshina IB, Kuyukina MS, Ritchkova MI, Philp JC, Cunningham CJ, Christofi N (2001) Oleophilic biofertilizer based on a *Rhodococcus* surfactant complex for the bioremediation of crude oil-contaminated soil. Contam Soil Sediment Water AEHS 4:20–24

Ivshina IB, Peshkur TA, Korobov VP (2002) Efficient uptake of cesium ions by *Rhodococcus* cells. Microbiology 71:357–361

Ivshina IB, Richkova MI, Vikhareva EV, Chekryshkina LA, Mishenina II (2006) Catalysis of the biodegradation of unusable medicines by alkanotrophic rhodococci. Appl Biochem Microbiol 42:392–395

Ivshina IB, Vikhareva EV, Richkova MI, Mukhutdinova AN, Karpenko JN (2012) Biodegradation of drotaverine hydrochloride by free and immobilized cells of *Rhodococcus rhodochrous* IEGM 608. World J Microbiol Biotechnol 28:2997–3006

Ivshina IB, Kuyukina MS, Kostina LV (2013) Adaptive mechanisms of nonspecific resistance to heavy metal ions in alkanotrophic actinobacteria. Russ J Ecol 44:123–130

Ivshina IB, Kuyukina MS, Krivoruchko AV, Barbe V, Fischer C (2014) Draft genome sequence of propane and butane oxidizing actinobacterium *Rhodococcus ruber* IEGM 231. Genome Announc 2:e01297–e01214. https://doi.org/10.1128/genomeA.01297-14

Ivshina I, Kostina L, Krivoruchko A, Kuyukina M, Peshkur T, Anderson P, Cunningham C (2016) Removal of polycyclic aromatic hydrocarbons in soil spiked with model mixtures of petroleum hydrocarbons and heterocycles using biosurfactants from *Rhodococcus ruber* IEGM 231. J Hazard Mater 312:8–17

Ivshina IB, Kuyukina MS, Krivoruchko AV (2017) Hydrocarbon-oxidizing bacteria and their potential in eco-biotechnology and bioremediation. In: Kurtböke I (ed) Microbial resources: from functional existence in nature to industrial applications. Elsevier, New York, pp 121–148

Jones N (2013) Food fuelled with fungi. Nature 504:199

Kalakoutskii LV (2016) In the flow of time. In: Ivshina IB, Kuyukina MS, Kamenskikh TN, Alfimova LA (eds) Microbial diversity: resource potential: Proceedings of the IV international conference. Institute of Ecology and Genetics of Microorganisms, Ural Branch, Russian Academy of Sciences; Perm State University. Moscow, Perm, Russia, pp 121 – 122

Kirsop BE (1992) WFCC endangered collections committee round table. In: Colwell RR (ed) Endangered culture collections. Proceedings of the first and second international symposia. WFCC. A Maryland Biotechnology Institute, Technical Publication, pp 31 – 38

Kirsop BE, Canhos VP (1998) The economic value of microbial genetic resources. In: Proceedings of the World Federation of Culture Collections. World Federation of Culture Collections, Campinas, Brazil pp 6 – 9

Korshunova IO, Pistsova ON, Kuyukina MS, Ivshina IB (2016) The Effect of organic solvents on the viability and morphofunctional properties of *Rhodococcus*. Appl Biochem Microbiol 52:43–50

Kuyukina MS, Ivshina IB (2010) Application of *Rhodococcus* in bioremediation of contaminated environments. In: Alvarez HM (ed) Biology of *Rhodococcus*. Springer, New York, pp 231–262

Kuyukina MS, Ivshina IB, Kamenskikh TN, Bulicheva MV, Stukova GI (2013) Survival of cryogel-immobilized *Rhodococcus* strains in crude oil-contaminated soil and their impact on biodegradation efficiency. Int Biodeter Biodegr 84:118–125

Kuyukina MS, Ivshina IB, Baeva TA, Kochina OA, Gein SV, Chereshnev VA (2015) Trehalolipid biosurfactants from nonpathogenic *Rhodococcus* actinobacteria with diverse immunomodulatory activities. New Biotechnol 32:559–568. https://doi.org/10.1016/j.nbt.2015.03.006

Kuyukina MS, Ivshina IB, Korshunova IO, Stukova GI, Krivoruchko AV (2016) Diverse effects of a biosurfactant from *Rhodococcus ruber* IEGM 231 on the adhesion of resting and growing bacteria to polystyrene. AMB Express 6. https://doi.org/10.1186/s13568-016-0186-z

Larkin MJ, Kulakov LA, Allen CR (2006) Biodegradation by members oh the *Rhodococcus*: biochemistry, physiology, and genetic adaptation. Adv Appl Microbiol 59:1–29

Larkum AWD, Ross IL, Kruse O, Hankamer B (2012) Selection, breeding and engineering of microalgae for bioenergy and biofuel production. Trends Biotechnol 30:198–205

Lederberg J (2000) Infectious history. Science 288:287–293

Lugtenberg B, Kamilova F (2009) Plant-growth-promoting rhizobacteria. Annu Rev Microbiol 63:541–556

Martínková L, Uhnáková B, Pátek M, Nešvera J, Křen V (2009) Biodegradation potential of the genus *Rhodococcus*. Environ Int 35:162–177

Medvedeva SE, Boyandin AN, Lankin YP, Kotov DA, Kargatova TV, Rodicheva EK, Popova LY (2005) BIOLUMBASE – the database of natural and transgenic bioluminescent organisms. Luminescence 20:90–96

Mendes R, Garbeva P, Raaijmakers JM (2013) The rhizosphere microbiome: significance of plant beneficial, plant pathogenic, and human pathogenic microorganisms. FEMS Microbiol Rev 37:634–663

Miyazaki M (2006) Economic value of microbial resources. Microbiol Cult Coll 22:15–19

MOSAICC (2011) MOSAICC micro-organisms sustainable use and access regulation international code of conduct. http://bccm.belspo.be/projects/mosaicc/docs/code2011.pdf; http://www.wdcm.org/workshop2014/one_one.pdf

Nogovitsina YM, Grishko VV, Ivshina IB (2011) Biocatalytic synthesis of pharmacology perspective stigmast-4-en-3-one using rhodococci cells. Rus J Bioorg Chem 37:626–633

OECD (2001) Biological resource centers: underpinning the future of life sciences and biotechnology. OECD Directorate for Science, Technology and Industry, Paris, p 68

OECD (2004) Guidance for the operation of biological resource centres (BRCs) [Biological Research Centres' is erroneously printed on the cover of this document], OECD. http://www.oecd.org/dataoecd/60/42/23547743.pdf

OECD (2007) OECD best practice guidelines for biological resource centres. OECD Directorate for Science, Technology and Industry, 115 pp. http://www.oecd.org/sti/biotech/oecdbestpracticeguidelinesforbiologicalresourcecentres.htm

OECD (2009) The bioeconomy to 2030. Designing a policy agenda. Main findings and policy conclusions. OECD, Paris, p 18

Overmann J (2015) Significance and future role of microbial resource centers. Syst Appl Microbiol 38:258–265

Overmann J, Scholz AH (2017) Microbial research under the Nagoya protocol: facts and fiction. Trends Microbiol 25:85–89

Ozerskaya SM, Kirillova NP, Vasilenko AN (2012) FungalDC: a database on fungal diversity in genetic resource collections. IMA Fungus 3:9–9

Pinevich AV, Mamkaeva KA, Titova NN, Gavrilova OV, Ermilova EV, Kvitko KV, Pljusch AV, Voloshko LN, Averina SG (2004) St. Petersburg Culture Collection (CFLU): Four decades of storage and research with microscopic algae, cyanobacteria and other microorganisms. Nova Hedwigia 79:115–126

Plotnikova EG, Solyanikova IP, Egorova DO, Shumkova ES, Golovleva LA (2012) Degradation of 4-chlorobiphenyl and 4-chlorobenzoic acid by the strain *Rhodococcus ruber* P25. Microbiology 81:143–153

Rubtsova EV, Kuyukina MS, Ivshina IB (2012) Effect of cultivation conditions on the adhesive activity of *Rhodococcus* cells towards *n*-hexadecane. Appl Biochem Microbiol 48:452–459

RU 2193464 C1 Ivshina IB, Kostarev SM, Kuyukina MS, Zakshevskaja LV Method of bioremediation of soils and grounds contaminated with oil and petroleum products. Application: 2001130686/13, 14.11.2001. Effective date for property rights: 14.11.2001. Date of publication: 27.11.2002. Bull 33

RU 2180276 C1 Kuyukina MS, Ivshina IB Oleophilic biological preparation useful for cleaning oil-polluted soil. Application: 2001104629/13, 19.02.2001. Effective date for property rights: 19.02.2001. Date of publication: 10.03.2002. Bull 7

RU 2216525 C1 Solovykh GN, Ushakova EI, Ivshina IB, Raimova EK Method for microbiological treatment of sewage from industrial plants from heavy metal ions: zinc, cadmium and lead. Application: 2002106289/13, 11.03.2002. Effective date for property rights: 11.03.2002. Date of publication: 20.11.2003. Bull 32

RU 2298033 C2 Podorozhko EA, Kuyukina MS, Ivshina IB, Filp JC, Lozinskij VI Composition for production of carrier for immobilized carbohydrate cleaving microorganisms, and method for carrier production. Application: 2005111376/13, 19.04.2005. Effective date for property rights: 19.04.2005. Date of publication: 27.04.2007. Bull 12

RU 2477316 C1 Ivshina IB, Grishko VV, Elkin AA Biocatalytic method to produce (r)-phenylmethyl sulfoxide. Application: 2011154122/10, 29.12.2011. Effective date for property rights: 29.12.2011. Date of filing: 29.12.2011. Date of publication: 10.03.2013. Bull 7

RU 2472857 C1 Ivshina IB, Grishko VV, Nogovitsina EM Method of producing stigmast-4-en-3-one from beta-sitosterol. Application: 2011154121/10, 29.12.2011. Effective date for property rights: 29.12.2011. Date of filing: 29.12.2011. Date of publication: 20.01.2013. Bull 2

RU 2475542 C1 Ivshina IB, Kuyukina MS, Krivoruchko AV, Naimark OB, Plekhov OA, Fedorova AY Method and plant to determine efficiency of adsorption immobilisation of microorganisms and monitoring of functional condition of biocatalysts based on immobilised microbial cells. Application: 2011154123/10, 29.12.2011. Effective date for property rights: 29.12.2011. Date of filing: 29.12.2011. Date of publication: 20.02.2013. Bull 5

RU 2496866 C1 Ivshina IB, Vikhareva EV, Richkova MI, Mukhutdinova AN, Karpenko JN Method of biodegradation of drotaverine hydrochloride. Application: 2012120923/10, 23.05.2012. Effective date for property rights: 23.05.2012. Date of filing: 23.05.2012. Date of publication: 27.10.2013. Bull 30

RU 2529365 C1 Grishko VV, Tarasova EV, Ivshina IB Method of production of betulon. Application: 2013137269/10, 08.08.2013. Effective date for property rights: 08.08.2013. Date of filing: 08.08.2013. Date of publication: 27.09.2014. Bull 27

RU 2525943 C1 Kuyukina MS, Ivshina IB, Serebrennikova MK, Rubtsova EV, Krivoruchko AV Method of specific differentiation of viable *Rhodococcus* immobilised in gel carrier. Application: 2013121968/10, 13.05.2013. Effective date for property rights: 13.05.2013. Date of filing: 13.05.2013. Date of publication: 20.08.2014. Bull 23

Sharma SK, Saini S, Verma A, Sharma PK, Lal R, Roy M, Singh UB, Saxena AK, Sharma AK (2017) National agriculturally important microbial culture collections in the global context of microbial culture collection centres. Proc Natl Acad Sci India Sect B Biol Sci 1–14. https://doi.org/10.1007/s40011-017-0882-8

Shimura J (2003) A regional strategy to achieve key information provision for implementation of the convention on biological diversity. In: Shimura J (ed) Building capacity in biodiversity information sharing. National institute for environmental studies, Tsukuba, pp 13–25. ISBN: 900626-03-1C3045

Sierra-García IN, Alvarez JC, de Vasconcellos SP, Pereira de Souza A, dos Santos Neto EV, de Oliveira VM (2014) New hydrocarbon degradation pathways in the microbial metagenome from Brazilian petroleum reservoirs. PLoS One 9(2):e90087. https://doi.org/10.1371/journal.pone.0090087

Smith D (2012) Culture collections. Adv Appl Microbiol 79:73–118

Smith D, Ryan MJ, Stackebrandt E (2008) The *ex situ* conservation of microorganisms: aiming at a certified quality management. Biotechnology. In: Doelle HW, DaSilva EJ (eds) Encyclopedia of life support systems (EOLSS), developed under the auspices of the UNESCO. Eolss, Oxford, 9 pp. http://wwweolss.net

Stackebrandt E (2010) Diversification and focusing: strategies of microbial culture collections. Trends Microbiol 18:283–287

Stackebrandt E, Schüngel M, Martin D, Smith D (2015) The microbial resource research infrastructure MIRRI: strength through coordination. Microorganisms 3:890–902

State Coordination Program for the Development of Biotechnology in the Russian Federation until 2020 (BIO-2020) (2012) Moscow VP-P8 322 http://owwz.de/fileadmin/Biotechnology/Information_Biotech/BIO2020_Programme_full.pdf

Stonik VA, Mikhailov VV, Bulgakov VP, Zhuravlev JN (2007) Biotechnological studies in the Far-Eastern Region of Russia. Biotechnol J 2:818–825

Suzuki K (2017) An overview of biological resource center-maintenance of microbial resources and their management. In: Kurtböke I (ed) Microbial resources: from functional existence in nature to industrial applications. Elsevier, New York, pp 257–274

Tarasova EV, Grishko VV, Ivshina IB (2017) Cell adaptations of *Rhodococcus rhodochrous* IEGM 66 to betulin biotransformation. Process Biochem 52:1–9

Ten Kate K (1999) Biotechnology in fields other that healthcare and agriculture. In: ten Kate K, Laird SA (eds) The commercial use of biodiversity. Earthscan, London, pp 228–261

The Lisbon European Council (2000) An agenda of economic and social renewal for Europe. European Council, Brussels

Tikhonovich IA, Provorov NA (2011) Microbiology is the basis of sustainable agriculture: an opinion. Ann Appl Biol 159:155–168

Trchounian A, Netrusov A (2016) Improving teaching approaches and algorithm of academic microbiology. In: Ivshina IB, Kuyukina MS, Kamenskikh TN, Alfimova LA (eds) Microbial diversity: resource potential: Proceedings of the IV International conference. Institute of Ecology and Genetics of Microorganisms, Ural Branch, Russian Academy of Sciences; Perm State University. Moscow, Perm, Russia, pp 163–164

Ufarté L, Laville É, Duquesne S, Potocki-Veronese G (2015) Metagenomics for the discovery of pollutant degrading enzymes. Biotechnol Adv 33:1845–1854

Vasilenko AN, Ozerskaya SM (2012) Fungal taxonomic diversity and bioinformatics-related issues. In: Proceedings of the 6th International conference in bioinformatics and biomedical engineering (iCBBE 2012). New York, IEEE Service Center, pp 633–634

Vasilenko A, Ozerskaya S, Stupar O (2011) Current WFCC-CC catalogues as a starting ground for networking efforts. WFCC Newslett 50:5–15

WDCM (2017) World Data Center of microorganisms. http://www.wdcm.org

WFCC (2010) World Federation for culture collections guidelines for establishment and operation of collections of cultures of microorganisms, 3rd edn. http://www.wfcc.info/guidelines/

WFCC (2017) World Federation of culture collections. http://www.wfcc.info/ccinfo/index.php/home/statistica/#m5

Whipps JM (2001) Microbial interactions and biocontrol in the rhizosphere. J Exp Bot 52:487–511

Wu L, Sun Q, Desmeth P, Sugawara H, Xu Z, McCluskey K, Smith D, Vasilenko A, Lima N, Ohkuma M, Robert V, Zhou Y, Li J, Fan G, Ingsriswang S, Oserskaya S, Ma J (2017) World data centre for microorganisms: an information infrastructure to explore and utilize preserved microbial strains worldwide. Nucleic Acids Res 45:D611–D618

Yamasato K (1992) Deteriorated/deteriorating culture collections. In: Colwell RR (ed) Endangered culture collections. Proceedings of the first and second international symposia. WFCC. A Maryland Biotechnology Institute, Technical Publication, New York, pp 45–50

Zavarzin GA (1995) The microorganisms: a concept in need of clarification or one now to be rejected. In: Allsopp D, Colwell RR, Hawksworth DL (eds) Microbial diversity and ecosystem function. CAB International, Wallingford, pp 17–18

Chapter 5
Agricultural Microbial Genetic Resources: Application and Preservation at Microbial Resource Centers

Tayebeh Soltanighias, Rajesh K. Vaid, and Praveen Rahi

Abstract To meet the increasing global demand for agriculture produce, the productivity of crops needs to be increased under declining soil quality and changing climate conditions. Emergence and widespread occurrence of plant pathogens and pests also pose a serious threat to agriculture. Plant–microbe interactions involve a beneficial, neutral, or negative effect on one or both partners. The plant-associated microorganisms perform complex roles in nutrient cycling and control of pathogens and induce stress tolerance. Several microbial strains have been isolated, screened, and evaluated for multiple plant growth-promoting activities and are rich resources to improve crop health and productivity. All these microbial resources are very important for agriculture research, and collection of such microbes provides researchers a way to look into the past and form a base for new scientific discoveries. Plant beneficial microorganisms have been preserved and maintained for a long time in a series of ex situ repositories across the globe, for their potential applications in sustainable agriculture. Similarly, a huge diversity of plant pathogens including bacteria and fungi have been isolated and preserved to be used as type specimens to breed disease-resistant crop varieties. Microbial resource centers (mBRCs) act as a repository of the diversity of microorganisms. The next-generation sequencing technologies have exhibited that only a small portion of microorganisms could be cultivated until now. It is expected that with high-throughput microbial cultivation approach, "culturomics" in conjunction with high-speed identification technique based on MALDI-TOF MS will lead to the cultivation of the not-yet-cultivated microorganisms, which can be harnessed to improve crop health and productivity. Adoption of culturomics in agriculture microbiology research will definitely increase the number of microbial strains, which will ultimately increase the load of mBRCs

T. Soltanighias · P. Rahi (✉)
National Centre for Microbial Resource (NCMR), National Centre for Cell Science, Pune, Maharastra, India
e-mail: praveen@nccs.res.in

R. K. Vaid
National Centre for Veterinary Type Cultures (NCVTC), ICAR-National Research Centre on Equines, Hissar, Haryana, India

S. K. Sharma, A. Varma (eds.), *Microbial Resource Conservation*, Soil Biology 54,
https://doi.org/10.1007/978-3-319-96971-8_5

maintaining these cultures. In this chapter, we discussed various aspects of agriculturally important microbes and their documentation and preservation at mBRCs.

5.1 Introduction

The nature of agricultural soils is shaped by management practices, which influences the development of microbial community composition and biomass. The shift in agriculture land use affects the structure and function of microbial communities. Consequently to the industrial revolution, the agriculture sector also witnessed several innovations and inventions, leading to the beginning of revolution in agriculture. The agricultural revolution had many positive impacts like increased yield and better crop management, but there were negative aspects as well. The alteration in the composition of the farm soil and compromised natural flow of nutrients are the most dire impacts of intensive agriculture. To minimize these undesirable consequences, the importance to understand the role of soil biology especially soil microorganisms has been realized (Carey 2016; Foo et al. 2017). The agriculture at present supports the large-scale monoculture to facilitate ease of mechanization in cultivation, fertilizer application, weed and pest control, and harvest. In addition to this, several million tonnes of nitrogen is turned into nitrogenous fertilizers through human activities; a large portion of this fixed nitrogen is not captured by plant and becomes a major pollutant. Intensification of such agricultural practices has posed huge burden on the fragile agroecosystems (Tsiafouli et al. 2015). The development of ecologically safe, efficient, and cheap biological alternatives to improve crop productivity has been desired for a long time. Plant-associated microorganisms have been proposed as an alternative to the application of chemical fertilizers and pesticides for sustainable agriculture (Busby et al. 2017; Rahi 2017; Singh and Trivedi 2017; van Lenteren et al. 2017; Trivedi et al. 2017). Exploitation of beneficial microorganisms (especially nitrogen-fixing rhizobia) can be traced back to the early centuries. Previously, experienced farmers mixed legume crops with nonlegume crops to improve yields and were unknowingly using the benefits of bacteria that fix nitrogen, a necessary nutrient, to the plants.

Microorganisms associated with plants play an important role in the availability of major nutrients including, nitrogen, phosphorus, and potassium and/or enhancing their uptake (Zhang et al. 2016; Oliveira et al. 2017; Wolińska et al. 2017). Plant-associated microbes can also decompose organic matter by releasing various enzymes. Several microorganisms exhibit multiple plant growth-promoting activities and help plants by phytohormone production, inhibition of plant pathogens, and inducing abiotic stress. In search of potential strains to support plant growth and productivity, a large number of bacteria have been isolated and evaluated for plant growth promotion (Glick et al. 1999; Bent et al. 2001; Rodríguez et al. 2008; Gulati et al. 2010; Vyas et al. 2010; Bashan et al. 2014; Rinu et al. 2014). Majority of these strains belonging to genera *Acinetobacter*, *Arthrobacter*, *Azospirillum*, *Azotobacter*, *Bacillus*, *Brachybacterium*, *Delftia*, *Enterobacter*, *Escherichia*, *Micrococcus*,

Paenibacillus, *Pantoea*, *Providencia*, *Pseudomonas*, *Rhanella*, *Rhizobium*, *Serratia*, *Staphylococcus*, *Stenotrophomonas*, and *Streptomyces* have been reported for performing plant growth-promoting activities in the rhizosphere (Arcand and Schneider 2006; Chen et al. 2006; El-Tarabily 2008; Gulati et al. 2008, 2010, 2011; Shoebitz et al. 2009; Upadhyay et al. 2009; Ali et al. 2010; Khare and Arora 2010; Mamta et al. 2010, Siddikee et al. 2010; Vyas et al. 2010; Eastman et al. 2014; Egamberdieva et al. 2016). Similar to other prokaryotic bacteria, the cyanobacteria have been used to improve soil fertility as they can fix the atmospheric nitrogen in free-living conditions (Mishra and Pabbi 2004; Singh et al. 2016). A huge diversity of fungi including *Beauveria*, *Discosia*, *Eupenicillium*, *Lecanicillium*, *Metarhizium*, *Paecilomyces*, *Penicillium*, and *Trichoderma* has also been reported to perform multiple plant growth-promoting activities (Vyas et al. 2007; Rahi et al. 2009; Contreras-Cornejo et al. 2014; Rinu et al. 2014; Jaber and Enkerli 2017; Nieto-Jacobo et al. 2017). Contrary to the beneficial effects to the plant, microorganisms also pose harmful effect on plants by causing various diseases. Plant pathogens can severely impact plant growth and productivity, leading to serious agricultural and economic losses. Understanding the plant disease cycle and diagnosis of plant pathogens is very critical; the absence of information on these plant pathogens can devastate crops and forests and disturb natural ecosystems. Major diseases caused by plant pathogens are rot, wilt, stunting, and seedling damping-off (Haas and Défago 2005). Information on the plant–pathogen interactions is also valuable in developing newer disease-resistant varieties (Nejat et al. 2017). Recently, a shift has been observed from the conventional to organic farming, which involves application of organic matter and nutrient-rich farmyard manure and sewage water to cultivate crop plants. Under-processed farmyard manure and sewage water could be a potential source of human pathogens (Ibenyassine et al. 2007; Gallegos-Robles et al. 2008; Teplitski et al. 2011).

High-throughput sequencing of community DNA from various stages of plant development exhibited the involvement of huge diversity of microorganisms in plant growth and development, the majority of which is still not cultivated (Da Rocha et al. 2013; Gomba et al. 2017). The culture-independent approach based on microbial community analysis helped to discover potentially beneficial microbial taxa that are not yet cultivated and represent difficult-to-cultivate group of microorganisms (Bergottini et al. 2017). Particularly the members of *Acidobacteria*, *Verrucomicrobia*, and *Planctomycetes* are considered to be difficult to culture but represent the most abundant groups of plant rhizosphere microbial community (Da Rocha et al. 2013; Foesel et al. 2016; Kalam et al. 2017). In addition to these groups, the fraction of cultured members is quite low with regard to the total representatives estimated by culture-independent methods in each prokaryotic group (Da Rocha et al. 2013). The involvement of several unknown and not-yet-cultivated members of plant microbiome in improving plant growth has been demonstrated by different cultivation-independent approaches like stable isotope probing-RNA analysis, proteomics, and transcriptomics (Vandenkoornhuyse et al. 2007; Mattarozzi et al. 2017; White et al. 2017). The cultivation-independent approaches generated sufficient evidences on the diversity of uncultivable microorganisms associated with plants and their potential role in improving growth and health, and taking clues from these studies, researchers are devising new strategies to cultivate such microbes. The not-yet-cultivated microorganisms could be cultivated in pure cultures by mimicking their natural habitat and by providing the chemical components of their natural environment (Kaeberlein et al. 2002; Zengler et al. 2002). Fast and low-cost microbial identification technique based on protein profiles generated by MALDI-TOF MS allowed the reliable identification and de-replication of large collections of microorganisms (Rahi et al. 2016). MALDI-TOF MS-based identification has been employed in combination with the high-throughput microbial cultivation approach "culturomics" to cultivate several not-yet-cultivated microorganisms (Lagier et al. 2016). It is imperative to store and preserve these newly generated microorganisms for long term, so that they can be used as reference for new discoveries. The collections of microorganisms serve as a catalogue of diversity of microbial life. Agriculturally important microorganisms hold several other reasons like these microorganisms can be used to understand the mechanisms of plant growth promotion, as genetic resource for plant improvement, and track the evolution of plant pathogens for their long-term preservations. Microbial resource centers (mBRCs) involved in the ex situ conservation of agriculturally important microorganisms are vital collections, which offer well-characterized and curated microbial cultures for various applications including biofertilizer and biopesticides for sustainable agriculture. Realizing the value of agriculturally important microbial culture collections, all major countries of the world have at least one microbial culture collection, which deals mainly with microorganisms with importance in agriculture. In this chapter, we highlighted the role of microorganisms in agriculture specifically for nutrient uptake, plant diseases, and soil improvement and discussed the importance of long-term preservation of microbial genetic resources in different mBRCs of the world.

5.2 Microbial Genetic Resources of Agricultural Importance

5.2.1 Soil–Plant–Microbe Interactions

Agriculture is the basis of all major civilization in the past and is one of the most important components of today's societies. Understanding of basic physical, chemical, and biological properties of soils is critical for successful agriculture. Soil microorganisms play a key role in regulating different processes that are involved in ecosystem functioning. In the very beginning of agriculture, the growth of less desirable plant species was suppressed to allow the growth of plant species for forage and food. Application of different biological inputs has been followed to improve the productivity. Later, with the development of chemical-based fertilizers and pesticides, the agriculture system witnessed a significant shift and led to the irreversible change in natural nutrient cycle in the soil. The land-use change due to the present-day intensive agriculture has led to consequent loss of soil organic matter and soil biodiversity. To minimize the damage of agricultural intensification under the climate change scenario, there is an upsurge to use microorganisms to replace chemical fertilizer and pesticides. Plant growth-promoting microorganisms have been used to improve crop productivity since the late nineteenth century. Especially the legume microsymbionts (i.e., rhizobia) have been used for almost 100 years around the world. Chemical-based fertilizers have not been used in Brazil for the cultivation of legumes, like soybean, and crops are only inoculated with their compatible rhizobia (Bashan and Halguin 1998). Similarly, inoculation of legume crops using rhizobia has also been made in countries like Australia, the USA, Eastern Europe, Egypt, Israel, South Africa, and New Zealand (Bashan 1998). The introduction of compatible rhizobia has been considered necessary to ensure the success of legume crops. Inoculation of rhizobia is critical in the absence of compatible rhizobial strains and in presence of ineffective or less effective indigenous rhizobia than the selected inoculants for a particular legume host variety (Giller 2001). The introduction of soybean cultivation in the USA has been reported to depend on the deliberate inoculation with *Bradyrhizobium japonicum* (Lohrke et al. 1996). Similarly, compatible symbionts have been required during the introduction of European *Lotus corniculatus*, a forage legume in New Zealand, and Asian *Cicer arietinum* as a grain legume in Australia (Sullivan et al. 1995; Howieson et al. 2000). In addition to symbiotic microorganisms, inoculation with free-living and plant-associated bacteria has also been used on a large scale, especially in Russia in the 1930s and 1940s (Bashan 1998). Enhanced growth and direct effect on metabolism of nonlegume plants in response to the inoculations with *Azospirillum* have been reported in the 1970s (Döbereiner and Day 1976; Bashan and Holguin 1997). This followed development of biocontrol agents, mainly based on the members of *Pseudomonas fluorescens* and *P. putida* groups (Kloepper and Schroth 1981; Défago et al. 1992; Glick 1995; Glick and Bashan 1997). Several fungal entomopathogens have been studied for over a hundred years as potential tools to control the harmful insects but have not been exploited to their full potential as effective biocontrol agents, especially under field conditions (Vega et al. 2009; Lopez et al. 2014; Zhou et al. 2016; Bamisile et al. 2018; Clifton et al. 2018). Diverse genera of entomopathogenic fungi have been reported as plant endophytes from a variety of crop plants, antagonists to plant pathogens (Vega et al. 2009; McKinnon et al. 2018; Jaber and Ownley 2017; Yun et al. 2018), plant growth promoters (Kabaluk and Ericsson 2007; Lopez and Sword 2015), and beneficial rhizosphere colonizers (St. Leger 2008; Bruck 2005).

With recent development in next-generation sequencing (NGS) technologies, plant microbiome has received substantial attention. Community DNA-based studies revealed the diversity of the phytomicrobiome, which represents microorganisms that are associated with plants (Berendsen et al. 2012). High-throughput sequencing has ignited the speed of research on microbes associated with plants, as traditionally the plant microbiome was characterized by isolating and culturing microbes on different media and growth conditions. In addition to the speed, NGS techniques also enhanced our understanding about plant microbiome and exhibited the potential role of uncultured majority in plant growth and health (Bulgarelli et al. 2015; Edwards et al. 2015; Rascovan et al. 2016). This information about the not-yet-cultivated majority pose a challenge for microbiologist to cultivate these microorganisms and test them for plant growth-promoting activities. The culture-based techniques missed the vast majority of microbial diversity in plant-associated habitats, which is now detectable by culture-independent techniques, but our understanding on the real potential of these not-yet-cultivated microorganisms is insufficient. Cultivation of microorganisms is advantageous over detection by cultivation-independent techniques, as the cultures can be evaluated for various biotechnological applications and can also be employed to prove the role of microorganisms in various ecosystems. Exploiting the information generated on the microbial diversity using cultivation-independent methods to cultivate the not-yet cultivated majority could lead to the discovery of several previously unknown microorganisms. Devising newer growth cultivation techniques by mimicking natural ecosystems has been proven effective strategy for the culturing difficult-to-cultivate microorganisms (Stewart 2012; Lagier et al. 2016; Wolfe 2018). In recent years, several microbial inoculant-based biofertilizers were formulated, produced, and used by farmers worldwide (Bhardwaj et al. 2014). Various studies have shown that plants also interact with a variety of microorganisms that are capable of increasing photosynthetic capacity (Xie et al. 2009; Zhang et al. 2008), conferring drought and salt tolerance (Dimkpa et al. 2009; Xie et al. 2009; Vurukonda et al. 2016) and increasing disease suppression (Chithrashree et al. 2011; Jetiyanon and Kloepper 2002; Sha et al. 2016; Okubara and Bonsall 2008) and plant growth (Hayat et al. 2010). In addition to the agriculture crops, other economical important crops like biofuel crops also need effective microbial partners as such crops are generally cultivated in areas unsuitable for agricultural production, where plants are exposed to stress condition including drought and salt (Tilman et al. 2009). Supplementing compost or farmyard manure with plant beneficial microorganisms has shown the ability to suppress plant pathogens and opportunistic human pathogens (Pugliese et al. 2011). Similarly microorganisms play an important role to clean up contaminated and polluted soils by their synergistic actions with plants. Realizing the immense potential of microorganisms in various activities, especially in agriculture, a large number of mBRCs have been established throughout the world (Tables 5.1 and 5.2). Utilization of agriculturally important microorganisms could help achieving the goal of *maximum gain by minimum change* for the sustainable agriculture.

5.2.2 Animal Gut to Plant Roots

The key to improve the agricultural productivity in a sustainable manner lies in the understanding and making use of information and knowledge gained from investigations on plant–microbe interactions, which play out at various spatial levels, i.e., below and above the soil (Vorholt 2012), host–microbe relations levels such as those beneficial or pathogenic to the plant (Newton et al. 2010), and environmental levels,

Table 5.1 Top ten countries based on microbial strain holdings and their major microbial culture collections

Country (holdings)	Major culture collections (total holdings)	Website
U.S.A (336,637)	American Type Culture Collection (ATCC: 75,079)	http://www.atcc.org/
	Agricultural Research Service Culture Collection (NRRL: 96,200)	http://nrrl.ncaur.usda.gov/
	Fungal Genetics Stock Center (FGSC:29,230)	http://www.fgsc.net/
Japan (254,830)	AHU Culture Collection (AHU: 2635)	http://www.agr.hokudai.ac.jp/oukin/index.html
	Microbial Culture Collection at National Institute for Environmental Studies (NIEZ: 2219)	http://mcc.nies.go.jp/
	Laboratory Culture Collection (IID: 1720)	http://www.ims.u-tokyo.ac.jp/bac/kinkabu00309/home.html
	NBRC Culture Collection (NBRC: 127,694)	http://www.nite.go.jp/en/index.html
	IAM Culture Collection (IAM: 3726)	http://www.iam.u-tokyo.ac.jp/misyst/ColleBOX/IAMcollection.html
	Japan Collection of Microorganisms (JCM: 24,784)	http://jcm.brc.riken.jp/
China (200,101)	China Center of Industrial Culture Collection (CICC: 10,925)	http://www.china-cicc.org
	China Center for Type Culture Collection (CCTCC: 22,071)	http://www.cctcc.org/
	China General Microbiological Culture Collection Center (CGMCC: 53,906)	http://www.cgmcc.net
India (197,658)	Indian Type Culture Collection (ITCC: 3820)	http://www.iari.res.in/
	Microbial Type Culture Collection & Gene Bank (MTCC: 3029)	http://www.imtech.ernet.in/mtcc/
	National Collection of Industrial Microorganisms (NCIM: 2970)	http://www.ncl-india.org/files/NCIM/Default.aspx
	National Centre for Microbial Resource (MCC: 164,652)	http://210.212.161.138/
	National Agriculturally Important Microbial Culture Collection (NAIMCC: 6300)	http://nbaim.org.in/pages/services-culture-collectionnaimccculture-collectionnaimcc
Rep. of Korea (167,127)	Korean Collection for Type Cultures (KCTC: 23,175)	http://kctc.kribb.re.kr/English/index.aspx
	National Culture Collection for Pathogens (NCCP: 2355)	http://nccp.cdc.go.kr

(continued)

Table 5.1 (continued)

Country (holdings)	Major culture collections (total holdings)	Website
Thailand (118,728)	Culture Collection, Bangkok MIRCEN (TISTR: 2425)	http://www.tistr.or.th/tistr_culture
	BIOTEC Culture Collection (BCC: 78,202)	http://www.biotec.or.th/bcc/
	Department of Medical Sciences Culture Collection (DMST: 10,000)	http://geocities.com/dmst_cc
Brazil (114,514)	Colecao de Culturas Tropical (CCT: 4656)	http://www.fat.org.br
	Collection of Microorganisms, DNA and Cells of Universidade Federal de Minas Gerais (CM-UFMG: 18,305)	carlrosa@icb.ufmg.br
Denmark (102,066)	Culture Collection of Fungi (IBT: 36,500)	http://www.bio.dtu.dk/
	The International *Escherichia* and *Klebsiella* Centre (IEKC, SSI: 63,500)	http://www.ssi.dk/sw1397.asp
	Scandinavian Culture Collection of Algae and Protozoa (SCCAP: 2066)	http://www.sccap.dk/
Belgium (100,044)	Belgian Coordinated Collections of Microorganisms/MUCL Agro-environmental Fungi Collection (MUCL: 22,227)	http://bccm.belspo.be/
	Belgian Coordinated Collections of Microorganisms/ Bacteria Collection (BCCM/LMG: 25,850)	http://bccm.belspo.be/index.php
Germany (95,593)	Leibniz-Institut DSMZ-Deutsche Sammlung von Mikroorganismen und Zellkulturen GmbH (DSMZ: 34,534)	http://www.dsmz.de/
	Jena Microbial Resource Collection (JMRC: 50,011)	http://www.jmrc.hki-jena.de
	Medical Culture Collection Marburg (MCCM: 3000)	mutters@mailer.uni-marburg.de

Data from WFCC website accessed on 24 February 2018

i.e., external climatic and soil factors (Hatfield and Prueger 2015; Carey 2016). But before examining the different levels of plant–microbe relations microscopically, it would be useful to focus our attention macroscopically on the level of interfaces between plants and animals that directly or indirectly affect the microbial milieu of various levels of plant–microbe interactions in either direction. The interaction between land plants and land animals, which depend on plants for their survival, is very deep. However, modern extensive agriculture which relies on intensive inputs of chemical fertilizers, and pesticides, may be considered to have broken the link of opportunities which various plant-animal interactions have to offer in terms of "microbial inputs" a plant might require, or acquire, from animals.

The role of animals has been appreciated in all traditional agricultural societies in sustainment of agriculture production by utilization of animal dung and draft power for crop cultivation (Chayanov 1966). Animals have been considered to help sustain agriculture production by increasing the rate of nutrient flows in the mixed crop-

Table 5.2 Major agricultural microbial genetic resource centers

Country	Culture collection	Holding (nos.)	Preservation	Services	Website/e-mail
Australia	Plant Pathology Herbarium (DAR)	Bacteria (2000), Fungi (4000), Virus: Plant (80)	Freeze-dry; L-dry; Subculturing	Distribution; Identification	http://www.dpi.nsw.gov.au/aboutus/services/collections/cultures
	Victorian Plant Disease Herbarium (VPRI)	Bacteria (20), Fungi (1000), Virus: Plant (100)	Freeze-dry; Subculturing	Identification	Ian.Pascoe@dpi.vic.gov.au
	Department of Agriculture and Food Western Australia Plant Pathology Collection (WAC)	Bacteria (379), Fungi (8000), Virus: Plant (200)	Freeze-dry; Liquid N2; Subculturing	Storage	neyres@agric.wa.gov.au
Brazil	Culture Collection of Phytopathogenic Fungi Prof. Maria Menezes (CMM)	Fungi (1780)	Mineral oil; Silica gel; Water	Storage; Distribution; Identification	http://www.pgfitopat.ufrpe.br/cmm.html
	Culture Collection of Diazotrophic and Plant Growth Promoting Bacteria of Embrapa Soja (CNPSo)	Bacteria (4125)	Freeze-dry; Freezer (−80 °C′ −50 °C); Subculturing	Storage; Distribution; Identification	http://alelomicro.cenargen.embrapa.br
China	Agricultural Culture Collection of China (ACCC)	Bacteria (8721), Fungi (6469), Yeast (609), Archaea (315)	Liquid N2; Freeze-dry; Subculturing; Mineral oil	Storage; Identification; Distribution; Patent deposits	http://www.accc.org.cn/show.asp?uver=cn
	Culture Collection, Beijing Agricultural University (CCBAU)	Rhizobium (5060)	Freeze-dry; Freezer; Subculturing	–	wenxin_chen@263.net
Egypt	Egypt Microbial Culture Collection (EMCC)	Bacteria (1398), Fungi (211), Yeast (176), Virus: Plant (15), Archaea (8)	Freeze-dry; Freezer; Under paraffin; Under glycerol	Storage; Identification; Distribution; Patent deposits	cairomircen78@gmail.com

(continued)

Table 5.2 (continued)

Country	Culture collection	Holding (nos.)	Preservation	Services	Website/e-mail
France	French Collection for Plant-associated Bacteria (CFBP/INRA)	Bacteria (6800)	Freeze-dry	Storage; Distribution	http://www.angers-nantes.inra.fr/cfbp/
	Pasteur Culture collection of Cyanobacteria (PCC)	Cyanobacteria (500)	–	Storage; Distribution; Patent deposits	http://www.crbip.pasteur.fr
Greece	Benaki Phytopathological Institute Collection (BPIC)	Bacteria (1556), Fungi (412)	Freeze-dry; Subculturing	Identification	k.elena@bpi.gr
Rep. of Korea	Korean Agricultural Culture Collection (KACC)	Bacteria (7783), Fungi (9829)	Freeze-dry; Freezer; Mineral oil	Patent deposits; Distribution; Identification	http://www.genebank.go.kr
	Plant Virus Gen Bank (PVGB)	Plant: Virus (1004), Mutant virus clones (301), Plant virus antisera (81)	Freeze-dry; Freezer; Liquid N2; $CaCl_2$ dried	Storage; Distribution; Identification	http://www.virusbank.org
India	National Agriculturally Important Microbial Culture Collection (NAIMCC)	Bacteria (1341), Fungi (3068), Cyanobacteria (120)	Freeze-dry; Subculturing; Glycerol Stock; Liquid N2; Live cultures	Storage; Distribution Identification for CIB &RC	http://www.mgrportal.org.in
The Netherlands	Culture Collection of Plant Pathogenic Bacteria (PD)	Bacteria (3500)	–	Storage; Distribution	d.janse@pd.agro.nl
	Plant Virus Collection (RIPO)	Virus: Plant (125)	Liquid N2	Distribution	http://www.plant.wageningen-ur.nl/products
New Zealand	International Collection of Microorganisms from Plants (ICMP)	Bacteria (9305), Fungi (9370)	Freeze-dry; Liquid N2	Storage; Identification; Distribution	http://www.landcareresearch.co.nz/resources/collections/icmp

Poland	Collection of Plant Pathogens (CPPIPP)	Bacteria (70), Fungi (800), Virus: animal plants (19)	Freeze-dry	Storage; Identification; Distribution; Patent deposits	–
Russia	Russian Collection of Agricultural Microorganisms (RCAM)	Bacteria (4227), Fungi (877), Yeast (204)	Freeze-dry; Freezer; Subculturing	Storage; Distribution; Patent deposits	http://www.arriam.spb.ru
	Collection for plant protection, All-Russian Institute of Plant Protection (VIZR)	Bacteria (765), Fungi (11), Yeast (9)	Freeze-dry; Freezer; Liquid N2; L-drying	Patent deposits; Identification	http://www.cl.spb.ru/vizrspb/
	Collection of Nonpathogenic Microorganisms for Agriculture (CIAM)	Bacteria (3926), Fungi (671), Yeast (184)	Freeze-dry; Freezer	Storage; Distribution; Patent deposits; Identification	http://www.arriam.spb.ru/
South Africa	South African Plant Pathogenic and Plant Protecting Bacteria (PPPPB)	Bacteria (860)	Freeze-dry; Freezer	_	nipbtg@plant1.agric.za
	South African Rhizobium Culture Collection (SARCC)	Bacteria (58)	Freeze-dry; Liquid N2; Subculturing	Storage; Identification	http://www.arc.agric.za
Switzerland	Swiss Collection of Arbuscular Mycorrhizal fungi (SAF)	Fungi (202)	Subculturing	Storage; Identification	www.agroscope.ch
U.K	Rhizobium Collection (WPBS)	Bacteria (650)	_	Distribution	
	National Collection of Plant Pathogenic Bacteria (NCPPB)	Bacteria (4500)	Freeze-dry; Freezer	Patent deposits; Storage; Identification	http://www.ncppb.com/
	CABI Genetic Resource Collection (IMI)	Bacteria (1739); Fungi (27,000); Yeast (261)	Freeze-dry; Liquid N2; Subculturing	Patent deposits; Storage; Identification	http://www.cabi.org/_

(continued)

Table 5.2 (continued)

Country	Culture collection	Holding (nos.)	Preservation	Services	Website/e-mail
U.S.A	Agricultural Research Service Culture Collection (NRRL)	Archaea (2), Bacteria (22,496), Fungi (55,733), Yeast (17,969)	Freeze-dry; Liquid N2	Patent deposits; Identification	http://nrrl.ncaur.usda.gov/
	USDA-ARS Rhizobium Germplasm Resource Collection (BRCC)	Bacteria (4016)	L-drying	Storage; Distribution	http://www.ars.usda.gov/is/np/systematics/rhizobium.htm
	University of Minnesota Rhizobium Collection (UMRC)	Soybean, bean, prairie legume rhizobia (2000)	Freeze drying	Distribution	http://www.Rhizobium.umn.edu/

Data from WFCC website accessed on 24 February 2018

livestock systems (Slicher van Bath and Ordish 1963). The recent interest toward advantages of organic, as against conventional farming, has been observed, as organic farming involves application of organic matter, farmyard manure, and even sewage water as agriculture inputs. Studies have shown the potential advantages of mixed crop-livestock farming over specialized systems with regard to the sustainable development of agriculture (Oomen et al. 1998; Hendrickson et al. 2008) and have been reported to enhance biodiversity through spatial heterogeneity (Fahrig et al. 2011). In a study, the addition of manure to the soil increased the nitrogen content in the sorghum shoots, 1.57 times higher than the control. The application of farmyard cattle manure to sorghum, the study further elucidated, resulted in increased concentration of amino acids in root compartment, greater microbial activity and diversity in the sorghum rhizosphere, and higher amino acid consumption near the roots (Chu et al. 2016).

Such studies, in which comparative analysis of soil and root microbiome in relation to application of conventional chemical fertilizers, and organic or farmyard manure or compost are few. Thus, the microbial niche functions of plants in which animals' derived microbes might be directly or indirectly involved have not been studied in detail. In addition, the evidence has been presented that participation of animals in agriculture system has allowed farmers to include crops that either increase soil fertility by fixing atmospheric nitrogen and release of bound phosphorus or enhance soil organic matter (Hoffland 1991). This might have link in the role of microbes that organic matter produced from livestock waste might provide to root microbial ecosystem. The plant root systems are directly exposed to soil microbes, and plants rely on root microbiomes to help garner the supply of nutrients from the soil and to fight pathogens. The plant microbiomes are mainly constituted from the microbes present in the soil, and the soil microbiomes itself may be positively influenced by the microbial constituents and/or animal–microbe-derived micronutrients originating from animal-derived organic and dung-matter. The plant-eating ruminant domestic animals, the most prominent animal components of a mixed farming approach, show a very rich microbial diversity in their gut microbiome (Ley et al. 2008), including probably plant-associated bacteria, especially endophytes (Rosenblueth and Martínez-Romero 2006). It is thus of interest to look for and ex situ conserve the animal–gut microbe and related consortia to understand their role in environmentally sustainable agriculture in general by influence of root microbe in a beneficial manner.

In the direction of documentation and conservation of microbial genetic resources, the Indian Council of Agricultural Research (ICAR), New Delhi, took giant steps when it established, among other culture collections, the Veterinary Type Culture facility in 2005 at the National Research Centre on Equines, Hisar, Haryana State, India, for conservation of the microbial diversity of animal origin. The activities of Veterinary Type Culture Collection (VTCC), which has been rechristened as the National Centre for Veterinary Type Cultures (NCVTC) in December 2015, comprise of acquisition, authentication, preservation, documentation, and repository database management system of animal microbes (http://ncvtc.org.in/). NCVTC has networking with numerous agricultural universities, ICAR institutes, and traditional universities to source its collection from veterinary, dairy, and ruminant components of animal microbial ecology. The conservation of microbes from rumen of different ruminant species like cattle, buffalo, sheep, goat, camel, and yak offers profound possibilities in application of such microbes to elucidate their role not only in rumen fermentation greenhouse effect and rumen energy metabolism but also in investigations on the role of these microbes in their shared niche functions in plant root milieu.

Many microbial genera which are reported to be players in their role as constituents of plant–root microbiome are also commonly found in the gut of animals including humans. Clostridia, and members of the family *Ruminococcaceae*, are prevalent in anaerobic root environment in water-saturated roots akin to anaerobic environments of the gut (Timmers et al. 2012). *Burkholderia cepacia* strain colonizes root regions (Sharma et al. 2008) which are otherwise important pathogens causing cystic fibrosis in humans (Mahenthiralingam et al. 2008). *Enterobacter asburiae*, which is found in maize kernels, colonizes the rhizosphere (Johnston-Monje and Raizada 2011); this bacterium is also a cellulase-producing bacterium in tilapia gut (Saha et al. 2006). *Chromobacterium violaceum* and *Pantoea agglomerans*, which produce root-stimulating riboflavin and lumichrome, are opportunistic animal and human pathogens, apart from *Pseudomonas* spp. and *Streptomyces* spp. strains.

As a result of high-throughput sequencing-based metagenomic studies, there is unraveling of the realm of the soil, root, human gut, and animal gut microbiome (Jami et al. 2013; Bulgarelli et al. 2015; Edwards et al. 2015). Novel philosophical transactions have been recently published on the topic dealing with putative similarities in structure, functions, and microbiota of gut and root microbiome (Ramírez-Puebla et al. 2013; Hacquard et al. 2015; Mendes and Raaijmakers 2015). The investigations on root and gut microbiomes have revealed a subset of common microbiome, called "core microbiome," which is important for both plant and animal health, and standardized core-microbiome transfer therapy can be a solution of plant disease treatment in the future (Gopal et al. 2013). The soil metagenomic study has detected a group of plant–disease-suppressive bacterial communities related to *Proteobacteria*, *Firmicutes*, and *Actinobacteria*, as core microbiome. In contrast, out of 15 phyla detected in the bovine rumen samples, the *Firmicutes*, *Bacteroidetes*, and *Proteobacteria* were detected as the dominant phyla regardless of age group (Jami et al. 2013). The characteristic similarity between the gut and root microbiota (Berendsen et al. 2012; Ramírez-Puebla et al. 2013) means that utilization of a consortium of microorganism, sourced from a combination of sources, soil, plant, and gut microbiota, for diseases control in plants can be followed. The suggestion has been mooted for pure-culture isolation of corresponding community members of microbes after an initial culture-independent survey of the plant microbiota (Lebeis et al. 2012). It would be pertinent to consider the sourcing of consortium of pure-culture microbes for soil treatment or plant disease treatment from myriad sources including animal and human gut.

In the process of investigation of disease outbreaks, the samples for isolation of bacteria are taken from various sources, which can be broadly classified as animal proper and its surrounding environment. The samples sourced from live or dead animal may include nasal, vaginal, skin, or milk sample or samples from inside the body after postmortem to look for typical disease lesions. In addition, samples are also obtained from surroundings like water or feed trough. Many genera and species of bacteria isolated unusually from animal samples are close to taxa represented in the NCBI taxonomy for all taxonomically classified OTUs from the rhizosphere samples (16S rRNA gene and metagenome survey) (Bulgarelli et al. 2015). Different strains like *Acinetobacter variabilis*, *Microbacterium imperiale*, *Flavobacterium* spp., *Brevibacillus agri*, *Brevibacterium* spp., and *Lysinibacillus fusiformis* have been isolated from equines. The isolation of plant-related taxa from animal sources indicates a probable shared environmental relatedness. Such taxa may represent endophytes of plants which are also involved in microbial community establishment in animal gut or may play some other function. The application of farmyard cattle manure to sorghum, which resulted in increased concentration of nutrients in root compartment and a greater microbial activity and diversity, is an indicator (Chu et al. 2016) that encourages us to explore the application of animal origin microbial strains in the plant rhizosphere microbial niche functions.

5.2.3 Plant Growth Promotion by Biosurfactant-Producing Microbes

Surfactants are surface-active compounds and amphiphilic (contain both hydrophobic and hydrophilic moieties) in nature (Moldes et al. 2013). These surface-active agents reduce surface and interfacial tension at the air/water and oil/water interfaces, respectively (Banat et al. 2010). Chemical surfactants mostly are derived from petroleum products that are noxious and non-biodegradable and create other types of pollutions in the environment (Oliveira et al. 2013). In recent decades, microbial surfactants produced by a variety of microorganisms such as bacteria, fungi, algae, and yeasts have attracted a huge interest in diverse industries like medicine, pharmaceutical, food, petroleum, and agriculture (Uzoigwe et al. 2015). Biosurfactants became interesting compounds due to their low toxicity, biodegradability, biocompatibility, effectiveness, and unique structures in the environment (Chen et al. 2015). Biosurfactants mostly are glycolipids, lipopeptides, phospholipids, natural lipids, and polymeric in nature (Pacwa-Plociniczak et al. 2011).

Biosurfactants are beneficial biomolecules with multifunctional applications in agriculture. Production of biosurfactants by soil microbes can influence plant growth, crop protection, and agricultural productivity. Moreover, biosurfactants play an important role in signaling, motility, and biofilm formation in plants. Ruth et al. (2006) reported a double function of quorum-sensing acyl homoserine lactone (AHLs) in control of swarming behavior and their action as biosurfactant. It has been reported the significant role of sensing molecules in the production of antifungal compounds by rhizobacteria. In addition to that, quorum-sensing molecule is required for regulation of essential exopolysaccharides in biofilm formation. Specific groups of biosurfactant producing-pseudomonads support root colonization (Tran et al. 2008). Many biosurfactants also have insecticidal activity and play a role in plant disease suppression. Yang et al. (2017) showed the capability of *Bacillus subtilis* Y9 to produce biosurfactant and its effectiveness against *Myzus persicae* causing decreased growth, shriveling of the leaves, and the death of various tissues. Biosurfactants like rhamnolipid enhanced the herbicidal activity of glyphosate (Liu et al. 2016). High concentration of heavy metals in soils led to root tissue necrosis and purpling of foliage and severely damage crop plants. Microbial biosurfactants help in remediation of organic and inorganic pollutants such as heavy metals and hydrocarbons from soils in order to improve the quality of agricultural soils ultimately improving the crop plants productivity. Biosurfactants enhance solubilization and bioavailability of contaminants and improve the process of biodegradation (Zhang et al. 2011).

Microbial surfactants also are capable to degrade and remove pesticides and insecticides from agricultural soil effectively. Phytoremediation of DDT-polluted soil was effectively enhanced with assistance of *Pseudomonas* sp. SB (Wang et al. 2017). Biosurfactant-producing *Actinobacteria* sp. has been effectively used for bioremediation of pesticides and heavy metal-impacted agricultural soils (Alvarez et al. 2017). Khedher et al. (2017) suggested that biosurfactant produced by *Bacillus amyloliquefaciens* AG1 could be a great insecticide against *Spodoptera littoralis* larvae and also prevent the resistance improvement of this pest. The antifungal and antibacterial activities of biosurfactants have been studied by many researchers (Kim et al. 2000; Zhao et al. 2014). The application of rhamnolipids (RLs) has been beneficial in controlling blight disease caused by *Phytophthora* in pepper plants and infection of *Colletotrichum orbiculare* on leaves of cucumber plants (Kim et al. 2000). Borah et al. (2016) found that the potential application of rhamnolipid biosurfactant produced by *Pseudomonas aeruginosa* SS14 could be an efficient biocontrol approach against maize plant pathogen *Fusarium verticillioides* FS7 which causes stalk and ear rot disease in maize. Apart from that, presence of biosurfactants can be beneficial for improvement of microbes–plants interaction. Microbes establish their association with plants by biofilm formation on plant roots, motility, and releasing of quorum-sensing molecules. Several plant growth-promoting rhizobacteria (PGPR), along with biosurfactant production, form biofilm, which restricts the essential nutrient uptake for phytopathogenic microorganisms from the environment (Rekadwad and Khobragade 2017). Biosurfactants or biosurfactant-producing microbes can be an excellent substitute to chemically available pesticides and insecticides.

5.3 Microbial Bioresource Centers and Microbial Preservation

The importance and influence of microbes on human life have been proven in many kinds of literature. Application of diverse microorganisms has been well established in many fields such as agriculture, environment, industries, medicine, and biotechnology (Prakash et al. 2013). Microbes and their primary and secondary metabolites such as insecticide, fungicide, growth hormones, vitamins, antibiotics, and novel therapeutical agents have gained huge interest for various researches (Demain and Sanchez 2009; Senni et al. 2011). In addition to their beneficial effects, microbes may cause many diseases to humans, plants, and animals. Approximately all aspects of human health gain benefit from microbial metabolites (Stewart 2012). Therefore, isolation, characterization, and preservation of diverse microbes are crucial. To study and verify the potential application of microorganisms, their cultivation is essential (Prakash et al. 2013). Culture-dependent approach is the traditional technique for isolation of microorganisms to investigate their functions. Till now just 1% of the microbial population could be cultivated (Pei et al. 2017). Recently molecular microbial ecology has a significant jump due to the development of several molecular techniques, such as sequencing of SSU ribosomal DNA, random amplified polymorphic DNA (RAPD), single-strand conformation polymorphism (SSCP), and gradient gel electrophoresis (DGGE/TGGE) and metagenomics (Caporaso et al. 2011). All these techniques are based on 16S rRNA genes and have become popular approaches to study microbial communities especially in environmental samples (Tringe and Hugenholtz 2008).

Metagenomics (culture independent) is a successful technology and carries great improvement in microbial diversity and their potential metabolites. However, the presence of only sequence data is not sufficient to study fundamental features of microbes, and it can be a serious challenge of metagenomics especially in habitats with highly diverse microbial communities (Prakash et al. 2013). The information related to microbial morphology, microbial physiology, ecology, and their natural products are achieved only with microbial cultivation in the laboratory (Stewart 2012). The clear and simple answer for why many bacteria are not cultivable in laboratory conditions is the lack of essential factors (nutrients, PH, temperature, etc.) of their natural environment. Mimicking the environmental conditions with synthetic growth media could allow cultivation of enormous diversity of not-yet-cultured microorganisms (Stewart 2012). Co-culturing is another approach to cultivate the not-yet cultivated microorganisms. Lewis et al. (2010) showed that the isolated bacteria from intertidal sand biofilm were capable of growing only in presence of other bacteria (helper culture) from the same environment. They also found that a diffusion of growth factor produced by the helper strain into media allowed the growth of previously uncultivable strain. There is a hidden ocean of bacterial taxa, representing genetic and biochemical diversity which have not been cultivated in a laboratory (Achtman and Wagner 2008). Plant–microbial diversity is extremely underrated; the data obtained by next-generation sequencing (NGS) technology have shown that only a small share of microbes associated with the plant is cultivated. The matrix-associated laser deionization time-of-flight mass spectrophotometry (MALDI-TOF MS)-based microbial identification approach, which offers low-cost and high-speed identification of microbes, provides an advantage to design high-throughput cultivation experiments to cultivate the not-yet-cultivated microorganisms (Rahi et al. 2016). It is an obvious fact that culture-based and culture-independent techniques are complementary. Combining culturing, MALDI-TOF MS, and 16S rRNA gene sequencing to create a high-throughput analysis will lead to finding what might otherwise be missed (Marx 2017).

Isolation and identification of microorganisms have their own importance; it also is necessary to preserve the newly isolated and identified strains for their potential applications in the future and to use them as a reference strain. Many researchers avoid depositing and preserving of reference cultures due to several reasons including lack of funding, interest, and incomplete projects, which led to the loss of huge microbial resource (Stackebrandt et al. 2014). A large number of microbial strains have been reported in various articles in research journals; a small portion of these strains were deposited in public culture collections. There is notable gap in the availability of reference cultures, especially in rare and less dominant taxa. Making the deposition of strains as essential part of publication process might help and stop the increase in gap between strains reported in literature and those available with mBRCs (Stackebrandt et al. 2014). mBRCs provide authentic cultures, long-term preservation, and identification of microbial strains. Similarly, patent deposition and training and distribution of reference cultures and database to international scientific groups are other services of culture collections. mBRCs also play an extensive role in microbial taxonomy, diversity, genome, and ecology of microorganisms. Hence it is critical responsibilities for mBRCs to keep their scientific level and services updated (Stackebrandt et al. 2014). Several valuable microbial cultures have been lost due to lack of a place for their storage and preservation, before the establishment of the mBRCs. To address this problem, almost a major country across the globe has one of more culture collections. Based on the present reports on WDCM (World Data Center for Microorganisms) website, there are 723 culture collections from 75 countries with 2,596,498 holding strains (www.wcdm.org). There are 11 culture collections from Africa, 153 from the USA, 223 from Asia, 220 from Europe, and 41 from Oceania, which have been registered with WDCM (www.wfcc.info). Each culture collection follows the particular strategy in deposition and type of holding strains. Based on the services and the number of microorganisms preserved, several culture collections have established them and are known to the microbiologists around the world (Table 5.1). Most of the mBRCs are publicly funded, established with the objective to provide quality-controlled and well-characterized microbial resources and data, at low cost to researchers. The preservation and maintenance of microbial cultures require very high amounts of money, and several mBRCs are facing severe monetary issues. To generate the funds to supplement the public funding, mBRCs are making serious efforts, and entering the field of bioprospecting in addition to their traditional work appears a promising strategy (Overmann and Smith 2017).

5.3.1 Agriculturally Important Microbial Resource Centers

Agriculture is the mainstay for human civilization, and agriculturally important microorganisms are the key players in developing strategies and technologies for sustainable agriculture systems. Understanding the value of agriculturally important microorganisms, all major countries in the world have established culture collections, which deal specifically with the microorganisms important in agriculture (Table 5.2). Resource centers for agriculturally important microorganisms have become important centers preserving the microorganisms, which have a role in agriculture or related sectors like fishery and veterinary. These culture collections are valuable resource centers for agriculture scientists, agriculture-based industry, and farmers by many ways, including:

- Maintain authentic strains of microorganisms beneficial to different agriculture and plantation crops.
- Maintain authentic reference strains for research.
- Maintain strains of pathogenic microorganisms.
- Host human resource of experts in understanding plant–microbe interactions.

These resource centers also preserve multiple strains of a single species to maintain the species diversity of particular microorganisms. Nitrogen-fixing members of the genus *Rhizobium* are preserved by many culture collections, as many species of this group have shown genetic flexibility and biogeographic patterns of distribution (Rahi et al. 2012; Sprent et al. 2017). There are several culture collections, which specifically maintain the strains of rhizobia, like the CB Rhizobium Collection (CB), CSIRO Canberra Rhizobium Collection (CC), Australia; Colecao de Culturas de Rhizobium da Fepagro (SEMIA), Brazil; CIAT Rhizobium Collection (CIAT), Colombia; Volcani Center Rhizobium Collection (VCRC), Israel; Rhizobium Collection (UPMR), Malaysia; South African Rhizobium Culture Collection (SARCC), South Africa; NifTAL Rhizobium Collection (CISM); Culture Collection of Soybean Rhizobia and Green Microalgae (KCCC), Thailand; WPBS Rhizobium Collection (WPBS), UK; USDA-ARS Rhizobium Germplasm Resource Collection (BRCC), University of Minnesota Rhizobium Collection (UMRC), Rhizobium Culture Collection (UPRM), USA; Novi Sad Collection of Nitrogen Fixing Bacteria (NSCNFB), Yugoslavia; and Grasslands Rhizobium Collection (MAR), Zimbabwe (http://www.wfcc.info/index.php/collections/display/). The culture collections of *Rhizobium* have been established since the isolation of these bacteria (Freire and Kolling 1986). With the availability and low cost of chemical nitrogenous fertilizers, the interest in these nitrogen-fixing bacteria reduced, but it is revamping once again. Rhizobia-based technologies are becoming important especially in the developing countries, which leads to isolation of locally adapted or native strains and developing reliable inoculants. The need for sustainable agriculture throughout the world is revitalizing the interest in biological nitrogen fixation and rhizobia–legumes symbioses especially the economically important legume crops (Laranjoa et al. 2014).

Cyanobacteria represents morphologically diverse group of photosynthetic prokaryotes, which occur almost in every illuminated habitat (Whitton 2012). They are the key players in the ecosystems ranging from the warmer oceans to many Antarctic sites, nutrient-rich lakes, rice fields, and many soils. Understanding of cyanobacterial ecology is very crucial as it helps to deal with different ecological problems such as the control of nuisance blooms and the use of cyanobacteria to manage semidesert soils. As most of the cyanobacteria grow very fast and have very little nutrient requirement, they have been considered as an organism of choice for bioenergy (Kumar and Singh 2016). Polyphasic approach based on the assessment of morphological and genotypic features has been used to estimate the cyanobacterial diversity (Komárek 2015). The lack of cultures of several cyanobacterial morphospecies and inadequate morphological data of strains with sequence information limits the comparative studies of morphological and genotypic features (Rajaniemi et al. 2005). Cyanobacteria are unique to all common members of prokaryotes and hence require special expertise for their preservation and maintenance. Considering the uniqueness and importance of cyanobacteria, there are dedicated culture collections like Brazilian Cyanobacteria Collection—University of Sao Paulo (BCCUSP), Brazil; Culture Collection of (sub)polar cyanobacteria (BCCM/ULC), Belgium; Canadian Phycological Culture Centre (formerly University of Toronto Culture Collection of Algae and Cyanobacteria) (CPCC), Canada; Culture Collection of Algae of Charles University in Prague (CAUP), Czech; Scandinavian Culture Collection of Algae and Protozoa (SCCAP), Denmark; Pasteur Culture Collection of Cyanobacteria (PCCC), France; Sammlung von Algenkulturen at the University of Göttingen (SAG), Germany; National Facility for Marine Cyanobacteria (BDU), India; and Culture Collection of Algae and Protozoa (CCAP), UK.

Similar to rhizobia and cyanobacteria, there are many other groups of microorganisms, which require special expertise for cultivation and preservation. All microorganisms cannot be preserved for a long time using the most common methods; therefore it is necessary to develop methodologies to maintain the viability of each strain and to retain their functional attributes (Caleza et al. 2017). In most of the fungal cultures, especially the non-sporulated, long-term preservation and maintenance are difficult to achieve in comparison to other microbes. Nearly all flowering plant species enter into an association with mycorrhizal fungi (Williams et al. 2017). The plant–mycorrhizal association is highly beneficial for crop plants as it helps in the uptake of nutrients, protection against pathogens, and maintenance of soil structure (Smith and Smith 2011). In addition to mycorrhizal fungi, several other members of fungi including yeast and filamentous fungi also influence plant growth and productivity, by direct and indirect mechanisms. Recent studies have demonstrated that certain entomopathogenic fungi can also perform endophytism, plant disease antagonism, plant growth promotion, and rhizosphere colonization (Jaber and Ownley 2017). All such studies indicated that fungi play critical and very unique roles in plant nutrient uptake and plant protection, and these strains should be preserved in mBRCs. Other than beneficial fungi, there is a huge diversity of fungi, which cause serious diseases to plants, and maintaining these pathogenic fungi is also important for developing disease-resistant crops.

5.3.2 Indian Microbial Resource Centers

India has been considered as an agrarian country as a large majority of its population depends on agriculture until today. Agriculture microbiology has always been an important area of research in all agriculture universities and institutes involved in agriculture-related research. There are 32 culture collections in India registered with WFCC (http://www.wfcc.info/ccinfo/). Collectively the Indian culture collections stand in fourth position in holding the number of microbial strains after the USA, Japan, and China (http://www.wfcc.info/ccinfo/statistics/). Many of Indian culture collections were maintained by single individuals, institutions, or companies and became nonfunctional due to several constrains. Among the major functional culture collections include National Centre for Microbial Resource (formerly Microbial Culture Collection), Pune; Microbial Type Culture Collection and Gene Bank, Chandigarh; National Fungal Culture Collection of India, Pune; National Collection of Industrial Microorganisms, Pune; Indian Type Culture Collection (ITCC), New Delhi; and ICAR-National Bureau of Agriculturally Important Microorganisms, Maunath Bhanjan (Sharma et al. 2017). Most of the Indian culture collections hold strains of some specific groups of microorganisms depending upon the expertise available and interest of their staff. It is expected that networking among all functional microbial culture collection will provide a common platform to access information about all type of microorganisms available with participating culture collections. Such networking will improve the functioning of culture collections and also led toward the sustainability of many collections, which don't have necessary equipment and specialized manpower.

At present, Indian culture collections are facing legal obligations while sharing their strains with international researchers especially for taxonomic purposes (Jayaraman 2016). As after signing as party to CBD, the Government of India brought Biological Diversity Act, 2002, with an extensive and intensive consultation to provide for conservation of biological diversity, sustainable use of its components, and fair and equitable sharing of the benefits arising out of the use of biological resources and knowledge and for matters connected therewith or incidental thereto. The National Biodiversity Authority (NBA) was established in 2003 to implement India's Biological Diversity Act (2002), which is a statutory, autonomous body, and it performs facilitative, regulatory, and advisory function for the Government of India on issues of conservation, sustainable use of biological resources, and fair and equitable sharing of benefits arising out of the use of biological resources. It is mandatory to seek approval of NBA before any transfer of the accessed biological resources of India. Consequent to these liabilities, the culture collections of India are not able to share microbial strains globally, including for taxonomic proposes. The National Biodiversity Authority rules include microorganisms under biological resources, and any non-Indian researcher or end user needs to take permission from the authority. This rule is contradictory to the International Code of Nomenclature of Prokaryotes (ICSP) rules, which require the deposit of type strains in a collection to be independent of other collections, other institutions, or other persons (Tindall and

Garrity 2008). The National Biodiversity Authority of India had made a provision for conducting noncommercial research or research for emergency purposes outside India by Indian researchers/Government institutions, as per regulation 13 of ABS Guidelines, 2014 (http://nbaindia.org/content/26/59/1/forms.html). Though individual researchers have the privilege to share microbial strains of Indian origin with non-Indian researchers, it is disappointing that Indian mBRCs are deprived of such privilege and they cannot supply Indian origin cultures without prior approval of the NBA. The refusal by one of the Indian culture collections (i.e., MTCC, Chandigarh) to provide type strains to non-Indians has been pointed out by Liu et al. (2015). It was also suggested that ICSP prevent such situations by rejecting certificates of deposit from Indian collections, as they are not able to follow the ICSP guidelines on type strains supply. Responding to these issues, the chairperson of the NBA insisted that the Indian bioresources are not inaccessible to non-Indians but can be accessed by taking approval, which one can get by following a simple procedure and paying a nominal fee (Jayaraman 2016). The National Biodiversity Authority of India has also assigned four culture collections (i.e., NCMR-NCCS, Pune; MTCC-IMTECH, Chandigarh; NAIMCC-NBIAM, Maunath Bhanjan; and IARI, New Delhi) as designated repositories for safe deposit of holotypes/isotypes/paratypes of new taxa discovered in India and samples of biological resources accessed by foreign citizens/entities for research or sent/carried abroad by Indian citizens/institutions for research (http://nbaindia.org/blog/675/3//archive.html). The culture collections agreed to this responsibility and are maintaining the new taxon deposited for indefinite period. The deposited biological resources are made available for research/academic purposes within the country. However, the right to approve the access to deposited materials (including type strains) by non-Indian researchers remains with the NBA. It is clear that the culture collections are the important institutions involved in preserving and maintaining the microbial resources and also have requisite human resource expertise to understand the scientific and economic importance of microbial resources; this makes the culture collection better judge to deal with the sharing of microbial resources to non-Indians. All major international culture collections are functioning independently and do not follow such restriction on sharing of microbial resources; the supply of microbial strains is done by signing a well-defined material transfer agreement, which restricts the undisclosed use of microbial resource by the end user.

5.4 Conclusion

The role of microorganisms in developing strategies and technologies for sustainable agriculture is crucial, and consequently the resource centers for agriculturally important microorganisms are key institutions for all countries. At present very few countries are supporting the mBRCs especially those dealing with agriculturally important microorganisms. In the majority of countries in the world, agriculture is still a cultural practice, and it is not developed as an industry, and a large human population is associated with agriculture. This population often includes farmers with smallholdings, who are generally not very educated, hence adopting the practices of conventional agriculture and employ chemical inputs extensively. All these contribute to the irreversible deterioration of agriculture soils and making them unfit for future applications. In such condition, it becomes vital to maintain the soil biodiversity. Soil microorganisms represent the major portion of soil biodiversity and are involved in many biogeochemical processes. Maintaining a healthy soil microbiome could contribute to enhance nutrient use efficiency, reduce disease outbreaks, and enrich soils by reducing greenhouse gas emissions and restoring nutrient. Several companies are developing microbes-based formulations targeted to improve soil nutrient status and controlling pest and pathogens. mBRCs preserve and maintain the microorganisms and host human resource expert to lead the research in this field. After the implementation of biodiversity laws and related treaties, governments are now more concerned about the conservation and protection of biological diversity. However, these regulations have also placed few hurdles for the research community involved in microbial taxonomy research, especially in developing countries like India, South Africa, and Brazil. It is expected that the governments of these countries will respect the intention of researchers and will come up with amicable solutions and regulations. In the ear of omics technologies, like metagenomics, proteomics, and culturomics, it is clear that the mBRCs have to match the speed to handle data and microbial strains generated by these high-throughput technologies. Solutions based on microorganisms have the potential to transform conventional agriculture, but this is only possible with a combined effort to understand the basics of plant–microbe interactions, product development, efficient delivery systems, regulatory mechanisms, and economic viability. mBRCs dealing with agriculture microbiomes could play an important role in finding solutions to all these issues and come up with microorganisms-based technologies for sustainable agriculture. Generating funds to supplement the public funding mBRCs is a big challenge, and it can be achieved by initiating new programs like bioprospection of cultures preserved with mBRCs. This functional characterization will allow industry to select strains based on their specific functions, and in return industries can also support the mBRCs.

Acknowledgments This work was supported by the Department of Biotechnology, Government of India (BT/Coord.II/01/03/2016), and the Department of Science and Technology, Government of India (YSS/2015/000149).

References

Achtman M, Wagner M (2008) Microbial diversity and the genetic nature of microbial species. Nat Rev Microbiol 6:431–440. https://doi.org/10.1038/nrmicro1872

Ali B, Sabri AN, Hasnain S (2010) Rhizobacterial potential to alter auxin content and growth of *Vigna radiata* (L.). World J Microbiol Biotechnol 26:1379. https://doi.org/10.1007/s11274-010-0310-1

Alvarez A, Saez JM, Costa JS, Colin VL, Fuentes MS, Cuozzo SA, Benimeli CS, Polti MA, Amoroso MJ (2017) Actinobacteria: current research and perspectives for bioremediation of pesticides and heavy metals. Chemosphere 166:41–62. https://doi.org/10.1016/j.chemosphere.2016.09.070

Arcand MM, Schneider KD (2006) Plant- and microbial-based mechanisms to improve the agronomic effectiveness of phosphate rock: a review. Anais da Academia Brasileira de Ciências SciELO Bra 78:791–807. https://doi.org/10.1590/S0001-37652006000400013

Bamisile BS, Dash CK, Akutse KS, Keppanan R, Wang L (2018) Fungal endophytes: beyond herbivore management. Front Microbiol 9:544

Banat IM, Franzetti A, Gandolfi I, Bestetti G, Martinotti MG, Fracchia L, Smyth TJ, Marchant R (2010) Microbial biosurfactants production, applications and future potential. Appl Microbiol Biotechnol 87:427–444. https://doi.org/10.1007/s00253-010-2589-0

Bashan Y (1998) Inoculants of plant growth-promoting bacteria for use in agriculture. Biotechnol Adv 16:729–770. https://doi.org/10.1016/S0734-9750(98)00003-2

Bashan Y, Holguin G (1997) *Azospirillum*-plant relationships: environmental and physiological advances (1990–1996). Can J Microbiol 43:103–121. https://doi.org/10.1139/m97-015

Bashan Y, Holguin G (1998) Proposal for the division of plant growth-promoting rhizobacteria into two classifications: biocontrol-PGPB (plant growth-promoting bacteria) and PGPB. Soil Biol Biochem 30:1225–1228

Bashan Y, de-Bashan LE, Prabhu SR, Hernandez J (2014) Advances in plant growth-promoting bacterial inoculant technology: formulations and practical perspectives (1998–2013). Plant Soil 378:1–33. https://doi.org/10.1007/s11104-013-1956-x

Bent E, Tuzun S, Chanway CP, Enebak S (2001) Alterations in plant growth and in root hormone levels of lodgepole pines inoculated with rhizobacteria. Can J Microbiol 47:793–800. https://doi.org/10.1139/w01-080

Berendsen RL, Pieterse CM, Bakker PA (2012) The rhizosphere microbiome and plant health. Trends Plant Sci 17:478–486. https://doi.org/10.1016/j.tplants.2012.04.001

Bergottini VM, Hervé V, Sosa DA, Otegui MB, Zapata PD, Juniera P (2017) Exploring the diversity of the root-associated microbiome of *Ilex paraguariensis* St. Hil. (Yerba Mate). Appl Soil Ecol 109:23–31. https://doi.org/10.1016/j.apsoil.2016.09.013

Bhardwaj D, Ansari MW, Sahoo RK, Tuteja N (2014) Biofertilizers function as key player in sustainable agriculture by improving soil fertility, plant tolerance and crop productivity. Microb Cell Factories 13:66. https://doi.org/10.1186/1475-2859-13-66

Borah SN, Goswami D, Sarma HK, Cameotra SS, Deka S (2016) Rhamnolipid biosurfactant against *Fusarium verticillioides* to control stalk and ear rot disease of maize. Front Microbiol 7:1505–1515. https://doi.org/10.3389/fmicb.2016.01505

Bruck DJ (2005) Ecology of Metarhizium anisopliae in soilless potting media and the rhizosphere: implications for pest management. Biol Control 32(1):155–163

Bulgarelli D, Garrido-Oter R, Münch PC, Weiman A, Dröge J, Pan Y, McHardy AC, Schulze-Lefert P (2015) Structure and function of the bacterial root microbiota in wild and domesticated barley. Cell Host Microbe 17:392–403. https://doi.org/10.1016/j.chom.2015.01.011

Busby PE, Soman C, Wagner MR, Friesen ML, Kremer J, Bennett A, Morsy M, Eisen JA, Leach JE, Dangl JL (2017) Research priorities for harnessing plant microbiomes in sustainable agriculture. PLoS Biol e2001793:15. https://doi.org/10.1371/journal.pbio.2001793

Caleza V, Castillo S, Gomis MD, Kamah S, López R, Garcia-Seco D, Varma A, Akdi K (2017) Conservation strategies of new fungi samples in culture collections: *Piriformospora indica* case. In: Varma A, Sharma A (eds) Modern tools and techniques to understand microbes. Springer, Cham. https://doi.org/10.1007/978-3-319-49197-4_27

Caporaso JG, Lauber CL, Walters WA, Berg-Lyons D, Lozupone CA, Turnbaugh PJ, Fiererb N, Knight R (2011) Global patterns of 16S rRNA diversity at a depth of millions of sequences per sample. Proc Natl Acad Sci U S A 108:4516–4522. https://doi.org/10.1073/pnas.1000080107

Carey J (2016) Crucial role of belowground biodiversity. Proc Natl Acad Sci U S A 113:7682–7685. https://doi.org/10.1073/pnas.1609238113

Chayanov AV (1966) In: Thorner D, Kerblay B, Smith REF (eds) The theory of peasant economy, The American Economic Association Translation Series, Homewood IL, p 317 (reprint)

Chen YP, Rekha PD, Arun AB, Shen FT, Lai WA, Young CC (2006) Phosphate solubilizing bacteria from subtropical soil and their tricalcium phosphate solubilizing abilities. Appl Soil Ecol 34:33–41. https://doi.org/10.1016/j.apsoil.2005.12.002

Chen WC, Juang RS, Wei YH (2015) Applications of a lipopeptide biosurfactant, surfactin, produced by microorganisms. Biochem Eng 103:158–169. https://doi.org/10.1016/j.bej.2015.07.009

Chithrashree, Udayashankar AC, Chandra Nayaka S, Reddy MS, Srinivas C (2011) Plant growth-promoting rhizobacteria mediate induced systemic resistance in rice against bacterial leaf blight caused by *Xanthomonas oryzae* pv. *oryzae*. Biol Control 59:114–122. https://doi.org/10.1016/j.biocontrol.2011.06.010

Chu Q, Sha Z, Nakamura T, Oka N, Osaki M, Watanabe T (2016) Differential responses of soybean and sorghum growth, nitrogen uptake, and microbial metabolism in the rhizosphere to cattle manure application: A rhizobox study. J Agric Food Chem 64:8084–8094. https://doi.org/10.1021/acs.jafc.6b03046

Clifton EH, Jaronski ST, Coates BS, Hodgson EW, Gassmann AJ, Reddy GVP (2018) Effects of endophytic entomopathogenic fungi on soybean aphid and identification of Metarhizium isolates from agricultural fields. PLoS One 13(3):e0194815

Contreras-Cornejo HA, Macías-Rodríguez L, Alfaro-Cuevas R, López- Bucio J (2014) *Trichoderma* spp. improve growth of Arabidopsis seedlings under salt stress through enhanced root development, osmolite production, and Na+ elimination through root exudates. Mol Plant-Microbe Interact 27:503–514. https://doi.org/10.1094/MPMI-09-13-0265-R

Da Rocha UN, Plugge CM, George I, van Elsas JD (2013) The rhizosphere selects for particular groups of *Acidobacteria* and *Verrucomicrobia*. PLoS One 8:e82443. https://doi.org/10.1371/journal.pone.0094514

Défago G, Berling C H, Burger U, Haas D, Kahr, G, Keel C, Voisard C, Wirthner P, Wüthrich B (1992) Suppression of black root rot of tobacco and other root diseases by strains of *Pseudomonas fluorescens*: potential applications and mechanisms. In: Hornby D (ed) Biological control of soil-borne plant pathogens. CAB Int, UK, pp 93–108. isbn:0851986374

Demain AL, Sanchez S (2009) Microbial drug discovery: 80 years of progress. J Antibiot 62:5–16. https://doi.org/10.1038/ja.2008.16

Dimkpa C, Weinand T, Asch F (2009) Plant-rhizobacteria interactions alleviate abiotic stress conditions. Plant Cell Environ 32:1682–1694. https://doi.org/10.1111/j.1365-3040.2009.02028.x

Döbereiner J, Day JM (1976) Associative symbioses in tropical grasses: characterization of microorganisms and dinitrogen-fixing sites. In Proceedings of First International Symposium on Nitrogen Fixation. Citeulike 2:518–538

Eastman AW, Weselowski B, Nathoo N, Yuan Z-C (2014) Complete genome sequence of *Paenibacillus polymyxa* CR1, a plant growth-promoting bacterium isolated from the corn rhizosphere exhibiting potential for biocontrol, biomass degradation, and biofuel production. Genome Announc 2:e01218–e01213. https://doi.org/10.1128/genomeA.01218-13

Edwards J, Johnson C, Santos-Medellín C, Lurie E, Podishetty NK, Bhatnagar S, Eisen JA, Sundaresan V (2015) Structure, variation, and assembly of the root-associated microbiomes of rice. PNAS 112:E911–E920. https://doi.org/10.1073/pnas.1414592112

Egamberdieva D, Wirth S, Behrendt U, Abd_Allah EF, Berg G (2016) Biochar treatment resulted in a combined effect on soybean growth promotion and a shift in plant growth promoting rhizobacteria. Front Microbiol 7:209. https://doi.org/10.3389/fmicb.2016.00209

El-Tarabily KA (2008) Promotion of tomato (Lycopersicon esculentum Mill.) plant growth by rhizosphere competent 1-aminocyclopropane-1-carboxylic acid deaminase-producing streptomycete actinomycetes. Plant Soil 308:161. https://doi.org/10.1007/s11104-008-9616-2

Fahrig L, Baudry J, Brotons L, Burel FG, Crist TO, Fuller RJ, Sirami C, Siriwardena GM, Martin JL (2011) Functional landscape heterogeneity and animal biodiversity in agricultural landscapes. Ecol Lett 14:101–112. https://doi.org/10.1111/j.1461-0248.2010.01559.x

Foesel BU, Mayer S, Luckner M, Wanner G, Rohde M, Overmann J (2016) *Occallatibacter riparius* gen. nov, sp. nov. and *O. savannae* sp. nov. two novel acidobacterial species isolated from Namibian soils and emended description of the family *Acidobacteriaceae*. Int J Syst Evol Microbiol 66:219–229. https://doi.org/10.1099/ijsem.0.000700

Foo JL, Ling H, Lee YS, Chang MW (2017) Microbiome engineering: current applications and its future. Biotechnol J 12:1600099. https://doi.org/10.1002/biot.201600099

Freire JRJ, Kolling I (1986) World culture collections of *Rhizobium*. MIRCEN J 2:231–235

Gallegos-Robles MA, Morales-Loredo A, Alvarez-Ojeda G, Vega PA, Chew MY, Velarde S, Fratamico P (2008) Identification of *Salmonella* serotypes isolated from cantaloupe and Chile pepper production systems in Mexico by PCR-restriction fragment length polymorphism. J Food Prot 71:2217–2222. https://doi.org/10.4315/0362-028X-71.11.2217

Giller KE (2001) Nitrogen fixation in tropical cropping systems. CAB Int, UK doi: https://doi.org/10.1079/9780851994178.000. isbn:0851994172

Glick BR (1995) The enhancement of plant growth by free-living bacteria. Can J Microbiol 41:109–117. https://doi.org/10.1139/m95-015

Glick BR, Bashan Y (1997) Genetic manipulation of plant growth promoting bacteria to enhance biocontrol of phytopathogens. Biotechnol Adv 15:353–378. https://doi.org/10.1016/S0734-9750(97)00004-9

Glick BR, Patten CL, Holguin G, Penrose DM (1999) Biochemical and genetic mechanisms used by plant growth promoting bacteria. Imperial College Press, London. pp 1–267. isbn:1-86094-152-4

Gomba A, Chidamba L, Korsten L (2017) Effect of postharvest practices including degreening on citrus carpoplane microbial biomes. J Appl Microbiol 122:1057–1070. https://doi.org/10.1111/jam.13396

Gopal M, Gupta A, Thomas GV (2013) Bespoke microbiome therapy to manage plant diseases. Front Microbiol 4:355. https://doi.org/10.3389/fmicb/2013.00355

Gulati A, Rahi P, Vyas P (2008) Characterization of phosphate-solubilizing fluorescent pseudomonads from the rhizosphere of seabuckthorn growing in the cold deserts of Himalayas. Curr Microbiol 56:73. https://doi.org/10.1007/s00284-007-9042-3

Gulati A, Sharma N, Vyas P, Sood S, Rahi P, Pathania V, Prasad R (2010) Production of organic acid and plant growth promotion as a function of phosphate solubilization by *Acinetobacter rhizosphaerae* from the cold deserts of the trans-Himalayas. Arch Microbiol 192:975. https://doi.org/10.1007/s00203-010-0615-3

Gulati A, Sood S, Rahi P, Thakur R, Chauhan S, Chawla I (2011) Diversity analysis of diazotrophic bacteria associated with the roots of tea (*Camellia sinensis* (L.) O. Kuntze). J Microbiol Biotechnol 21:545–555. https://doi.org/10.4014/jmb.1012.12022

Haas D, Défago G (2005) Biological control of soil-borne pathogens by fluorescent *pseudomonas*. Nat Rev Microbiol 3:307–319. https://doi.org/10.1038/nrmicro1129

Hacquard S, Garrido-Oter R, Gonza'lez A, Spaepen S, Ackermann G, Lebeis S, McHardy AC, Dang JL, Knight R, Ley R, Schulze-Lefert P (2015) Microbiota and host nutrition across plant and animal kingdoms. Cell Host Microbe 17:603–616. https://doi.org/10.1016/j.chom.2015.04.009

Hatfield JL, Prueger JH (2015) Temperature extremes: effect on plant growth and development. Weather Clim Extrem 10:4–10. https://doi.org/10.1016/j.wace.2015.08.001

Hayat R, Ali S, Amara U, Khalid R, Ahmed I (2010) Soil beneficial bacteria and their role in plant growth promotion: a review. Ann Microbiol 60:579. https://doi.org/10.1007/s13213-010-0117-1

Hendrickson JR, Hanson JD, Tanaka DL, Sassenrath GF (2008) Principles of integrated agricultural systems: introduction to processes and definition. Renewable Agric Food Syst 23:265–271. https://doi.org/10.1017/S1742170507001718

Hoffland E (1991) Mobilisation of rock phosphate by rape (*Brassica napus L.)* Ph.D. Thesis. Wageningen Agricultural University. The Netherlands, 93 pp

Howieson JG, Malden J, Yates RJ, O'Hara GW (2000) Techniques for the selection and development of elite inoculant rhizobial strains in southern Australia. Symbiosis 28:33–48 ISSN 0334-5114

Ibenyassine K, Mhand RA, Karamoko Y, Anajjar B, Chouibani MM, Ennaji M (2007) Bacterial pathogens recovered from vegetables irrigated by wastewater in Morocco. J Environ Health 69:47–51

Jaber LR, Enkerli J (2017) Fungal entomopathogens as endophytes: can they promote plant growth? Biocontrol Sci Tech 27:28–41. https://doi.org/10.1080/09583157.2016.1243227

Jaber LR, Ownley BH (2017) Can we use entomopathogenic fungi as endophytes for dual biological control of insect pests and plant pathogens? Biol Control 116:36–45

Jami E, Israel A, Kotser A, Mizrahi I (2013) Exploring the bovine rumen bacterial community from birth to adulthood. ISME J 7:1069–1079. https://doi.org/10.1038/ismej.2013.2

Jayaraman KS (2016) Microbe hunters concerned over 'research in peril'. Nat India. https://doi.org/10.1038/nindia.2016.62

Jetiyanon J, Kloepper JW (2002) Mixtures of plant growth-promoting rhizobacteria for induction of systemic resistance against multiple plant diseases. BioControl 24:285–291. https://doi.org/10.1016/S1049-9644(02)00022-1

Johnston-Monje D, Raizada MN (2011) Conservation and diversity of seed associated endophytes in Zea across boundaries of evolution, ethnography and ecology. PLoS One 6:e20396. https://doi.org/10.1371/journal.pone.0020396

Kabaluk JT, Ericsson JD (2007) Seed treatment increases yield of field corn when applied for wireworm control. Agron J 99(5):1377

Kaeberlein T, Lewis K, Epstein SS (2002) Isolating "uncultivable" microorganisms in pure culture in a simulated natural environment. Science 296:1127–1129. https://doi.org/10.1126/science.1070633

Kalam S, Das SN, Basu A, Podile AR (2017) Population densities of indigenous Acidobacteria change in the presence of plant growth promoting rhizobacteria (PGPR) in rhizosphere. J Basic Microbiol 57:376–385. https://doi.org/10.1002/jobm.201600588

Khare E, Arora NK (2010) Effect of indole-3-acetic acid (IAA) produced by *Pseudomonas aeruginosa* in suppression of charcoal rot disease of chickpea. Curr Microbiol 61:64. https://doi.org/10.1007/s00284-009-9577-6

Khedher SB, Boukedi H, Dammak M, Kilani-Feki O, Sellami-Boudawara T, Abdelkefi-Mesrati L, Tounsi S (2017) Combinatorial effect of *Bacillus amyloliquefaciens* AG1 biosurfactant and *Bacillus thuringiensis* Vip3Aa16 Toxin on *Spodoptera littoralis* larvae. J Invertebr Pathol 144:11–17. https://doi.org/10.1016/j.jip.2017.01.006

Kim BS, Lee JY, Hwang BK (2000) In vivo control and in vitro antifungal activity of rhamnolipid B, a glycolipid antibiotic, against *Phytophthora capsici* and *Colletotrichum orbiculare*. Pest Manag Sci 56:1029–1035. https://doi.org/10.1002/1526-4998(200012)

Kloepper JW, Schroth MW (1981) Plant growth promoting rhizobacteria under gnotobiotic conditions. Am Phytopathol Soc 71:642–644

Komárek J (2015) Review of the cyanobacterial genera implying planktic species after recent taxonomic revisions according to polyphasic methods: state as of 2014. Hydrobiologia 764:259–270. https://doi.org/10.1007/s10750-015-2242-0

Kumar A, Singh JS (2016) Microalgae and cyanobacteria biofuels: a sustainable alternate to crop-based fuels. In: Singh JS, Singh DP (eds) Microbes and environmental management. Studium Press Pvt. Ltd., New Delhi, pp 1–20

Lagier JC, Khelaifia S, Alou MT, Ndongo S, Dione N, Hugon P, Caputo A, Cadoret F, Traore SI, Seck EH, Dubourg G, Durand G, Mourembou G, Guilhot E, Togo A, Bellali S, Bachar D, Cassir N, Bittar F, Delerce J, Mailhe M, Ricaboni D, Bilen M, Dangui Nieko NP, Dia Badiane NM, Valles C, Mouelhi D, Diop K, Million M, Musso D, Abrahão J, Azhar EI, Bibi F, Yasir M, Diallo A, Sokhna C, Djossou F, Vitton V, Robert C, Rolain JM, La Scola B, Fournier PE, Levasseur A, Raoult D (2016) Culture of previously uncultured members of the human gut microbiota by culturomics. Nat Microbiol 1:16203. https://doi.org/10.1038/nmicrobiol.2016.203

Laranjoa M, Alexandrea A, Oliveira S (2014) Legume growth-promoting rhizobia: an overview on the *Mesorhizobium* genus? Microbiol Res 169:2–17. https://doi.org/10.1016/j.micres.2013.09.012

Lebeis SL, Rott M, Dangl JL, Schulze-Lefert P (2012) Culturing a plant microbiome community at the cross-Rhodes. New Phytol 196:341–344. https://doi.org/10.1111/j.1469-8137.2012.04336.x

Lewis K, Epstein S, D'Onofrio A, Ling LL (2010) Uncultured microorganisms as a source of secondary metabolites. J Antibiot 63:468–476. https://doi.org/10.1038/ja.2010.87

Ley RE, Hamady M, Lozupone C, Turnbaugh PJ, Ramey RR, Bircher JS, Schlegel ML, Tucker TA, Schrenzel MD, Knight R, Gordon JI (2008) Evolution of mammals and their gut microbes. Science 320:1647–1651. https://doi.org/10.1126/science.1155725

Liu Y, Kumar NR, Lai Q, Du J, Dobritsa AP, Samadpour M, Shao Z (2015) Identification of strains *Bacillus aerophilus* MTCC 7304T as *Bacillus altitudinis* and *Bacillus stratosphericus* MTCC 7305T as a *Proteus* sp. and the status of the species *Bacillus aerius* Shivaji et al. 2006. Request for an Opinion. Int J Syst Evol Microbiol 65:3228–3232. https://doi.org/10.1099/ijsem.0.000408

Liu H, Shao B, Long X, Yao Y, Meng Q (2016) Foliar penetration enhanced by biosurfactant rhamnolipid. Colloids Surf B Biointerfaces 145:548–554. https://doi.org/10.1016/j.colsurfb.2016.05.058

Lohrke SM, Orf JH, Sadowsky MJ (1996) Inheritance of host controlled restriction of nodulation by *Bradyrhizobium japonicum* strain USDA 110. Crop Sci 36:1271–1276. https://doi.org/10.2135/cropsci1996.0011183X003600050033x

Lopez DC, Zhu-Salzman K, Ek-Ramos MJ, Sword GA, Wilkinson TL (2014) The entomopathogenic fungal endophytes Purpureocillium lilacinum (Formerly Paecilomyces lilacinus) and Beauveria bassiana negatively affect cotton aphid reproduction under both greenhouse and field conditions. PLoS One 9(8):e103891

Lopez DC, Sword GA (2015) The endophytic fungal entomopathogens Beauveria bassiana and Purpureocillium lilacinum enhance the growth of cultivated cotton (Gossypium hirsutum) and negatively affect survival of the cotton bollworm (Helicoverpa zea). Biol Control 89:53–60

Mahenthiralingam E, Baldwin A, Dowson CG (2008) *Burkholderia cepacia* complex bacteria: opportunistic pathogens with important natural biology. J Appl Microbiol 104:1539–1551. https://doi.org/10.1111/j.1365-2672.2007.03706.x

Mamta RP, Pathania V, Gulati A, Singh B, Bhanwra RK, Tewari R (2010) Stimulatory effect of phosphate-solubilizing bacteria on plant growth, stevioside and rebaudioside-A contents of *Stevia rebaudiana* Bertoni. Appl Soil Ecol 46:222–229. https://doi.org/10.1016/j.apsoil.2010.08.008

Marx V (2017) Microbiology: the return of culture. Nat Methods 14:37–40. https://doi.org/10.1038/nmeth.4107

Mattarozzi M, Manfredi M, Montanini B, Gosetti F, Sanangelantoni AM, Marengo E, Careri M, Visioli G (2017) A metaproteomic approach dissecting major bacterial functions in the rhizosphere of plants living in serpentine soil. Anal Bioanal Chem 409:2327–2339. https://doi.org/10.1007/s00216-016-0175-8

McKinnon AC, Glare TR, Ridgway HJ, Mendoza-Mendoza A, Holyoake A, Godsoe WK, Bufford JL (2018) Detection of the entomopathogenic fungus Beauveria bassiana in the rhizosphere of wound-stressed Zea mays plants. Front Microbiol 9:1161

Mendes R, Raaijmakers JM (2015) Cross-kingdom similarities in microbiome functions. ISME J 9:1905–1907. https://doi.org/10.1038/ismej.2015.7

Mishra U, Pabbi S (2004) Cyanobacteria: a potential biofertilizer for rice. Resonance 9:6–10. https://doi.org/10.1007/BF02839213

Moldes AB, Paradelo R, Vecino X, Cruz JM, Gudina E, Rodrigues L, Teixeria JA, Domínguez JM, Barral MT (2013) Partial characterization of biosurfactant from *Lactobacillus pentosus* and comparison with sodium dodecyl sulphate for the bioremediation of hydrocarbon contaminated soil. Biomed Res Int 2013:1–26. https://doi.org/10.1155/2013/961842

Nejat N, Rookes J, Mantri NL, Cahill DM (2017) Plant-pathogen interactions: toward development of next-generation disease-resistant plants. Crit Rev Biotechnol 37:229–237. https://doi.org/10.3109/07388551.2015.1134437

Newton AC, Fitt BDL, Atkins SD, Walters DR, Daniell TJ (2010) Pathogenesis, parasitism and mutualism in the trophic space of microbe-plant interactions. Trends Microbiol 18:365–373. https://doi.org/10.1016/j.tim.2010.06.002

Nieto-Jacobo MF, Steyaert JM, Salazar-Badillo FB, Nguyen DV, Rostás M, Braithwaite M, De Souza JT, Jimenez-Bremont JF, Ohkura M, Stewart A, Mendoza-Mendoza A (2017) Environmental growth conditions of *Trichoderma* spp. affects indole acetic acid derivatives, volatile organic compounds, and plant growth promotion. Front Plant Sci 8:102. https://doi.org/10.3389/fpls.2017.00102

Okubara PA, Bonsall RF (2008) Accumulation of *Pseudomonas*-derived 2,4-diacetylphloroglucinol on wheat seedling roots is influenced by host cultivar. BioControl 46:322–331. https://doi.org/10.1016/j.biocontrol.2008.03.013

Oliveira DWF, Franca IWL, Felix AKN, Martins JJL, Giro MEA, Melo VMM (2013) Kinetic study of biosurfactants production by *Bacillus subtilis* LAMI005 grown in clarified cashew apple juice. Colloids Surf B Biointerfaces 101:34–43. https://doi.org/10.1016/j.colsurfb.2012.06.011

Oliveira RS, Carvalho P, Marques G, Ferreira L, Nunes M, Rocha I, Ma Y, Carvalho MF, Vosátka M, Freitas H (2017) Increased protein content of chickpea (*Cicer arietinum L.)* inoculated with arbuscular mycorrhizal fungi and nitrogen-fixing bacteria under water deficit conditions. J Sci Food Agric 97:4379–4385. https://doi.org/10.1002/jsfa.8201

Oomen GJM, Lantinga EA, Goewie EA, Van Der Hoek KW (1998) Mixed farming systems as a way towards a more efficient use of nitrogen in European Union agriculture. Environ Pollut 102:6697–6704. https://doi.org/10.1016/S0269-7491(98)80101-2

Overmann J, Smith D (2017) Microbial resource centers contribute to bioprospecting of bacteria and filamentous microfungi. In: Paterson R, Lima N (eds) Bioprospecting. Topics in Biodiversity and Conservation, vol 16. Springer, Cham. https://doi.org/10.1007/978-3-319-47935-4_4

Pacwa-Plociniczak M, Plaza GA, Piotrowska-Seget Z, Cameotra SS (2011) Environmental applications of biosurfactants: recent advances. Int J Mol Sci 12:633–654. https://doi.org/10.3390/ijms12010633

Pei C, Mi C, Sun L, Liu W, Li O, Hu X (2017) Diversity of endophytic bacteria of *Dendrobium officinale* based on culture-dependent and culture-independent methods. Biodiversity Ecosys 31:112–119. https://doi.org/10.1080/13102818.2016.1254067

Prakash O, Shouche Y, Jangid K, Kostka JE (2013) Microbial cultivation and the role of microbial resource centers in the omics era. Appl Microbiol Biotechnol 97:51–62. https://doi.org/10.1007/s00253-012-4533-y

Pugliese M, Liu B, Gullino ML, Garibaldi A (2011) Microbial enrichment of compost with biological control agents to enhance suppressiveness to four soil-borne diseases in greenhouse. J Plant Dis Protect 118:45–50. https://doi.org/10.1007/BF03356380

Rahi P (2017) Phytomicrobiome: a reservoir for sustainable agriculture. In: Kalia V, Shouche Y, Purohit H, Rahi P (eds) Mining of microbial wealth and MetaGenomics. Springer, Singapore. https://doi.org/10.1007/978-981-10-5708-3_7

Rahi P, Vyas P, Sharma S, Gulati A, Gulati A (2009) Plant growth promoting potential of the fungus *Discosia* sp. FIHB 571 from tea rhizosphere tested on chickpea, maize and pea. Indian J Microbiol 49:128–133. https://doi.org/10.1007/s12088-009-0026-9

Rahi P, Kapoor R, Young JPW, Gulati A (2012) A genetic discontinuity in root-nodulating bacteria of cultivated pea in the Indian trans-Himalayas. Mol Ecol 21:145–159. https://doi.org/10.1111/j.1365-294X.2011.05368.x

Rahi P, Sharma OP, Shouche YS (2016) Matrix-assisted laser desorption/ionization time-of-flight mass-spectrometry (MALDI-TOF MS) based microbial identifications: challenges and scopes for microbial ecologists. Front Microbiol 7:1359. https://doi.org/10.3389/fmicb.2016.01359

Rajaniemi P, Hrouzek P, Kastovská K, Willame R, Rantala A, Hoffmann L, Komárek J, Sivonen K (2005) Phylogenetic and morphological evaluation of the genera Anabaena, Aphanizomenon, Trichormus and Nostoc (Nostocales, Cyanobacteria). Int J Syst Evol Microbiol 55:11–26. https://doi.org/10.1099/ijs.0.63276-0

Ramírez-Puebla ST, Servín-Garcidueñas LE, Jiménez-Marín B, Bolaños LM, Rosenblueth M, Martínez J, Rogel MA, Ormeño-Orrillo E, Martínez-Romero E (2013) Gut and root microbiota commonalities. Appl Environ Microbiol 79:2–9. https://doi.org/10.1128/AEM.02553-12

Rascovan N, Carbonetto B, Perrig D, Díaz M, Canciani W, Abalo M, Alloati J, González-Anta G, Vazquez MP (2016) Integrated analysis of root microbiomes of soybean and wheat from agricultural fields. Sci Rep 6:28084. https://doi.org/10.1038/srep28084

Rekadwad BN, Khobragade CN (2017) Microbial biofilm: role in crop productivity. In: Kalia V (eds) Microbial applications, vol. 2. Springer, Cham, pp 107–118. doi:https://doi.org/10.1007/978-3-319-52669-0_5 isbn:978-3-319-52669-0

Rinu K, Sati P, Pandey A (2014) *Trichoderma gamsii* (NFCCI 2177): A newly isolated endophytic, psychrotolerant, plant growth promoting, and antagonistic fungal strain. J Basic Microbiol 54:408–417. https://doi.org/10.1002/jobm.201200579

Rodríguez H, Vessely S, Shah S, Glick BR (2008) Effect of a nickel-tolerant ACC deaminase-producing *Pseudomonas* strain on growth of nontransformed and transgenic canola plants. Curr Microbiol 57:170. https://doi.org/10.1007/s00284-008-9181-1

Rosenblueth M, Martínez-Romero E (2006) Bacterial endophytes and their interactions with hosts. Mol Plant-Microbe Interact 19:827–837. https://doi.org/10.1094/MPMI-19-0827

Ruth D, Sven R, Hans H, Christel V, Joost J, Kristien B, Maarten F, Serge B, Christophe H, Ivo L, Drik EDV, Jos V, Jan V, Jan M (2006) Quorum signal molecules as biosurfactants affecting swarming in *Rhizobium etli*. Proc Natl Acad Sci U S A 103:14965–14970. https://doi.org/10.1073/pnas.0511037103

Saha S, Roy RN, Sen SK, Ray AK (2006) Characterization of cellulase-producing bacteria from the digestive tract of tilapia, *Oreochromis mossambica* (Peters) and grass carp, *Ctenopharyngodon idella* (Valenciennes). Aquac Res 37:380–388. https://doi.org/10.1111/j.1365-2109.2006.01442.x

Senni K, Pereira J, Gueniche F, Delbarre-Ladrat C, Sinquin C, Ratiskol J, Godeau G, Fischer AM, Helley D, Sylvia CJ (2011) Marine polysaccharides: a source of bioactive molecules for cell therapy and tissue engineering. Mar Drugs 9:1664–1681. https://doi.org/10.3390/md9091664

Sha Y, Wang Q, Li Y (2016) Suppression of *Magnaporthe oryzae* and interaction between *Bacillus subtilis* and rice plants in the control of rice blast. Springer Plus 5:1238. https://doi.org/10.1186/s40064-016-2858-1

Sharma S, Sharma S, Singh RK, Vaishampayan A (2008) Colonization behavior of bacterium *Burkholderia cepacia* inside the *Oryza sativa* roots visualized using green fluorescent protein reporter. World J Microbiol Biotechnol 24:1169. https://doi.org/10.1007/s11274-007-9589-y

Sharma SK, Saini S, Verma A, Sharma PK, Lal R, Roy M, Singh UB, Saxena AK, Sharma AK (2017) National agriculturally important microbial culture collection in the global context of microbial culture collection centres. Proc Natl Acad Sci India Sect B Biol Sci. https://doi.org/10.1007/s40011-017-0882-8

Shoebitz M, Ribaudo CM, Pardo MA, Cantore ML, Ciampi L, Curá JA (2009) Plant growth promoting properties of a strain of *Enterobacter ludwigii* isolated from *Lolium perenne* rhizosphere. Soil Biol Biochem 41:1768–1774. https://doi.org/10.1016/j.soilbio.2007.12.031

Siddikee MA, Chauhan PS, Anandham R, Han GH, Sa T (2010) Isolation, characterization, and use for plant growth promotion under salt stress, of ACC deaminase-producing halotolerant bacteria derived from coastal soil. J Microbiol Biotechnol 20:1577–1584. https://doi.org/10.4014/jmb.1007.07011

Singh BK, Trivedi P (2017) Microbiome and the future for food and nutrient security. Microb Biotechnol 10:50–53. https://doi.org/10.1111/1751-7915.12592

Singh JS, Kumar A, Rai AN, Singh DP (2016) Cyanobacteria: a precious bio-resource in agriculture, ecosystem, and environmental sustainability. Front Microbiol 7:529. https://doi.org/10.3389/fmicb.2016.00529

Slicher van Bath BH, Ordish O (1963) The Agrarian History of Western Europe, A.D. 500–1850, translated by O. Ordish. Arnold London

Smith FA, Smith SE (2011) What is the significance of the arbuscular mycorrhizal colonisation of many economically important crop plants? Plant Soil 348:63–79. https://doi.org/10.1007/s11104-011-0865-0

Sprent JI, Ardley J, James EK (2017) Biogeography of nodulated legumes and their nitrogen-fixing symbionts. New Phytol 215:40–56. https://doi.org/10.1111/nph.14474

Stackebrandt E, Smith D, Casaregola S, Varese GC, Verkleij G, Lima N, Bridge P (2014) Deposit of microbial strains in public service collections as part of the publication process to underpin good practice in science. Springer Plus 3:208. https://doi.org/10.1186/2193-1801-3-208

Stewart EJ (2012) Growing unculturable bacteria. J Bacteriol 194(16):4151–4160

St. Leger RJ (2008) Studies on adaptations of Metarhizium anisopliae to life in the soil. J Invertebr Pathol 98(3):271–276

Sullivan JT, Patrick HN, Lowther WL, Scott DB, Ronson CW (1995) Nodulating strains of *Rhizobium loti* arise through chromosomal symbiotic gene transfer in the environment. Proc Natl Acad Sci U S A 92:8985–8989

Teplitski M, Warriner K, Bartz J, Schneider KR (2011) Untangling metabolic and communication networks: interactions of enterics with phytobacteria and their implications in produce safety. Trends Microbiol 19:121–127. https://doi.org/10.1016/j.tim.2010.11.007

Tilman D, Socolow R, Foley JA, Hill J, Larson E, Lynd L, Pacala S, Reilly J, Searchinger T, Somerville C, Williams R (2009) Beneficial biofuels- the food, energy, and environment trilemma. Science 325:270–271. https://doi.org/10.1126/science.1177970

Timmers RA, Rothballer M, Strik DP, Engel M, Schulz S, Schloter M, Hartmann A, Hamelers B, Buisman C (2012) Microbial community structure elucidates performance of *Glyceria maxima* plant microbial fuel cell. Appl Microbiol Biotechnol 94:537–548. https://doi.org/10.1007/s00253-012-3894-6

Tindall BJ, Garrity GM (2008) Proposals to clarify how type strains are deposited and made available to the scientific community for the purpose of systematic research. Int J Syst Evol Microbiol 58:1987–1990. https://doi.org/10.1099/ijs.0.2008/006155-0

Tran H, Kruijt M, Raaijmakers JM (2008) Diversity and activity of biosurfactant-producing *Pseudomonas* in the rhizosphere of black pepper in Vietnam. J Appl Microbiol 104:839–851. https://doi.org/10.1111/j.1365-2672.2007.03618.x

Tringe SG, Hugenholtz P (2008) A renaissance for the pioneering 16S rRNA gene. Curr Opin Microbiol 11:442–446. https://doi.org/10.1016/j.mib.2008.09.011

Trivedi P, Delgado-Baquerizo M, JeffriesT C, Trivedi C, Anderson IC, McNee M, Flower K, Pal Singh B, Minkey D, Singh BK (2017) Soil aggregation and associated microbial communities modify the impact of agricultural management on carbon content. Environ Microbiol 19:3070–3086. https://doi.org/10.1111/1462-2920.13779

Tsiafouli MA, Thébault E, Sgardelis SP, de Ruiter PC, van der Putten WH, Birkhofer K, Hemerik L, de Vries FT, Bardgett RD, Brady MV, Bjornlund L, Jørgensen HB, Christensen S, Hertefeldt TD, Hotes S, Gera Hol WH, Frouz J, Liiri M, Mortimer SR, Setälä H, Tzanopoulos J, Uteseny K, Pižl V, Stary J, Wolters V, Hedlund K (2015) Intensive agriculture reduces soil biodiversity across Europe. Glob Change Biol 21:973–985. https://doi.org/10.1111/gcb.12752

Upadhyay SK, Singh DP, Saikia R (2009) Genetic diversity of plant growth promoting rhizobacteria isolated from rhizospheric soil of wheat under saline condition. Curr Microbiol 59:489–496. https://doi.org/10.1007/s00284-009-9464-1

Uzoigwe C, Burgess JG, Ennis CJ, Rahman PKSM (2015) Bioemulsifiers are not biosurfactants and require different screening approaches. Front Microbiol 6:245. https://doi.org/10.3389/fmicb.2015.00245

Van Lenteren JC, Bolckmans K, Köhl J, Ravensberg WJ, Urbaneja A (2017) BioControl using invertebrates and microorganisms: plenty of new opportunities. BioControl 63:39–59. https://doi.org/10.1007/s10526-017-9801-4

Vandenkoornhuyse P, Mahe S, Ineson P, Staddon P, Ostle N, Cliquet JB, Francez AJ, Fitter AH, Young JPW (2007) Active root inhabiting microbes identified by rapid incorporation of plant-derived carbon into RNA. Proc Natl Acad Sci U S A 104:16970–16975. https://doi.org/10.1073/pnas.0705902104

Vega FE, Goettel MS, Blackwell M, Chandler D, Jackson MA, Keller S, Koike M, Maniania NK, Monzón A, Ownley BH, Pell JK, Rangel DEN, Roy HE (2009) Fungal entomopathogens: new insights on their ecology. Fungal Ecol 2(4):149–159

Vorholt JA (2012) Microbial life in the phyllosphere. Nat Rev Microbiol 10:828–840. https://doi.org/10.1038/nrmicro2910

Vurukonda SSKP, Vardharajula S, Shrivastava M, Ali S k Z (2016) Enhancement of drought stress tolerance in crops by plant growth promoting rhizobacteria. Microbiol Res 184:13–24. https://doi.org/10.1016/j.micres.2015.12.003

Vyas P, Rahi P, Chauhan A, Gulati A (2007) Phosphate solubilization potential and stress tolerance of *Eupenicillium parvum* from tea soil. Mycol Res 111:931–938. https://doi.org/10.1016/j.mycres.2007.06.003

Vyas P, Joshi R, Sharma KC, Rahi P, Gulati A, Gulati A (2010) Cold-adapted and rhizosphere-competent strain of *Rahnella* sp. with broad-spectrum plant growth-promotion potential. J Microbiol Biotechnol 20:1724–1734. https://doi.org/10.4014/jmb.1007.07030

Wang B, Wang Q, Liu W, Liu X, Hou J, Teng Y, Luo Y, Christie P (2017) Biosurfactant-producing microorganism *Pseudomonas* sp. SB assists the phytoremediation of DDT-contaminated soil by two grass species. Chemosphere 182:137–142. https://doi.org/10.1016/j.chemosphere.2017.04.123

White RA, Borkum MI, Rivas-Ubach A, Bilbao A, Wendler JP, Colby SM, Köberl M, Jansson C (2017) From data to knowledge: the future of multi-omics data analysis for the rhizosphere. Rhizosphere 3:222–229. https://doi.org/10.1016/j.rhisph.2017.05.001

Whitton BA (2012) Ecology of Cyanobacteria II. Springer, Netherlands, p 533. https://doi.org/10.1007/978-94-007-3855-3

Williams A, Manoharan L, Rosenstock NP, Olsson PA, Hedlund K (2017) Long-term agricultural fertilization alters arbuscular mycorrhizal fungal community composition and barley (*Hordeum vulgare*) mycorrhizal carbon and phosphorus exchange. New Phytol 213:874–885. https://doi.org/10.1111/nph.14196

Wolfe BE (2018) Using cultivated microbial communities to dissect microbiome assembly: challenges, limitations, and the path ahead. mSystems 3(2):e00161-17

Wolińska A, Kuźniar A, Zielenkiewicz U, Banach A, Izak D, Stępniewska Z, Błaszczyk M (2017) Metagenomic analysis of some potential nitrogen-fixing bacteria in arable soils at different formation processes. Microb Ecol 73:162. https://doi.org/10.1007/s00248-016-0837-2

Xie X, Zhang H, Paré PW (2009) Sustained growth promotion in Arabidopsis with long-term exposure to the beneficial soil bacterium *Bacillus subtilis* (GB03). Plant Signal Behav 4:948–953. https://doi.org/10.4161/psb.4.10.9709

Yang SY, Lim DJ, Noh MY, Kim JC, Kim YC, Kim IS (2017) Characterization of biosurfactants as insecticidal metabolites produced by *Bacillus subtilis* Y9. Entomol Res 47:55–59. https://doi.org/10.1111/1748-5967.12200

Yun H-G, Kim D-J, Gwak W-S, Shin T-Y, Woo S-D (2018) Entomopathogenic fungi as dual control agents against both the pest and phytopathogen. Mycobiology 45(3):192–198

Zengler K, Toledo G, Rappe M, Elkins J, Mathur EJ, Short JM, Keller M (2002) Cultivating the uncultured. Proc Natl Acad Sci U S A 99:15681–15686. https://doi.org/10.1073/pnas.252630999

Zhang H, Xie X, Kim MS, Kornyeyev DA, Holaday S, Paré PW (2008) Soil bacteria augment Arabidopsis photosynthesis by decreasing glucose sensing and abscisic acid levels. Plant J 56:264–273. https://doi.org/10.1111/j.1365-313X.2008.03593.x

Zhang C, Wang S, Yan Y (2011) Isomerization and biodegradation of beta-cypermethrin by *Pseudomonas aeruginosa* CH7 with biosurfactant production. Bioresour Technol 102:7139–7146. https://doi.org/10.1016/j.biortech.2011.03.086

Zhang L, Xu M, Liu Y, Zhang F, Hodge A, Feng G (2016) Carbon and phosphorus exchange may enable cooperation between an arbuscular mycorrhizal fungus and a phosphate-solubilizing bacterium. New Phytol 210:1022–1032. https://doi.org/10.1111/nph.13838

Zhao P, Quan C, Wang Y, Wang J, Fan S (2014) *Bacillus amyloliquefaciens* Q-426 as a potential biocontrol agent against *Fusarium oxysporum* f. sp. *spinaciae*. J Basic Microbiol 54:448–456. https://doi.org/10.1002/jobm.201200414

Zhou W, Starr JL, Krumm JL, Sword GA, Sessitsch A (2016) The fungal endophyte negatively affects both above and belowground herbivores in cotton. FEMS Microbiol Ecol 92(10):fiw 158

Chapter 6
Rhizobia: Culture Collections, Identification, and Methods of Preservation

Manvika Sahgal and Vandana Jaggi

Abstract Across the globe nitrogen is limiting primary productivity. Although fertilizer nitrogen could supplement soil nitrogen depletion, efforts should be to augment soil biological nitrogen fixation mediated by microbes. This could be addressed by conservation and sustainable use of rhizobia that can fix nitrogen in soil in association with legumes. Rhizobia commonly occur in soils but often fail to produce effective nodulation either because their population in soil is low or those present cannot effectively nodulate the particular legume. Rhizobia present in various ecosystems are considerably diverse; at present 130 species within 15 genera are reported. Rhizobial inoculation is almost always needed when certain new leguminous crops are introduced to new areas or when ineffective and incompatible rhizobia are present in the soils. For this, depository of authentic microbial gene pool is a must. Culture collection centers act as repository of valuable microbial strains. In the changing global scenario, these centers are becoming Biological Resource Centres (BRCs), carry out research, enhance value of strains, and control access to dangerous microorganisms. For adoption and popularizing use of rhizobium-legume technology, their isolation, identification, preservation, and deposition in biological banks are imminent. Moreover, there is a need to identify the compatible rhizobium-legume symbioses for sustainable agriculture.

6.1 Introduction

The major challenge of modern agriculture is to achieve a food production level sufficient to feed the rapidly increasing global population. Current world population of 7.6 billion is projected to reach 8.6 billion by 2030, 9.8 billion in 2050, and 11.2 billion in 2100 (Anonymous 2017). Half of this population will be shared by nine countries: India, Nigeria, Congo, Pakistan, Ethiopia, Tanzania, the United States,

M. Sahgal (✉) · V. Jaggi
Department of Microbiology, G. B. Pant University of Agriculture and Technology, Pantnagar, Uttarakhand, India
e-mail: m_sahgal@gbpuat-cbsh.ac.in

S. K. Sharma, A. Varma (eds.), *Microbial Resource Conservation*, Soil Biology 54,
https://doi.org/10.1007/978-3-319-96971-8_6

Uganda, and Indonesia. Hence, India will be the most populated country. The other challenges for increasing food production are the sharp decline in cultivable land area and the increased deterioration of soil health and environmental quality as a result of intense agriculture practices, particularly since the green revolution of the 1960–1970s (Tilman et al. 2001; Trewavas 2001). High agriculture productivity will put immense pressure on soil nitrogen (N) and phosphorous (P) demands. Human N requirements are fulfilled directly or indirectly from plants. Plants acquire N from (a) soil, through commercial fertilizers, manure, and/or mineralization of organic matter; (b) from the atmosphere through biological N_2 fixation; and (c) minor contributions from other biological processes. However, since the green revolution, management of N inputs into agricultural system has become a contentious issue. It is already known that cultivation of legumes can enhance N acquisition and use. However with the goal of enhanced grain production, increased awareness of organic production, health issues, importance of biodiversity, and maintenance of soil fertility, the role of legumes in the sustainable management of N, as well as of P, is reaffirmed. It has been estimated that properly managed alfalfa (*Medicago sativa* L.) in rotation with corn (*Zea mays* L.) will reduce fertilizer inputs by up to 25%. Further legume green manures can replace more than 100 kg N ha^{-1} for the subsequent crop (Peterson and Russelle 1991). The use of legumes accompanied with good agronomic practices could save up to 20 TgN $year^{-1}$. Legumes can be used for reclaiming acid soils low in P and N, by the incorporation of residues, resulting in higher soil organic matter content and increased P and N availability. Legumes with low harvest index also improve P and N sustainability in low technified agriculture. Intercropping of pigeon pea (*Cajanus cajan*) with cereals is a proven management strategy for increasing N and P. Thus, in addition to providing an immediate source of dietary N, incorporation of pigeon pea residues after seed harvest increases the availability of P and N, contributing to agriculture sustainability. In another example, the development of rhizobial inoculants for low fertility acidic soils of the Brazilian Cerrado has allowed Brazil to become a leading exporter of soybeans.

In India research on legume crops is being undertaken through All India Coordinated Research Projects (AICRP) of Indian Council of Agricultural Research (ICAR). There are five AICRPs undertaking research on pigeon pea, arid legumes, groundnut, soybean, and MULLaRP [mung bean (*Vigna radiata*), urad bean (*Vigna mungo*), lentil (*Lens culinaris*), lathyrus (*Lathyrus sativus*), rajma (*Phaseolus vulgaris*), pea (*Pisum sativum*)]. Under the aegis of AICRPs, there is large gene pool of rhizobia specific to chickpea, lentil, and other legumes from various agroclimatic regions in India which have been screened for high N_2-fixing efficacy and various plant growth-promoting attributes (www.aicrpchickpea.res.in; www.icar.org.in/content/aicrps-network-projects). However, there are no records to prove the submission of these rhizobial isolates in culture collection centers. Similarly Centre for Research on Bacteria and Archaea (CRBA) operational at the Department of Microbiology, Govind Ballabh Pant University of Agriculture and Technology, Pantnagar, India, as part of the Ministry of Environment and Forests funded All India Coordinated Project on Taxonomy during 2000–2012 have characterized and identified rhizobia from nodules of tree (*Dalbergia sissoo*), medicinal

(*Mucuna pruriens*), pulse legumes (*Lens culinaris*, *Cicer arietinum*), and spice legume (*Trigonella foenum-graecum*). A few rhizobia nodulating *Dalbergia sissoo* have been deposited at the culture collection center at IMTECH, Chandigarh, India (Sahgal et al. 2004). The development of high-quality rhizobial inoculants is not possible unless the authentic rhizobial strains are available for advanced research, efficacy trial in fields, and for industrial applications. Thus ex situ conservation of microbial wealth is essential under Convention on Biological Diversity (CBD). Hence, establishment of culture collection centers is imminent (Smith 2003).

6.2 Global Directory of Culture Collection

The culture collection centers around the globe are registered under the World Federation of Culture Collection (WFCC) founded in 1963 and based at University of Queensland, Australia. In 1972, it published the first printed volume of world directory of culture collection centers of microorganisms. WFCC has been relocated in 1986 to RIKEN, Saitama, Japan, and again in 1999 to the National Institute of Genetics, Japan (Smith 2003; Sly 2010). All the microbial resource centers registered under WFCC are networked under WFCC-MIRCEN World Data Centre for Microorganisms (WDCM) (www.wdcm.org). WDCM constitutes the *World Directory of Culture Collections* (sixth version, 2014) and has various constituent databases such as CCINFO http://www.wfcc.info/ccinfo/ and STRAIN http://www.wfcc.info/ccinfo/search/strain_search/. Culture Collections Information Worldwide (CCINFO) is the directory of all registered culture collection in world, whereas STRAIN is the database that includes the list of holdings from registered culture collections. CCINFO allows the search through culture collection centers, whereas STRAIN allows the search through name and type of strain. Nowadays, there are 728 culture collection centers registered globally. They hold 29,63,164 microbial cultures, of which 12,20,838 represent bacteria, 8,14,206 fungi 38,002 viruses and 32,128 cell lines (www.wfcc.info/ccinfo/).

6.3 Global Catalogue of Microorganisms

The global catalogue of microorganisms (GCM) is the database of microbial culture deposits in the collection centers worldwide (gcm.wfcc.info). The main objective of the GCM is to develop a data bank of information on biological holdings of registered culture collection centers (s). It provides the description of microbial strains and a friendly user system to manage and disseminate the information. Besides, it also provides the information of strains to scientific and industrial communities for their usage. All WFCC members are required to publish online or a printed catalogue of microbial holdings with them. However, as per recent statistics, only 1/6 of the total collection centers registered under CCINFO have published their online catalogue. Nonpublication of an online catalogue by all WFCC members greatly hinders the visibility and accessibility of strains. Only 112 culture collections representing 43 countries or regions have participation in the GCM (gcm.wfcc.info). From these, only 77 collection centers have rhizobial culture deposits (Fig. 6.1).

6.3.1 Rhizobial Culture Collections: Past to Present

There is increased interest in *Rhizobium* research because of its symbiotic biological nitrogen fixation (BNF). The research on rhizobia and BNF is almost 125 years old. During this period numerous rhizobial strains have been identified and tested for nitrogen fixing and plant growth-promoting potential. In spite of being researched since long, rhizobium/legume technology is not widely used as yet. The major constraint has been availability of well-characterized and identified authentic cultures. Here comes the importance and role of culture collection centers. They are depository of authentic rhizobial strains available for advanced research and application, as patent deposits and as microorganisms cited in research publications for confirmation of results.

The status and progress of rhizobial culture collection centers have been reviewed by Freire and Kolling (1986). The first rhizobial culture collection was established as early as 1889. Five decades later O.N. Allen established the earliest and largest rhizobial culture repository at Department of Bacteriology, University of Wisconsin, Australia (Johnson and Allen 1952). Thereafter several other collection centers were established in Australia. They are Commonwealth Scientific and Industrial Research Organization (CSIRO) Cunningham Laboratory, Brisbane (https://www.csiro.au/en/Publications), and Australian Inoculant Research and Control Service, New South Wales Department of Agriculture, Melbourne (http://www.dpi.nsw.gov.au/). Several large rhizobial culture collections are present in South America mainly in Brazil, Uruguay, and Argentina. In Europe, the largest collections of rhizobium are in Belgium, Bulgaria, Czechoslovakia, Poland, and the United Kingdom. In Asia, China has the largest repository of rhizobia at Beijing Agriculture University. The first edition of the World Catalogue of Rhizobium Culture Collection (1973) listed 3000 strains from 59 collection centers spread over 29 countries, whereas its second edition (1983) listed the same number of strains from 64 repositories in 38 countries. The online printed catalogue is available for 112 collection centers, of which only 77 centers have rhizobial strain deposits (gcm.wfcc.info). Till today 15 genera representing rhizobia have been described. They are *Aminobacter*, *Azorhizobium*, *Bradyrhizobium*, *Burkholderia*, *Cupriavidus*, *Devosia*, *Ensifer* (earlier *Sinorhizobium*) *Mesorhizobium*, *Methylobacterium*, *Microvirga*, *Neorhizobium*, *Ochrobactrum*, *Phyllobacterium*, *Rhizobium*, and *Shinella* (Weir 2016). Within 15 genera, a total of 130 species of rhizobia have been identified (Table 6.1). World over all rhizobial culture collection centers have several strains that have only genus designation. They are not yet identified at the species level. The majority of culture collections, of the 77 with online published catalogue, have deposits designated as *Bradyrhizobium* sp.,

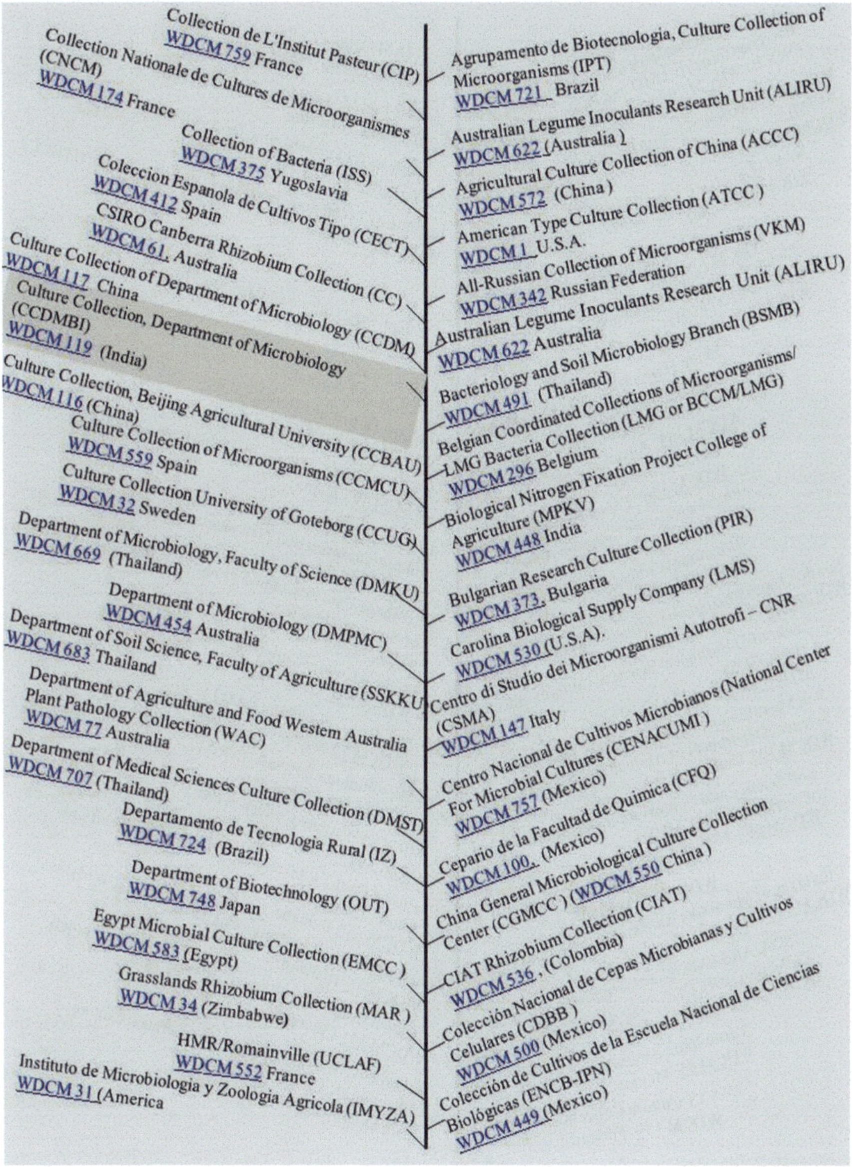

Fig. 6.1 The list of rhizobial culture collections that have participation in global catalogue of microorganisms and published online printed catalogue

Mesorhizobium sp., and *Rhizobium* sp. They have not been assigned species designation. The strains have been assigned to genus on the basis of alignment of their 16S rDNA region with that of completely identified strains. To identify species of these strains, a set of housekeeping genes (two to five) and symbiotic genes (*nod*, *nif*, and *fix*) must be sequenced and analyzed. The sequence analysis of these genes is used to infer

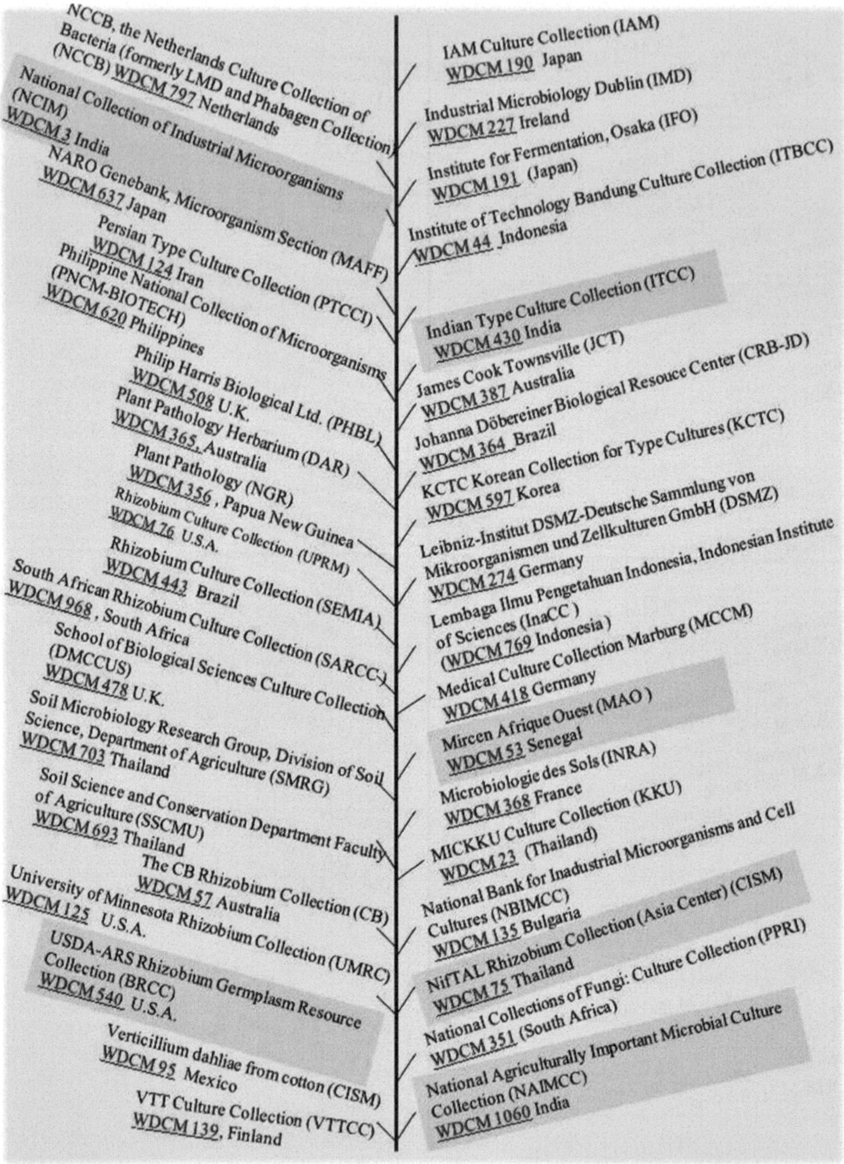

Fig. 6.1 (continued)

the phylogenetic relationship. In these culture collection centers, several strains are still designated by their former names, such as *R. japonicum*, *R. meliloti*, *R. phaseoli*, *R. trifolii*, *R. fredii*, *R. loti*, *and R. viceae*, although they have been reassigned to different groups.

Table 6.1 Genera of nodulating bacteria along with species described within each genus

Genus	Species within the genus
Aminobacter	This genus contains only one species *Aminobacter anthyllidis*
Azorhizobium	This genus currently contains two species. *Azorhizobium caulinodans, A. doebereinerae*
Bradyrhizobium	This genus contains 10 species. *Bradyrhizobium canariense, B. cytisi, B. denitrificans, B. elkanii, B. iriomotense, B. japonicum, B. jicamae, B. liaoningense, B. pachyrhizi, B. yuanmingense.* They are slow growing bacteria
Devosia	This genus contains only one species. *Devosiane ptuniae*, fast growing rhizobia
Mesorhizobium	This genus contains 31 species. *Mesorhizobium abyssinicae, M. albiziae, M. alhagi, M. amorphae, M. australicum, M. camelthorni, M. caraganae, M. chacoense, M. ciceri, M. erdmanii, M. gobiense, M. hawassense, M. huakuii, M. jarvis*ii, *M. loti, M. mediterraneum, M. metallidurans, M. muleiense, M. opportunistum, M. plurifarium, M. qingshengii, M. robiniae, M. sangaii, M. shangrilense, M. shonense, M. silamurunense, M. septentrionale, M. tamadayense, M. tarimense, M. temperatum and M. tianshanense*
Methylobacterium	This genus contains only one species *Methylobacterium nodulan.* Colonies 0.5–1 mm diameter after 4–7 days incubation
Microvirga	This genus contains only three species. *Microvirga lupine, M.lotononidis* and *M. zambiensis*
Neorhizobium	This genus contains only one species *Neorhizobium galegae*
Ochrobactrum	This genus contains two species. *Ochrobactrum cytisi* and *O. lupine*
Phyllobacterium	This genus contains three species. *Phyllobacterium trifolii, P. ifriqiyense* and *P. leguminum*
Rhizobium	This genus contains 49 species namely *Rhizobium alamii, R. alkalisoli, R. azibense, R.calliandrae, R.cauense, R. cellulosilyticum, R.daejeonense, R.endophyticum, R. etli, R. fabae, R. freirei, R. galegae, R. gallicum, R. giardinii, R. grahamii, R. hainanense, R. halophytocola, R. herbae, R. huautlense, R. indigoferae, R.jaguaris, R. laguerreae, R. leguminosarum, R. leucaenae, R. loessense, R. lusitanum, R.mayense, R. mesoamericanum, R.mesosinicum, R. miluonense, R.mongolense, R. multihospitium, R. oryzae, R. paranaense, R. petrolearium, R. phaseoli, R. pisi, R. tibeticum, R. sophorae, R. sophoriradicis, R. sphaerophysae, R.sullae, R. taibaishanense, R. tropici, R. tubonense, R. undicola, R. vallis, R. vignae* and *R. yanglingense.* They are fast growing bacteria
Ensifer (Sinorhizobium)	This genus contains 17 species. *Ensifer abri, E. americanum, E. arboris, E. fredii, E.garamanticus, E.indiaense, E. kostiensis, E. kummerowiae, E. medicae, E. meliloti, E. mexicanus, E. morelense', E. adhaerens, E. numidicus, E.saheli, sojae* and *E. terangae*
Shinella	This genus contains one species. *Shinella kummerowiae*
Burkholderia	This genus contains seven species. *Burkholderia caribensis, B. cepacia, B. mimosarum, B. nodosa, B. phymatum, B. sabiae* and *B. tuberum*
Cupriavidus	This genus contains only one species. *Cupriavidus taiwanensis*

6.3.2 MIRCENs' Culture Collection

In 1974 the concept of the Microbial Resource Centers (MIRCENs) was coined by the Microbiology Panel of UNEP/UNESCO/ICRO with specific objectives like (1) world network of regional and interregional cooperating laboratories, (2) efforts for conservation of microorganisms specifically *Rhizobium*, (3) technology development for strengthening rural economics, and (4) trained manpower. All five MIRCENs with BNF as the main aim have been established during 1977–1982. These are the (1) University of Nairobi, Kenya (1977); (2) Porto Alegre, Brazil (1978), jointly at the Department of Soils, University of Rio Grande do Sul, and Institute of Agronomic Research, and State Department of Agriculture; (3) NifTAL at the University of Hawaii (1981); (4) Cell Culture and Nitrogen Fixation Laboratory, USDA, Beltsville, USA (1981); and (5) at the Centre National de Recherches Agronomiques de Bambey, Senegal (1982). All rhizobia MIRCENs hold 4000 strains. The NifTAL MIRCEN has about 2000 strains from 50 countries and 283 legume species. Of these 41 strains are recommended for economically important legumes and 55 have been characterized for lesser known potential legumes. FEPAGRO/UFRGE-MIRCEN has a collection of 700 strains of which 150 are *Bradyrhizobium japonicum*. About 106 strains from this collection have proven high N_2 fixation efficiency for 50 legume species. The USDA-Beltsville MIRCEN has around 1000 strains.

6.3.2.1 NifTAL MIRCEN

Nitrogen fixation of tropical agricultural legume (NifTAL) was established at the University of Hawaii under United States Agency for International Development (USAID) contract to promote use of legume-rhizobia technology for increasing food production in developing countries. It has a collection of 1774 rhizobial strains for legume cultivated at both high and low elevation in tropics for grains, forage, fodder, firewood, green manure as cover crop, erosion control, and shade-providing trees. The rhizobium germplasm resource at NifTAL is a member of WFCC. It contains complete information for rhizobial strains tested as inoculant for 18 agriculturally important tropical legumes that were part of the International Network of Legume Inoculation Trials (INLIT) (NifTAL catalogue of strains 1994).

6.3.3 Rhizobial Repositories in India

6.3.3.1 Indian Type Culture Collection

It is an oldest culture collection established at the Division of Mycology and Plant Pathology, Indian Agricultural Research Institute, New Delhi in 1936. Indian Type

Culture Collection (ITCC) is a member of the WFCC and registered as WDCM 430. The main objectives of ITCC are to act as a repository, to supply authentic fungal/ bacterial cultures as well as identification and provide related services to farmers and scientists working in research institutions/universities and industries. It has 3800 bacterial and fungal deposits.

6.3.3.2 National Microbial Resource Centre

The Department of Biotechnology (DBT), Government of India, funded national facility affiliated to National Centre for Cell Science (NCCS), Pune, India. It is a member of WFCC and is registered as WDCM 930. It was recognized by the World Intellectual Property Organization (WIPO), Geneva, Switzerland as an International Depository Authority (IDA) in April 2011. The total fungal and bacterial deposits including rhizobia at National Microbial Resource Centre (*NMRC*) (formerly MCC), Pune, is 15338 and 149314, respectively.

6.3.3.3 The International Crops Research Institute for the Semiarid Tropics

It is an international organization established in 1972 at Patancheru, India, by the Ford and the Rockefeller foundations with two regional hubs, Nairobi, Kenya, and Bamako, Mali. International Crops Research Institute for the Semiarid Tropics (ICRISAT's) research agenda is achieved through CGIAR Research Program on Grain Legumes. The rhizobial strains deposited in the collection are associated to various drought-tolerant legumes such as chickpea (*Cicer arietinum* L.), groundnut (*Arachis hypogaea* L.), and pigeon pea (*Cajanus cajan* (L.) Millsp.) The total deposits in this collection include 819 strains, of which 259 are of chickpea, 150 of groundnut, and 410 of pigeon pea collected from 6 countries including Niger, Nigeria, Zimbabwe, Malawi, Ethiopia, and Mozambique.

6.3.3.4 National Agriculturally Important Microbial Culture Collection

NAIMCC was established in the year 2004 by the National Bureau of Agriculturally Important Microorganisms (NBAIM), Mau, Uttar Pradesh India. It is recognized as one of national repositories of India by the National Biodiversity Authority of India and has storage facility of 10,000 agriculturally important microorganisms (AIMs). The AIMs comprising of fungi, bacteria, actinomycetes, and cyanobacteria are being preserved at National Agriculturally Important Microbial Culture Collection (NAIMCC). The accessioned microorganisms at NAIMCC are 6327 including fungi (3809), bacteria and actinomycetes (2327), and cyanobacteria (228). It has published a *Catalogue of Microbial Cultures: Supplement-2014* giving detailed information on microbial deposits.

6.4 Authentication

Authentication is the ability of rhizobial isolates to form a nodule on a legume and fix nitrogen (effectiveness). This is done through in vivo plant infection assays. The various methods for authentication of rhizobia are described and compiled in a recent manual (Howeison and Dilworth 2016). The effective nodulation of legume with a rhizobial strain is a result of genetic compatibility between both partners. This also gives clue to the host range of a specific strain. Rhizobia host range is often the main criterion for the release of strain in the field. Nodulation of a legume with rhizobia may either be effective or noneffective (Terpolilli et al. 2008).

6.5 Identification of Rhizobia

Rhizobia are Gram-negative bacteria present in soil. They are distinguished from other prokaryotes by the ability to elicit nodules on roots and stems of leguminous plants and fixing atmospheric nitrogen. Initially one species of legume root-nodulating bacterium, *Rhizobium leguminosarum*, was known (Frank 1889). Nearly five decades later, it was observed that rhizobia considerably differed in the range of host plants; they could nodulate effectively and thus emerged the concept of cross-inoculation group in rhizobia. Hence *R. leguminosarum* (symbiont of peas and vetches), *R. trifolii* (clover rhizobium), *R. phaseoli* (bean rhizobium), *R. lupini* (host plant lupin), *R. japonicum* (nodulated soybean in Japan), and *R. meliloti* (sweet clover) were identified and described (Fred et al. 1932). Another 50 years later, Jordan proposed that growth rate and biochemical properties along with host range should be the basis of taxonomy (Jordan 1982). Hence second rhizobial genus, *Bradyrhizobium*, was described, and *R. leguminosarum*, *R. trifolii*, and *R. phaseoli* were amalgamated into single species *R. leguminosarum* with three different "biovars" denoting their symbiotic preferences. Since then emerged the concept of polyphasic taxonomy which is currently followed for identification of rhizobia. Consequently, till date 15 genera (13 in α-proteobacteria and 2 in β-proteobacteria) of rhizobia with 120 species have been described (Table 6.1). Since authentication of rhizobia for nodulation and nitrogen fixation on legume plant is the key step prior to their identification. Thus the concept of symbiovar (earlier biovar) is still useful.

6.5.1 Polyphasic Taxonomy of Rhizobia

In the 1990s the concept of defining prokaryotic species using the combination of phenotypic, biochemical, and genotypic methods, collectively known as polyphasic taxonomic approach, emerged. Almost a decade later in 2002, the International Committee on Systematics of Prokaryotes (ISCP) defined prokaryotic species as a collection of strains that have phenotypic and/or morphological similarity; genome similarity of >70% in DNA–DNA hybridization; G+C content similarity, as indicated by melting temp of DNA within 5 °C; and less than 3% divergence of 16S rRNA sequence (Stackebrandt et al. 2002). The description of any novel prokaryotic species must be first published in the *International Journal of Systematic and Evolutionary Microbiology* (http://ijs.sgmjournals.org/). For rhizobia, polyphasic taxonomy has been used by several workers (Graham et al. 1991; Vandamme et al. 1996) and is overseen by the ICSP Subcommittee on the taxonomy of *Rhizobium* and *Agrobacterium* (Lindström and Young 2011). Under this scheme, the species designation to rhizobia is based on the combination of morphological, biochemical, physiological, and genetic fingerprinting methods, along with host range for nodulation. The genotypic fingerprinting is performed using several techniques, viz., restriction fragment length polymorphism (RFLP), pulse-field gel electrophoresis (PFGE), PCR-fingerprinting, and gene sequencing. The sequence analysis of 16S rRNA gene has been used in phylogeny and taxonomy of rhizobia, but this does not reflect the symbiotic features including legume host range. Moreover, there are multiple copies of the 16S rRNA gene, with up to 5% intragenomic differences (Kampfer and Glaeser 2012); the 16S rRNA gene sequence is highly conserved and has ability to resolve closely related strains but only up to genus level, and it is also vulnerable to horizontal gene transfer (Willems et al. 2001; Gevers et al. 2005). Stackebrandt and co-workers (2002) have proposed the16S–23S rRNA ITS region and genes located in the core genome (housekeeping) as alternative phylogenetic markers. According to them, at least five housekeeping genes are necessary for reliable taxonomy. Although housekeeping genes have evolved at a rate faster than the 16S rRNA, they are considerably conserved to retain genetic information. The housekeeping genes frequently used in taxonomic characterization are *dnaK*, *dnaJ*, *glnA*, *gyrB*, *gltA*, *glnII*, *recA*, *rpoA*, *rpoB*, and *atpD* (Menna et al. 2009; Menna and Hungria 2011; Azevedo et al. 2015). There are several criteria for selecting housekeeping genes for MLSA such as that they should be (1) present in single copies in a genome; (2) distributed throughout the genome, spaced by distance of at least 100 kb between two genes; (3) with nucleotide length sufficient for sequencing; (4) carry information sufficient for the analysis; and (5) be broadly distributed among taxa. The combination of genes for use in taxonomic analysis is also important and should be carefully selected. The increased use of MLSA with rhizobia suggests that it may soon be an accepted tool to define new species. It is further proposed that 94–96% identity for multilocus sequences may replace the 70% homology in DNA–DNA hybridization. In case of rhizobia, analysis of nodulation (*nod*) and nitrogen-fixing (*nif, fix*) genes can provide the information on evolution of the symbioses and its host range for nodulation.

Since whole genome sequences for bacterial species are available, average nucleotide identity (ANI) of the whole genome has also been increasingly used for the identification of novel species (Konstantinidis and Tiedje 2005a, b, Ormeño-Orrillo et al. 2015). For the genome comparison, 94% identity of the shared genes between two strains would be equivalent to the 70% level of DNA–DNA hybridization. This criterion might be reevaluated as more rhizobial genomes are sequenced and available for analysis.

6.5.2 Criteria for Description of Rhizobia

Taxonomy includes three elements: characterization, classification, and nomenclature. The characterization of one strain is a key step in prokaryotic systematic and precedes classification and nomenclature. Over the past century, characterization methodologies have considerably improved. Presently, both modern and traditional methods are used for assigning a strain to a taxon. The scheme of currently available methods for characterization of the strain comprehensively has been outlined by Tindall and co-workers (2010). The scheme of tests for rhizobial species demarcation is as follows:

1. *Number of strains used to describe a new rhizobial species.* The new strain descriptions are preferentially based on a minimum of at least three different isolates as revealed by IGS-PCR RFLP, rep-PCR, and AFLP fingerprints or sequence data. Sampling should be made from different ecological settings, and more than 12 isolates should preferentially represent each ecological setting. Hence, *new species descriptions based on a single isolate are strongly discouraged!*
2. *Number and type of molecular markers. The high-resolution molecular typing method appropriate to reveal diversity within species (i.e., rep-PCR or AFLP genomic fingerprints) combined with MLSA of at least three protein-coding loci should be used.* For rhizobia it *would be appropriate to generate full-length 16S rDNA sequences for a few carefully selected strains, along with the partial sequencing of three protein-coding loci (e.g. glnII, recA, and rpoB) and at least one symbiotic (nifH, nodA, and nodC) locus.*
3. *Group of phenotypic tests.* New isolates should be screened for host range in vivo which gives insights on biovars and symbiotic varieties. Other commonly used and potentially relevant attributes are pH and temperature growth range, salt tolerance, growth on different C and N sources, as well as antibiotic resistance profiling and fatty acid methyl ester analysis (FAME). *The selection of phenotypic and chemotaxonomic tests is based on the niches from where isolate(s) was recovered.*
4. *DNA–DNA hybridization vs MLSA. For delineation of species, a thorough MLSA must be combined with key phenotypic tests. They should further be supported by full-length 16S rDNA sequencing of the type strain and two other strains, along with DNA–DNA hybridization data. DNA–DNA hybridization has been increasingly replaced by MLSA and genomic comparisons (ANI).*

6.5.3 *Classification of Rhizobia on Plant Host Range for Nodulation*

Rhizobia are characterized by inherent ability to nodulate legume plants effectively. Another characteristic property of rhizobia is host specificity. At genetic level host specificity is determined by nodulation (e.g., *nod*, *rhi*) and nitrogen fixation (*fix*, *nif*) genes that form accessory genome and/or present on plasmids. The stable taxonomy is based on "core" or housekeeping genes. Hence, bacteria of same species can have different host specificity. The host specificity or symbiotic preferences in rhizobia were denoted through biovar description (Jordan 1984). In subsequent years several biovars have been described within rhizobia (Table 6.2). Biovars can be differentiated on the basis of different biochemical and physiological properties. Considering the polyphasic taxonomy, in rhizobia the biovar must be supported with sequence data of symbiotic genes. Thus the use of "symbiovar" is more appropriate to reflect symbiotic preferences of a particular rhizobium species (Rogel et al. 2011).

6.6 The Need to Preserve Cultures

Preservation of the cultures in a genetically stable form is highly important, as experimentation with strains of rhizobia can last for many decades. It has been more than 100 years since the first bacteria from the nodules were isolated, and several decades have elapsed since the symbiosis was scientifically understood, yet very few rhizobial cultures are available for exploitation. The probable reason is that strains were commonly stored on agar. Agar slope-borne cultures have a relatively short shelf life. For this reason, methods for long-term preservation of valuable cultures are required to ensure survival over long period of inattention.

6.7 Methods of Preservation

The main goal of any method of preservation is to maintain the purity and viability of the culture for the longest duration possible. In addition, the method should be easy to implement, of low cost, and easily accessible (Romeiro 2001). The requirements for short-term storage after the strain has been isolated from a nodule and passed through the purification process and long-term storage to preserve the integrity of the strain after authentication are different. A short-term system is based on agar or glycerol storage (with strains labelled with a temporary code) and long-term preservation (freeze-drying or lyophilization) after authentication (Table 6.3). The strains may lose desirable properties during storage or after repeated subculture. Hence, storage methods must be such that minimize the opportunity for variation or mutation besides keeping cultures viable for long duration. Several methods are

Table 6.2 Description of different biovars within rhizobial species and their legume host

Biovars	Rhizobial species	Legume host	References
acaciae	*S. terangae* *S. sahelense* *S. meliloti*	*Acacia* *Acacia* *A. tortilis*	Lortet et al. (1996) Haukka et al. (1998) Ba et al. (2002)
acaciellae	*S. chiapanecum* *S. mexicanum*	*Acaciella angustissima* *A. angustissima*	Rogel et al. (2011)
biserrulae	*M. opportunistum*	*Biserrula pelecinus*	Nandasena et al. (2007)
ciceri	*M. amorphae* *M. tianshanense* *M. ciceri* *M. mediterraneum* *S. meliloti*	*Cicer arietinum* *C. arietinum* *C. arietinum* *C. arietinum* *C. arietinum*	Rivas et al. (2007) Rivas et al. (2007) Nandasena et al. (2007) Nour et al. (1995) Maatallah et al. (2002)
gallicum	*R. gallicum* *R. giardinii*	*Phaseolus vulgaris, Leucaena leucocephala* *P. vulgaris, L. leucocephala*	Amarger et al. (1997) Amarger et al. (1997)
genistearum	*B. japonicum*	*Genisteae, Loteae*	Vinuesa et al. (2005)
giardinii	*R. giardinii*	*P. vulgaris, L. leucocephala*	Amarger et al. (1997)
glycinearum	*B. japonicum*	*Glycine*	Vinuesa et al. (2005)
lancerottense	*S. meliloti*	*Lotus lancerottense*	Leon-Barrios et al. (2009)
medicaginis	*S. meliloti*	*Medicago laciniata*	Villegas et al. (2006)
mediterranense	*S. fredii* *S. meliloti*	*P. vulgaris* *P. vulgaris*	Mnasri et al. (2007) Mnasri et al. (2007)
meliloti	*S. meliloti*	*Medicago sativa, M. truncatula*	Villegas et al. (2006) Mnasri et al. (2007)
mimosae	*R. etli*	*P. vulgaris, L. leucocephala, Mimosa affinis*	Mnasri et al. (2007) Wang et al. (1999)
officinalis	*R. galegae*	*Galega officinalis*	Radeva et al. (2001)
orientalis	*R. galegae*	*Galega orientalis*	Quispel (1988)
orientale	*R. mongolense, Rhizobium* spp.	*Medicago ruthenica, P. vulgaris*	Silva et al. (2005) Amarger et al. (1997)
phaseoli	*R. gallicum P* *R. giardini* *R. leguminosarum* *R. etli* *R. phaseoli*	*P. vulgaris* *P. vulgaris* *P. vulgaris* *P. vulgaris* *P. vulgaris*	Amarger et al. (1997) Amarger et al. (1997) Jordan (1984) Segovia et al. (1993) Ramírez-Bahena et al. (2008)

(continued)

Table 6.2 (continued)

Biovars	Rhizobial species	Legume host	References
sesbaniae	*S. terangae* *S. sahelense* *Agrobacterium* sp.	*Sesbania* *Sesbania* *Sesbania*	Lortet et al. (1996) Lortet et al. (1996) Cummings et al. (2009)
trifolii	*R. leguminosarum* *R. leguminosarum*	*Trifolium* *Vicia sativa*	Jordan (1984) Jordan (1984)
viciae	*R. fabae* *R. pisi*	*Viciafaba* *V. sativa*	Tian et al. (2008) Ramírez-Bahena et al. (2008)

Table 6.3 Cell viability of various methods for preserving rhizobia

Methods	Characteristics and conditions	Period of viability
Agar slopes	YEMA, stored at 5–7 °C Low cost and transfer is simple	1 year
Agar slopes covered with mineral oil or paraffin wax	As above. In addition the agar slopes were covered with sterilized mineral or paraffin oil	2 years
Porcelain beads	Suspension of rhizobium cells is poured on sterilised porcelain beads, air dried and kept in a tube with dehydrated silica. Stored at 5–7 °C	2 years
Soil, peat or clay	The material is finely ground, corrected for chemical properties and sterilized. The culture to be stored is inoculated into the measured amount of soil, peat or clay. Stored at 5–7 °C	2–4 years
Freezing	Storage in temperatures ranging from −70 to −190 °C in deep freeze or liquid nitrogen. Viability depends on the culture medium; freezing speed, freezing temperature, type of cryoprotectant used	Months to several years
Lyophilization	Viability depends on the physiological state of the culture, cell concentration, and medium and lyophilization rate; can be kept at room temperature for decades but little information available	Several decades

(Source: modified from Hungria et al. 2016)

available for the preservation of bacterial cultures such as immersing in mineral oil, ordinary freezing, drying, in situ preservation in soil, in sterile distilled water (SDW), glycerol, synthetic polymers, liquid nitrogen, and lyophilization (van Elsas 2001; Campos et al. 2004; Denardin and Freire 2000; Fernandes Júnior et al. 2009). Carboxymethyl cellulose (CMC) and glycerol are commonly used for preserving bacterial cultures. Carboxymethyl cellulose (CMC) is cellulose-derived ester and a highly hygroscopic and viscous polymer, nontoxic to humans (Sanz et al. 2005). Glycerol is a cryoprotectant commonly used as osmoregulator in freezing processes (Campos et al. 2004).

6.7.1 Storage on Agar Slopes

1. Select preferred growth medium (yeast extract mannitol agar or tryptone yeast agar). Heat to dissolve the agar and ensure it is thoroughly mixed.
2. Dispense mixture with a screw-capped bottle or test tube to fill 33% of the volume (i.e., 10 into 30 ml McCartney bottle or 1 into 3 ml plastic vial).
3. Place the bottles and tubes into autoclavable basket. After autoclaving place the bottles and tubes at an angle of 45–60 °C until set.
4. Take a loopful of culture, and streak across the surface of agar slope, allow growing until visible.
5. Store at 5–7 °C.
6. Alternatively cover the agar slope with sterile paraffin or mineral oil to decrease the rate of desiccation.

6.7.2 Storage on Porcelain Beads

1. The desiccated silica gel is placed in the autoclavable vial occupying up to 30% of the volume. The vial is covered with cotton wool, topped with cleansed porcelain beads.
2. The vial is screw capped on loosely then autoclaved.
3. The culture is grown in selected liquid medium to visible turbidity. Alternatively, culture suspension is prepared by washing from solid medium with diluents.
4. Transfer the sterilized beads aseptically to broth culture in the tubes and re-plug. Allow the beads to soak for 1–2 h. The beads impregnated with rhizobia culture are placed into storage tubes aseptically; replace and tighten the caps; store at 5–7 °C. The tubes are observed for moisture. If silica gel becomes pink or colorless, either too much moisture was absorbed during transfer or there is an improper seal permitting entry of moisture.
5. For reviving a culture, one bead is removed and dropped into a yeast mannitol broth medium. This is allowed to grow till visible turbidity is observed; a loopful is then streaked on YM agar with bromothymol blue (BTB) or congo red (CR) indicator dye to check purity.

6.7.3 Long-Term Preservation

(a) *Lyophilization*

This method allows removal of moisture from the culture sample without concurrent change in their physiological, biochemical, or genetic properties. It is based on the principle of removing moisture from culture under strong vacuum; the evaporation of moisture results in drop of the temperature. Although cultures are desiccated, cells in sufficient numbers survive. Freeze-drying of cultures is carried out in two stages, primary stage where 90–95% of the water is removed and secondary stage where additional 4% of water is removed; retention of 1% of water is essential for survival of bacteria. This is achieved by suspending the cells in medium that will not permit complete removal of water. For example, a 50–50 mixture of 10% w/v peptone with 10% w/v Na-glutamate or 5% w/v peptone with 10% w/v sucrose is used. Cultures are then sealed in glass ampoules under very low vacuum. Various cryoprotectants such as dimethyl sulfoxide, glycerol, albumin, skim milk, and peptone (Hubálek 2003; Day and Stacey 2007) are used.

(i) *Preparation of Ampoules*

1. Glass ampoules, dimensions 4 mm × 50 mm open at one end, are made of high-quality glass.
2. A small cotton wool swab is placed at the bottom of the tube with the help of inoculating loop. Then the labels are placed into ampoules above the cotton wool, plugged with non-absorbent cotton wool and autoclaved at 121 °C for 30 min.

(ii) *Preparation of Cultures*

1. The rhizobial strains are grown in the preferred agar medium to check for purity.
2. Dispense 2 ml of the lyophilization mixture (e.g., 1:1 mixture of 50% peptone and 10% Na-glutamate) into screw-capped vials and autoclave.
3. A loopful of the rhizobial strain is added into the cooled lyophilization mixture, and lid replaced and vortexed. Alternatively, take 1 ml of a broth suspension of rhizobia and mix well into the lyophilization medium.
4. The suspension (0.1 ml) is transferred to the ampoule using a sterile pasture pipette. Usually per strain 10 ampoules are prepared. If less than 0.1 ml culture suspension is added, then primary drying process can be omitted.
5. The cotton wool plug is pushed halfway down the ampoule, below the constriction point, to close the vial and protect the culture during the lyophilization process.

(iii) *Primary Drying Process*

1. Turn on the lyophilizer, introduce the ampoules, and follow the manufacturer's instructions. The first drying stage (90–95% of dehydration) usually takes about 90 to 120 min in most lyophilizers.

(iv) *Ampoule Constriction*

1. After the culture is lyophilized, ampoules are constricted in the middle above the inserted plug by gently turning ampoules while holding the center point over a flame, allowing the glass to melt.

2. During the process ampoules are rolled. Consequently glass flows into the middle of the ampoule. At this point ampoule is removed from the flame while continuing to gently stretch the middle section. The ampoules are not sealed during this process.

(v) *Secondary Drying*

1. Constricted ampoules are applied to a manifold by pushing over into rubber sleeve so that they are held tightly.
2. The vacuum is started again, and drying is continued for approximately 1 h until a vacuum reaches 6.7 pascals. Drying over a desiccant, for example, phosphorus pentoxide, is done for the removal of remaining moisture. It is noteworthy that phosphorus pentoxide is poisonous.
3. Once the vacuum has been achieved, ampoules may be sealed by holding a flame to the constricted part of the ampoule. After the narrow neck has melted, ampoule is held at the end and twisted, so that it is separated from the top half of the ampoule.
4. The ampoules should retain a vacuum. The vacuum within sealed ampoule is checked using a high-frequency spark tester. If correct level of vacuum is present, a faint glow like a neon light appears inside ampoule.
5. Viability of cell should be ascertained in at least one ampoule per culture per batch immediately

(vi) *Recovery from Lyophilized Ampoules*

1. Break the ampoule by hitting the glass above the cotton plug with a steel file, heat the score mark, and gently bend the glass allowing it to crack.
2. The cotton from ampoule is removed with sterile forceps, and 0.1–0.3 ml growth medium or lyophilizing mixture is added with a micropipette.
3. Homogenize and transfer a drop or two to a Petri plate containing growth medium. Spread and incubate at optimum temperature for growth.

(b) *Cryopreservation*

It is the process of maintenance of living organisms at low temperature (−80 °C or below) so that they survive thawing. Generally at −80 °C temperature, the viability of cultures decreases with time. For long-term preservation, bacteria should be stored at temperature below −140 °C (OECD 2007). For preservation, rhizobia are grown in glycerol-peptone medium (Gerhardt et al. 1981).

(i) *Preparation of Glycerol Stock*
Method 1

1. The pure culture is inoculated into liquid glucose-peptone broth till log phase is achieved.

2. 150 μl of sterile 80% (v/v) glycerol is added to the cryotube. The final volume is made up to 1 ml by adding broth culture and mixed thoroughly.
3. The temperature of the tube is decreased at a rate of 1.0 °C per minute till it reaches −50 °C and then stored at the final temperature.

Method 2

1. The glucose-peptone medium without indicator dyes is mixed with 80% v/v glycerol in the ratio 85:15 (final concentration = 12% glycerol) and autoclaved.
2. The above mixture is dispensed into sterile cryotubes (1–2 ml).
3. Add loopful of culture from solid agar after confirming purity to the cryotube, and vortex to suspend the culture.
4. Alternatively the cryotubes can be dipped into liquid nitrogen allowing a rapid drop of temperature.

Method 3

1. The glycerol cell suspensions may also be added to vials containing small sterile glass beads. The excess liquid is removed; the beads get coated with cell suspensions. For revival, individual coated beads are removed and inoculated into the standard medium.

Method 4

1. The glucose-peptone medium without indicator dyes is mixed with 100% glycerol in the ratio1:1.
2. The loopful of culture is inoculated into glucose-peptone broth (10 ml) in a test tube and grown till log phase is attained.
3. Now the mixture of glucose-peptone broth and glycerol, 4.5 ml in 10 ml of log phase culture broth is added and thoroughly mixed. The mixture is transferred to cryotubes in aliquots of 1 ml each. The cryotubes were labelled properly and stored at −80 °C

(c) *Revival of Preserved Cultures*

The preserved cultures should be checked every 5 years.

1. The frozen cultures should be kept on ice, if they are to be returned to cold storage or deep freezer.
2. A loopful of frozen culture is transferred on to agar plate with the help of warm sterile loop.
3. Once thawed, the culture is streaked on to the plate. The purity is checked for which the cultures should be streaked till single colonies appear.
4. The growth of rhizobial cultures streaked from frozen stocks may be slower and required longer incubation period than those streaked out from routine slopes or plates. There can be loss of nod character during storage/frequent transfers.

6.8 Conclusion and Future Scenario

There is network of 728 culture collection centers registered under WFCC-MIRCEN and coordinated by WDCM and of these 77 have published online catalogues. The Convention on Biological Diversity and Budapest Treaty has been adopted globally. The use and peruse of rhizobium-legume technology is well established. Although advent and advancement of sequencing techniques have led to increased number of rhizobial strains being described and identified, till date 130 rhizobial species have been described. However, for rhizobia plant infection, tests are very useful and relevant. Presently, hundreds of rhizobial strains have been recovered, explored, exploited, and preserved. But still 90% of the legumes spread across the globe remain to be explored for their microbial partner. More and more microbial repositories should be established around the world especially in regions that are hotspots of biodiversity.

Biogeographically India has characteristic elements from all three realms—Afro-tropical, Indo-Malayan, and Paleo-Arctic realms. Therefore it harbors 2 of the 18 hotspots of biodiversity spread across the globe. These include floristically rich areas of Northeast India, the Western Ghats, northwest Himalayas, and the Andaman and Nicobar Islands. Considering rich biodiversity, very few culture collection centers are established across India, and only six have published online catalogues.

The exploration of legumes for their microbial partner, its characterization, description, and conservation should be taken up exhaustively and systematically so that we do not lose out rich biodiversity. This will further help in strict implementation of the Convention on Biological Diversity.

Acknowledgments The authors thank Indian Council of Agricultural Research (ICAR), and Ministry of Environment & Forests (MoEF), and GOI for financial assistance to MS. The senior author MS thanks Mariangela Hungria, Brazil for critical suggestions.

References

Amarger N, Macheret V, Laguerre G (1997) *Rhizobium gallicum* sp. nov. and *Rhizobium giardinii* sp. nov., from *Phaseolus vulgaris* nodules. Int J Syst Bacteriol 47:996–1006

Anonymous (2017) India to overtake China's population in 7 yrs. A UN report published in Times of India

Australian Inoculant Research and Control Service, New South Wales Department of Agriculture Melbourne. http://www.dpi.nsw.gov.au/

Azevedo H, Lopes FM, Silla PR, Hungria M (2015) A database for the taxonomic and phylogenetic identification of the genus *Bradyrhizobium* using multilocus sequence analysis. BMC Genomics 16:S10

Ba S, Willems A, de Lajudie P, Roche P, Jeder H, Quatrini P, Neyra M, Ferro M, Prome JC, Gillis M, Boivin-Masson C, Lorquin J (2002) Symbiotic and taxonomic diversity of rhizobia isolated from *Acacia tortilis* subsp. Raddiana in Africa. Syst Appl Microbiol 25:130–145

Campos AK, de Andrade Mota M, de Araújo JV, Cecon PR (2004) Predatory activity, radial growth and sporulation of fungi predating *Monacrosporium* nematode submitted to cryopreservation. Sci Rural 34:465–469

Commonwealth Scientific and Industrial Research Organization's (CSIRO) Cunningham Laboratory, Brisbane. https://www.csiro.au/en/Publications

Cummings SP, Gyaneshwar P, Vinuesa P, Farruggia FT, Andrews M, Humphry D, Elliott GN, Nelson A, Orr C, Pettitt D, Shah GR, Santos SR, Krishnan HB, Odee D, Moreira FMS, Sprent JI, Young JPW, James EK (2009) Nodulation of *Sesbania* species by *Rhizobium (Agrobacterium)* strain IRBG74 and other rhizobia. Environ Microbiol 11:2510–2525

Day JG, Stacey GN (eds) (2007) Cryopreservation and freeze-drying protocols, 2nd edn. Methods in molecular biology series 368. Humana, Totowa

Denardin ND, Freire JRJ (2000) Assessment of polymers for the formulation of legume inoculants. World J Microbiol Biotechnol 16:215–217

Fernandes Júnior PI, Rohr TG, de Oliveira PJ, Xavier GR, Rumjanek NG (2009) Polymers as carriers for rhizobia inoculant formulations. Pesq Agropec Bras Brasília 44:1184–1190

Frank B (1889) Über die Pilzsymbiose der Leguminosen. Ber Dtsch Bot Ges. 7:332–346

Fred EB, Baldwin IL, McCoy E (1932) Root nodule bacteria and leguminous plants. University of Wisconsin. Digitized19 Feb 2008

Freire JRJ, Kolling I (1986) World culture collections of Rhizobium. World J Microbiol Biotechnol 2:231–235

Gerhardt P, Murray RGE, Costilow RN, Nester EW, Wood WA, Krieg NR, Phillips GB (eds) (1981) Manual of methods for general bacteriology. American Society of Microbiology, Washington

Gevers D, Cohan FM, Lawrence JG, Spratt BG, Coenye T, Feil EJ, Stackebrandt E, de Peer YV, Vandamme P, Thompson FL, Swings J (2005) Re-evaluating prokaryotic species. Nat Rev Microbiol 3:733–739

Graham PH, Sadowsky MJ, Kersters HH, Barnet YM, Bradley RS, Cooper JE, De Ley DJ, Jarvis BDW, Roslycky EB, Strijdom BW, Young JPW (1991) Proposed minimal standards for the description of new genera and species of root-and stem-nodulating bacteria. Int J Syst Bacteriol 41:582–587

Haukka K, Lindstrom K, Young JPW (1998) Three phylogenetic groups of nodA and nifH genes in *Sinorhizobium* and *Mesorhizobium* isolates from leguminous trees growing in Africa and Latin America. Appl Environ Microbiol 64:419–426

Howeison JG, Dilworth MJ (2016) Working with rhizobia. Australian Centre for International Agricultural research, Canberra

Hubálek Z (2003) Protectants used in the cryopreservation of microorganisms. Cryobiology 46:205–229

Hungria M, Ardley J, O'Hara GW, Howeison JG (2016) Preservation of rhizobia. In: Howeison JG and Dilworth MJ (eds) Working with rhizobia. Australian Centre for International Agricultural Research, Canberra, pp 61–72

Johnson MD, Allen ON (1952) Cultural reactions of rhizobia with special reference to strains isolated from *Sesbania* species. Ant Leewenhoek 18:1–12

Jordan DC (1982) Notes: transfer of Rhizobium japonicum Buchanan 1980 to Bradyrhizobium gen. Nov., a genus of slow-growing, root nodule bacteria from leguminous plants. Int J Syst Bacteriol 32(1):136–139

Jordan DC (1984) Family III. Rhizobiaceae Conn (1938). In: Krieg NR, Holt JG (eds) Bergey's manual of systematic bacteriology, vol 1. Williams and Wilkins, Baltimore, pp 234–254

Kampfer P, Glaeser SP (2012) Prokaryotic taxonomy in the sequencing era the polyphasic approach revisited. Environ Microbiol 14:291–317

Konstantinidis K, Tiedje JM (2005a) Genomic insights that advance the species definition for prokaryotes. Proc Natl Acad Sci USA 102:2567–2572

Konstantinidis KT, Tiedje JM (2005b) Towards a genome-based taxonomy for prokaryotes. J Bacteriol 187:6258–6264

Leon-Barrios M, Lorite MJ, Donate-Correa J, Sanjuan J (2009) *Ensifer meliloti* bv. lancerottense establishes nitrogen-fixing symbiosis with Lotus endemic to the Canary Islands and shows distinctive symbiotic genotypes and host range. Syst Appl Microbiol 32:413–420

Lindström K, Young JP (2011) International Committee on Systematics of Prokaryotes Subcommittee on the taxonomy of *Agrobacterium* and *Rhizobium*: minutes of the meeting, 7 September 2010, Geneva, Switzerland. Int J Syst Evol Microbiol 61:3089–3093

Lortet G, Mear N Lorquin J, Dreyfus B, de Lajudie P, Rosenberg C, Boivin C (1996) Nod factor thin-layer chromatography profiling as a tool to characterize symbiotic specificity of rhizobial strains: application to *Sinorhizobium saheli, S. teranga*, and *Rhizobium sp.* strains isolated from Acacia and *Sesbania*. Mol Plant–Microbe Interact 9:736–747

Maatallah J, Berraho E, Munoz S, Sanjuan J, Lluch C (2002) Phenotypic and molecular characterization of chickpea rhizobia isolated from different areas of Morocco. J Appl Microbiol 93:531–540

Menna P, Hungria M (2011) Phylogeny of nodulation and nitrogen fixation genes in *Bradyrhizobium*: supporting evidence for the theory of monophyletic origin, and spread and maintenance by both horizontal and vertical transfer. Int J Syst Evol Microbiol 61:3052–3067

Menna P, Barcellos FG, Hungria M (2009) Phylogeny and taxonomy of a diverse collection of *Bradyrhizobium* strains based on multilocus sequence analysis of the 16S rRNA gene, ITS region and glnII, recA, atpD and dnaK genes. Int J Syst Evol Microbiol 59:2934–2950

Mnasri B, Mrabet M, Laguerre G, Aouani ME, Mhamdi R (2007) Salt-tolerant rhizobia isolated from a Tunisian oasis that are highly effective for symbiotic N2-fixation with *Phaseolus vulgaris* constitute a novel biovar (bv. mediterranense) of *Sinorhizobium meliloti*. Arch Microbiol 187:79–85

Nandasena KG, O'Hara GW, Tiwari RP, Sezmis E, Howieson JG (2007) In situ lateral transfer of symbiosis islands results in rapid evolution of diverse competitive strains of mesorhizobia suboptimal in symbiotic nitrogen fixation on the pasture legume *Biserrula pelecinus L.* Environ Microbiol 9:2496–2511

Nour SM, Cleyet Marel JC, Normand P, Fernandez MP (1995) Genomic heterogeneity of strains nodulating chickpeas (*Cicer arietinum L.*) and description of *Rhizobium mediterraneum* sp. nov. Int J Syst Bacteriol 45:640–648

OECD (2007) Promoting pro-poor growth: policy guidance for donors. OECD Publishing. https://doi.org/10.1787/9789264024786-en

Ormeño-Orrillo E, Servín-Garcidueñas LE, Rogel MA, González V, Peralta H, Mora J, Martínez-Romero J, Martínez-Romero E (2015) Taxonomy of *Rhizobia* and *Agrobacteria* from the Rhizobiaceae family in light of genomics. Syst Appl Microbiol 38:287–291

Peterson TA, Russelle MP (1991) *Alfalfa* and the nitrogen cycle in the Corn Belt. J Soil Water Conserv 46:229–235

Quispel A (1988) Hellriegel and Wilfarth's discovery of (symbiotic) nitrogen fixation one hundred years ago. In: Bothe H, de Bruijn FJ, Newton WE (eds) Nitrogen fixation: one hundred years after. Gustav Fisher, Stuttgart, pp 3–12

Radeva G, Jurgens G, Niemi M, Nick G, Suominen L, Lindstrom K (2001) Description of two biovars in the *Rhizobium galegae* species: *biovar orientalis* and *biovar officinalis*. Syst Appl Microbiol 24:192–205

Ramírez-Bahena MH, García Fraile P, Peix A, Valverde A, Rivas R, Igual JM, Mateos PF, Martínez Molina E, Velázquez E (2008) Revision of the taxonomic status of the species *Rhizobium leguminosarum* (Frank 1879) Frank 1889AL, *Rhizobium phaseoli* Dangeard 1926AL and *Rhizobium trifolii* Dangeard 1926AL. *R. trifolii* is a later synonym of *R. leguminosarum*. Reclassification of the strain *R. leguminosarum* DSM 30132 (=NCIMB 11478) as *Rhizobium pisi* sp. nov. Int J Syst Evol Microbiol 58:2484–2490

Rivas R, Laranjo M, Mateos PF, Oliveira S, Martínez-Molina E, Velázquez E (2007) Strains of *Mesorhizobium amorphae* and *Mesorhizobium tianshanense*, carrying symbiotic genes of common chickpea endosymbiotic species, constitute a novel biovar (*ciceri*) capable of nodulating *Cicer arietinum*. Lett Appl Microbiol 44:412–418

Rogel MA, Ormeño-Orrillo E, Martinez Romero E (2011) Symbiovars in rhizobia reflects bacterial adaptation to legumes. Syst. Appl Microbiol 34:96–104. https://doi.org/10.1016/j.syapm.2010.11.015

Romeiro RS (2001) Methods in plant bacteriology. UFV, Viçosa, p 279

Sahgal M, Sharma A, Johri BN, Prakash A (2004) Selection of growth promotory rhizobia for *Dalbergia sissoo* from diverse soil ecosystems of India. Symbiosis 36:83–96

Sanz T, Fernández MA, Salvador A, Muñhoz J, Fiszman SM (2005) Thermogelation properties of methylcellulose (MC) and their effect on a batter formula. Food Hydrocoll 19:141–147

Segovia L, Young JPW, Martinez-Romero E (1993) Reclassification of American *Rhizobium leguminosarum* biovar *phaseoli* type I strains as *Rhizobium etli* sp. nov. Int J Syst Bacteriol 43:374–377

Silva C, Vinuesa P, Eguiarte LE, Souza V, Martinez-Romero E (2005) Evolutionary genetics and biogeographic structure of *Rhizobium gallicum* sensu lato, a widely distributed bacterial symbiont of diverse legumes. Mol Ecol 14:4033–4050

Sly LI (2010) Biodiversity and the role of microbial resource centres. In: Brown AG (ed) Proceedings of conference "biodiversity and world food security: nourishing the planet and its people". Conducted by the Crawford Fund for International Agricultural Research, Parliament House, Canberra, Australia, 30 August–1 September, 2010, pp 60–67

Smith D (2003) Culture collections over the world. Int Microbiol 6:95–100

Stackebrandt E, Frederiksen W, Garrity GM, Grimont PAD, Kampfer P, Maiden X, Nesme MCJ, Rossello-Mora R, Swings J, Truper HG, Vauterin L, Ward AC, Whitman WB (2002) Report of the adhoc committee for the re-evaluation of the species definition in bacteriology. Int J Syst Evol Microbiol 52:1043–1047

Terpolilli J, O'Hara GW, Tiwari RP, Dilworth MJ, Howieson JG (2008) The model legume *Medicago truncatula* A17 is poorly matched for N2 fixation with the sequenced microsymbiont *Sinorhizobium meliloti* 1021. New Phytol 179:62–66

Tian CF, Wang ET, Wu LJ, Han TX, Chen WF, Gu CT, Gu JG, Chen WX (2008) *Rhizobium fabae* sp. nov., a bacterium that nodulates *Vicia faba*. Int J Syst Evol Microbiol 58:2871–2875

Tilman D, Reich PB, Knops J, Wedin D, Mielke T, Lehman C (2001) Diversity and productivity in a long-term grassland experiment. Sci 294:843–845

Tindall BJ, Rosselló-Móra R, Busse H-J, Ludwig W, Kämpfer P (2010) Notes on the characterization of prokaryote strains for taxonomic purposes. Int J Syst Evol Microbiol 60:249–266

Trewavas A (2001) Urban myths of organic farming. Nature 410:409–410

Van Elsas JD (2001) Effects of ecological factors on the survival and physiology of *Ralstonia solanacearum* bv.2 in irrigation water. Can J Microbiol 47:842–854

Vandamme P, Pot B, Gillis M, De Vos P, Kersters K, Swings J (1996) Polyphasic taxonomy; a consensus approach to bacterial systematics. Microbiol Mol Biol Rev 60:407–438

Villegas MC, Rome S, Maure L, Domergue O, Gardan L, Bailly X, Cleyet- Marel JC, Brunel B (2006) Nitrogen-fixing *sinorhizobia* with *Medicago laciniata* constitute a novel biovar (bv. *medicaginis*) of *S. meliloti*. Syst Appl Microbiol 29:526–538

Vinuesa P, Leon-Barrios M, Silva C, Willems A, Jarabo-Lorenzo A, Perez-Galdona R, Werner D, Martinez-Romero E (2005) *Bradyrhizobium canariense* sp. nov., an acid-tolerant endosymbiont that nodulates endemic genistoid legumes (Papilionoideae: Genisteae) from the Canary Islands, along with *Bradyrhizobium japonicum* bv. *genistearum*, *Bradyrhizobium* genospecies alpha and *Bradyrhizobium* genospecies beta. Int J Syst Evol Microbiol 55:569–575

Wang ET, Rogel MA, Garcia-De los Santos A, Martinez-Romero J, Cevallos MA, Martinez-Romero E (1999) *Rhizobium etli* bv. *mimosae*, a novel biovar isolated from *Mimosa affinis*. Int J Syst Bacteriol 49:1479–1491

Weir BS (2016) The current taxonomy of rhizobia. NZ Rhizobia website. https://www.rhizobia.co.nz/taxonomy/rhizobia. Last updated: 10 Jan 2016

Willems A, Coopman R, Gillis M (2001) Comparison of sequence analysis of 16S–23S rDNA spacer regions, AFLP analysis and DNA–DNA hybridizations in *Bradyrhizobium*. Int J Syst Evol Microbiol 51:623–632

World Directory of Culture Collections, sixth version (2014) www.wfcc.info/ccinfo/index.php/home/statistics/#m2

Chapter 7
Biobank for Conservation of Arbuscular Mycorrhizal (AM) Fungi

Richa Agnihotri, Hemant S. Maheshwari, Anil K. Sharma, and Mahaveer P. Sharma

Abstract The development of microbial inocula is a tedious process that requires labor, scientific know-how, and huge amount of patience to ensure the availability of viable and contamination-free inocula. Arbuscular mycorrhizal fungi (AMF) have been routinely used as biofertilizers as a part of sustainable agriculture strategy, but due to their obligate symbiotic nature, the widespread cultivation under in vitro is hindered. However, over the years several methods have been developed for AMF cultivation. Preservation of AMF gene pool has become one of the most challenging aspects of mycorrhizal research. Mycorrhizal fungi conservation banks have aimed to preserve and provide viable and pure AMF starter cultures to researchers. The in vitro technology for AMF culture is a breakthrough in mycorrhizal propagation studies, although not all germplasm conservation banks have used this technology; therefore, trap cultures using appropriate hosts for maintaining and propagating AM strain (s) are still being largely practiced and followed. This chapter summarizes prerequisites for AMF germplasm conservation, AMF inoculum production, preservation techniques, and current scenarios on regulations and constraints being faced in maintaining mycorrhizal germplasm conservation agencies.

7.1 Introduction

Arbuscular mycorrhizal fungi (AMF) being obligate biotrophs form mutualistic relationship with most of the terrestrial plant species. AMF have potential role in nutrients acquisition, plant growth promotion, as well as mitigation of biotic and abiotic stresses (Smith and Read 2008). The ability of AMF to grow well in soils having low fertility thus minimizing synthetic fertilizer inputs to maximize yield and soil health is one of

R. Agnihotri · H. S. Maheshwari · M. P. Sharma (✉)
ICAR-Indian Institute of Soybean Research, Indore, Madhya Pradesh, India
e-mail: Mahaveer.Sharma@icar.gov.in

A. K. Sharma
Department of Biological Sciences, CBSH, GB Pant University of Agriculture and Technology, Pantnagar, Uttrakhand, India

S. K. Sharma, A. Varma (eds.), *Microbial Resource Conservation*, Soil Biology 54,
https://doi.org/10.1007/978-3-319-96971-8_7

the factors that make AMF a choice for sustainable agriculture. A reduction in mycorrhizal colonization is observed in soils having elevated P levels (Hepper 1983) and tillage (Jansa et al. 2002). Host species for AMF chiefly include angiosperms, pteridophytes, few gymnosperms, lycopods, and mosses (Smith and Read 1997). AMF life cycle has three phases, i.e., asymbiotic, presymbiotic, and symbiotic phase (Akiyama 2007). Taxonomically, AMF belong to the phylum *Glomeromycota* with 3 classes, 5 orders, 14 families, and 29 genera (Oehl et al. 2011). The classification system was further revised by Redecker et al. (2013).

Mycorrhizal host plants, i.e., trap plants employed for pot culture, include *Sorghum halepense*, *Paspalum notatum*, *Panicum maximum*, *Cenchrus ciliaris*, *Zea mays*, *Trifolium subterraneum*, and *Allium cepa* (Chellappan et al. 2001). AMF and plant communities correlate in terms of community composition (Yang et al. 2012). Yang et al. (2012) via ITS rDNA metadata revealed that host plants mediate a number of factors such as biogeographical, continental, climatic, and ecosystem, all of which are drivers of AMF community composition as AMF distribution was found to be different among the abovementioned factors. Phylogenetic clustering observed within sites depicts habitat filtering, thereby exerting strong effect on assemblage of AMF communities (Kivlin et al. 2011). However, a particular habitat might play a more important role than host species as different plant species in a given habitat harbored similar AMF communities as compared to species of the same plant grown in different habitats (Li et al. 2010). It is worth to know as to how do AM species respond differently with host plants (host specific and site specific); therefore, maintaining specialized type of AM species becomes very important to get desired response for particular edapho-climatic conditions and host plants.

To employ AMF for agricultural purpose for increasing plant growth and yield and to improve soil health, a pure and viable culture is needed. AMF cannot be cultured in a laboratory, but a number of methods have been developed for mass production of AMF. The obligate symbiotic nature of AMF limits its smooth production under laboratory conditions and hence remains a major bottleneck for propagating all the AMF species. Nevertheless, the AMF propagation under in vitro employing hairy root cultures in bioreactor/root organ culture has made significant progress (Fortin et al. 2002) and found to be commercially viable technology worldwide. However, due to some factors, e.g., germination ability of some AM species, all species cannot be propagated under in vitro; therefore, raising AMF species in pot cultures becomes very important. AMF culture once obtained by such methods needs to be preserved till its application. Smith and Onions (1994) reviewed prerequisites to preserve fungal species and categorized them into three major groups, i.e., method involving continuous growth and the ones which exempt the requirement of serial subculturing, i.e., fungal isolates are preserved on growth medium at low temperature, covered with oil/under water, drying by means of air/silica gel/freezing, and arresting metabolism by reducing the water availability by means of dehydration or cryopreservation at ultra-low temperature/liquid nitrogen. However, the obligate symbiotic nature of AMF makes their cultivation difficult, as some AM species are tedious to grow and revive, and therefore, preservation becomes a point of paramount importance for conserving AM fungal species. Conservation and maintenance of AMF at a place ensures availability of desired AMF species, and culture collection banks aim to provide authentic AMF cultures, i.e., appropriately identified and aseptically maintained. The prerequisite of such cultures according to Declerck et al. (2005) are VIPS, i.e., Viability-Identity-Purity-Stability. Microbial germplasm culture collections give access to the cultures as well as provide passport data of all the cultures, ensuring their preservation and distribution, consultancy, contracts, safe deposit, and publications and also give sponsorship for organizing training camps and workshops.

7.2 Biobanks of Mycorrhiza and Their Affiliations and Regulations

Germplasm conservation biobanks are recognized by the World Federation for Culture Collections (WFCC) which defines their objective as "the promotion and development of collections of cultures of microorganisms and cultured cells." WFCC aims at regulation of some of the important criteria such as involvement of skilled personnel, preservation, proper identification, distribution, detailed documentation, biosafety monitoring, research, training, etc. to ensure the reliability and authenticity of culture collections. A list of the most renowned AMF culture collections is given below, and details are also provided in Table 7.1.

- International Culture Collection of Arbuscular Mycorrhizal Fungi (INVAM)
- International Bank of the *Glomeromycota* (IBG/BEG)
- *Glomeromycota* In Vitro Collection (GINCO)
- International Culture Collection of *Glomeromycota* (CICG)
- Centre for Mycorrhizal Culture Collection (CMCC)
- Swiss Collection of AMF (SAF)

Out of these, INVAM, BEG, CICG, CMCC, and SAF are the trap-/pot culture-based banks that preserve AMF using trap plants via continuous culture under greenhouse conditions, while GINCO is maintaining cultures under in vitro/monoaxenic conditions on in vitro RiT-DNA hairy root/root organ culture system. However, other culture collections also exist, but due to unavailability of information, they are not mentioned here. The aim of different culture collections is to preserve the mycorrhizal diversity, i.e., to increase the availability of a pure and stable form of present AMF strains that have been identified with the latest technologies to fulfill research purpose and industrial use (Lalaymia et al. 2012).

Table 7.1 AMF germplasm collections across the globe

Culture collection	Type	Location	Website (hyperlink)
Glomeromycota in vitro collection (GINCO)	In vitro	GINCO-BEL in Belgium and GINCO-CAN in Canada	http://www.mycorrhiza.be/ginco-bel/
The International Bank for the Glomeromycota (IBG) [formerly called Banque Européenne de Glomales (BEG)]	Trap/Pot culture based	Dijon, France	www.i-beg.eu
International Culture Collection of (Vesicular) Arbuscular Mycorrhizal Fungi (INVAM)	Trap/Pot culture based	1090 Agricultural Sciences Building, West Virginia University, Morgantown, West Virginia 26506-6018, USA	http://invam.wvu.edu
International Collection of Glomeromycota (CICG)	Trap/Pot culture based	Blumenau Regional University Department of Natural sciences Santa Catarina	www.furb.br/cicg/
Centre for Mycorrhiza Culture Collection (CMCC)	Trap/Pot culture based/In vitro	India	http://mycorrhizae.org.in/cmcc/
Swiss Collection of AMF (SAF)	Trap/Pot culture based	Agrospace ART, Zurich, Switzerland	www.agroscope.admin.ch

7.3 Methods of Culturing AMF

Traditional methods to culture AMF mainly include trap-based (pot and on farm), aeroponic, and hydroponic system (Ijdo et al. 2011).

(a) *Aeroponic technique*: In aeroponic culture AMF pre-colonized seedlings are placed inside controlled chambers and misted with nutrient solutions (Hung and Sylvia 1988). The benefits provided by aeroponic culture include adequate root development, AMF infection, sporulation attained due to absence of a physical substrate (Singh et al. 2012; Abdul-Khaliq et al. 2001), control over contamination, and eliminated need of carrier material, whereas the disadvantages include the special efforts required for the establishment of setup (Gianinazzi and Vosátka 2004).

(b) *Nutrient film technique (NFT)*: Another technique used to produce soil-deficient AMF inoculum is the nutrient film technique (NFT) introduced by Cooper (1975). In this technique, plant roots are supplied with a thin layer/film of swiftly flowing nutrient solution allowing root growth in the form of a mash where by the virtue of upper liquid layer a moisture coat is formed around the roots. This system like aeroponic system uses pre-colonized seedlings. A more concentrated and effortlessly harvested inoculum is the advantage provided by this technique

Fig. 7.1 Raising of starter culture of AMF using pots

over other substrate (soil/solid media)-based methods of inoculum production (Abdul-Khaliq et al. 2001; Chellappan et al. 2001).

(c) *Pot-/trap-based culture*: Pot-/trap-based culture of AMF is the most widely used method employed till date. The method entails the use of potting mix in the form of different substrates notably organic ones with AMF compatible host plants under greenhouse conditions (Fig. 7.1).

AMF inoculum procured from various sources can be multiplied by on-farm production for successive use (Dodd et al. 1990) using raised beds (Sharma et al. 1996). The use of nursery plots with soil (Sieverding 1991) has also been practiced for a long time which is a low-cost method but might present the risk of easier contamination. Substrates that have been used as potting mix include combination of compost with vermiculite and perlite/horticulture potting medium (Douds et al. 2010). Advantages offered by this method include cost-effectiveness, absence of a complicated setup/easier establishment, and reduced risk of contamination in case of separate pots, whereas disadvantages include lack of complete purity and limited scope for use at industrial scale (Gianinazzi and Vosátka 2004). The widespread use of pot culture is limited due to certain drawbacks which include problems associated with transportation, its bulky nature and limited amount, and issues with inoculums' genetic stability, and the purity of inoculum produced via this method is questionable where chances of cross-contamination of the nearby pots also exist (Abdul-Khaliq et al. 2001).

In vitro cultivation techniques: Biodiversity conservation of AMF is achieved by germplasm collection, where in vitro technology provides access to contamination-free isolates. AMF culture under in vitro is also known as root organ culture where AMF are co-cultivated on hairy roots obtained from transformation by Ri T-DNA of

Agrobacterium rhizogenes (Mosse and Hepper 1975; Bécard and Fortin 1988; Fortin et al. 2002). In this technique AMF infectious propagules mainly spores, infected root fragments are grown with an excised transformed root on a synthetic growth medium under sterile growth conditions (Fortin et al. 2002; Declerck et al. 2005).

In vitro grown spores can be harvested and examined irrespective of growth stage (Dalpé 2001). Under in vitro system, it is possible to closely observe the dynamics of extramatrical mycelium, auxiliary cells (Declerck et al. 2004), and sporulation (Declerck et al. 2001). However, the genetic stability of AMF species maintained monoxenically remains questioned. The sub-cultivation generates risk of somaclonal variation (Plenchette et al. 1996; Cárdenas-Flores et al. 2010). There could be successful sporulation in the first generation of spores during the monoxenic cultivation which might not occur in successive generations (Declerck et al. 2005). Prerequisite for making in vitro culture effective is that it should be continuous, i.e., capable of smooth maintenance and subculture for enhanced production of propagules for all the AMF species (Bécard and Piche 1992). In vitro cultivation of AMF, i.e., root organ culture, is being used at *Glomeromycota* In Vitro Collection (GINCO). GINCO aims to provide high-quality impurity-free AMF inocula for scientific research purpose. At GINCO for in vitro cultivation, "gel plugs" taken from monoxenic cultures of AMF are transferred to new root organ cultures, whereas for in vivo gel plug is placed in the rhizosphere for multiplication under pot conditions (Fig. 6.4). Although continuous culture is widely accepted, its prolonged maintenance with repeated subculturing could reduce infectivity (Plenchette et al. 1996). Storage at lower temperatures could provide a solution for the same as lower temperature slows down the metabolic processes. For some specific cultures particularly of safe deposit, GINCO follows cryopreservation at −130 °C.

7.4 Germplasm Conservation Banks

Maintenance of separate greenhouses and workstations, restricted assess of field soil in pure culture chambers and refrigerators, proper documentation of analyzed samples, systematic labeling of zip bags, proper alphabetic arrangement of samples, and regular cleaning of rooms, racks, and bags are some of the prerequisites that germplasm repositories follow. Under this section, we have provided the prerequisites of a germplasm conservation bank and technologies being followed at different germplasm collections (Fig. 6.2). A list of AMF germplasm conservation banks has been provided in Table 7.1. The facilities present at INVAM and process of culture deposition are outlined in Fig. 7.2a.

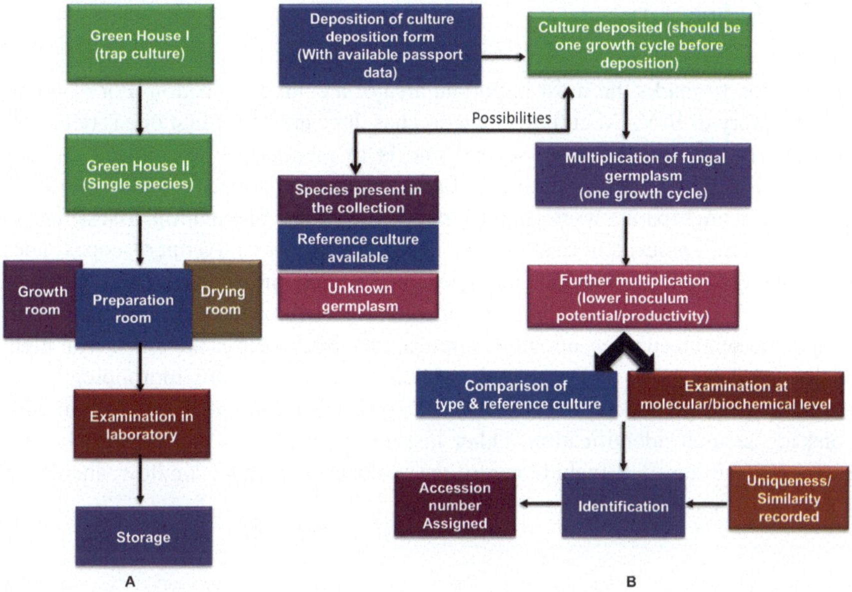

Fig. 7.2 (**a**) Existing trap/pot based germplasm facilities at INVAM. (**b**) Culture deposition process being followed in biobank of AMF

7.4.1 Greenhouse and Associated Chambers

A trap-/pot culture-based gene bank essentially requires greenhouse facilities for growing mycorrhizal plants. At CICG, there are two greenhouses where one is being used to grow trap cultures and the other one is devoted to single-species cultures of AMF. INVAM has a number of growth rooms across the greenhouse that provide controlled conditions of temperature and light and provision for collecting water from cultures to check and prevent cross-contamination. A drying room and a preparation room are also housed at INVAM where the former marks the storage area, pot culture drying in the absence of natural/artificial light, and recording observations, whereas the latter marks the area which restricts the movement of field soil, thus acting as separate space for contamination-free operations like establishment of new cultures, transplantation, and quantitative and qualitative estimation of spores. All the sections have provision to regulate temperature and light, and regular cleaning by bleach solutions is also routinely done to render chambers insect- and pest-free and also devoid of any other biotic and abiotic contamination.

7.4.2 Laboratory

The laboratory marks the most important area of a culture collection. For example, the laboratory of INVAM culture collection has three areas: the first one is dedicated to active culture examination, second one is to microscopy and image capture/visualization, and third one is to the study of molecular aspects of the culture. At CICG, there are separate work stations for active culture examination, assessment of root infection, spore identification, etc. There are high-quality microscopes aided with imagers, and all the data obtained are entered into computers. Therefore, computers become an important tool for storing documents, webpages, database library and search engines, and other applications. Microscopes are aided with high-quality images which help to study all the aspects of spore morphology, i.e., subtending hyphal attachments, shape, size, wall layers, septation, and ornamentations for accurate identification. Other instruments that must be required for any basic microbiology lab include centrifuge, autoclave, laminar air flow, incubator, refrigerators, etc.

7.4.3 Storage Space

Cold room marks the storage area in INVAM where there is provision of precise temperature control, restricted access of field soil, and shiftable racks for convenient retrieval of cultures. At CICG, refrigerators are used to separate species-specific culture as well as for storing the material for regular use. Keeping in view of longevity of AMF starter cultures, storage of inoculum of some species at INVAM is also being done through freeze drying/lyophilization and cryopreservation.

7.4.4 Determination of Inoculum Potential and Quality Assurance

As part of AM quality assurance, ensuring the vigor and colonization ability of any AMF culture is one of the most significant steps that is essentially needed for any gene bank maintaining AMF cultures. The methods used to quantify AMF biomass have been summarized in Table 7.2. However, these methods are microscopic skill oriented (Gange et al. 1999) and often lack reproducibility of quantified data by different observers.

Table 7.2 Methods for assessing AMF biomass in soil and root

S. no.	Method	References
1.	Wet sieving and decanting	Gerdemann and Nicolson (1963)
2.	Root staining	Phillips and Hayman (1970)
3.	Gridline intersect method for percent AM root colonization	Giovannetti and Mosse (1980)
4.	Frequency distribution method for AM root colonization	Baermann and Linderman (1981)
5.	Gridline intersect method (for assessing the root length)	Tennant (1975)
6.	Most probable number (MPN) method	Porter (1979)
7.	Mycorrhizal soil infectivity	Plenchette et al. (1989)
8.	Infection unit method	Franson and Bethlenfalvay (1989) Sharma et al. (1996)
9.	Mean Percentage Infection (MPI)	Moorman and Reeves (1979)
10.	Inoculum potential	Liu and Luo 1994

7.4.5 Acquisition, Deposition, and Regulations of Cultures

At germplasm culture collections, fungal isolates are provided to interested researchers in a prescribed MTA form consisting details and purpose. As per the policy of banks, cultures are also provided to all the contributors/depositors/researchers where accession codes are assigned to the deposited cultures (Fig. 7.2b). As per INVAM and CICG banks, all the depositors are required to provide passport data/information on the culture, growth, generation time, site characteristics, climate, biogeography, symbiotic partner (plant host) and surrounding flora, soil edaphic factors, etc. If the isolate has been multiplied in pots, the second set of information which includes the detailed history of pot culture (i.e., important dates of beginning and harvest of the culture and other information, nature, i.e., characteristics of the culture medium) should also be provided. The culture deposition forms are available at germplasm collections or on their websites. Once the culture is received, after verification, curator assigns the accession number which depends on the choice of the depositor, i.e., if the culture deposition is publicly accessible or should be considered for safe deposit. At GINCO, deposits have been divided into three categories, i.e., public, safe, and patent which gives right to the depositor to choose whether a culture has to be cataloged or not.

A material transfer agreement (MTA), i.e., the contract meant for the acquisition of research material, has to be executed. The document defines the criteria, i.e., terms and conditions regarding supply, distribution, handling, and transfer of the material in acceptance with all applicable laws and regulations that have to be fulfilled. After identification, a specific code is assigned in accordance with the International Code of Nomenclature for Algae, Fungi, and Plants which is based on specimen (holotype), i.e., spores. Both CICG and INVAM use a similar system where an alphanumeric code is followed that contains three letters (three in case of CICG and two at INVAM) representing geographic/country/location of origin, a hundred is followed by a letter (initial coding corresponds to region and mycorrhizal community and letter represents species), and a stroke is followed by a number (represents crop generation). A comparative study is performed between type and reference culture, and depending upon the uniqueness of characters, the culture is categorized as a reference culture (CICG).

7.4.6 Ordering of Cultures and Other Services Being Rendered by Banks

Depending upon the mode of culture maintenance and propagation, cultures can be ordered. For example, GINCO uses in vitro technology and provides cultures in vials or Petri dishes containing, i.e., spores immersed/submerged in growth medium (gel). Recommended storage temperature is 4 °C with suggested subculturing to keep the culture pure and viable. INVAM provides dried inoculum (infectious propagules in growth medium), extracted spores (in sterile distilled water/sand), or bulk inoculum (multiple species).

Besides culture acquisitions, gene banks also provide know-how on AMF collection, culture maintenance methodologies, taxonomy, etc. For example, INVAM has emerged as the largest platform for providing support data leading to species identification from reference cultures. INVAM maintains a database of accession numbers on available AM cultures and also provides services including supply of cultures (bulk inoculum, whole inoculum, i.e., root/hyphae/spores/growth medium and clean spores), identification at morphological and molecular level, AMF biomass quantification (spore, root colonization, infectivity, etc.), high-resolution images, visits for students and researchers, trainings, etc. GINCO aims to provide strain characterization and preservation and recently updated developments in the in vitro technology with expert supervision to carry out research work. International trainings are also provided for the same.

The International Bank of the *Glomeromycota* (IBG), apart from maintaining germplasm repository and providing reference cultures, also develops molecular probes, makes germplasm lines commercially available, and provides technical training.

In India, the conservation of microorganisms is being administered through the Ministry of Agriculture and Farmers' Welfare; Ministry of Environment, Forest and Climate Change; Department of Biotechnology; Department of Science and Technology; and Department of Agricultural Research and Education. For example, Microbial Type Culture Collection (MTCC) has been established and operated from the Institute of Microbial Technology (CSIR-IMTECH), Chandigarh; National Centre for Microbial Resource (DBT-NCCs), Pune; National Agriculturally Important Microbial Culture Collection (NAIMCC), ICAR-NBAIM, Mau; etc. However these microbial culture collection banks are not maintaining AM cultures but are mainly dealing with actinomycetes, bacteria, fungi, etc.

For AM conservation, a DBT-Ministry of Science and Technology-funded initiative Centre for Mycorrhizal Culture Collection (CMCC) has been established at TERI, New Delhi. At CMC culture deposition; assigning of accession number; development of single-spore cultures and their identification, i.e., at microscopic (morphotaxonomic), molecular (DNA sequencing), and biochemical (fatty acid methyl esters) level; and supply of cultures are being undertaken.

7.5 Factors Affecting AMF Inoculum Maintenance

7.5.1 Host Plant

Highly mycotrophic plants are chosen to multiply AMF in trap-based cultures. Trap plant species might exhibit some sort of selective effect on community composition which is the possible reason why community composition differs when field spores are multiplied using trap plants (Jansa et al. 2002). Egerton-Warburton et al. (2007) noted that in terms of increasing hyphal biomass, the combination of N fertilization and C_4 plants proved a better strategy for *Glomus* species, whereas in case of *Gigasporaceae* N fertilization and C_3 plants proved an effective combination. INVAM suggests some of the important points to be remembered while choosing a suitable host for AMF, i.e., the plant should be strictly mycotrophic, be suited for greenhouse conditions and a variety of soils, have average root growth rate, have ample branching of roots, be photosynthetically efficient, and have high P requirement as well as resistance to pathogens, insect, pests, and nematodes.

7.5.2 Forms of Inocula and Colonizing Ability

AM inoculum supplied in the form of spores, infected roots, and hyphae should have the ability to infect the plant. Spores of AMF have been shown in Fig. 7.3. Besides spores, AMF infectious propagules such as hyphae and vesicles could prove to be an effective component of AMF for initiating symbiosis with plants (Fig. 7.4). The presence of vesicle in the intraradical phase reduces the risk of contamination; however the colonization efficiency of spores and vesicles could vary (Fortin et al. 2002). Declerck et al. (1998) successfully cultured *Glomus versiforme*, *Glomus intraradices*, *Glomus macrocarpum*, and *Glomus fasciculatum* in vitro and using colonized root bits and single vesicles. Continuous culture of these species was possible with this approach, so it was suggested to be effective for germplasm collection. Considerable variation could exist between species producing large dark-colored spores and those with light-colored spores, where in most of the

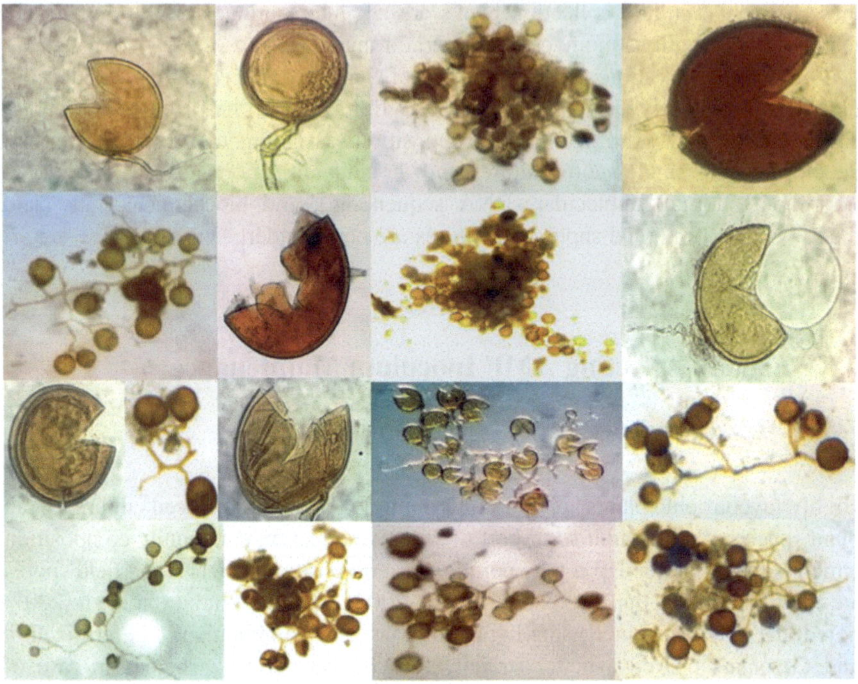

Fig. 7.3 Diversity of AMF spores and sporocarps

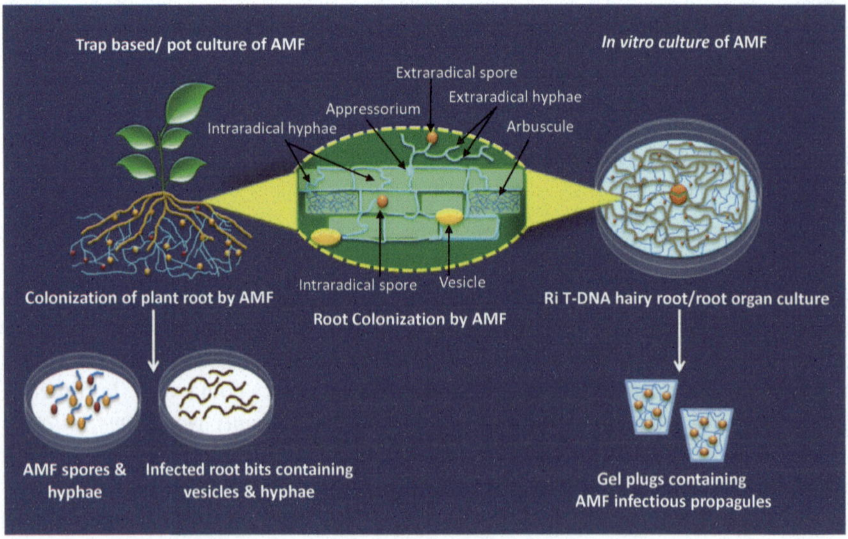

Fig. 7.4 Forms of AMF inocula obtained via trap based and in vitro culture

cases the former exhibit degradation, parasitism, and loss of viability to a greater extent than the latter (INVAM).

7.5.3 Sporulation Issue

According to CICG, some species of AMF like *Acaulospora* and *Entrophospora* need to undergo a dormancy period prior to germination, so trap cultures have to be kept in zip pouches before isolating spores for setting up of single-spore cultures. In order to keep spores in dormant form, spores have to be maintained on ice while performing important steps from spore extraction to rinsing in antibiotic solution (Fortin et al. 2002). Before germination, spores of some species need cold stratification at 4 °C (Smith and Read 1997), which might not be needed in case of *Glomus* (Fortin et al. 2002).

7.5.4 Subculturing and Growth Cycles

Declerck et al. (2005) observed that under monoxenic cultivation only the first generation had successful sporulation and subsequent cycles failed to grow. Under such conditions, it becomes necessary to submit the culture to germplasm collections after the culture has been grown for at least two cultivation cycles. INVAM accepts germplasm with the prerequisite that the germplasm has undergone at least one growth cycle under pot conditions. Similarly, in case of germplasm identification, both INVAM and CICG take 3–4 months as one cultivation cycle has to be completed to get sufficient number of healthy spores.

7.5.5 Storage Conditions

Storage is one of the important factors that contribute to the maintenance of infectivity of AMF inocula. Storage conditions include, i.e., temperature, humidity, duration, and carrier material. An increase in the storage duration and temperature tends to decrease infectivity potential of AMF inoculum (Daft et al. 1987; Mugnier and Mosse 1987).

7.6 Conservation and Preservation of AMF

AMF are being preserved by immobilizing their infectious propagules through polymer-based encapsulation. This involves the formation of a porous gel matrix that allows the biological material viable. Most commonly used compounds include kappa-carrageenan, agar, and alginates. However, in most of the studies, calcium alginate is employed for the entrapment of AMF infectious propagules (Lalaymia et al. 2012, 2014). One of the most promising techniques for prolonged preservation is cryopreservation. Cryopreservation at ultra-low temperature has largely been applied to filamentous fungi to keep the genetic stability intact (Smith and Onions 1994). Alginate bead entrapment followed by cryopreservation at ultra-low temperature has been identified as a proven strategy to preserve AMF. Bead used for encapsulation could provide a microenvironment providing the infectious propagules held inside with essential nutrients (Strullu et al. 1991). Encapsulation of monoxenically grown spores of *Glomus intraradices* in alginate beads followed by cryopreservation at -100 °C was found to be successful in the experiments of Declerck and Angelo-van Coppenolle (2000) where encapsulated spores showed successful germination, infection, and completed fungal life cycle. One of the important factors that drive the success of cryopreservation is the age of spore, i.e., mature spores have marked capacity to undergo cryopreservation and successive germination (Lalaymia et al. 2012). When grown in vitro, growth of sporulation follows a sigmoid pattern comprising of a lag, log, and plateau phase (Declerck et al. 2001). This phase plateau/stationary phase is characterized by conditions of nutrient limitation, i.e., spores are exposed to stress during this phase, i.e., mature spores are produced (Lalaymia et al. 2012). Mature spores have multiple and thicker spore layers as compared with young ones (Declerck et al. 2000) thus conferring resistance to cryopreservation-induced stress (Lalaymia et al. 2012).

Unlike extracted spores, encapsulated spores possess mechanical strengthful, thereby making their physical handling easier (Declerck and Angelo-van Coppenolle 2000). In the process of cryopreservation, encapsulation/entrapment of biological material is followed by osmotic dehydration in sucrose, subsequent drying in sterile air/silica gel, and cryopreservation at ultra-low temperature (Fabre and Dereuddre 1990). The method eliminates the need of routine sub-cultivation hence the risk of somaclonal variation. The factors which contribute to the success of cryopreservation include ice formation, entrapment in alginate beads, and the age of culture (Lalaymia et al. 2012). Lalaymia et al. (2012) demonstrated the preservation of 12 AMF isolates belonging to the genus *Rhizophagus via* alginate bead encapsulation for 6 months at -130 °C. Here, the germination ability of cryogenically preserved isolates was examined by PIB%, i.e., percentage of potentially infected beads which was found to be more than 70% after storing many months. However, the questions on the genetic stability and morphological as well as phenotypic aspects were answered in the experiments of Lalaymia et al. (2013) where the genetic stability remained unaffected determined based on amplified fragment length polymorphism profiles and enzymatic activities. Later on, Lalaymia et al. (2014)

Table 7.3 Forms of AMF inoculum and their storage

Component, forms of inoculum	Method of storage	References
Spores	Ringer's solution	Schenck and Perez (1990)
AMF spores	Single stage lyophilisation	Dalpé (1987)
Spores, dry hyphae and mother cells	L-drying	Tommerup (1988)
Spores of several AM fungi belonging to *Glomus*, *Gigaspora*, *Acaulospora*	Slow drying of soil of pot culture followed by in situ freezing of spores for cryoprotection and cryopreservation at 60–70 °C	Douds and Schenck (1990)
Vesicles and Hyphae of *Glomus* species (pot culture based)	Alginate bead encapsulation and storage at 4 °C	Strullu and Plenchette (1991)
Spores of *Glomus intraradices* species cultured in vitro	Cryopreservation (alginate beads encapsulation followed by trehalose incubation before cryopreservation at −100 °C in a two steps decrease	Declerck and Van Coppenolle (2000)
AMF spores and root bits produced in vitro (*Rhizophagus* species)	Cryopreservation following alginate bead encapsulation, trehalose incubation, drying	Lalaymia et al. (2012)
In vitro or in vivo produced i.e., spores, vesicles, auxiliary cells, and colonized root pieces having *Rhizophagus*, *Glomus*, *Claroideoglomus*, *Paraglomus*, *Septoglomus*, *Gigaspora*		Lalaymia et al. (2014)

showed that regardless of cultivation strategy, i.e., in vitro/in vivo, the encapsulation drying followed by cryopreservation is useful for *Rhizophagus*, *Glomus*, *Claroideoglomus*, *Septoglomus*, *Paraglomus*, and *Gigaspora*. Methods of storage of AMF are summarized in Table 7.3.

7.7 Commercial Production of AMF

AMF are being produced commercially through pot cultures and by using RiT-DNA hairy roots as bioreactor popularly known as root organ culture (ROC) technique. ROC system is an attractive technology which is rapid and contamination-free and delivers a pure and viable product within limited space (Fortin et al. 2002; Cranenbrouck et al. 2005). Douds (2002) reported monoxenic culture of *G. irregularis* with Ri T-DNA transformed roots in two-compartment Petri dishes as a very useful technique for physiological studies and the production of clean fungal tissues. Recently, Rosikiewicz et al. (2017) modified the standard dual compartment system given by St-Arnaud et al. (1996) and provided a cost-effective alternative for contamination-free in vitro production of *Rhizophagus irregularis* for molecular studies. Various forms of AM fungi are commercially produced and available in various formulations for sale throughout the world (Table 7.4). Ijdo et al. (2011)

Table 7.4 Sources of commercial AM products/inocula[a]

S. no.	Name of AMF product	Forms/Composition	Other details	Manufacturer
1.	Symbivit®	Fragments of colonized roots, spores and mycelium fragments of six AM species *Glomus etunicatum* *Glomus microaggregatum* *Glomus intraradices* *Glomus claroideum* *Glomus mosseae* *Glomus geosporum* 100 prop/g (minimum content) Inert carrier material and bioadditive components Natural clay, Humates, ground mineral, extracts from sea organisms Naturally degradable granules of a water-retaining gel	NA	*Symbiom, sro* Sázava 170, 563 01 Lanškroun, Czech Republic www.symbiom.cz http://www.inoculumplus.eu/nos-produits-en/symbivit
2.	Shubhodaya	Spores with hyphae in carrier Multiple *Glomus* species	NA	Cosme Biotech Private Limited, Vatadev Sarvona, Bicholim 403 529 Goa http://fabcoders.com.cs-mum-2.bluehostwebservers.com/shubhodaya/#section-team
3.	MYCOgold	*Beauveria bassiana*, AMF, *Azospirillum*/*Azotobacter* and micronutrients	NA	MycoGold LLC 3828 Round Bottom Rd., Cincinnati, OH 45244 http://www.mycogold.com/
4.	Mycormax™	*Glomus* species spores, colonized root fragments and other propagules in clay based carrier	NA	JH Biotech, Inc. 4951 Olivas Park Dr, Ventura, CA 93003, USA http://jhbiotech.com
5.	Micofert®	A mixture of AMF colonized soil with colonized root fragments Produced on sorghum *bicolour* in a 3:1 (v/v) mixture of soil/cachaza (sugarcane residue)	NA	Institute of Ecology and Systematics (IES), Cuba; www.ecosis.cu

(continued)

Table 7.4 (continued)

S. no.	Name of AMF product	Forms/Composition	Other details	Manufacturer
6.	AM120™ (Standard)	*Glomus intraradices* propagules (one or more species of AMF at a minimum rate of 120 total propagules; of which 20 (spores per gram), 100 (as hyphal fragments and infective root component)	NA	www.reforest.com Reforestation Technologies International 5355 Monterey Frontage Rd. Gilroy, Ca 95020
7.	Mykos™ Pure Mycorrhizal Inoculant	*Rhizophagus intraradices* 300 prop/g		
8.	MycoApply®	Ecto and Endomycorrhizal fungi (available in variety of formulations including liquid, powder and granular)	NA	Mycorrhizal Applications 710 NW E St, Grants Pass, OR 97526, USA www.mycorrhizae.com
	MycoApply Micronized Endo	*Glomus intraradices, Glomus mosseae, Glomus aggregatum, and Glomus etunicatum* Wettable powder with a particle >300 microns		
	MycoApply Micronized Endo/Ecto	Endomycorrhizal fungi (*Glomus intraradices, Glomus mosseae, Glomus aggregatum*, and *Glomus etunicatum*) Ectomycorrhizal fungi species (*Rhizopogon villosulus, Rhizopogon luteolus, Rhizopogon amylopogon, Rhizopogon fulvigleba, Pisolithus tinctorius, Scleroderma cepa, Scleroderma citrinum*) Wettable powder with a particle >300 microns		
	MycoApply Ultrafine Endo (Suitable for 'water in' or 'spray' application)	*Glomus intraradices, Glomus mosseae, Glomus aggregatum* and *Glomus etunicatum* Wettable powder with a particle >300 microns		
9.	SymBean™	AMF and *Rhizobium* (for soybean) Spores of *Glomus intraradices*: 7500 prop/g	NA	SyMyco Inc. 1005 North Warson Road, Suite #402, Creve Coeur, Missouri-63132

(continued)

Table 7.4 (continued)

S. no.	Name of AMF product	Forms/Composition	Other details	Manufacturer
		Glomus etunicatum: 3750 prop/g *Glomus clarum*: 3750 prop/g *Bradyrhizobium japonicum*: 2×10^4 cells/g		USA http://www.symyco.com
10.	SYMPLANTA-001	Spores of *Rhizophagus irregularis* water-insoluble fine calcined attapulgit clay based dry powder (liquid/gel available on request) One million spores/100 g inoculum + fewer root segments	In vitro	SYMPLANTA GmbH & Co. KG Milanweg 46, D-26127 Oldenburg Germany www.symplanta.com
11.	MycoUp	*Glomus iranicum var. tenujhypharum var. nov.* (Mycorrhizal inoculant based on *Glomus iranicum* var. *tenuihypharum* var. nov. 1.2×10^4 NMO in 100 ml substrate 0.2×10^4 prop in 100 ml of product (according to the Most Probable Number)	NA	Symborg Polígono Industrial Cabezo Cortado Avenida Jesús Martínez Cortado, 51, 30100-Espinardo, Murcia (Spain) www.symborg.com
	MycoUpActiv	*Glomus iranicum var. tenujhypharum var. nov.* with 1.2×10^4 prop/100 ml and 3.14% (w/v) of iron		
12.	MycoBloom	Endomycorrhizal fungi (diverse mycorrhizae species mix) (7 mycorrhizal species mix)	NA	519 W DODDS ST, Bloomington, IN, 47403, USA http://www.mycobloom.com

[a]Information taken from websites of respective biofertilizers

categorized AMF production into three groups, i.e., the "classical" system that uses soil/sand and other substrate combinations, hydroponic/aeroponic system producing sheared root inocula, and in vitro system employing transformed roots. Gianinazzi and Vosátka (2004) outlined some prerequisites to be followed by inoculum manufacturers, i.e., the plant that has to be inoculated should be capable of forming mycorrhizal association; no agent in AMF inoculum should demote normal plant growth and presence of a sufficiently higher shelf life of the inocula. They also recommended some standard criteria regarding the details on inoculum's physical and chemical properties. Moreover, a highly reproducible, rapid, and universal protocol for determining the quality of AMF has to be adopted globally to ensure the standard and effectiveness of any inocula produced by a particular firm. It is also suggested that the inoculum should be devoid of other microbial agents and storage conditions, shelf life, and dilution strategy should be precisely mentioned on the packet. There are reports of commercial inoculum carrying unwanted fungal species such as *Olpidium brassicae*, i.e., which is a species that forms intraradical structures similar to AMF was reported by Tarbell and Koske (2007). Therefore, maintaining the quality of AM inocula as per standards of organisms prevailing in a country has to be adhered and regulated accordingly.

7.8 Conclusions and Prospects

The germplasm conservation biobanks provide an easier approach to make the cultures available for research purpose as well as for the appropriate identification of AMF isolate form different regions. Different germplasm banks work under expert supervision that ensures the quality of AMF inocula as examination at both molecular and microscopic basis is critically performed. The inoculum thus obtained is pure, viable, and free from contaminants.

In India, for large-scale application, the AMF have been commercialized after being notified by the *Gazette* [The fertilizer control order (1985), Govt. of India] as "mycorrhizal biofertilizer" in 2010. The progress of AMF biofertilizer in India is comparatively slow as AMF are less accessed and less popular as compared to bacterial fertilizers. There is still a need of rigorous quality control measures for production of pure and viable AMF inocula for delivery. In order to promote AMF application for specialized agroecosystems where some of them are vulnerable stressed ecosystems, e.g., saline-alkaline soils, heavy metal-polluted sites, wetlands, etc., there is a need of native AMF from these ecosystems. Therefore besides AMF of crop-farming systems, concerted efforts are required to develop a gene pool of AMF for stressed ecosystems. Moreover, for quality of assessment, the existing conventional methods (spore count, infectivity potential/unit, MPN) are based on person's skills and often lack reproducibility; hence an analytical, molecular probe-based method for assessing the quality of AM inocula is required. With regard to availability of AM inocula, the existing germplasm conservation banks need to be dedicated to undertake mycorrhizal production using in vitro technology and render training for capacity building for mycorrhiza researchers at affordable cost so as to increase the availability of viable and pure AMF cultures globally. Currently, around 250 AMF species have been identified (Schübler and Walker 2010), of which majority of species with a few exceptions remain confined to the genus *Rhizophagus* maintained under trap-/pot-based system at IBG and INVAM and maintained under in vitro at GINCO (*Glomeromycota* In Vitro Collection). Ohsowski et al. (2014) noted that many AMF species colonize natural habitats, i.e., wild plants remain uncultured. Dalpé et al. (2005) showed that species with large spores might not be successfully maintained after two to three cycles of sub-cultivation as the nutrient profile/conditions present in the culture might be less appropriate; hence there is a need to conserve and develop methods to propagate these AMF species. Nevertheless, to attempt in vitro propagation and study symbiosis of novel and specialized niche AMFs, pot-/trap-based system still provides a more natural environment to propagate such AMFs. Therefore, it becomes pertinent that germplasm repositories encourage the submission of new species from different geographic locations to maximize the availability of strains. Owing to the benefits exerted by AMF on crop plants, more germplasm repositories are needed to explore all the available species and to promote the use of AMF as potential biofertilizers.

Acknowledgment The authors would like to thank the director of ICAR-Indian Institute of Soybean Research, Indore, for providing the necessary facilities. Funding from DBT-North Eastern Region Twinning Project, Govt. of India, to MPS and fellowship to RA is gratefully acknowledged.

References

Abdul-Khaliq, Gupta ML, Alam A (2001) Biotechnological approaches for mass production of arbuscular mycorrhizal fungi: current scenario and future strategies. In: Mukerji KG, Manoharachary C, Chamola BP (eds) Techniques in mycorrhizal studies. Kluwer Academic, Dordrecht, pp 299–312. https://doi.org/10.1007/978-94-017-3209-3_16

Akiyama K (2007) Chemical identification and functional analysis of apocarotenoids involved in the development of arbuscular mycorrhizal symbiosis. Biosci Biotechnol Biochem 71:1405–1414. https://doi.org/10.1271/bbb.70023

Baermann B, Linderman RG (1981) Quantifying vesicular-arbuscular mycorrhizae: a proposed method towards standardization. New Phytol 87:63–67. https://doi.org/10.1111/j.1469-8137.1981.tb01690.x

Bécard G, Fortin JA (1988) Early events of vesicular-arbuscular mycorrhiza formation on Ri T-DNA transformed roots. New Phytol 108:211–218. https://doi.org/10.1111/j.1469-8137.1988.tb03698.x

Bécard G, Piche Y (1992) Establishment of vesicular arbuscular mycorrhizal in root organ culture: review and proposed methodology. In: Norris JR, Read DJ, Varma AK (eds) Methods in microbiology: techniques for study of mycorrhiza. Academic, London, pp 89–108. https://doi.org/10.1016/S0580-9517(08)70089-8

Cárdenas-Flores A, Draye X, Bivort C, Cranenbrouck S, Declerk S (2010) Impact of multispores in vitro subcultivation of Glomus sp. MUCL 43194 (DAOM 197198) on vegetative compatibility and genetic diversity detected by AFLP. Mycorrhiza 20:415–425. https://doi.org/10.1007/s00572-009-0295-5

Chellappan P, Christy SAA, Mahadevan A (2001) Multiplication of arbuscular mycorrhizal fungi on roots. In: Mukerji KG, Manoharachary C, Chamola BP (eds) Techniques in mycorrhizal studies. Kluwer Academic, Dordrecht, pp 285–297. https://doi.org/10.1007/978-94-017-3209-3_15

Cooper AJ (1975) Crop production in the re-circulating nutrient solutions. Sci Horticult 3:251–258

Cranenbrouck S, Votes L, Bivort C, Renard L, Strullu D, Declerck S (2005) Methodologies for in vitro cultivation of arbuscular mycorrhizal fungi with root organs. In: Declerck S, Strullu DG, Fortin JA (eds) In vitro culture of mycorrhizas. Springer, Heidelberg, pp 341–348. https://doi.org/10.1007/3-540-27331-x_18

Daft MJ, Spencer D, Thomas GE (1987) Infectivity of vesicular arbuscular mycorrhizal inocula after storage under various environmental conditions. Trans Br Mycol Soc 88:21–27. https://doi.org/10.1016/S0007-1536(87)80181-X

Dalpé Y (1987) Spore viability of some Endogonaceae submitted to a single stage lyophilisation. In: Sylvia DM, Hung LL, Graham JH (eds) Proceedings of the seventh North American conference on mycorrhiza. Gainesville, p 279

Dalpé Y (2001) In vitro monoxenic culture of arbuscular mycorrhizal fungi: a major tool for taxonomical studies. In: Olalde V (ed) Proceedings of the 3rd national symposium on mycorrhizal symbioses. Guanajuato, Mexico, October, 2000. ComiteNacional de Investigacion y enseñanza de la micorriza

Dalpé Y, Cranenbrouck S, Seguin S, Declerck S (2005) The monoxenic culture of arbuscular mycorrhizal fungi as a tool for germplasm collections. In: Soil biology. In vitro culture of mycorrhizas, vol 4. Springer Berlin, pp 32–43

Declerck S, Angelo-Van Coppenolle MG (2000) Cryopreservation of entrapped monoxenically produced spores of an arbuscular mycorrhizal fungus. New Phytol 148:169–176. https://doi.org/10.1046/j.1469-8137.2000.00740.x

Declerck S, Strullu DG, Plenchette C (1998) Monoxenic culture of the intraradical forms of *Glomus* sp. isolated from a tropical ecosystem: a proposed methodology for germplasm collection. Mycologia 90:579–585. https://doi.org/10.2307/3761216

Declerck S, Cranenbrouck S, Dalpé Y, Séguin S, Grandmougin-Ferjani A, Fontaine J, Sancholle M (2000) Glomus proliferum sp. Nov.: a description based on morphological, biochemical. molecular and monoxenic cultivation data. Mycologia 92:1178–1187. https://doi.org/10.2307/3761485

Declerck S, D'or D, Cranenbrouck S, Boulengé LE (2001) Modelling the sporulation dynamics of arbuscular mycorrhizal fungi in monoxenic culture. Mycorrhiza 11:225–230. https://doi.org/10.1007/s005720100124

Declerck S, D'or D, Bivort C, de Souza FA (2004) Development of extraradical mycelium of *Scutellospora reticulata* under root-organ culture: spore production and function of auxiliary cells. Mycol Res 108:84–92. https://doi.org/10.1017/S0953756203008761

Declerck S, Seguin S, Dalpe Y (2005) The monoxenic culture of arbuscular mycorrhizal fungi as a tool for germplasm collection. In: Declerck S, Strullu DG, Fortin JA (eds) In vitro culture of mycorrhizas. Springer, Heidelberg, pp 18–30. https://doi.org/10.1007/3-540-27331-X_2

Dodd JC, Arias I, Koomen I, Hayman DS (1990) The management of populations of vesicular arbuscular mycorrhizal fungi in acid-infertile soils of a savanna ecosystem. I. The effect of pre-cropping and inoculation with VAM fungi on plant growth and nutrition in the field. Plant Soil 122:229–240. https://doi.org/10.1007/BF02851980

Douds DD (2002) Increased spore production by *Glomus intraradices* in the split-plate monoxenic culture system by repeated harvest, gel replacement, and resupply of glucose to the mycorrhiza. Mycorrhiza 12:163–167. https://doi.org/10.1007/s00572-002-0174-9

Douds DD, Schenck NC (1990) Cryopreservation of spores of vesicular–arbuscular mycorrhizal fungi. New Phytol 115:667–647. https://doi.org/10.1111/j.1469-8137.1990.tb00498.x

Douds DD, Nagahashi G, Hepperly PR (2010) On-farm production of inoculum of indigenous arbuscular mycorrhizal fungi and assessment of diluents of compost for inoculum production. Bioresour Technol 101:2326–2330. https://doi.org/10.1016/j.biortech.2009.11.071

Egerton-Warburton LO, Johnson NC, Allen EB (2007) Mycorrhizal community dynamics following nitrogen fertilization: a cross-site test in five grasslands. Ecol Monogr 77:527–544. https://doi.org/10.1890/06-1772.1

Fabre J, Dereuddre J (1990) Encapsulation-dehydration: a new approach to cryopreservation of Solanum shoot-tips. Cryo-Letters 11:413–426

Fortin JA, Bécard G, Declerck S, Dalpe Y, St-Arnaud M, Coughlan AP, Piché Y (2002) Arbuscular mycorrhiza on root-organ cultures. Can J Bot 80:1–20. https://doi.org/10.1139/b01-139

Franson RL, Bethlenfalvay GJ (1989) Infection unit method of vesicular-arbuscular mycorrhizal propagule determination. Soil Sci Soc Am J 53:754–756. https://doi.org/10.2136/sssaj1989.03615995005300030020x

Gange AC, Bower E, Stagg PG, Aplin DM, Gillam AE, Bracken M (1999) A comparison of visualization techniques for recording arbuscular mycorrhizal colonization. New Phytol 142:123–132. https://doi.org/10.1046/j.1469-8137.1999.00371.x

Gerdemann J, Nicolson T (1963) Spores of mycorrhizal endogone species extracted from soil by wet sieving and decanting. Trans Br Mycol Soc 46:235–244. https://doi.org/10.1016/S0007-1536(63)80079-0

Gianinazzi S, Vosátka M (2004) Inoculum of arbuscular mycorrhizal fungi for production systems: science meets business. Can J Bot 82:1264–1271. https://doi.org/10.1139/b04-072

Giovannetti M, Mosse B (1980) An evaluation of techniques for measuring vesicular arbuscular mycorrhizal infection in roots. New Phytol 84:489–500. https://doi.org/10.1111/j.1469-8137.1980.tb04556.x

Hepper CM (1983) The effect of nitrate and phosphate on the vesicular arbuscular mycorrhizal infection of lettuce. New Phytol 92:389–399. https://doi.org/10.1111/j.1469-8137.1983.tb03439.x

Hung LL, Sylvia DM (1988) Production of vesicular-arbuscular mycorrhizal fungus inoculum in aeroponic culture. Appl Environ Microbiol 54:353–357

Ijdo M, Cranenbrouck S, Declerck S (2011) Methods for large-scale production of AM fungi: past, present, and future. Mycorrhiza 21:1–16. https://doi.org/10.1007/s00572-010-0337-z

Jansa J, Mozafar A, Anken T, Ruh R, Sanders IR, Frossard E (2002) Diversity and structure of AMF communities as affected by tillage in a temperate soil. Mycorrhiza 12:225–234. https://doi.org/10.1007/s00572-002-0163-z

Kivlin SN, Hawkes CV, Treseder KK (2011) Global diversity and distribution of arbuscular mycorrhizal fungi. Soil Biol Biochem 43:2294–2303. https://doi.org/10.1139/cjm-2014-0416

Lalaymia I, Cranenbrouck S, Draye X, Declerck S (2012) Preservation at ultra-low temperature of in vitro cultured arbuscular mycorrhizal fungi via encapsulation e drying. Fungal Biol 116:1032–1041. https://doi.org/10.1016/j.funbio.2012.07.007

Lalaymia I, Declerck S, Cranenbrouck S (2013) Cryopreservation of in vitro-produced *Rhizophagus* species has minor effects on their morphology, physiology, and genetic stability. Mycorrhiza 23:675–682. https://doi.org/10.1007/s00572-013-0506-y

Lalaymia I, Declark S, Naveau F, Cranenbrouck S (2014) Cryopreservation of arbuscular mycorrhizal fungi from root organ and plant cultures. Mycol Res 24:233–237. https://doi.org/10.1007/s00572

Li LF, Li T, Zhang Y, Zhao ZW (2010) Molecular diversity of arbuscular mycorrhizal fungi and their distribution patterns related to host-plants and habitats in a hot and arid ecosystem, southwest China. FEMS Microbiol Ecol 71:418–427. https://doi.org/10.1111/j.1574-6941.2009.00815.x

Liu RJ, Luo XS (1994) A new method to quantify the inoculum potential of arbuscular mycorrhizal fungi. New Phytol 128:89–92. https://doi.org/10.1111/j.1469-8137.1994.tb03990.x

Moorman T, Reeves FB (1979) The role of endomycorrhizae in revegetation practices in the emi-Arid West. II. A bioassay to determine the effect of land disturbance on endomycorrhizal populations. Am J Bot 66:14–18

Mosse B, Hepper CM (1975) Vesicular–arbuscular infections in root-organ cultures. Physiol Plant Pathol 5:215–223. https://doi.org/10.1016/0048-4059(75)90088-0

Mugnier J, Mosse B (1987) Spore germination and viability of a vesicular-arbuscular mycorrhizal fungus, *Glomus mosseae*. Trans Br Mycol Soc 88:411–414

Oehl F, Sieverding E, Palenzuela J, Ineichen K, Alves da Silva G (2011) Advances in Glomeromycota taxonomy and classification. IMA Fungus 2:191–199. https://doi.org/10.5598/imafungus.2011.02.02.10

Ohsowski BM, Zaitsoff PD, Öpik M, Hart MM (2014) Where the wild things are: looking for uncultured Glomeromycota. New Phytol 204:171–179. https://doi.org/10.1111/nph.12894

Phillips J, Hayman D (1970) Improved procedures for clearing roots and staining parasitic and vesicular-arbuscular mycorrhizal fungi for rapid assessment of infection. Trans Br Mycol Soc 55:158–160. https://doi.org/10.1016/S0007-1536(70)80110-3

Plenchette C, Perrin R, Duvert P (1989) The concept of soil infectivity and a method for its determination as applied to Endomycorrhizas. Can J Bot 67:112–115. https://doi.org/10.1139/b89-016

Plenchette C, Declerck S, Diop TA, Strullu DG (1996) Infectivity of monoaxenic subcultures of the arbuscular mycorrhizal fungus Glomus versiforme associated with Ri-T-DNA-transformed carrot root. Appl Microbiol Biotechnol 46:545–548. https://doi.org/10.1007/s002530050858

Porter WM (1979) The "most probable number" method for enumerating infective propagules of vesicular-arbuscular mycorrhizal fungi. Aust J Soil Res 17:515–519. https://doi.org/10.1071/SR9790515

Redecker D, Schüßler A, Stockinger H, Stürmer SL, Morton JB, Walker C (2013) An evidence-based consensus for the classification of arbuscular mycorrhizal fungi (Glomeromycota). Mycorrhiza 23:515–531. https://doi.org/10.1007/s00572-013-0486-y

Rosikiewicz P, Bonvin J, Sanders IR (2017) Cost-efficient production of in vitro *Rhizophagus irregularis*. Mycorrhiza 27:477–486. https://doi.org/10.1007/s00572-017-0763-2

Schenck NC, Perez Y (1990) Manual for the identification of VA mycorrhizal fungi, 3rd edn. Synergistic, Gainesville

Schübler A, Walker C (2010) The Glomeromycota. A species list with new families and new genera. The Royal Botanic Garden/BotanischeStaatssammlung Munich/Oregon State University, Edinburgh & Kew/Munich/Oregon

Sharma MP, Gaur A, Bhatia NP, Adholeya A (1996) Growth responses and dependence of *Acacia nilotica* var. cupriciformis on the indigenous arbuscular mycorrhizal consortium of a marginal wasteland soil. Mycorrhiza 6:441–446. https://doi.org/10.1007/s005720050144

Sieverding E (1991) Vesicular-arbuscular mycorrhiza management in tropical agrosystems. Deutsche GesellschaftfürTechnischeZusammenarbeit (GTZ), Eschbom, p 371

Singh S, Srivastava K, Badola JC, Sharma AK (2012) Aeroponic production of AMF inoculum and its application for sustainable agriculture. Wudpecker J Agric Res 1:186–190

Smith D, Onions AHS (1994) The preservation and maintenance of living fungi. IMI Technical Handbooks, N2. International Mycological Institute. An Institute of CAB International, pp 50–65. https://doi.org/10.1111/j.1365-3164.1994.tb00033.x

Smith SA, Read DJ (1997) Mycorrhizal symbiosis, 2nd edn. Academic, San Diego. ISBN: 0-12-652840-3

Smith SE, Read DJ (2008) Mineral nutrition, toxic element accumulation and water relations of arbuscular mycorrhizal plants. In: Smith SE, Read DJ (eds) Mycorrhizal symbiosis, 3rd edn. Academic, London, pp 145–118. https://doi.org/10.1016/B978-012370526-6.50007-6

St.-Arnaud M, Hamel C, Vimard B, Caron M, Fortin JA (1996) Enhanced hyphal growth and spore production of the arbuscular mycorrhizal fungus *Glomus intraradices* in an *in vitro* system in the absence of host roots. Myc Res 100:328–332

Strullu DG, Plenchette C (1991) The entrapment of *Glomus* sp. in alginate beads and their use as root inoculum. Mycol Res 95:1194–1196

Strullu DG, Romand C, Plenchette C (1991) Axenic culture and encapsulation of the intraradical forms of *Glomus* spp. W J Microbiol Biotechnol 7:292–297. https://doi.org/10.1016/B978-012370526-6.50007-6

Tarbell TJ, Koske RE (2007) Evaluation of commercial arbuscular mycorrhizal inocula in a sand/peat medium. Mycorrhiza 18:51–56. https://doi.org/10.1007/s00572-007-0152-3

Tennant D (1975) A test of a modified line intersect method of estimating root length. J Ecol 63:995–1001. https://doi.org/10.2307/225861799

Tommerup IC (1988) Long-term preservation by L-drying and storage of vesicular arbuscular mycorrhizal fungi. Trans Br Mycol Soc 90:585–591

Yang H, Zang Y, Yuan Y, Tang J, Chen X (2012) Selectivity by host plants affects the distribution of arbuscular mycorrhizal fungi: evidence from ITS rDNA sequence metadata. BMC Evol Biol 12:50. https://doi.org/10.1186/1471-2148-12-50

Chapter 8
Conservation of Fungi: A Review on Conventional Approaches

Sanjay K. Singh, P. N. Singh, S. B. Gaikwad, and D. K. Maurya

Abstract Being an essential component of biodiversity, fungi extend various ecosystem services in nature. However, rich fungal diversity is diminishing at accelerated pace in most of the ecosystems requiring conservation strategies and appropriate methods for maintaining them on sustained basis. Many of the rare groups of fungi need suitable techniques for long-term preservation and maintenance at bioresource centre(s). Different short- and long-term methods of preserving fungi have been introduced from time to time in relation to the morphological, biochemical and genetic stability. Maintenance of fungi in a viable state is the strategic requirement of active research and development in various disciplines of basic and applied sciences. It is therefore repositories/bioresource centres the world over are practicing conventional and advanced techniques in preserving fungi on long-term basis. Generally, these methods depend on their practical relevance and cost inputs. No single method has been applied efficiently to all groups of fungi. Some techniques are inexpensive, while others are extremely expensive. The purpose of this article is to make a review and document scattered information on various conventional methods/techniques for fungal preservation, especially those which consist of natural substrates, introduced from time to time. This would help in providing ready information to users intending to practice preservation strategies for fungi on long-term basis.

8.1 Introduction

Fungi are known for centuries, but the history of identifying them dates back to their systematic study from the beginning of the eighteenth century after invention of the microscope (Ainsworth 1976). These fungi were treated under the kingdom *Plantae*. However, recent advances on understanding of the evolution have separated fungi

---

S. K. Singh (✉) · P. N. Singh · S. B. Gaikwad · D. K. Maurya
National Fungal Culture Collection of India (NFCCI), Biodiversity and Palaeobiology Group, MACS' Agharkar Research Institute, Pune, Maharashtra, India
e-mail: sksingh@aripune.org

S. K. Sharma, A. Varma (eds.), *Microbial Resource Conservation*, Soil Biology 54,
https://doi.org/10.1007/978-3-319-96971-8_8

Fig. 8.1 Illustrates tree of life, showing relationships between eukaryotic groups of fungi, and animals are sister groups, with plants as next closest relative [Taken from Tree of Life Web Project: http://tolweb.org/tree/]

from other living organisms and treated under 'The Kingdom Fungi' which consists of tremendous biological diversity and various lifestyles, forms, habitats and sizes. Recent study revealed that fungi are sister to animals; it includes numerous lineages, from yeasts to mushroom-forming fungi, rusts, smuts, moulds, etc. (Fig. 8.1). They play indispensable ecological role, in decomposition processes, in addition to their important symbiotic associations and known noteworthy parasitism (Alexopoulos et al. 1996). While much has been discovered, fungi still remain a cryptic and understudied group of organisms. Advancement in conventional and molecular techniques especially fast development of next-generation sequencing technologies promises further progress towards a better understanding of diversity and function.

As such fungi are poorly documented and the phylogenetic relationships within the kingdom are not yet conclusive, though recent attempts have enlightened the evolutionary history of the fungi (James et al. 2006; Hibbett et al. 2007).

Though fungi have established their credentials as promising gold mines for a variety of basic and applied studies, less than 1% of the fungal diversity available in nature has received serious attention for biotechnology investigations (Heywood 1995). Dimension of fungal biology is changing fast and can play important role in mainstream biology as model organism. However, increased incidences of new diseases and resistance to existing drugs lead to actual worry, require effective strategies to deal with. Systematic exploration, documentation and conservation of fungi are important for future research activities. Conservation of fungal genetic resources on a sustained basis has become a strategic requirement for supporting basic and applied research. Developing strategies to overcome the ongoing destruction of natural habitats is essential in order to compensate the loss of diverse gene pool. The conservation of fungal genetic resources by preserving dried/exsiccate specimens in mycological herbaria (Hawksworth 1974) and live cultures in germplasm banks (Singh and Baghela 2017) has been practiced. As such, conserving fungal genetic resources has become essential requirement for supporting basic, applied researches and for biotechnological innovation.

Primary objectives of preserving and storing of fungi are to maintain them in a viable state ensuring their morphological, physiological and genetic stability under laboratory conditions, which require a thorough understanding of the biology, physiology and nutritional requirement of organism to be preserved. In addition to the several advanced (unconventional) methods used in preservation of fungi, several other conventional methods consisting of different types of substrates, some of them being cost-effective methods, like agar strips, cereal grains, soil, wood chips, etc., have been reported from time to time to suit short and long-term preservation of specific group of fungi while retaining their viability and other characteristic properties. Therefore, it is intended to document different conventional methods of preservation which are simple and cost effective, but less in practice.

8.2 Preservation and Maintenance of Fungi on Organic Substrata

Review of literature reveals the fact that over the years, various effective and ingenious methods have been developed by researchers. In general, some of them are well in practice, while others are less in practice due to obvious reasons. The preservation of fungi on various organic substrata such as wood chips, cereal grains, straw, filter paper, etc. has already attracted the attention of researchers in various fields. However, several methods developed to suit long-term preservation of a specific group of fungi were not rigorously tested for nontarget group of fungi. Some of the interesting methods are detailed hereunder.

8.3 Preservation on Wood Chips

This is an interesting method for preserving wood-inhabiting fungi. Some wood-inhabiting Basidiomycetes and Ascomycetes have been reported to be maintained and stored on wood chips for about 10 years (Johnson and Martin 1992). However, in the case of fungi, they do not vigorously colonize the wood chips, and the method sometimes fails. Similarly, these fungi can successfully be stored on toothpicks as long as the colony is growing vigorously (Delatour 1991; Singleton et al. 1992). Croan et al. (1999) report preservation of tropical wood-inhabiting basidiomycetes.

Protocol

- Take about 50–60 small pieces (12-mm diam. × 6-mm thick) of beechwood and add to flask.
- Sterilize flask with wood pieces at 121 °C for 20 min twice after 24 h (Delatour 1991).
- About 15 sterilized wood chips are taken out of the flask aseptically.
- Place wood pieces on pre-grown colony on MEA in Petri plates.
- Seal the Petri plates with Parafilm and keep for incubation.
- After 10–15 days of incubation, colonized wood chips are taken out of Petri plates.
- Transfer colonized wood chips to sterile test tubes (18 × 180 mm) containing 5–7 ml of 2% malt agar. Tubes are plugged with cotton; after a week, cotton is replaced with Parafilm and aluminium foil, if required.
- Tubes may be stored at 4 °C for 10 years (*Basidiomycetes* and *Ascomycetes*).

Retrieval/revival To retrieve the culture, a piece of wood chip is removed and placed on fresh agar medium and incubated at optimal temperature. The tube is then resealed and returned to the refrigerator.

8.4 Preservation of Microfungi on Cereal Grains

This is a simple and low-cost method to maintain the fungi on cereal grains, like pearl millet, barley, wheat, maize, etc. Different fungal genera like *Trichoderma*, *Beauveria*, *Metarhizium*, *Aspergillus*, *Penicillium*, *Cladosporium*, *Rhizopus* and *Rhizoctonia* can be usually preserved, and they can survive for longer duration without changing their inherited properties (Onions 1971; Sneh et al. 1991). Singleton et al. (1992) report preservation and storage of *Sclerotinia*, *Magnaporthe*, *Leptosphaeria* and *Rhizoctonia* up to 10 years on seeds of oats, barley, rye, millet and sorghum.

Protocol

- Soak the selected cereal grains overnight in water supplemented with chloramphenicol (250 μg mL^{-1}).
- Decant the water and distribute 4 g of seeds in a screw cap vial.
- Autoclave the vial containing cereal grains for 2 consecutive days at 121 °C for 1 h. Seeds were checked on the medium for sterility before inoculation.
- Mycelial disc cut out from margins of actively pre-grown cultures was transferred to vials having cereal grains.
- Incubate the vials containing inoculated grains at optimal condition (25 ± 2 °C) for 10 days and store at −20 °C.
- Viability of the cultures stored on grains should be checked periodically, though reported to be stored for 5–10 years.

Retrieval/revival A few grains colonized by the fungus can be taken out of vials and transferred to suitable agar plate, which can be kept for incubation at optimal temperature. After the initiation of growth, purity and identity may be rechecked, and then revived culture can be used.

8.5 Preservation of Mushrooms on Cereal Grains

This is an interesting method for long-term preservation and maintenance of pure culture of several commercial mushrooms. The in vitro axenic culture of target fungus can be raised in laboratory following standard procedure (Veena and Pandey 2010). Cereal grains like jowar/sorghum (*Sorghum vulgare*) can be used to maintain culture in viable and stable form for longer periods without any change in its genotypic, phenotypic and physiological traits. Commonly mushroom culture is stored at room temperature (28–30 °C) for a limited period of 1–2 months, and hence it is not considered as a good method. However, maintenance of mushroom culture on cereal grains is an inexpensive and simple method of preservation for an extended period. The general protocol is as follows:

Protocol

- Thoroughly clean and wash cereal grain with sufficient water, and boil for 1–2 h.
- Remove the extra moisture by mixing chalk powder (calcium carbonate) with boiled grains at 4% and put in vials.
- Autoclave the vials filled with grains at 15 lb. at 121 °C for 2 h and then cool it. Seeds were checked on the medium for sterility before inoculation.
- Inoculate the autoclaved grains with 2–3 mycelial discs of 5 mm cut out from actively growing mushroom cultures in Petri plates.
- Incubate the inoculated vials in BOD incubator at optimally desired temperature.
- After complete colonization of the grains in 7–10 days, shift the vials in refrigerator (4–5 °C) for long-term storage (more than 1 year).

Retrieval/revival A few grains colonized by the fungus can be taken out of vials and kept at room temperature for some time and then transferred to suitable agar plate, which can be kept for incubation at optimal temperature. After the initiation of growth, purity and identity may be rechecked and then can be used.

8.6 Preservation on Filter Papers

The filter paper method is a good, cost-effective and efficient method. From time to time, certain modification in protocol has been reported. The modified method was found to be very effective for preservation of species of *Fusarium*, *Marasmiellus* and *Ganoderma* (Fong et al. 2000). It has been reported that 75% *M. inoderma* isolates were found to be viable after 2 years of storage at −19 °C, 81% of *Ganoderma* isolates after 5 months and 100% of *F. oxysporum* isolates after 4 years of preservation by this method. For successful recovery, fungal cultures must be incubated under optimum growth conditions prior to preservation, and mycelial mats on filter papers must be properly dried. As such this preservation method is simple, which requires a vacuum system consisting of a desiccator and a vacuum pressure station and a domestic refrigerator. Comparatively, this modified method is cost-effective than liquid nitrogen and is good for small laboratory.

Protocol

- Prepare a full-strength PDA at pH 5.6 and pour in 90-mm Petri plate aseptically.
- Cut a piece of about 70-mm filter paper (Whatman No. 1) and sterilize it.
- Filter paper can be sterilized by putting it in aluminium bag in an autoclave for 60 min at 121 °C and 15 p.s.i.
- After sterilization filter paper must be subjected to drying at 150 °C in oven for 2–3 h.
- Place this sterile and dried filter paper in the centre of a 90-mm plastic Petri dish containing full-strength PDA.
- Cut out 5 mm of agar punch from margin of actively growing colony of target fungus.
- Place five to six pieces of agar punch on PDA at equal distances apart, slightly away from the circumference of the filter paper.
- Seal the inoculated Petri dish with Parafilm, and incubate at 26–28 °C in an incubator.
- Peel off the filter paper completely covered with mycelia after 7–8 days of growth; using sterile forceps, separate it completely from PDA medium.
- Place this filter paper covered with mycelial mat inside a sterile 90-mm Petri plate, separately one filter paper per Petri plate.
- Place this Petri plate inside simple vacuum desiccator, completely filled with dry silica gel at the base.
- Connect the desiccator to a vacuum pressure station for drying out the filter paper under vacuum.
- Silica gel in desiccator may have to be replaced/recharged from time to time as it becomes wet (turns from blue to pink).

- Stop the drying process when filter papers appear dry. It takes about 15 h to dry ten filter papers.
- Ensure that the filter papers are completely dried before placing them in the freezer for storage. Too much moisture remaining in the filter papers can affect the viability of the fungal cultures destined for storage below freezing temperatures.
- Remove Petri dishes with the dried filter papers aseptically from desiccator. Take out dried filter paper with forcep and cut into small pieces, and place it inside a dry sterile screw cap bottle aseptically.
- Each fungal isolate was stored in a separate bottle. Care should be taken to ensure that the cap of the storage bottle is screwed tightly.
- Shift this bottle inside the freezer compartment of a normal household refrigerator.
- It can be stored for 8–20 years (but not all isolates can be stored this way).

Retrieval/revival A few pieces of the filter paper from individual isolate may be taken out and placed on full-strength PDA at pH 5.6 and incubated at 26–28 °C in an incubator. Confirm the viability, purity and identity before it is used.

8.7 Preservation in Soil

This method is also of low cost and effective. Several groups of fungi can be preserved successfully for many years in dry, sterile soil (Booth 1971). The shelf life of the cultures varies from one (1) to twenty (20) years depending upon the species (Smith and Onions 1994). Certain fungi such as *Rhizoctonia*, *Septoria* and *Pseudocercosporella* (Reinecke and Fokkema 1979; Sneh et al. 1991) were successfully tested. In many cases 100% of viability and retention of virulence have been reported. However, in the case of other fungi, storage in soil has been discouraged because of the risk of mutation and loss of sporulation, like *Fusarium* sp. (Windels et al. 1993; Atkinson 1954). The general protocol is as follows (Singh 2017):

Protocol

- Collect garden soil, and fill to one-third in tubes/bottles (20–30 ml) with 20% water.
- Autoclave the tubes/bottle containing soil twice on consecutive days at 121 °C for 20 min.
- Prepare mycelial/spore suspension from pre-grown culture in sterile d-water.
- Add about 1 ml of suspension to two tubes/bottles of sterile soil for making two sets.
- Label both the sets as reserve and working stocks.
- Incubate the inoculated tube/bottle at suitable temperature and time depending on the type of fungus to be stored.
- Screw the tube/bottle caps down tightly and store in a refrigerator (4–7 °C).

Retrieval/revival Take out a pinch of soil from the working stock, and sprinkle it on to a suitable agar medium and incubate under appropriate growth conditions.

Alternatively, small pinch of soil may be dissolved in sterile water, and about 0.5 ml of suspension may be inoculated on PDA medium. After 3–5 days of incubation at optimal temperature, initiation of growth may be observed. After checking the purity and identity, the revived culture can be used.

8.8 Preservation in Sterile Distilled Water

This is a low-cost and easy-to-apply technique described in the 1930s. The principle lies in extending the life of an agar culture which is extremely successful. Many filamentous moulds and yeasts are reported to survive for more than a decade in sterile water (Marx and Daniel 1976; Castellani 1967). Zygomycetous fungi are reported to be less stable in water storage, surviving only for a few months, but fungi commonly encountered in clinical and plant pathology laboratories remain viable for years and become important for those group of fungi that generally failed to freeze-drying, e.g. *Oomycetes* (Clark and Dick 1974). In general this method has been used successfully for a wide range of fungi including human or plant pathogens. Onions and Smith (1984) reported 58% viability in water after 5 years of preservation of *Pythium* and *Phytophthora* strains. Several other researchers have also used sterile water for long-term storage (Ellis 1979; Richter and Bruhn 1989; Burdsall and Dorworth 1994). Recently, Singh (2017) has reviewed in detail ex situ preservation of different groups of fungi in sterile distilled water. The length of storage is often variable. Some fungi are reported to remain viable even up to 20 years (Hartung de Capriles et al. 1989), while others lose viability much sooner. McGinnis et al. (1974) reported maintenance of stock cultures of filamentous fungi, yeasts and aerobic *Actinomycetes* in sterile distilled water. In a nutshell this method is considered a suitable storage technique for most of the fungal pathogens. Protocol as per Singh (2017) is as follows:

Protocol

- Use young and vigorously growing culture as inoculum.
- Grow on an agar slope either in a cryovial/universal bottle/test tube.
- Autoclave distilled water in separate bottle/tubes.
- Pour sterile distilled water into tube containing agar culture to cover completely the agar surface.
- Alternatively, cut out agar blocks (5 mm^3) from the growing edge.
- Transfer agar blocks (10–12) to 10 ml of sterile distilled water in bottles/tubes.
- Alternatively, sporulating or nonfilamentous organisms can be harvested without agar and can be suspended in presterilized distilled water.
- Label one bottle/tube as reserve stock and the other as working stock.
- Screw the cap of the bottle/tube tightly and store between 20 and 25 °C (Boeswinkel 1976).

Retrieval/revival Remove an agar block from a working stock, and inoculate on a suitable agar medium. Alternatively, organism can be recovered by transferring a small amount of the suspension onto suitable agar medium and incubating under optimum growth conditions. After the initiation of growth, the culture may be monitored for viability and contamination and then may be used.

8.9 Preservation/Storage in Mineral Oil

Mineral oil (paraffin oil) has been traditionally used as an effective method of long-term preservation of live cultures (Lumiere and Chevrotier 1914). Several modifications are found in literature to suit the specific requirement of a particular isolate. This method has been well recognized for preserving diverse group of microbes, and economics were also studied and reaffirmed. In principle sterile mineral oil prevents desiccation and is reported to diminish gas exchange which substantially reduces the metabolism of fungal strains to be stored. By applying this method, culture(s) can be maintained for years together. This method is space intensive and needs periodic inspection for contaminations. This method has been reviewed in detail, and the protocol provided here is as per Singh (2017).

Protocol

- Grow desired culture on agar slan and in glass test tube.
- Autoclave medical-grade white mineral oil to 2 consecutive days, to kill bacterial spores activated after first autoclave.
- Cover the culture slant with sterile mineral oil to the depth of 1 cm.
- Cover tube with caps/cotton plugs, and then apply couple of layers of paraffin film.
- Store overlaid culture tube in upright position either in refrigerator or at room temperature.

Retrieval/revival A small mass of mycelium with spores is taken out with needle/loop, and excess oil is drained off, before streaked on a suitable agar medium. Growth of the culture initially is slow due to the presence of oil, but after two to three transfers, original growth rate can be restored. Oil can be removed either by washing the mycelial mass in sterile water or by inoculating an agar slant and incubating upright, so that the oil can drain to the bottom (Singh 2017).

8.10 Preservation on Agar Strips

This method basically applies vacuum-drying fungal cultures on agar strips described by Nuzum (1989). Several difficult groups of fungi tested like *Pythium*, *Rhizoctonia* and basidiomycetous fungi have been reported to survive for 18 months. Ascomycetes and their mitosporic forms were able to survive by this method for 3–5 years.

Protocol

- Grow the fungus on suitable agar medium in Petri plate.
- After full growth cut out 1-cm-long strip from the growing edge of the colony.
- Place in an empty sterile Petri plate for air drying at room temperature.
- Transfer the dried strips of agar into sterile ampoule.
- Vacuum-dry, seal and label the ampoule.
- Place the ampoule at optimal temperature (as per requirement depending on the type of fungus used).

Retrieval/revival Remove an agar strip from ampoules, and place/inoculate on freshly prepared suitable agar medium. Alternatively, organism can be recovered by transferring a small portion of agar strip into sterile distilled water. Inoculate the suspension on to suitable agar medium, and incubate it under the optimum growth conditions. After the initiation of growth, the culture may be monitored for viability and contamination and then may be used.

8.11 Preservation on Perlite (Aluminosilicate Volcanic Mineral Particles)

This method was introduced by Homolka and Lisá (2008) for maintaining fungal cultures on perlite and cryovials. Compared to agar slant, preservation and maintenance of culture on perlite is reported to be advantageous in various ways like viability, micro- and macromorphology and growth. Besides, extracellular laccase-producing property of basidiomycetous strains after 3-week-old colonies was also checked and reported unchanged after storing on perlite. Moisture-holding property of perlite (agriculture grade) is good which can be released when required.

Protocol

- Grow the target culture on agar medium.
- Add about 200 mg of perlite particle (agriculture grade) moistened with 1 ml of wort in cryovials (1.5–2 ml).
- Inoculate the perlite in cryovials with agar plugs (5–6 mm) cut out from freshly grown colony to be preserved.
- Incubate inoculated perlite vials for about 15 days at 24–25 °C (as per type of fungus to be preserved).
- Cryovials with perlite particles overgrown by the mycelium can be stored for 4 years at 4 °C.

Retrieval/revival Small quantity of perlite culture is taken out of cryovials and subjected to 2–3 aliquots and then inoculated on suitable agar Petri dishes. After attaining the appropriate growth in about 15 days at 24–25 °C, check for purity and identity before further use. The culture meant to enzyme production can be checked for stability after revival following the method of Niku-Paavola et al. (1990).

8.12 Maintenance and Preservation of Mycological Herbarium Specimens

Dried and correctly pressed plant specimens and their ancillary collections (e.g. high-resolution photographs) and passport data are remarkable and irreplaceable sources of information about a particular plant and its inhabitants. Herbarium specimens serve as the comparative material essential for studies in taxonomy, systematics, ecology, anatomy, morphology, conservation biology, biodiversity, ethnobotany and palaeobiology. They are being used for teaching and are a veritable gold mine of information and serve as the foundation of comparative biology. This is one of the most important and culture-independent ways to maintain variety of fungi on its natural hosts as exsiccate. There are various ways of preparing herbarium samples suitably for long-term storage in herbaria. Since each group of fungi has its own requirements and limitations, they need to be handled differently. Considering the life cycles of different groups of fungi perpetuating at different times in a year, it is important to collect samples in different seasons and consecutive years in order to get the different fruiting structures important for taxonomic study, e.g. rust and smut infections are found prominent during autumn, while mushrooms are collected during or soon after the rainy season. Different fertile fruiting bodies play important roles in identification of Ascomycetes and Basidiomycetes and require special attention for their exploration and collection. It is not always expected that different stages like asexual and sexual morphs are found in the same location and in the same season. Hence a systematic planning is required for making comprehensive and referral collections. The longevity of preserved specimens especially various fruiting structures present along with symptoms on aerial part of the plants largely depends on the methods followed correctly in preparation of mycological herbarium. It is advised to refer to Hawksworth (1974) for correct preparation of mycological herbarium and their long-term storage.

8.13 Importance and Basic Functions of Mycological Herbarium

Herbaria serve as natural collection of preserved specimens for various purposes, like the following:

- Serve as a secure repository of exsiccate required for taxonomy and systematics of fungal specimens.
- Serve as reference collection for comparative taxonomy and systematics of associated fungi.
- Provide material for making morphological measurements as part of taxonomy and systematic studies.

- Provide passport information including locality which is important for planning field exploratory trips.
- Verify names/nomenclature revision and monographic works.
- Promote exchange of specimens among bonafied institutions.
- Provide information regarding fungal fruiting times, seasons, temperatures, climatic conditions, etc.
- Provide information for GIS studies of past and future collecting expeditions.
- Provide basis for an illustration of taxonomy and general publishing.
- Provide material for DNA analysis of associated fungi.
- Provide vouchers, photographs used during lectures and publications of taxonomic account.
- Maintain information about common, rare or extinct fungi that can no longer be found in nature.
- Provide material for microscopic observations.
- Provide information for studies of expeditions and explorers (history of science).
- Serve as an archive for related material (field notebooks, letters, reprints, etc.)
- Provide information on effect of climate change on life cycle and fruiting of fungi.
- Organize photographs of different taxonomic groups of fungi associated with voucher collection.
- Provide material for teaching and promote diversity by making specimens available for viewing by students, researchers and the public.
- Attract students to systematic research and help in running education courses for teachers.
- Serve as centre of identification of various groups of fungi used for various purposes.

8.14 In Situ Conservation

It is now well recognized that fungi are the essential component of biodiversity requiring serious attention. Even after implementation of convention of biological diversity status of fungal conservation, in general, is yet to improve and require legal protection. Apart from ex situ, these fungi can be conserved in situ. Different types of strategies have been recognized like (a) conservation of natural habitats, (b) establishment of mycological reserves and (c) use of ecological corridors (Courtecuisse 2001). It is essential to compile red list fungi as per the guidelines of IUCN, which will help in deciding appropriate strategies of protection/in situ conservation by concerted efforts of national policy, like protection of many forests by Indian Forest Act, 1927, in different categories, like reserve forests, protected forests, national parks, etc. In this regard, a detailed MycoAction Plan is very much essential for the protection of fungi at various levels and their sustainable utilization (Hawksworth 2003).

8.15 Conclusion and Future Prospects

Fungi have always been put to human uses, playing significant roles in advances of molecular genetics, fermentation technology and modern biotechnology. Bioresource centres (BRCs)/culture collections (CC) preserve, maintain and manage microbial genetic resources complying OECD guidelines of best practices. BRCs/CC can play a capacity-building role to help better understand and utilize microbial genetic resources of the country. They can help initiate/establish ex situ and/or in situ conservation programs using preservation technologies in order to retain properties of microbial strains by practicing various simple and cost-effective methodologies. These efforts would make the availability of authentic and high-quality research materials (strains) to research community on sustained basis. Well-preserved indigenous strains are the important genetic materials for screening programs. This would lead to the biodiversity-rich country, like India, to benefit by sustainable utilization of the microbial genetic resources. Also collection can protect sovereign rights, ownership and IPR and generate bio-economy on sustained basis through collaborative approaches and partnerships established with intergovernmental agencies. Key objectives of the Convention on Biological Diversity (CBD) also mandate and support these views, especially (1) conservation of biodiversity, (2) sustainable development of genetic resources and (3) fair and equitable sharing of resultant benefits. Though fungal conservation has never been treated as part of conservation biology, short- and long-term preservation and maintenance of fungal strains/isolates using classical, cost-effective and convenient methodologies would help in popularizing conservation efforts as regular practice which helps conserve wild strains in gene banks for future research and development.

Acknowledgement Authors thank the director of MACS-Agharkar Research Institute (ARI) for providing laboratory facilities and Department of Science and Technology (DST), Government of India for establishing National Facility for Culture Collection of Fungi at MACS'ARI.

References

Ainsworth GC (1976) Introduction to the history of mycology. Cambridge University Press, Cambridge, UK

Alexopoulos CJ, Mims CW, Blckwell M (1996) Introductory mycology, 4th edn. Wiley, New York, p 870

Atkinson RG (1954) Quantitative studies on the survival of fungi in five-year-old dried soil cultures. Can J Bot 32:673–678

Boeswinkel HJ (1976) Storage of fungal cultures in water. Trans Br Mycol Soc 66:183–185

Booth C (1971) The genus *Fusarium*. CMI Kew, England, p 237

Burdsall HH, Dorworth EB Jr (1994) Preserving cultures of wood-decaying Basidiomycotina using sterile distilled water in cryovials. Mycologia 86:275–280

Castellani A (1967) Maintenance and cultivation of common pathogenic fungi of man in sterile distilled water. Further researched. J Trop Med Hyg 70:181–184

Clark G, Dick MW (1974) Long-term storage and viability of aquatic Oomycetes. Trans Br Mycol Soc 63:611–612

Courtecuisse R (2001) Current trends and perspectives for the global conservation of fungi. In: Moore D, Nauta MM, Evans SE, Rotheroe M (eds) Fungal conservation. Issues and solutions. Cambridge University Press, Cambridge, UK, pp 7–18

Croan SC, Bardsall-Jr HH, Rentmeester RM (1999) Preservation of tropical wood- inhabiting basidiomycetes. Mycologia 91:908–916

Delatour CA (1991) A very simple method for long-term storage of fungal cultures. Eur J Forest Pathol 21:444–445

Ellis J (1979) Preserving fungus strains in sterile water. Mycologia 71:1072–1075

Fongs YK, Anuar S, Lim HP, Tham FY, Sanderson FR (2000) A modified filter paper technique for long-term preservation of some fungal cultures. Mycologist 14:127–130

Hartung de Capriles C, Mata S, Middelveen M (1989) Preservation of fungi in water (Castellani): 20 years. Mycopathologia 106:73–80

Hawksworth DL (1974) Mycologists' hand book. CMI Kew, Surrey, p 231

Hawksworth DL (2003) Monitoring and safeguarding fungal resources worldwide: the need for an international collaborative MycoAction Plan. Fungal Divers 13:29–45

Heywood VH (1995) Global biodiversity assessment. Cambridge University Press, Cambridge, UK

Hibbett DS, Binder M, Bischoff JF, Blackwell M, Cannon PF, Eriksson OE, Huhndorf S, James T, Kirk PM, Lucking R, Thorsten LH, Lutzoni F, Matheny PB, McLaughlin DJ, Powell MJ, Redhead S, Schoch CL, Spatafora JW, Stalpers JA, Vilgalys R, Aime MC, Aptroot A, Bauer R, Begerow D, Benny GL, Castlebury LA, Crous PW, Day YC, Gams W, Geiser DM, Griffith GW, Gueidan C, Hawksworth DL, Hestmark G, Hosaka K, Humber RA, Hyde KD, Ironside IE, Koljalg U, Kurtzman CP, Larsson KH, Lichtwardt R, Longcore J, Miądlikowska J, Miller A, Moncalvo J-M, Mozley-Standtridge S, Oberwinkler F, Parmasto E, Reeb W, Rogers J-D, Roux C, Ryvarden L, Sampaio JP, Schussler A, Sugiyama J, Thorn RG, Tibell L, Untereiner WA, Walker C, Wang Z, Weir A, Weiss M, White MM, Winka K, Yao Y-J, Zhang N (2007) A higher level phylogenetic classification of the fungi. Mycol Res 111:509–547

Homolka L, Lisá L (2008) Long-term maintenance of fungal cultures on perlite in cryovials – an alternative for agar slants. Folia Microbiol 53:534–536

James TY, Kauff F, Schoch CL, Matheny PB, Hofstetter V, Cox CJ, Celio G, Gueidan C, Fraker E, Miadlikowska J, Lumbsch HT, Rauhut A, Reeb V, Arnold AE, Amtoft A, Stajich JE, Hosaka K, Sung GH, Johnson D, O'Rourke B, Crockett M, Binder M, Curtis JM, Slot JC, Wang Z, Wilson AW, Schussler A, Longcore JE, O'Donnell K, Mozley-Standridge S, Porter D, Letcher PM, Powell MJ, Taylor JW, White MM, Griffith GW, Davies DR, Humber RA, Morton JB, Sugiyama J, Rossman AY, Rogers JD, Pfister DH, Hewitt D, Hansen K, Hambleton S, Shoemaker RA, Kohlmeyer J, Volkmann-Kohlmeyer B, Spotts RA, Serdani M, Crous PW, Hughes KW, Matsuura K, Langer E, Langer G, Untereiner WA, Lucking R, Budel B, Geiser DM, Aptroot A, Diederich P, Schmitt I, Schultz M, Yahr R, Hibbett DS, Lutzoni F, McLaughlin DJ, Spatafora JW, Vilgalys R (2006) Reconstructing the early evolution of fungi using a six-gene phylogeny. Nature 443:818–822

Johnson GC, Martin AK (1992) Survival of wood inhabiting fungi stored for 10 years in waters and under oil. Can J Microbiol 38:861–864

Lumiere A, Chevrotier J (1914) Sur la vitalite des cultures de gonocoques. Compt Rend 158:1820–1821

Marx DH, Daniel WJ (1976) Maintaining cultures of ectomycorrhizal and plant pathogenic fungi in sterile water cold storage. Can J Microbiol 22:338–441

McGinnis MR, Padye AA, Ajello L (1974) Storage of stock cultures of filamentous fungi, yeast and some aerobic actinomycetes in sterile distilled water. Appl Microbiol 28:218–222

Niku-Paavola ML, Raaska L, Itavaara M (1990) Detection of white-rot fungi by a non-toxic stain. Mycol Res 94:27–31

Nuzum C (1989) A simple method for the preservation of some non-sporing fungi. Australian Plant Pathol 18:104–105

Onions AHS (1971) Preservation of fungi. In: Booth C (ed) Methods in microbiology, vol 4. Academic, London, pp 113–151
Onions AHS, Smith D (1984) Current status of culture preservation and technology. In: Batra LR, Iigima T (eds) Critical problems of culture collections. Institute of Fermentation, Osaka
Reinecke P, Fokkema NJ (1979) *Pseudocercosporella herpotrichoides*: storage and mass production of conidia. Trans Br Mycol Soc 72:329–331
Richter DL, Bruhn JN (1989) Revival of saprotrophic and mycorrhizal basidiomycete cultures from cold storage in sterile water. Can J Microbiol 35:1025–1060
Singh SK (2017) Ex situ conservation of fungi: a review. In: Satyanarayana T, Deshmukh SK, Johri BN (eds) Developments in fungal biology and applied mycology. Springer, Singapore, pp 543–562
Singh SK, Baghela A (2017) Cryopreservation of microorganisms. In: Varma A, Sharma AK (eds) Modern tools and techniques to understand microbes. Springer, Cham, Switzerland, pp 321–333
Singleton LL, Mihail JD, Rush CM (1992) Methods for research on soilborne phytopathogenic fungi. APS, St Paul, Minnesota, p 266
Smith D, Onions AHS (1994) The preservation and maintenance of living fungi. In: IMI technical handbook, 2nd edn. CMI, Kew, p 122
Sneh BL, Burpee L, Ogoshi A (1991) Identification of *Rhizoctonia* species. APS, St Paul, Minnesota, p 133
Veena SS, Pandey M (2010) A simple method for culture conservation of some commercial mushrooms. Mycosphere 1:191–194
Windels CE, Burnes PM, Kommedahll T (1993) *Fusarium* species stored on silica gel and soil for ten years. Mycologia 85:21–23

Chapter 9
Veterinary Type Cultures and Their Preservation: Status and Challenges

Rajesh K. Vaid, Taruna Anand, K. Shanmugasundaram, and Bhupendra N. Tripathi

Abstract Microbial repositories are seminal to the development of biotechnology and its myriad applications in animal and agriculture bio-economy. These are the specialized storehouses of authenticated microbial cultures and accessory products needed for research and development in the animal agriculture. The public fund managed microbial culture collections like National Centre for Veterinary Type Cultures (NCVTC) and National Bureau of Agriculturally Important Microorganisms (NBAIM), initiated by the Indian Council of Agricultural Research (ICAR), are the future of the emerging agriculture bio-economy as they cement their positions as premier sources of authenticated microbial resources from ex situ preserved biodiversity of microbial strains of veterinary, dairy, rumen, and agriculture origins. Apart from this, they can fuel the development of taxonomy and exploration of animal microbial biodiversity and act as source of new knowledge, data, and skill in the field of microbiology. Their role as managers and disseminators of IPR issues is related to safekeeping of various categories of strain deposits. Although the demands and expectations from repositories are increasing, the challenges faced in the day-to-day functioning and maintenance of culture collection are also enormous.

9.1 Microbe Resource Centers: Bedrock of Modern Bio-economy

Culture collections (CC) or microbial resource centers (MRC) are needed for preservation and provision of biological resources for scientific, industrial, agricultural, environmental, and veterinary-medical research and development (R&D) applications. The essential functions of such centers consist of doing research on these biological resources with an aim for ex situ conservation of biodiversity and authentication and characterization of strains in order to make them available for

R. K. Vaid (✉) · T. Anand · K. Shanmugasundaram · B. N. Tripathi
National Centre for Veterinary Type Cultures (NCVTC), ICAR-National Research Centre on Equines, Hissar, Haryana, India
e-mail: rk_vaid@yahoo.com

S. K. Sharma, A. Varma (eds.), *Microbial Resource Conservation*, Soil Biology 54,
https://doi.org/10.1007/978-3-319-96971-8_9

future application in biotechnology and enhancing food security (Sly 2010). These centers act as repositories of critical microbial reference materials in laboratory research, testing, and diagnosis (Stackebrandt 2010). The creation of new knowledge of genetic engineering and molecular biology in the last century is giving way to modern omic sciences including genomics, proteomics, and transcriptomics, the tools which hold promise to unlock the useful information inherent in versatile microbial genetic resources (MGR). In order to capture and utilize such benefits of developing bio-economy, culture collections provide the well-described living resources (OECD 2001). They also function as repositories of microbial resources for protection of intellectual property. These will act as resources for public information and policy formulation also. The governments world over are actively establishing, nurturing, and sustaining MRCs because by making available microbes, their products and information of guaranteed identity and quality, a MRC will serve an essential infrastructural function for scientific investigation and R&D (Sasson and Malpica 2017).

It is conventional wisdom that a scientific research experiment requires reproducibility with statistical robustness. Scientific experiments, products, and vaccines must be able to replicate the responses and results from one geographical setting to another (McCluskey et al. 2017). Similarly, it is important that the microbial resources are also purified, properly authenticated, and genetically stable as is required of the purity of chemical reagents and the precision of equipments used to conduct scientific research (Prakash et al. 2013). The availability of known, validated, and precisely identified microbial resources is essential for biomedical and biotechnological research (Overmann 2015). Molecular-based assays like PCR, qPCR which are used to target specific genes of virulence factor or taxonomic indicator gene (*rpo*B) from a given microbial analyte, need to be demonstrating a diagnostic validity, for the target pathogen, for which positive and negative control strains are required (USFDA 2015).

Antimicrobial sensitivity testing is a routine procedure in clinical laboratories. The emergence of antimicrobial-resistant organisms and multidrug-resistant strains (MDR) makes it important for a more judicious use of antimicrobials, as newer drugs are rare to come by (May 2014). Therefore, antimicrobial drug testing is an important part of treatment for proper guidance to clinician in selection of an antimicrobial agent; however, a quality-controlled drug testing requires reference strains for different group of infections and antibiotics (CLSI 2008). Microbial strains collected in different space and time are useful resources for evaluation of temporal and spatial trend studies of their drug susceptibility profile, as evaluated by measurement of minimal inhibitory concentrations (MIC) against given antimicrobials (Jorgensen and Ferraro 2009). Moreover, as the discovery of novel antibiotics by Waksman platform method, which requires screening and evaluation of a large number of potential strains, is tapering off (Lewis 2012), the genomic tools are offering innovative ways to produce novel antibiotics by synthetic biology techniques. For instance, the sequence of *Streptomyces coelicolor* revealed many clusters of genes which if manipulated can lead to production of novel antimicrobial products (Bentley et al. 2002; Braff et al. 2016).

9.2 Wider Veterinary Research and Withered Microbial Germplasm

The history of establishment of research infrastructure in the field of veterinary sciences in India starts in 1893 with the establishment of Imperial Veterinary Research Institute (IVRI) by British colonial rulers at Himalayan hilly location of Mukteshwar and Gangetic plains of Izzatnagar near Bareilly, in United Provinces, the present Uttar Pradesh (UP) (Sinha 2010). In order to prepare vaccine and hyperimmune sera for use in prophylaxis of cattle plague or rinderpest (RP), the additional facility was created in 1901 in the plain region of Izzatnagar, Bareilly. The preparation of sera for bacterial diseases like anthrax and hemorrhagic septicemia (HS), vaccine for black quarter (BQ), and mallein reagent was started in the first decade of the twentieth century. The research ambit of IVRI has, since its inception, gradually increased from doing investigation and treatment of domestic animals to research and technological development on all animal endemic diseases of India. However, a systematic repository of the microbial strains, which might have been isolated in late eighteenth century and in early nineteenth century up to 1947, when Imperial Veterinary Research Institute was rechristened Indian Veterinary Research Institute, was either not created or was probably lost. Old Indian origin cultures, NCTC 3708 (mule isolate, 1932, India) and NCTC 3709 (horse isolate, 1932, India), of *Burkholderia mallei*, a highly pathogenic zoonotic bacteria causing glanders in equines, a disease still endemic in India, are part of National Centre for Type Cultures (NCTC), UK, but such or other strains are sadly not available in India. These and other cultures of Indian origin are being utilized for studying their biology (Gee et al. 2003). Similarly, another *B. mallei* strain (Genome Online database ID Go0000225; NCBI taxonomy ID 320388), which is identified as a strain (SAVP1) "that had previously caused disease in a mule (in India)," lacks in information on isolation metadata (Schutzer et al. 2008).

Although, Civil Veterinary Departments were created in the provinces in 1889, veterinary colleges were started at Babugarh (1877), Lahore (1882), Bombay (1890), and Madras and Calcutta (1893), and investigation of the diseases of the animals in India began in 1889–1890 with appointment of Dr. Alfred Lingard as imperial bacteriologist at the College of Science at Poona; however, any initiation to curation of cultures of veterinary-medical significance isolated during that period, or subsequently, has been lost in time. Meanwhile, National Collection of Type Cultures, UK, was established in 1920 (Russell 2014) and the American Type Culture Collection (ATCC) took its root in 1925 (Buchanan 1966), which are presently functional and cover a wide variety of reference microbes for use by industry. Similar facilities, however, were lacking in India, especially in the agricultural and livestock sectors, until recently. The Convention on Biological Diversity (CBD), which came into force from December, 1993, prompted signatory national governments to establish and maintain facilities for the ex situ conservation of different facets of biological diversity in the country of origin, considering that microbes are integral part of biodiversity.

9.3 Culture Collection for Value Addition

Enormous sums have continuously been spent in veterinary science institutions of teaching and research in our country toward isolation of microbes and their genes from nature. The elucidation of the genetic and functional molecular elements of microbes is capital intensive (OECD 2011). It is not only essential for these veterinary microbial resources to be preserved but also to be used further, so that research findings can be strengthened and corroborated. Without such a microbial repository, every user would have to reinvent the wheel and invest innumerable hours and money in the recovery of organisms and genes and their characterization, again and again. International studies which elucidate the impact of MRCs have categorically shown that microbial repositories and other biological resource centers (BRCs) significantly increase the value of research if the research materials are deposited and thus shared (Furman and Stern 2011).

The MRCs are safe heavens where researchers can deposit their well-characterized strains so that these are available for further research. It will also foster a culture of utilization of deposited strains for research work. The impact of publications increases if well-curated cultures are utilized in studies (Furman and Stern 2011). Moreover, as the cultures are isolated, identified, characterized, and preserved ex situ, they gain in value as compared to the in situ clone, which is considered to be of negligible monetary value (WFCC 1997). Although it is not always possible to ascertain the true value of a given microbial isolate, many factors, such as time and expertise needed to grow and catalogue a strain, its fastidiousness on culture media, and its level of characterization and preservation, all add value to a cultured strain. Consequently, it has been calculated that a microbial strain, which has been isolated and deposited in a European CC, attains a value of approximately 1000 €, which is roughly equal to 752,528 ₹, whereas same strain will be of 5042 €, approximately equal to 379,424 ₹ (2017 prices) in India (Overmann 2015). It thus implies that by ex situ isolation, characterization, and preservation of veterinary microbial resources, we can look forward to development of animal agriculture bio-economy through various value additions, in contrast to uncharacterized wild isolates remaining in situ.

The value added to a characterized isolate also indirectly indicates the quantum of economic loss incurred by our nation when researchers either do not deposit or fail to deposit their isolated cultures in public repositories. Internationally also, the deposition of "microbial strains cited in publications" into microbial repositories for their smoother publication is poor (Stackebrandt 2010). In a cursory back of the page survey done on Indian Journal of Animal Sciences publications related to the microbes, it was observed that out of total number of 16 papers published in 12 issues (2016–2017), in which research related to various aspects of microbes was reported, only 2 articles utilized authenticated reference or accredited accessioned cultures, and none of the papers indicated deposition of the cultures in any of the public-funded microbial collections (unpublished data). This loss can be reversed if public-funded institutions, funding agencies, and journals can formulate their own coherent policy regarding the mandatory deposition of microbial cultures isolated in a research in a MRC. In addition to serve as repository of such strains as deposited by principal investigators (PIs) of projects and authors, a veterinary MRC will have opportunities to carry out R&D on the microbial resources they house including research on identification, characterization, and preservation of biological resources. Their R&D activities can contribute to the advancement of the veterinary-medical life sciences. However, MRC need to balance their R&D function with their culture distribution service function, providing and preserving microbial resources for the wider scientific communities.

The Bureau of Animal Industry of the USA in the decades of 1920–1940 embarked upon an ambitious program to control brucellosis, a dreaded zoonotic disease, which is caused by pathogen *Brucella abortus* in cattle. Research was conducted with one of the specific aims to formulate an effective vaccine. Incidentally, the bureau veterinarian, Dr. John M. Buck, had maintained a group of *B. abortus* cultures for "well over a year" on his desk at room temperature. In a subsequent evaluation of "stored cultures" for immunogenicity and stability, the 19th culture evaluated was found to be significantly less pathogenic. It was also stable even after many animal transmission studies, thus eventually becoming an optimal vaccine strain (Graves 1943). This live, attenuated strain S19 had been in use all over the world including India since the early 1930s as an effective vaccine to prevent brucellosis in cattle (Nicoletti 1990).

The anecdote illustrates the importance of well-documented microbial strains as a tool of biology. Microbial strains or MGRs obtained from animals, such as bacteria, pathogenic or commensal strains, viruses including bacteriophages, fungi and their genes, plasmids, nuclear material, cDNAs, and the related information are the essential raw materials for the advancement of animal biotechnology, veterinary and human health, and research and development in the life sciences in general and veterinary-medical sciences and public health in particular. Clinical microbial strains of bacteria, viruses, and fungi isolated from disease outbreaks and preserved with relevant epidemiological metadata are useful tools for analysis and understanding of patterns of pathogen transmission and modes of dispersion and taxonomical studies (Holmes et al. 1978; Plainvert et al. 2014). The data attached to strains can give information about time and location of the host from which the pathogen was isolated.

Culture collections are the custodians of microbial diversity and play a key role in storage and supply of authentic reference material for research and development. The importance of collection of biological specimens has always held its forte. However, the recent past has witnessed a new increased awareness of the value of culture collections. This has been partly due to emergence of new genomic and proteomic level molecular technological breakthroughs and also due to the realization of the value of conservation of genetic resources and maintenance of biodiversity, especially after the comity of consenting nations signed CBD. The breakthrough in the science of sequencing and gene hunting has opened up a new and exciting field of discovery. It is thus imperative to understand and nurture MRCs or CC which are going to be the support source houses for scientific development in the field of agriculture, health, and environment biotechnology.

9.4 Overview of the National Centre for Veterinary Type Cultures

The National Centre for Veterinary Type Cultures (NCVTC) was established as a microbial repository for the collection, characterization, and distribution of animal microbes at ICAR-NRCE, Hisar. The NCVTC was sanctioned in April 2004 and its actual functioning began in 2006. The National Agriculturally Important Microbial Culture Collection (NAIMCC), dedicated to the microbes in agriculture, was established at NBAIM, Maunath Bhanjan, in Uttar Pradesh by ICAR in agriculture sector and with a similar mandate applicable in microbes of plant origin.

The activities of NCVTC comprise of acquisition, authentication, preservation, documentation, and repository database management system of microbes of animal origin, so that resource generated can be utilized for application in control and prevention of animal disease, application in dairy science, and in understanding the rumen function. Apart from these functions, NCVTC also caters to requests of taxonomic identification of culture isolates of veterinary background. In order to rapidly develop the repository for representation of microbial strains from different animal species and environments, a plan of carrying forward the collection work was mooted by formation of a Network of State Agricultural Universities and ICAR throughout the country. As India is a large country, so initially the states of Haryana, Rajasthan, Uttar Pradesh, Himachal Pradesh, Assam, Jammu and Kashmir, Tamil Nadu, Gujarat, Uttarakhand, Karnataka, Arunachal Pradesh, and Nagaland were included in Network.

A culture collection must meet the high standards of quality and expertise expected from a researcher for the delivery of biological information and materials. A veterinary type culture setup has to provide access to microbial resources on which R&D in the veterinary-medical sciences, animal sciences, and the advancement of animal biotechnology depends. Precisely in this direction, the DARE, Indian Council of Agricultural Research, New Delhi, had sanctioned this new activity of veterinary type culture, during the tenth plan period with the following mandate and objectives.

9.4.1 Mandate

- National repository of veterinary, dairy, and rumen microorganisms and their identification, characterization, and documentation.

- Distribution of microbes for teaching, research, and development of new technologies.

9.4.2 Objectives

(a) Veterinary Microbes

- Exploration and collection of microbes of veterinary origin/significance/relevance
- To identify and characterize the isolated/collected microbial pathogens (viruses, bacteria, fungi, mollicutes, protozoa, etc.) isolated from different disease conditions and carrier animals by conventional and molecular techniques
- Collection and central storage of animal microbes from existing culture collection centers, institutions, and universities
- To clone and sequence important genes to generate epidemiological information and to understand disease pathogenesis
- To create a genome/gene bank of different microbes by storing genomic DNA/RNA, recombinant plasmid clones, transfected cell lines, transformed bacteria, etc. for conservation of microbial biodiversity

(b) Rumen Microbes

- To isolate cellulolytic rumen bacteria from Indian cattle, sheep, goats, and buffaloes
- To isolate rumen fungi from domesticated and wild animals
- To determine the best fibre digesting organism

(c) Dairy Microbes

- To isolate indigenous strains of lactic acid bacteria including the ones with probiotic attributes from ethnic fermented milk and other foods prepared in various parts of India
- To identify and characterize the isolated strains using conventional and molecular techniques
- To analyze the plasmid pool encoding commercially important traits of selected cultures and their characterization at molecular level
- To identify the commercial potential of these diversified strains for designing novel starter cultures with functional/probiotic properties
- To conserve these strains and their germplasm ex situ and create a well-catalogued collection

To date, 726 culture collections from 75 countries have been registered in World Federation of Culture Collection (WFCC 2017). However, out of these, only a small percentage caters to animal and human pathogenic microbe culture collection.

Although the number of veterinary and animal microbe-related CC is unknown, a cursory survey of WFCC data indicates that more than 20 repositories are listed in World Data Center for Microbes (WDCM) (Table 9.1). There are many specialized bacteria culture collections also, and at least ten of these are dedicated to collection and research on strains of *Leptospira*, *Brucella*, *Salmonella*, *Yersinia*, and even antimicrobial-resistant bacteria (Table 9.2).

Table 9.1 International culture collections dedicated to veterinary microbes, dairy and related strains

Culture collection/ (status)	Acronym	WDCM number	Category of cultures	Date of institution	Number of cultures
Centro de Investigaciones en Ciencias Veterinarias, Buenos Aires, Argentina	CICV	WDCM33	Animal bacteria, protozoa, hybridoma, viruses, plant viruses	1981	200–500
Center for Veterinary Culture Collection, Beijing, China	CVCC	WDCM129 WDCM876	Veterinary bacteria, cell lines and viruses	1985	1300
Pusat Veterinaria Farma, Surabaya, Indonesia	PVF	WDCM628	Veterinary and industrial cultures	1989	NA
Korean Veterinary Culture Collection, Gyeongsangbuk-do, South Korea	KVCC	WDCM954	Veterinary bacteria, viruses	2009	>6000
Veterinary Branch of National Strain Collection, Pulawy, Poland	DMVB	WDCM194	Bacteria	1981	450
Thai Collection of Medical Microorganism, Chulalongkorn University, Bangkok, Thailand	TCMM	WDCM661	Bacteria	1981	NA
Animal Health Division Culture Collection, Perth, Australia	AHLDA	WDCM334	Bacteria	1981	1350
Rumen Yeast Collection of Animal Science Institute, Mayabeque, Cuba	RYCASI	WDCM980	Yeast	2011	NA
Collection of Animal Pathogenic Microorganisms, Brno, Czech	CAPM	WDCM181	Veterinary bacteria/Viruses	1962	900
Collection of Animal Viruses, Baranya, Hungary	RCAT	WDCM425	Veterinary Viruses	1981	~50
Regional Collection of Animal Viruses and Tissue Cultures, Szeged, Hungary	IMMH	WDCM427	Animal Viruses	1981	36
National Institute of Animal Health, Ibaraki, Japan	NIAH	WDCM638	Veterinary microbes	1990	100

(continued)

Table 9.1 (continued)

Culture collection/ (status)	Acronym	WDCM number	Category of cultures	Date of institution	Number of cultures
Culture Collection of Animal Cells, Foot and Mouth Disease Institute, Ankara, Turkey	HUKUK	WDCM756	Animal cells and hybridoma	2008	100
Rumen Microorganisms, Grassland Research Centre, New Zealand	RM	WDCM764	Rumen microbes	1995	150
Culture Collection of Dairy Microorganisms Laktoflora, Tabor, Czech	CCDM	WDCM878	Dairy bacteria, yeasts and fungi	2005	1000
New Zealand Reference Culture Collection of Microorganisms, Dairy Section	NZRD	WDCM318	Bacteria, virus	1968	3000
Centro de Referencia para Lactobacilos, Argentina	CERELA	WDCM614	*Lactobacillus, Propionibacteria and Bifidobacteria*	1981	900
Greek Aquaculture Bacteria, Veterinary Medicine, University of Thessaloniki, Greece	GAB	–	Aquaculture	2006	100

Table 9.2 International culture collections dedicated to specific pathogenic microorganisms

Culture collection/(Status)	Acronym	WDCM number	Cultures	Date of institution	Number of cultures
Brucella ANSES Culture Collection, Maisons-Alfort, France	BACC	WDCM789	*Brucella*	1998	>3000
WHO/FAO/OIE Collaborating Centre for Reference and Research on Leptospirosis, Queensland, Australia	CPHS	WDCM14	*Leptospira*	1981	213
Leptospirotheque, Paris, France	LIPP	WDCM345	*Leptospira*	1981	1020
Collection of Leptospira Strains, Instituto Superiore di Sanita, Rome, Italy	RV	WDCM421	*Leptospira*	1981	600

(continued)

Table 9.2 (continued)

Culture collection/(Status)	Acronym	WDCM number	Cultures	Date of institution	Number of cultures
W.H.O./F.A.O. Collaborating Centre for Reference and Research on Leptospirosis, Royal Tropical Institute, Amsterdam, Netherlands	ITH	WDCM196	*Leptospira*	1981	250
Salmonella Genetic Stock Centre, Department of Biology Sciences, University of Calgary, Canada	SGSC	WDCM338	*Salmonella*	1981	10,000
International Salmonella Centre (W.H.O.), Paris, France	ISC	WDCM63	*Salmonella*	–	–
Collection of Salmonella Microorganisms, Medical University of Gdańsk, Poland	KOS	WDCM784	*Salmonella*	1998	2000
Centre des Yersinia, CCOMS Reference and Research Center for Yersinia, Pasteur Institute, France	CY	WDCM7	*Yersinia*	–	–
Collection de Champignons et Actinomycetes Pathogenes, Pasteur Institute France	PCC	WDCM481	Actinomycetes	–	–
Culture Collection of Antimicrobial Resistant Microorganisms, Seoul, S. Korea	CCARM	WDCM847	Antibiotic Resistant Microbes	1999	20,226
Colecao de Trypanosoma de Mamiferos Silvestres, Domesticos e Vetores, Avenida, Rio de Janeiro, Brazil	ColTryp	WDCM 949	*Trypanosoma* of Wild and Domestic Mammals and Vectors	2009	700–800

9.5 NCVTC and Its Role in Research and Development

9.5.1 Conservation of Biodiversity

Ex situ microbial culture collections help preserve biodiversity, which is threatened by unsustainable economic development, natural disasters, and global change. The benefits of the conservation of biological resources are emphasized by the CBD, which highlights the need for creation of ex situ conservatories of biodiversity for its uses and conservation. Repository should strive to be a source of certified, stable, and validated microbial materials.

Several collections, called International Depository Authorities (IDAs) in the Budapest Treaty (Budapest Treaty on the International Recognition of the Deposit of Micro-organisms for the Purposes of Patent Procedures), serve as repositories of biological resources for the purpose of implementing intellectual property rights (IPRs). A veterinary MRC should strive to attain the competency to act as IDA in relevant field; however, MRC functions are a set of many complex activities. Apart from the basic work of populating a collection through isolation of clinical outbreak strains, cultures of various taxa received from depositors need subsequent purification, identification, characterization, and long-term preservation. In addition to characterization of bacteria, viruses, fungi, protozoa, and cell lines of animal relevance/significance, development and evaluation of assays and techniques for validating research resources and preserving and distributing microbial biological materials to the public and private sector research communities need to be done in a quality-conscious and timely manner. In order to rise to the level of expertise inherent in the definition of a biological resource center (BRC) (OECD 2009), emphasis of NCVTC is needed on value addition of microbial cultures by advanced molecular characterization, economic operations, and competitive benchmarking for all areas of the MRC. In order to develop expertise on collection, preservation, and distribution in a quality-conscious standardized manner, it is imperative to establish the repository with standard set of protocols (like collection of microbes, collection of information on microbes from depositors, storage conditions of cultures of different type, inventory management of liquid nitrogen with backup, location and retrieval system of samples/cultures, power backup system, media and chemical quality control, security of collection, access to collection. etc.) and documentation of various nature (such as various data/deposit forms, declaration, identity proofs). Apart from serving as a bio-repository and distributor, NCVTC envisages to provide specialized services related to its overall mission as a MRC, for example, HRD in the area of biological repository management services. However, the plans of NCVTC to develop into a worthy leader in R&D expertise for identifying, characterizing, preserving, and distributing a wide range of microbes have its own set of challenges.

9.6 Animals and Microbial Life: What to Preserve?

For most people, microbes are out of sight and thus out of mind. However, the biodiversity of plants and animals is largely known to us. In this genetic biodiversity lie mines of precious genetic information (Fraser and Dujon 2000). Among the variety of environmental niches, the animals are also one of the important locations of microbial life. Animals harbor various forms of microorganisms in and on their body in different role of associations, which can be described as commensals, opportunists, mutualistic, beneficial, or pathogenic. The close association between humans and animals also leads to transmission of pathogenic bacteria directly or indirectly through animal products as zoonoses. Taylor et al. (2001) catalogued 1415 known human pathogens, out of which 62% were of zoonotic origin. Intermingling of commensals with pathogens has also led to emergence of pathogens as a result of horizontal gene transfer (Ochman and Moran 2001).

The term microorganisms comprise all prokaryotes (archaea and bacteria), some eukaryotic organisms (fungi, yeasts, algae, protozoa), noncellular entities (e.g., viruses), their replicable parts, and other derived materials, e.g., genomes, plasmids, cDNA, etc. In the veterinary-medical health and production systems, the diversity of microbial life forms, which have an economic impact on health and production of animal and public health, is enormous and wide ranging. The horizon of microflora ranges from a variety of bacteria with different growth and environment requirements. This ranges from aerobic facultative organisms to microaerophilic and strict anaerobes on one side and fastidious pathogens like *Mycobacterium* spp., *Leptospira* spp., *Chlamydiales*, Rickettsiae (*Rickettsia rickettsii*), *Anaplasmataceae*, and *Coxiella burnetii* on the other. The *Mycoplasma* of class Mollicutes are the smallest prokaryotes capable of replicating on their own. This taxonomic group with five out of eight genera found in animals and humans are important agents to warrant research on their disease aspects and as persistent contaminants of cell lines (Corral-Vázquez et al. 2017). *Mycoplasma* cause significant disease in domestic and companion animals with about 35 species recorded taxonomically to be involved in infections of poultry, pigs, cattle, goats, sheep, horses, dogs, cats, and laboratory animals. Each one of the myriad microbial forms, including categories of bacteria, fungi, and viruses, requires separate laboratory setup, biosafety environments, growth and preservation necessities, human expertise, and handling requirements.

India is the seventh largest country in the world, with a variety of geo-climatic zones and landforms. The two broad climatic regions, viz., temperate or continental zone in the north and tropical zone in the south, are further divided into 11 regions depending upon the amount of rainfall and temperature. The variety of climatic and physiographical features leads to a complex ecology of diseases in Indian subcontinent. National Institute of Veterinary Epidemiology and Disease Informatics (2014–2015) report indicates that common bacterial diseases prevalent in India are hemorrhagic septicemia (HS), black quarter (BQ), enterotoxemia (ET), brucellosis, anthrax, leptospirosis, and salmonellosis, and among viral disease, peste de petits ruminants (PPR), foot and mouth disease (FMD), classical swine fever (CSF), goat pox, sheep pox, orf, blue tongue, swine pox, and rabies are prevalent in domestic animal population (AICRP-ADMAS Annual Report 2014–2015) (Table 9.3). Among these, most of the diseases are vaccine preventable, and for devising robust prevention and control measures, an inventory of spatially and temporally indexed isolates for long-term preservation with serum samples in all categories of domestic animals would be an effective tool in future control of disease. It would lead to a better understanding of disease ecology and agent biology for effective control and diagnostic reagent development.

Table 9.3 Summary of animal diseases prevalent in some Indian States

State	Bacterial disease	Animals	Viral disease	Animals
Andaman and Nicobar Islands	Leptospirosis	–	Goat pox (GP), Classical Swine Fever (CSF)	Goat, pigs
Andhra Pradesh	Anthrax, Haemorrhagic septicaemia (HS), Enterotoxaemia (ET), Black Quarter (BQ)	Cattle, sheep, goat, buffalo	Ranikhet disease (RD), Pesti des petits Ruminants (PPR), Sheep-pox, Goat-pox	Poultry, sheep, goat
Telangana	HS, BQ, ET, Anthrax, Brucellosis	Goat, sheep, cattle, buffalo	PPR, CSF, goat-pox, sheep-pox	Sheep, goat, pig
Assam	BQ, HS, ET, Swine Erysipelas	Cattle, goat, pig	CSF, RD, PPR	Pig, poultry, goat
Gujarat	HS	–	Sheep-pox, rabies, PPR	Sheep
Jammu and Kashmir	Foot Rot, BQ	Sheep	FMD, sheep and Goat pox, PPR, RD, contagious ecthyma	Cattle, sheep, goat, poultry
Jharkhand	HS, Anthrax	Cattle	FMD, PPR, CSF, avian influenza (AI)	Cattle, sheep, pig, poultry
Karnataka	BQ, HS, ET, Anthrax	Cattle, buffalo, sheep, goat	Sheep and goat-pox, PPR, BT, FMD	Sheep, goat
Kerala	HS, Anthrax, ET, CCPP	Cattle, sheep, goat	CSF, PPR, swine-pox, rabies (in cattle), Orf, Kyasanur forest disease (KFD), AI (In ducks and turkeys)	Pig, sheep, cattle, monkey,
Punjab	HS, BQ, Anthrax	Cattle	FMD, PPR, rabies	Cattle
Maharashtra	HS, Fowl typhoid,	Cattle, poultry	PPR	Small ruminant
Manipur	BQ, HS, Brucellosis, CRD, Salmonellosis, Colibacillosis	Cattle, sheep	Blue tongue (BT), CSF, goat pox, rabies, RD, IBD, fowl-pox	Sheep, goat, dogs
West Bengal	Anthrax, BQ, HS	–	CSF, PPR, FMD, sheep and goat-pox	Pig, sheep, goat
Rajasthan	HS, ET	–	Rabies, PPR	–

Adapted from Annual Report, AICRP on ADMAS 2014–2015

9.7 Costly Endemic Diseases Need Strategic Collection

An important aspect of our understanding of host-microbe interaction is the ability to detect, identify, and isolate microorganisms and to recognize diseases caused by them. Our abilities and achievements in this aspect have so far been limited. This has been due to use of conventional methods of pathogen research like dependence on microbial propagation methods, nonspecific clinical and epidemiological indicators, insensitive imprecise and sometimes cross-reacting serological tests, and an almost nil understanding of the microbial "background" in the external and internal host environment. In order to control diseases in animals and humans, molecular understanding of pathogens and pathogenesis is imperative. Colibacillosis, for example, caused by various strains of *Escherichia coli* takes a heavy toll of domestic animal neonates. These bacteria also spread through water and food to community causing diarrhea (Vaid et al. 2003). Further, there are many strains of *E. coli*; among them are those not causing any disease and those causing high mortality in humans due to verocytotoxins. These are emerging pathogens like *E. coli* O157:H7 strains (Besser et al. 1999). The neonatal diarrhea causing enterotoxigenic strains of *E. coli* (ETEC) are an economic drain on animal industry as they lead to undermining of fresh animal crop by high mortality (Bandyopadhyay et al. 2011). It will, however, be a boost to the research on ETEC, if scientists working in this field can deposit the ETEC strains, which can serve as positive control strains for future research. Availability of such strains (e.g., K88 variants, K99 and 987P fimbrial antigen-positive strains) has been difficult, as, in author's experience, many efforts to locate standard control strains for ETEC research from public health institutions like AIHPH, Kolkata; NICED, Kolkata, and Central Research Institute, Kasauli, have in the past been unsuccessful, probably due to differences in institutional mandates. The NCVTC and other MRCs have the sacred duty as well as obligation to make a concerted effort with all stakeholders, funding agencies, and publishers of research findings to start a dialogue within a timeline to make sure that important microbial strains, which are a component of funded/published research, are available to all contemporary and future generations by their deposition, in order to optimize and economize on government-funded research output.

Pasteurella multocida is a heterogeneous species that produces septicemic or respiratory diseases in domesticated and wild large and small ruminants (Rimler and Rhoades 1989). Considerable variation has been observed among strains with respect to host predilection, pathogenicity, carbohydrate fermentation, colonial morphology, and antigenic specificity (Carter and Chengappa 1981). It is also an important economic disease of buffalo, goat, sheep, rabbits, and pigs and has zoonotic links also (Donnio et al. 1991; Vaid et al. 2012). The taxonomy of *Pasteurella* has undergone many changes since its first isolation from domestic animals and it continues to change as new characterization procedures are used (Jaworski et al. 1998). A systematic collection, preservation, and characterization of pathogens of this group are imperative to understand the molecular virulence factors and pathogenicity mechanism in order to conceptualize novel vaccines. NCVTC has been instrumental in obtaining *P. multocida* isolates from various geographical locations for preservation and distribution. In a recent outbreak investigation in buffaloes in N. India (Kumar et al. 2014), we have isolated a strain of *Mannheimia varigena* from pneumonic buffalo for the first time in India (NRCE Annual Report 2013–2014). This isolation is an indication for the need of fresh taxonomic investigation of *Pasteurellaceae* family members in domestic and wild ruminants of India as these constitute an important group of pathogens causing respiratory diseases in largest group of food animals.

Similarly, mastitis is one of the costliest diseases affecting the dairy industry. Mastitis as a disease may be caused by *Staphylococcus* spp., *Streptococcus* spp., *Mycoplasma* spp., *Corynebacterium* spp., and several Gram-negative bacteria (Nickesrson et al. 1995). The subclinical form of mastitis leads to maximum damage. *Staphylococcus aureus* is the most frequent infectious agent in subclinical manifestations representing an economic problem for the milk industry the world over (Baumgartner et al. 1984). During growth in milk, enterotoxigenic *S. aureus* strains are able to produce thermostable enterotoxins that, when ingested, cause nausea, vomiting, and diarrhea in humans (Jablonsky and Bohach 1997). In this way, *Staphylococcus* spp. become critical pathogens in causing mastitis, food-borne zoonoses, and clinical nosocomial infection in humans, especially by emerging antibiotic-resistant strains. Not much comparable investigations, however, have been initiated in *S. aureus* of mastitis origin. Workers have also isolated a new species of staphylococci, *Staphylococcus nepalensis* sp. nov. from goats of Himalayan region in Nepal (Spergser et al. 2003); however no such reports of isolation of any novel staphylococci or other veterinary pathogenic strains have been reported in India. As the microbial repositories are the central point for thorough characterization of strains, exploration of biodiversity by isolation, identification, and description of novel strains is facilitated in such scientific institutions. It is thus imperative that capacity for polyphasic characterization is accommodated in NCVTC.

The science of microbial systematic and approaches to the classification and identification of veterinary microorganisms is also, as in other prokaryotes, based on complete phenotypic and genotypic characterization of microbes which is based on collection and comparison of microbe's biochemical and molecular attributes with type strains (polyphasic characterization) (Rosselló-Móra and Amann 2015). Polyphasic characterization, apart from phenotypic and chemotaxonomic data, also includes results obtained from 16S rRNA gene sequence pair similarity and phylogenetic analysis, DNA-DNA hybridization, and DNA G + C content. Due to recent advancements in next-generation sequencing (NGS) platforms and reduction in cost of sequencing, now whole genome sequence (WGS) comparison is leading to rapid polyphasic characterization by replacing DNA-DNA hybridization techniques (Chun and Rainey 2014). Polyphasic characterization not only leads to identification of novel species but is also instrumental in further strain-level differentiation of established species. Workers in Spain were able to delineate a novel subspecies *Streptococcus phocae* subsp. *salmonis* on the basis of polyphasic characterization (Avendaño-Herrera et al. 2014). However, the discipline of taxonomy, which is

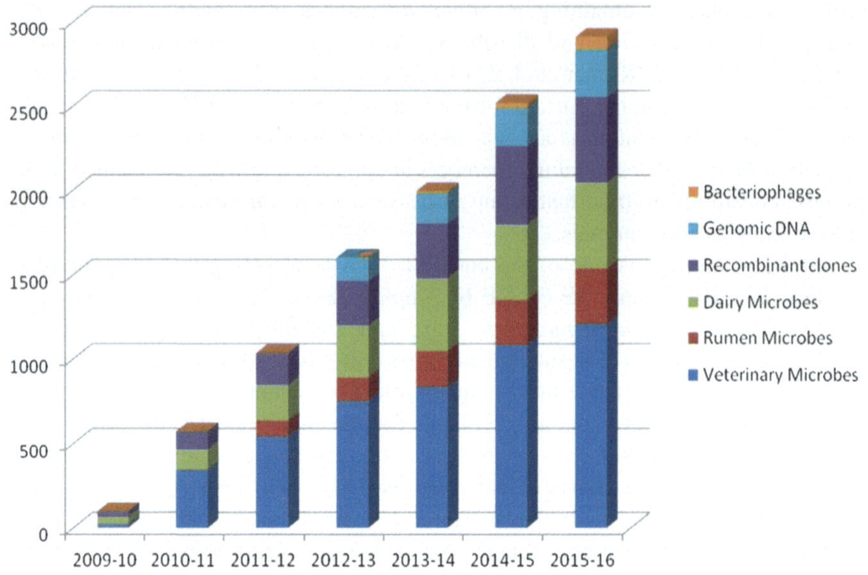

Fig. 9.1 The NCVTC repository has shown a steady progress in collection of different microbes

advancing at molecular level, also requires type strains as reference strains for comparison with under investigation novel strains for nomenclatural research and publications (Stackebrandt 2010; Tindall et al. 2010). However, such resources, especially the type strains of taxa relevant to veterinary-medical isolates, are hardly available in India, and efforts to get these strains get entangled in host of biodiversity regulations and administrative approvals (Overmann and Scholz 2017).

At present the NCVTC, with its limited scientific strength, initiated its CC activities in the field of aerobic, facultative anaerobic microbes and anaerobes, viruses, and bacteriophages. The biodiversity of taxa belonging to this category of microbes itself is huge. Within the ambit of this biodiversity, the available human resource capacity, capability, and aspirations are presently stretched. In spite of constraints, the NCVTC has, since its inception, reached a veterinary bacterial collection of more than 1000 strains representing greater than 50 genera. Apart from veterinary bacteria, numbers of various taxa of viruses, bacteriophages, dairy microbes, rumen bacteria and archaea, and recombinant clones preserved in NCVTC have shown a steady increase (Fig. 9.1). The facilities and funding requirements for enhancing and maturing the authentication and characterization skills need further appreciation. As the number of strains increases, the biodiversity, both at the species level and at higher taxa level, also increases. In order to aspire for a quality benchmark of BRC, a further investment in characterization technologies, automated preservation system, strain data management systems, and identification and nurturing of new talent in taxonomy of additional category of microbes, e.g., fungi, is urgently required. In addition, the NCVTC has also started distribution of bacterial strains to researchers in government and private network. Consequently, availability of technical staff is a very critical aspect in maintenance of cultures, distribution to customers, routine subculturing from seed repository, lyophilization, file work, packaging of strains, postage, and maintenance of records.

The establishment of NCVTC may also lead to progress in exploration of biodiversity of microbial cultures of animal origin in India and discovery of novel strains. However, the genetic studies can only take place if, after isolation, we develop a long-term preservation environment which includes methods of cataloguing the isolates, their DNA isolation, and preliminary 16S rRNA-based identification and classification (Woese et al. 1985, 2000; Harmsen and Karch 2004). Methods for rapid amplification of specific known sequences or the amplification of broad-range group sequences giving genus level information have improved our understanding of real state of etiology, detection, and exploration of microbial diversity. Among these approaches are BioLog MicroLog automated system based on metabolic characterization, ribotyping of isolates by RiboPrinter, and targeting of microbial small molecules such as fatty acid profile (Sherlock microbial identification system) of bacterial cell wall (Srinivasan et al. 2001). The characterization of accessioned isolates however is challenging without recourse to modern technologies utilized in the polyphasic characterization of isolates. Technologies such as Sanger sequencing platform to sequence smaller genetic elements such as 16S rRNA, and various housekeeping genes for particular taxa identification viz., *rpo*B, *inv*A, and NGS for quick measurement of Average Nucleotide Identity (ANI) (Kim et al. 2014) are instrumental in measurements of genomic identity comparison between strains, which is indispensable in the prokaryotic taxonomy.

Comparison of genome sequence (comparative genomics) of closely related pathogens and or non-pathogens (commensals/opportunistic pathogens) e.g., *E. coli* from goat/sheep/cattle/buffalo and environmental isolates has the potential to provide a rapid and effective method for understanding pathogenesis (Strauss and Falkow 1997). The enormous database of prokaryotic genomes now getting sequenced will prove a cornerstone of genome-based microbiology. The analysis of the entire available microbial sequences is already opening innovative research ideas and insights. What will be the potential impact of this database on human society when the projected database of over 2000 small genomes of each species will be available to the comity of scientists by the end of 2020? Certainly, NCVTC will be able to provide the pivot point from where suitable, critical bacterial isolates can be recognized for WGS work. It will be useful, because the cultures will be backed up by its disease metadata information. NCVTC has already embarked upon ambitious attempt to WGS some important pathogenic isolates (Vaid et al. 2014, 2015) (Table 9.4). However, bioinformatic analysis of curated genome data needs expertise in high-end genome data hosting, application of window-based analytical software, software skills in installation, and running of Linux OS platforms and computational analysis. The NCVTC needs to be either collaborating with expert bioinformatic institutions or achieve its goals of bioinformatic analysis by getting a human resource placed.

Table 9.4 Details of NCVTC accessioned strains with WGS

ACC.	DID	ID	Source	Genome size
BAA1	Eq24E	*Bordetella bronchiseptica*	Nasal swab	5,264,383 bp
BAA267	Bu5	*Trueperella pyogenes*	Buffalo pus	2,407,500 bp
BAA264	Bu1	*Pasteurella multocida*	Buffalo intestine	2,073,865 bp
BAA445	Eq28B	*Actinobacillus equilli*	Nasal swab foal	2,295,342 bp
BAA614	Sal40	*Salmonella gallinarum*	Poultry fowl typhoid	4,809,037 bp

The metadata collected during the collection of samples from which pathogens are isolated constitutes all the information concerned about the circumstances of isolation of pathogen thus providing a real epidemiological backdrop of the pathogen (Field et al. 2005). Example of such information is information on host, environment including meteorological data and type of samples. However, the epidemiological context of genes and organisms will be threaded only by an information system, which can give quick access to metadata based on key search criteria. Furthermore, information management is also important in day-to-day operation and use of culture collection (Casaregola et al. 2016). The information systems like Microbial Information Network Europe (MINE), which has been integrated across the networks of European collections and has evolved into recently developed Microbial Resource Research Infrastructure (MIRRI), encompass key feature of common minimal microbial metadata sets including taxonomy, growth, and biochemical and hazard information, among others (Stalpers et al. 1990; Smith et al. 2016). These are highly advanced online integrated systems networking the European BRCs. The optimal utilization and expansion of activities in collections like NCVTC and NBAIM will require advanced interactive websites, searchable electronic catalogue system, and tools of information system for curation of strain data. An interactive user interface for structured data dissemination and communication systems between users and staff and in between staff is important for operational efficiency. For optimization of utilization of microbial resources available in regional collections, the communication system's interoperability between collections, controls on regulatory issues and restrictions on data, ability to conduct simple searches within and across databases and to mine databases, etc. are some of the essential requisites that need to be developed.

9.8 Conclusion

The vision of our nation developing in leaps and bounds in the realm of bio-economy is very ambitious. That there is no dearth of competent human resource and biodiversity is well documented. However, biodiversity aspect of microbial variety can only be better utilized, if it is conserved ex situ, where its value can be enhanced by characterization and value addition. The current repositories of veterinary culture collection resources, including ex situ culture collections of veterinary pathogens, are inadequate to meet our needs for veterinary microbial resources because of the issues of standards, quality, and reliability. Microbial resource centers like NCVTC are the important institutions which act as source of biological material for reference and biological and biotechnological research, and they act as rallying point for development of taxonomy of microbes of animal origin. The effort of national government in responding to need for various biodiversity capturing institutions is timely and noteworthy, as historically, private participation in such endeavors is hardly feasible. Indian Council of Agricultural Research also embarked upon nurturing of various institutes, among which NCVTC caters to ex situ preservation of veterinary, dairy, and rumen microbes.

Isolation, preservation, and conservation of veterinary pathogens and use of molecular tools for their identification and their genomic and proteomic level characterization work can give impetus in the direction of combating newly emerging and reemerging veterinary pathogens, majority of which also cross-species barrier. This will also give impetus for strengthening taxonomic base and work in the fields of functional and comparative genomics. NCVTC is working toward becoming a veterinary MGR center to fulfill a long-felt resource gap in the fast developing knowledge-based society facing new challenges in the field of agriculture, health, and environment. Requirements such as new communication standards, technologies, Internet services, database and information management system, and web interfaces with efficient integration of veterinary microbial resources information in the bioinformatic network are needed. In the given limited resources, in which to carry out essential research and work to enhance the value and applications of strains and provide metadata information, the requirement and application of new technologies to aid in identifying and characterizing microbial diversity is a continuous challenge.

References

AICRP-ADMAS Annual Report (2014–15) In: Gajendragad MR, Reddy GBM, Suresh KP (eds) Annual Report, All India coordinated research project on animal disease monitoring and surveillance. Bengaluru

Avendaño-Herrera R, Balboa S, Castro N, González-Contreras A, Magariños B, Fernández J, Romalde JL (2014) Comparative polyphasic characterization of *Streptococcus phocae* strains with different host origin and description of the subspecies *Streptococcus phocae* subsp. *salmonis* subsp. nov. Int J Syst Evol Microbiol 64(Pt 5):1775–1781. https://doi.org/10.1099/ijs.0.056978-0

Bandyopadhyay S, Mahanti A, Samanta I, Dutta TK, Ghosh MK, Bera AK, Bandyopadhyay S, Bhattacharya D (2011) Virulence repertoire of Shiga toxin-producing *Escherichia coli* (STEC) and enterotoxigenic *Escherichia coli* (ETEC) from diarrhoeic lambs of Arunachal Pradesh, India. Trop Anim Health Prod 43(3):705–710. https://doi.org/10.1007/s11250-010-9757-1 Epub 2010 Nov 23

Baumgartener A, Nicolet J, Eggiman M (1984) Plasmid profile of *Staphylococcus aureus* causing bovine mastitis. J Appl Bacteriol 56:159–163

Bentley SD, Chater KF, Cerdeño-Tárraga AM, Challis GL, Thomson NR, James KD, Harris DE, Quail MA, Kieser H, Harper D, Bateman A, Brown S, Chandra G, Chen CW, Collins M, Cronin A, Fraser A, Goble A, Hidalgo J, Hornsby T, Howarth S, Huang CH, Kieser T, Larke L, Murphy L, Oliver K, O'Neil S, Rabbinowitsch E, Rajandream MA, Rutherford K, Rutter S, Seeger K, Saunders D, Sharp S, Squares R, Squares S, Taylor K, Warren T, Wietzorrek A, Woodward J, Barrell BG, Parkhill J, Hopwood DA (2002) Complete genome sequence of the model actinomycete *Streptomyces coelicolor* A3(2). Nature 417:141–147

Besser RE, Griffin PM, Skuster L (1999) *Escherichia coli* O157:H7 gastroenteritis and the hemolytic uremic syndrome: an emerging infectious disease. Annu Rev Med 50:355–367

Braff D, Shis D, Collins JJ (2016) Synthetic biology platform technologies for antimicrobial applications. Adv Drug Deliv Rev 105(Pt A):35–43. https://doi.org/10.1016/j.addr.2016.04.006

Buchanan RE (1966) History and development of the American Type Culture Collection. The Q Rev Biol 41:101–104

Carter GR, Chengappa MM (1981) Recommendations for a standard system of designating serotypes of *Pasteurella multocida*. Am Assoc Vet Lab Diagn 24:37–42

Casaregola S, Vasilenko A, Romano P, Robert V, Ozerskaya S, Kopf A, Glöckner FO, Smith D (2016) An information system for European culture collections: the way forward. Springerplus 5:772. https://doi.org/10.1186/s40064-016-2450-8

Chun J, Rainey FA (2014) Integrating genomics into the taxonomy and systematics of the Bacteria and Archaea. Int J Syst Evol Microbiol 64:316–324. https://doi.org/10.1099/ijs.0.054171-0

CLSI (2008) Performance standards for antimicrobial disk and dilution susceptibility tests for bacteria isolated from animals; approved standard 3rd edn (M31-A3). Clinical and Laboratory Standards Institute, Wayne, Pennsylvania

Corral-Vázquez C, Aguilar-Quesada R, Catalina P, Lucena-Aguilar G, Ligero G, Miranda B, Carrillo-Ávila JA (2017) Cell-lines authentication and *Mycoplasma* detection as minimum quality control of cell lines in biobanking. Cell Tissue Bank 18:271–280. https://doi.org/10.1007/s10561-017-9617-6 Epub 2017 Mar 2

Donnio PY, Avril JL, Andre PM, Vaucel J (1991) Dermonecrotic toxin production by strains of *Pasteurella multocida* isolated from man. J Med Microbiol 34:333–337

Field D, Garrity G, Morrison M, Selengut J, Sterk P, Tatusova T, Thomson N (2005) eGenomics: Cataloguing our complete genome collection. Comp Funct Genomics 6:363–368

Fraser CM, Dujon B (2000) The genomics of microbial diversity. Editorial overview. Curr Opin Microbiol 3:443–444

Furman JL, Stern S (2011) Climbing atop the shoulders of giants: the impact of institutions on cumulative research. Am Econ Rev 101:1933–1963. https://doi.org/10.1257/aer.101.5.1933

Gee JE, Sacchi CT, Glass MB, De BK, Weyant RS, Levett PN, Whitney AM, Hoffmaster AR, Popovic T (2003) Use of 16S rRNA gene sequencing for rapid identification and differentiation of *Burkholderia pseudomallei* and *B. mallei*. J Clin Microbiol 41:4647–4654

Graves RR (1943) The story of John M. Buck's and Matilda's contribution to the cattle industry. J Am Vet Med Assoc 102:193–195

Harmsen D, Karch H (2004) 16S rDNA for diagnosing pathogens: a living tree. ASM News 70:19–24

Holmes B, Snell JJS, Lapage SP (1978) Revised description, from clinical strains, of *Flavobacterium breve* (Lustig) Bergey et al. 1923 and proposal of the Neotype strain. Int J Syst Evol Microbiol 28:201–208 0020-7713/78/0028-0201$02.00/0

Jablonsky LM, Bohach GA (1997) *Staphylococcus aureus*. In: Doyle MP, Beauchat LR, Montville TJ (eds) Food microbiology: fundamentals and frontiers. ASM Press, Washington, pp 353–376

Jaworski MD, Hunter DL, Ward ACS (1998) Biovariants of isolates of *Pasteurella* from domestic and wild ruminants. J Vet Diagn Investig 10:49–55

Jorgensen JH, Ferraro MJ (2009) Antimicrobial susceptibility testing: a review of general principles and contemporary practices. Med Microbiol Clin Infect Dis 49:1749–1755

Kim M, Oh HS, Park SC, Chun J (2014) Towards a taxonomic coherence between average nucleotide identity and 16S rRNA gene sequence similarity for species demarcation of prokaryotes. Int J Syst Evol Microbiol 64:346–351

Kumar A, Maan S, Batra K, Yadav S, Mahajan NK, Ghosh A, Sunayna, Jindal N, Banger R, Boora A, Vaid R K et al (2014) Emerging infectious disease outbreaks at an organised buffalo farm in Haryana. In: XXVIII Annual Convention of Indian Association of Veterinary Microbiologists, Immunologists and Specialists in Infectious Diseases (IAVMI) 30 Oct.–1 Nov., 2014, Abstract No S3:P15, pp 24

Lewis K (2012) Antibiotics: recover the lost art of drug discovery. Nature 485:439–440

May M (2014) Time for teamwork. Nature 509:S4–S5

McCluskey K, Mills KB, Dye G, Ehmke E, Gunnell GF, Kiaris H, Richmond MP, Yoder AD, Zeigler DR, Zehr S, Grotewold E (2017) The challenges faced by living stock collections in the USA. eLife 2017 6:e24611. https://doi.org/10.7554/eLife.24611

Nickerson SC, Owens WE, Boddie RL (1995) Mastitis in dairy heifers: initial studies on prevalence and control. J Dairy Sci 78:1607–1618

Nicoletti P (1990) Vaccination against *Brucella*. Adv Biotechnol Process 13:147–168

NRCE Annual Report (2013–14) In: Virmani N, Vaid RK, Barua S, Anand T, Bera BC, Raut AA (eds) National Research Centre on Equines, Hisar, Haryana, India, pp 49

Ochman H, Moran NA (2001) Genes lost and genes found: evolution of bacterial pathogenesis and symbiosis. Science 292:1096–1099

OECD (2001) In: In: Wald S (ed) Biological Resource Centres –underpinning the future of life sciences and biotechnology. Organisation for Economic Development and Cooperation, Paris, France, pp 68

OECD (2009) Organisation for economic cooperation and development. The bioeconomy to 2030: designing a policy agenda. OECD, Paris, France. ISBN:978-92-64-03853-0

OECD (2011) Future prospects for industrial biotechnology. OECD Publishing https://doi.org/10.1787/9789264126633-en

Overmann J (2015) Significance and future role of microbial resource centers. Syst Appl Microbiol 38:258–265

Overmann J, Scholz AH (2017) Microbiological research under the Nagoya protocol: facts and fiction. Trends Microbiol 25:85–88

Plainvert C, Dinis M, Ravins M, Hanski E, Touak G, Dmytruk N, Fouet A, Poyart C (2014) Molecular epidemiology of *sil* locus in clinical *Streptococcus pyogenes* strains. J Clin Microbiol 52:2003–2010. https://doi.org/10.1128/JCM.00290-14

Prakash O, Nimonkar Y, Shouche YS (2013) Practice and prospects of microbial preservation. FEMS Microbiol Lett 339:1–9. https://doi.org/10.1111/1574-6968.12034

Rimler RB, Rhoades KR (1989) *Pasteurella multocida*. In: Adlam C, Rutter JM (eds) *Pasteurella* and pasteurellosis. Academic, London, United Kingdom, pp 37–73

Rosselló-Móra R, Amann R (2015) Past and future species definitions for bacteria and archaea. Syst Appl Microbiol 38:209–216. https://doi.org/10.1016/j.syapm.2015.02.001

Russell JE (2014) Historical perspective, the national collection of type cultures microbiologist, September, 25–27

Sasson A, Malpica C (2017) Bioeconomy in Latin America. New Biotechnol 40:40–45. https://doi.org/10.1016/j.nbt.2017.07.007

Schutzer SE, Schlater LRK, Ronning CM, DeShazer D, Luft BJ, Dunn JJ, Ravel J, Frasaer-Liggett CM, Nierman WC (2008) Characterization of clinically-attenuated *Burkholderia mallei* by whole genome sequencing: candidate strain for exclusion from select agent lists. PLoS One 3 (4):e2058. https://doi.org/10.1371/journal.pone.0002058

Sinha JN (2010) Veterinary science and animal husbandry in India: a case study of Indian Veterinary Research Institute at Mukteshwar-Izatnagar. Indian J Hist Sci 45(4):559–568

Sly LI (2010) Biodiversity and the Role of Microbial Resource Centres. Paper at "Biodiversity and world food security: nourishing the planet and its people" conference conducted by the

Crawford Fund for International Agricultural Research, Parliament House, Canberra, Australia, 30 August–1 September, 2010

Smith D, Martin D, Novossiolova T (2016) Microorganisms: good or evil, MIRRI provides biosecurity awareness. Curr Microbiol 74:299–308. https://doi.org/10.1007/s00284-016-1181-y

Spergser J, Wieser M, Täubel M, Rosselló-Mora RA, Rosengarten R, Busse HJ (2003) *Staphylococcus nepalensis* sp. nov., isolated from goats of the Himalayan region. Int J Syst Evol Microbiol 53(Pt 6):2007–2011

Srinivasan A, Kraus CN, DeShazer D, Becker PM, Dick JD, Spacek L, Bartlett JG, Byrne WR, Thomas DL (2001) Glanders in a military research microbiologist. N Engl J Med 345:287–289

Stackebrandt E (2010) Diversification and focusing: strategies of microbial culture collections. Trends Microbiol 18:283–287

Stalpers JA, Kracht M, Janssens D, De Ley J, van der Toorn J, Smith J, Claus D, Hippe H (1990) Structuring strain data for storage and retrieval of information on Bacteria in MINE, the Microbial Information Network Europe. Syst Appl Microbiol 13:92–103

Strauss E, Falkow S (1997) Microbial pathogenesis, genomics and beyond. Science 276:707–712

Taylor LH, Latham SM, Woolhouse ME (2001) Risk factors for human disease emergence. Philos Trans R Soc Lond Ser B Biol Sci 356:983–989

Tindall BJ, Rosselló-Móra R, Busse HJ, Ludwig W, Kampfer P (2010) Notes on the characterization of prokaryote strains for taxonomic purposes. Int J Syst Evol Microbiol 60:249–266. https://doi.org/10.1099/ijs.0.016949-0

USFDA (2015) Guidelines for the validation of analytical methods for the detection of microbial pathogens in foods and feeds. 2nd edn. US Food & Drug Administration Office of Foods and Veterinary Medicine

Vaid RK, Rana R, Kumar A, Vihan VS (2003) Isolation of *E. coli* of serogroups indicating probable presence of Verocytotoxic *E. coli* (VTEC) in diarrhoeal kids. Indian J Anim Sci 73:1197–1199

Vaid RK, Shanamugasundaram K, Boora A, Riyesh T, Bera BC, Shukla BN, Anand T, Virmani N, Barua S, Rana N, Singh KP, Malik P, Singh RK (2012) Sporadic outbreak of Haemorrhagic Septicaemia in buffalo calves in an organized farm. Vet Pract 13:326–329

Vaid RK, Shanmugasundaram K, Boora A, Bera BC, Shukla BN, Anand T, Singha H, Riyesh T, Virmani N, Barua S, Ahir VB, Koringa PG, Sajnani MR, Bhat VD, Rana N, Singh KP, Malik P, Singh RK, Joshi CG (2014) Draft genome sequence of *Pasteurella multocida* subsp. *multocida* B:2 strain VTCCBAA264 isolated from *Bubalus bubalis* in North India. Genome Announc 2 (4):e00755–e00714. https://doi.org/10.1128/genomeA.00755-14

Vaid RK, Jindal N, Anand T, Bera BC, Riyesh T, Virmani N, Barua S, Gupta R, Mahajan NK, Joshi CG, Singh RK (2015) First draft genome sequence of *Salmonella enterica* serovar Gallinarum strain VTCCBAA614, isolated from chicken in India. Genome Announc 3(5):e01221-15. https://doi.org/10.1128/genomeA.01221-15

WFCC (1997) In: Kirsop, B, Canhos VP (Eds.), The economic value of microbial genetic resources. Proceedings of the World Federation for Culture Collections (WFCC) Workshop at the eighth international symposium on microbial ecology (ISME-8). World Federation for Culture Collections, http://www.wfcc.info/pdf/halifax98.pdf, isbn:92 9109 043 3

WFCC (2017) World Federation for Culture Collections, http://www.wfcc.info/ccinfo/

Woese CR, Stackebrandt T, Macke J, Fox GE (1985) A phylogenetic definition of the major eubacterial taxa. Syst Appl Microbiol 6:143–151

Woese CR, Olsen GJ, Ibba M, Soll D (2000) Comparison of complete genome sequences allow the most objective and comprehensive descriptions possible of a lineage's evolution. Microbiol Mol Biol Rev 64:202–236

Chapter 10
Biodiversity Conservation of Phages and Microbial Populations

Galina Novik, Victoria Savich, and Elena Kiseleva

Abstract The article is dedicated to microbial diversity in nature, methods of investigation, identification and classification of bacteria and bacteriophages, as well as their applications in biotechnology. Microbial culture collections and the strategies for preservation of microbial diversity are considered.

10.1 Introduction

Microorganisms comprise huge numbers of individual organisms on Earth. Only a small part of bacteria (about 1%) was described. Disappearance of the natural habitats leads to the respective vanishing of microorganisms associated with the specific environment (Colwell 1997). Microbial culture collections are able to conserve these species as well as commercially valuable bacteria and bacteriophages. The preservation strategy demands preliminary taxonomic identification and studies of microbial characteristics.

G. Novik (✉) · V. Savich
Belarusian Collection of Non-pathogenic Microorganisms, Institute of Microbiology, National Academy of Sciences of Belarus, Minsk, Republic of Belarus
e-mail: galina_novik@mbio.bas-net.by

E. Kiseleva
Institute of Bioorganic Chemistry, National Academy of Sciences of Belarus, Minsk, Republic of Belarus

S. K. Sharma, A. Varma (eds.), *Microbial Resource Conservation*, Soil Biology 54,
https://doi.org/10.1007/978-3-319-96971-8_10

10.2 Biodiversity, Identification, and Conservation of Microorganisms

10.2.1 Microbial Resources in the Environment

Bacteria represent an abundant group of microorganisms occupying different ecological niches, even extreme habitats, and interacting with the related species and other taxa. Bacteria use different types of metabolic strategies allowing to adapt to the specific environment. These features make bacteria vital constituents of ecosystems. They act as symbionts and pathogens of plants and animals, regulators of biogeochemical cycles providing energy and carbon to the other organisms or serving as decomposers.

Majority of prokaryotes occur in the open ocean, in soil, and in marine and terrestrial subsurface media, where cell concentrations reach 1.2×10^{29}, 2.6×10^{29}, 3.5×10^{30}, and $0.25 - 2.5 \times 10^{30}$, respectively. Bacterial cells are estimated to contain overall 350–550 Pg of carbon, up to 60–100% of the total carbon found in plants, as well as 85–130 Pg of nitrogen and 9–14 Pg of phosphorous, exceeding approximately 10 times the corresponding levels in plants, and thus accumulate the largest pool of these nutrients in living organisms (Whitman et al. 1998). Despite a huge number of bacteria, so far only 7000-odd species have been described (Achtman and Wagner 2008).

Such diversity demands more thorough investigations. The development of molecular-genetic and biochemical methods allows to systemize anew and further study bacteria and their phages (Sidorenko et al. 2008; Rakhuba et al. 2010). The variety of metabolic pathways makes bacterial organisms attractive for application in medicine, agriculture, and industry, which emphasizes the need in detailed research and conservation of microbial resources.

10.2.2 Microbial Culture Collections and Gene Banks

Microbial culture collections play a crucial role in preservation and management of microbial resources. To maintain viability of stock entries and ultimately to secure microbial diversity are the primary goals of any microbial depository providing cultures to biotechnology sector, academic and research institutions. All collections can be divided into three categories: private collections with restricted access usually managed by individuals, laboratories, institutes, hospitals, and commercial firms; public collections containing massive funds to meet requirements of various clients; and specialized collections intended for taxonomic studies and conservation of specific groups of microorganisms (Mahilum-Tapy 2009). Gene banks are engaged in preservation of genetic material.

World Federation for Culture Collections (WFCC) coordinates the activities of microbial depositories around the world. The membership list tends to expand steadily and currently embraces 760 collections in 76 countries and regions registered in World Data Centre for Microorganisms (WDCM) database: 14 from Africa, 191 from America, 270 from Asia, 41 from Oceania, and 244 from Europe (WDCM 2018).

10.3 Belarusian Collection of Nonpathogenic Microorganisms

The intensive consumption of bioresources by academic research institutions and corporate sector has drawn keen interest to collections of microbial cultures. Nowadays in developed countries the problem of conservation and rational use of biological and genetic materials is emphasized as a top-priority national challenge. This trend was supported by the proclamation of the principle of state sovereignty over genetic resources in the International Convention on Biodiversity. In compliance with this principle, international legal mechanisms of access to genetic resources (MOSAICC, Micro-organisms Sustainable Use and Access Regulation International Code of Conduct; Bonn Guidelines on Access to Genetic Resources and Fair and Equitable Sharing of the Benefits Arising out of their Utilization; Nagoya Protocol on Access and Benefit-sharing), the transboundary relocation of genetically engineered organisms (GMOs, the Cartagena Protocol on Biosafety to the Convention on Biological Diversity), and the international patent procedure for the inventions involving microorganisms [the International Microorganism Deposit System, World Intellectual Property Organization (WIPO)] have been developed.

State collections of microorganisms, being the central depositories of typical and industrially valuable strains, act in accordance with national legislation, regulations, and rules and carry out a number of important functions. For example, many industrial microbial cultures deposited at the Belarusian collection of microorganisms could be used in manufacturing enzyme preparations, ferments ensiling plant substrates, producing preventive-therapeutic compositions enhancing immune potential in humans and animals, development of biological agents to control plant pathogens, bacterial preparations for degradation of toxic organic substances, and bioremediation of natural and industrial media. Principal research trends of the collection are:

- Depositing and maintenance of microbial cultures valuable for different areas of microbiology and biotechnology in a specialized collection fund
- Selection and genetic engineering of highly active strains to develop biopreparations of novel generation
- Molecular-genetic identification of bacteria, bacteriophages, filamentous fungi, yeasts and studies on physiology of microorganisms representing diverse taxonomic groups
- Optimization of long-term preservation methods for microbial cultures stored in collection depository; elaboration and introduction of procedures for washless

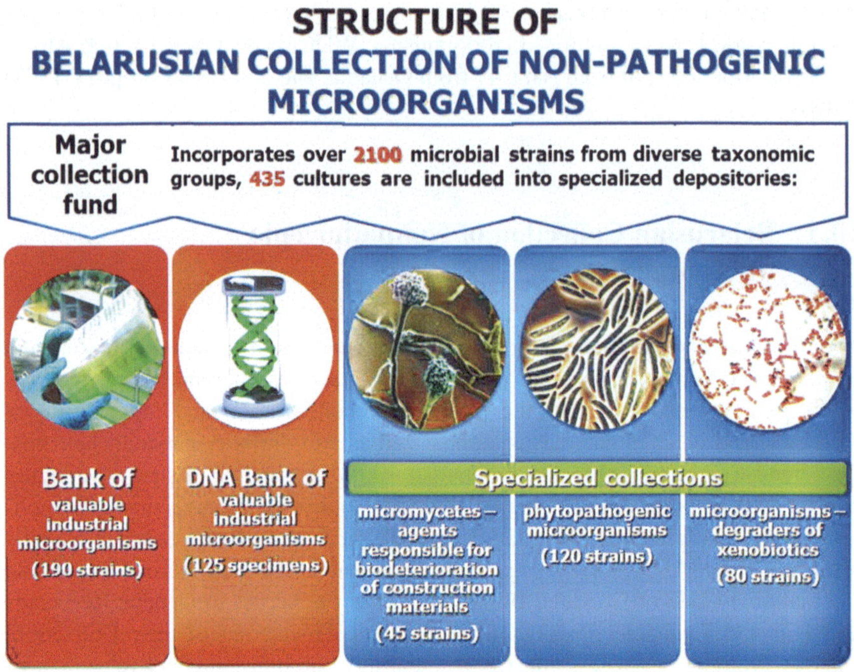

Fig. 10.1 Belarusian collection of non-pathogenic microorganisms—the National Asset of the Republic of Belarus

method of low-temperature conservation of bacteria, bacteriophages, filamentous fungi, and yeasts; optimization of refrigerating-defrosting and desiccation regimes for cell suspensions and bacterial concentrates

- Classification of information on deposited cultures, setting up data banks describing microbial collection stock, consulting on identification and preservation of microorganisms, depositing microbial strains for patent procedure, and provision of cultures ordered by the clients

Currently, over 2100 microbial strains are maintained at the Belarusian collection of nonpathogenic microorganisms. Catalogues and databases of microbial cultures comprising entries of type, reference, and industrial strains were compiled in Russian and English versions. Belarusian collection of nonpathogenic microorganisms is rated as the national asset of the Republic of Belarus (Fig. 10.1).

It was registered as a member of the World Federation of Culture Collections [WFCC-MIRCEN, World Data Centre for Microorganisms (WDCM), Global Catalogue of Microorganisms (GCM); acronym BIM, collection number 909]. The fund of microbial resources deposited at Belarusian collection of nonpathogenic microorganisms was described in the 6th edition of *World Directory of Collections of Cultures of Microorganisms*.

For further progress of innovative biotechnologies, close interaction and collaboration of the national culture collections with biotechnological industry are very important. This activity can be realized via continuous replenishment of microbial collections, banks of DNA, and industrial cultures used in various biotechnological processes; screening of promising microbial genes for subsequent genetic engineering of strains-hyperproducers of bioactive substances, like enzymes, vitamins, organic acids, etc.; safety evaluation of industrial microbial cultures, including recombinants; monitoring of legislation regulating turnover of microbial genetic resources; and commercial application of type, reference, and industrial microbial strains in different countries and regions of the world.

10.4 Modern Approaches to Taxonomy and Identification of Bacteria

10.4.1 Taxonomy of Bacteria

The basis of taxonomy was laid by Linnaeus in the eighteenth century, but only in the late nineteenth century bacterial taxonomy was shaped. Ferdinand Cohn in 1872 was the first to classify six genera of bacteria as members of the plant kingdom. At that time morphology, growth requirements and pathogenic potential were the most important taxonomic markers. Later, in addition to morphology, physiological and biochemical properties were introduced into classification procedure. However, for a long time, bacteria were still referred to plants (*Protophyta*, primitive plants). In 1962 a detailed and well-accepted division scheme of prokaryotic and eukaryotic organisms was proposed. In 1974 bacteria were no longer considered as plants and were recognized as members of the kingdom *Procaryotae* (Schleifer 2009).

Nevertheless, to date there is no official classification of prokaryotes. Taxonomy remains a matter of scientific judgment and general agreement. Bergey's Manual is merely the best agreed and balanced source for classification at the moment, although there are always controversial areas where the available data are lacking or confusing, resulting in different opinions and taxonomic arguments (Sneath and Brenner 1992).

Currently, the two prokaryotic domains *Bacteria* and *Archaea* are subdivided into 39 phyla: 34 and 5 phyla in domains *Bacteria* and *Archaea*, respectively. All in all domains incorporate about 89 classes, 1 subclass, 197 orders, 20 suborders, 446 families, and 2857 genera (LPSN 2018).

Modern classification divides microorganisms into groups or taxa on the basis of similarities and differences, demonstrates relationships between organisms, and serves as data storage and retrieval system. Taxonomy provides information useful for diverse theoretical and practical purposes and enables to compile and update most relevant databases of microorganisms. Classification is essential when pathogens of plants and animals must be identified. It allows to distinguish the known strains and recognize novel ones when choosing cultures for industrial processes. Taxonomy helps to make predictions and formulate research hypotheses based on knowledge of identical bacteria. Thus taxonomy of bacteria has important practical and theoretical implications (Goodfellow 2000).

10.4.2 Morphology and Physiology of Bacteria

Bacteria constitute a versatile group of organisms, so that various morphological and physiological characteristics may differ among representatives of the same species and can be used in identification and description of bacteria. Classical microbiology utilizes both gross and microscopic morphology to identify microorganisms. Gross morphology analyzes colony shape, size, surface parameters, color, and pigment production. Light and electron microscopy examines cellular morphology. Shape of cells can be readily seen via light microscopy with no sample preparation (Jackman 2012). Light and electron microscopy of specimens (Fig. 10.2) allows to visualize cells and their components which cannot be seen with the naked eye. Light microscopy makes use of visible light and a system of lenses to magnify images, while electron microscopy applies a beam of accelerated electrons as a source of illumination. The biological research frontiers that can be investigated with light microscopy are limited by the image resolution, the parameter which defines the minimum distance between two distinguishable points. Electron microscopy has higher image resolution and can be divided into scanning microscopy and transmission microscopy. In the former case, the electron beam is generally scanned in a raster scan pattern, and the beam position is combined with the detected signal to produce an image resolution better than 1 nm. In the latter case, the transmitted beam of electrons interacts with ultrathin specimen as it passes through it. Clear image is formed as a result of this interaction (Mishra and Chauhan 2016). Optical microscopes also can be manufactured in different versions designed for specialized purposes: bright-field, phase contrast, fluorescence, etc. (KEYENCE 2017).

Bacteria can show enormous diversity in cell size ranging from ~0.3 μm for obligate intracellular pathogenic members of the genus *Mycoplasma* to ~600 μm for *Epulopiscium fishelsoni*, a Gram-positive commensal inhabitant of surgeonfish guts, and 750 μm for *Thiomargarita namibiensis*, a chemilithotrophic Gram-negative bacterium native to coastal Namibia. While *T. namibiensis* has a large gas vesicle surrounded by a thin layer of cytoplasm and involved in sulfide oxidation under anaerobic conditions, *Epulopiscium* overcomes diffusion-dependent limitations by increasing genome number on the cell periphery, where the latter is thought to facilitate responses to local stimuli and thereby contribute to maintenance of the extremely large cell size (Razin and Cosenza 1966; Angert et al. 1993; Schulz and Jorgensen 2001; Mendell et al. 2008). Archaeal cells are usually a few μm in diameter. Shape of bacteria also differs in a wide range. Cells can be amorphous, ovoid, square, stellate, filamentous, or stalked. They may be grouped in pairs, clumps, chains, rosettes, cuboid packets, flat squares, networks, mycelia, or fruiting

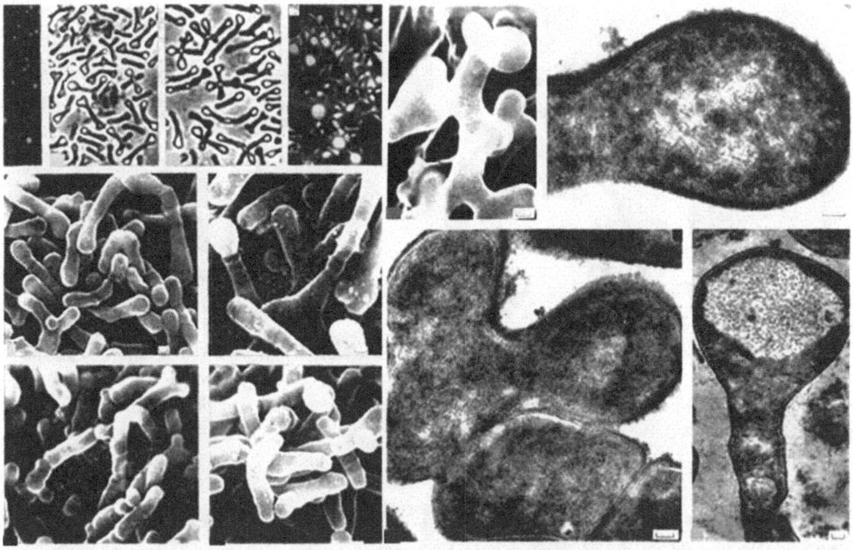

Fig. 10.2 Morphology of bacterial cells revealed by means of light and electron microscopy. Visualisation of pleomorphic rods with a particular cell morphology ranging from regular rods to various branched and club-shaped forms (photo Galina Novik)

bodies (Zinder and Dworkin 2006). Size and shape are determined by diffusion-mediated nutrient import. Alteration of nutrient fluxes or growth state may be accompanied by morphological changes, such as filamenting or extruding prosthecae, which increase the surface area available for nutrient import without increasing the surface-to-volume ratio. More specialized shapes cope with the nutritional requirements of inhabitants of atypical environmental niches (Young 2006).

Bacterial multiplication proceeds as binary transverse fission common for most studied bacteria. However, there is a variety of alternative division strategies (Fig. 10.3) (Novik et al. 1994, 2001; Novik 1998). Division of *Hyphomicrobium* and *Rhodomicrobium* is carried out by budding, while actinomycetes are characterized by mycelial extension and subsequent fragmentation. Pleurocapsalean cyanobacteria display multiple fissions (Zinder and Dworkin 2006).

Bacterial populations differ not only in size and shape but in some cases in cell architecture. Most prokaryotes have a rigid cell wall characterized by presence of peptidoglycan murein. Cell walls of Gram-positive and Gram-negative bacteria are distinguished by different composition. The cell wall of Gram-positive bacteria is thick, consisting of several layers of peptidoglycan and containing teichoic acids unique to this group. Gram-negative bacteria have thin layer of peptidoglycan surrounded by the outer membrane. The outer leaflet of this membrane is composed mainly of lipopolysaccharides comprising three domains: a lipophilic moiety termed lipid A, a hydrophilic glycan called the *O*-specific polysaccharide (*O*-chain or *O*-antigen), and linking core oligosaccharide (Knirel and Valvano 2011). The

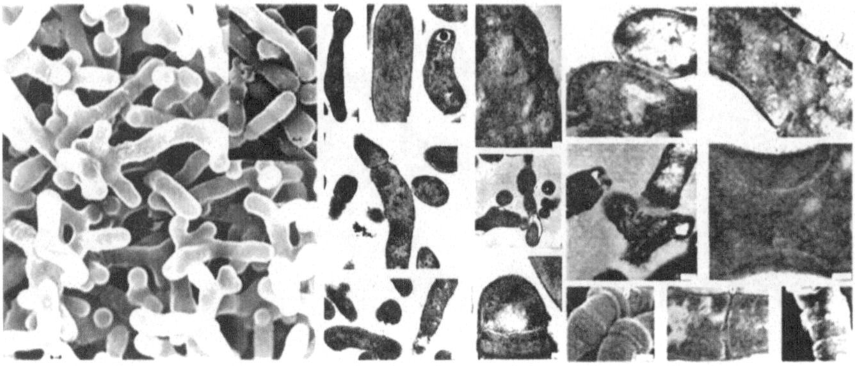

Fig. 10.3 Bacterial cell division at exponential and stationary phases of population development (photo Galina Novik)

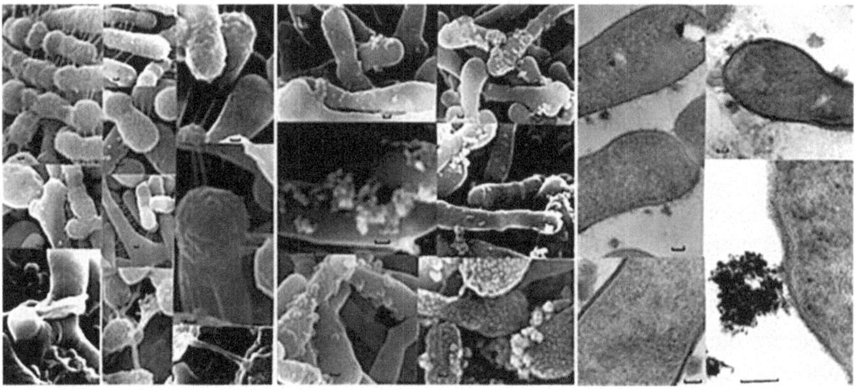

Fig. 10.4 Structure and aggregation of bacterial capsule. Cohesion and formation of microfibrils in the apical sections of cells (photo Galina Novik)

interaction of bacteria with the environment generated such additional structures as flagella, pili, or capsule (Fig. 10.4) taking part in movement, attachment to surface, DNA transfer during conjugation, and protection of cells (Novik and Vysotskiĭ 1995; Novik 1998). Bacterial cytoplasmic membrane consists of a phospholipid bilayer sufficient to perform major functions of a cell membrane. It limits inner space of the cell represented by a few intracellular organelles such as chromosome, plasmids, and ribosomes. Other structures are found in specialized groups of bacteria to promote adaptation to specific environmental conditions.

Bacteria exhibit a wide variety of metabolic pathways allowing them to adapt to the environment. Bacterial metabolism is classified into nutritional groups on the basis of the type of energy and the source of carbon used for growth. Bacteria can derive energy from the sun or chemical reaction and carbon—from utilization of organic compounds or directly by carbon fixation. Some bacteria are able to use

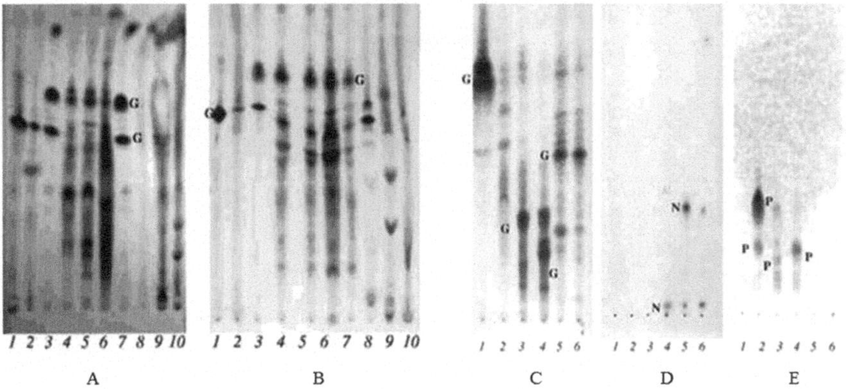

Fig. 10.5 Analysis and detection of bacterial lipids. A: 1—*St. mucilaginosus* PCM 2415^T, 2—*S. hirsute* PCM 2279 (=ATCC 27875^T), 3—*P. propionicum* PCM 2431^T, 4, 5, 6—*B. adolescentis* 94 BIM, 7—*P. propionicum* PCM 2431^T, 8, 9—*N. dassonvillei* PCM 2492, 10—*R. equi* PCM^T 559 (=ATCC 3969); B: 1—*St. mucilaginosus* PCM 2415^T, 2—*S. hirsute* PCM 2279 (=ATCC 27875^T), 3—*P. propionicum* PCM 2431^T; 4, 5—*B. longum* B 379 M, 6—*B. adolescentis* 94 BIM, 7—*B. bifidum* № 791, 8—*N. dassonvillei* PCM 2492, 9—*R. equi* PCM^T 559 (=ATCC 3969), 10—*G. bronchialis* PCM 2167; C, D, E: G—glycolipids, N—lipids containing free amino groups, P—phospholipids (1–6—extract of lipids of *B. adolescentis* 94 BIM). Solvent system: chloroform/methanol/H_2O (65:25:4, v/v/v). Detection: A, B, C—treatment with 0.5% orcinol in ethanol with 3% H_2SO_4; D—treatment with 0.5% solution of ninhydrin in butanol; E—treatment with molybdenum reagent (picture Galina Novik)

inorganic electron donors for energy conservation or nitrogen fixation (Zinder and Dworkin 2006).

Morphological and physiological properties have been applied in classification of bacteria since origination of these studies. Chemical composition of cell constituents is another useful classification feature (chemotaxonomy). Chemotaxonomic methods can be used for those groups of prokaryotes where morphological and physiological characters have largely failed or have not been plentiful to provide a satisfactory classification. Various traits can be used as markers: chemical composition of cell walls (capsular polysaccharides and teichoic acid of Gram-positive bacteria), guanine-cytosine content of DNA, and presence and diversity of compounds of different chemical composition (glycolipids, phospholipids, fatty acids, isoprenoid quinones, cytochromes, polyamines, etc.) (Figs. 10.5 and 10.6). Chemotaxonomic data are very useful for reliable classification and identification, but they are not sufficient for a comprehensive reconstruction of their phylogeny (Hamana and Matsuzaki 1992; Schleifer 2009; Novik et al. 2002, 2005a, b, 2006a, b; Zdorovenko et al. 2009; Valueva et al. 2011, 2013).

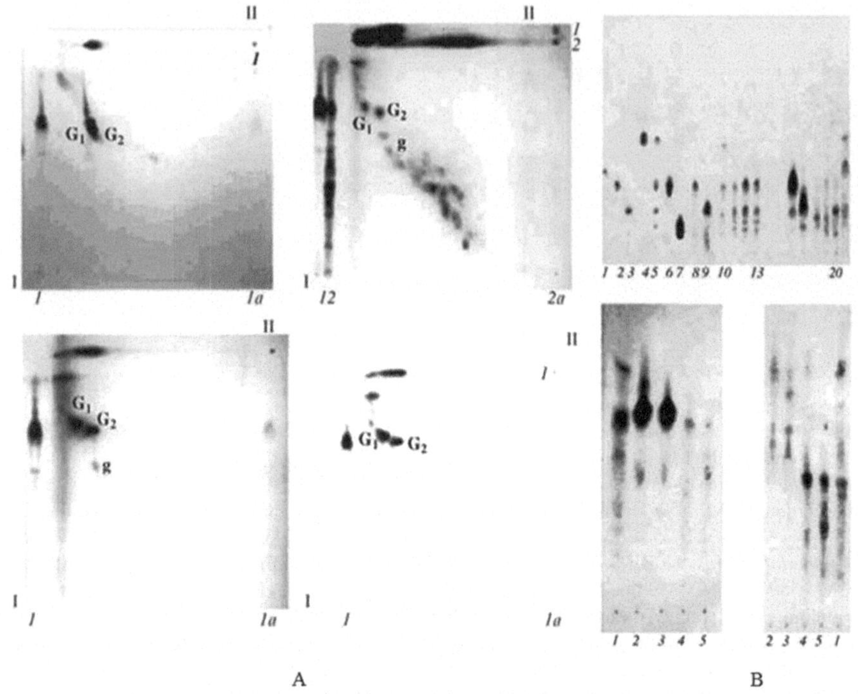

Fig. 10.6 Two dimensional thin-layer chromatography of glycolipids (A) and determination of bacterial polar lipids by using purchased standards of phospholipids (B). Solvent system: A—first dimension—chloroform/methanol/H_2O (65:25:4, v/v/v), second dimension—chloroform/acetic acid/methanol/H_2O (80:15:12:4, v/v/v/v) or *n*-butanol/acetic acid/H_2O (60:20:20, v/v/v); B—chloroform/methanol/H_2O (65:25:4, v/v/v). Detection: A—treatment with 0.5% orcinol in ethanol with 3% H_2SO_4; B—treatment with molybdenum reagent and the part with 0.5% orcinol in ethanol with 3% H_2SO_4. A: 1, 1a—methanol-soluble fraction of lipids; 2, 2a—extract of lipids; G_1, G_2—major glycolipids, g—minor glycolipids; B: 1—diphosphatidylglycerol, 2—phosphatidylglycerol, 3—phosphatidylinositol, 4—phosphatidylethanolamine, 5—mixture of phosphatidylethanolamine, phosphatidylcholine, phosphatidylinositol, lysophosphatidylcholine, 6—phosphatidylcholine, 7—lysophosphatidylcholine, 8—phosphatidic acid, 9—lysophosphatidylserine, 10–20—methanol-soluble fraction of lipids, 21—extract of lipids; orcinol-treated part: 1—extract of lipids, 2–5—methanol-soluble fraction of lipids (picture Galina Novik)

10.5 Molecular-Genetic Identification of Bacteria

Discrimination of closely related strains by phenotypic methods is not exhaustive, and unambiguous identification can take days to weeks. With this regard, various genetic methods have been developed in recent decades. In the 1960s, the technique of DNA–DNA hybridization (DDH) was introduced to evaluate genetic relationship, but wide application of molecular-genetic methods for identification has started since 1980s. At that time, the development of PCR and sequencing of the 16S rRNA gene triggered major changes in prokaryotic taxonomy resulting in key role of 16S rRNA gene as classification parameter (Sentausa and Fournier 2013).

Now 16S rRNA gene sequence comparison is the most common tool to elucidate the phylogenetic position of prokaryotes. The main features of this method determining its popularity are:

1. This gene is present in all bacteria making it a universal target for bacterial identification.
2. The function of the 16S rRNA gene over time has not changed, suggesting that random sequence changes are a relatively accurate measure of evolution in time.
3. The 16S rRNA gene is large enough (approximately 1500 bp) to provide distinctive and statistically authentic measurements (Patel 2001).

Nevertheless, application of this methodology faces some problems. 16S rRNA gene sequencing has low phylogenetic power at the species level and below and poor discriminatory power for some genera, like *Bacillus*. Species can share over 99% sequence similarity with regard to their 16S rRNA genes, but DDH method may not exhibit close relationship leading to classification of objects as distinct species (Patel 2001; Janda and Abbott 2007). The 16S rRNA gene sequence is composed of both variable and conserved regions, but not all genera possess the same hypervariable region. Strains of the same species may differ in genotype and phenotype, but sequence distinctions may be minimal (Clarridge 2004). Some species, like *Aeromonas veronii*, could contain up to six copies of the 16S rRNA gene varying by up to 1.5% which might cause identification problems. The challenges aroused during 16S rRNA identification can also be related with poor quality of nucleotide sequences deposited in databases, although the data are constantly revised (Janda and Abbott 2007).

The 5S and 23S rRNA genes don't have such a widespread application as 16S rRNA. The 5S rRNA is considered to contain not sufficiently long sequences for statistically reliable comparisons. The23S rRNA is rarely used in taxonomic classification due to the lack of established broad-range sequencing primers and the difficulty of sequencing larger genes with early sequencing technology. However, compared to 16S rRNA, 23S rRNA genes contain more characteristic sequence stretches determined by greater length and unique insertions and/or deletions and perhaps display better phylogenetic resolution because of higher sequence variation. A recent study indicated that 23S rRNA genes also contain conserved regions suitable for design of broad-range primers with a similar degree of universality to the broad-range primers for 16S rRNA genes and their phylogenetic trees are in good agreement with 16S rRNA, so that in the latest decades 23S rRNA came into use as a taxonomic research tool (Bavykin et al. 2004; Pei et al. 2009).

The 16S, 23S, and 5S rRNA genes are found in the same genetic locus, and they are separated by noncoding regions called internal transcribed spacers (ITS). The 16S–23S rRNA ITS sequence is not subject to the same selective pressure as the rRNA genes and consequently shows a 10 times greater evolution rate apparently able to overcome the evident limitations of these genes (Barry et al. 1991; Gürtler and Stanisich 1996). Phylogenetic analysis based on ITS approach additionally reveals the relationship similar to 16S rRNA genes (Wang et al. 2008).

The degree of intrageneric resolution achieved with 16S rRNA sequence analysis is not always sufficiently discriminatory because of the extremely slow rate of 16S rRNA evolution. Thus sequence of housekeeping genes was considered as the alternative to 16S rRNA. Housekeeping genes are constitutive, involved in various processes (metabolism, gene expression, etc.), and often encode proteins. As a rule, protein-encoding genes evolved much faster and ensured higher resolution than 16S rRNA sequences (Yamamoto and Harayama 1998). Studies revealed that sequence of housekeeping genes provided better resolution on species and subspecies levels as compared to 16S rRNA, and the results matched DDH data (Christensen et al. 2004; Case et al. 2007; Wang et al. 2007).

Multilocus sequence analysis (MLSA) is also a widely used method to establish phylogenetic relationships of species within a genus or genera within a family. MLSA studies usually employ partial sequences of four to five housekeeping genes. They must be in single copies in the genome, as well as be homologous and widely distributed in the studied taxa. Genes frequently analyzed are those coding for subunits of ubiquitous enzymes, such as the β-subunit of DNA gyrase (*gyrB*), the β-subunit of RNA polymerase (*rpoB*), the sigma 70 (sigma D) factor of RNA polymerase (*rpoD*), recombinase A (*recA*), the β-subunit of ATP synthase F0F1 (*atpD*), translation initiation factor IF-2 (*infB*), tRNA modification GTPase ThdF or TrmE (*thdF*), or the chaperonin GroEL (*groEL*). Gene fragments from 400 to under 600 nucleotides are most often applied in the analysis. The MLSA is successfully used in bacterial taxonomy to differentiate species and to estimate the phylogenetic relationships of different species or genera (Glaeser and Kämpfer 2015).

The abovementioned methods are usually engaged in identification to genus or species levels, and in some cases, subspecies and strains can be determined. Methods of molecular typing are applied to discriminate different bacterial isolates of the same species. Pulsed-field gel electrophoresis (PFGE) is the technique used for separation of larger pieces of DNA by applying electric current that periodically changes direction (three directions) in a gel matrix, unlike the conventional gel electrophoresis where the current flows only in one direction. The technique can be modified by using restriction endonucleases recognizing few digestion sites in the chromosome and generating large DNA fragments further separated by PFGE. The method allows to perform sample typing with high discriminatory power (Hansen et al. 2002; He et al. 2014).

The restriction fragment length polymorphism (RFLP) involves fragmentation of DNA sample by the restriction enzyme that can recognize and cut DNA whereby a specific short sequence occurs in the process known as restriction digestion. The resulting DNA fragments are then separated in length by agarose gel electrophoresis and transferred to membrane via the Southern blot procedure. Hybridization of the membrane to labeled DNA probe determines the length of the fragments complementary to the probe. Depending on the restriction enzyme, the pattern of hybridizing bands can be species- or strain-specific. Two or more different restriction enzymes in separate digestions can be used to generate different patterns from the same isolates and better demarcate similar strains (Lin et al. 2014; Ranjbar et al. 2014). Ribotyping is based on RFLP analysis, but it applies ribosomal DNA generating restriction fragment pattern by enzymatic digestion of rRNA genes of the chromosomal DNA (Ranjbar et al. 2014).

Cleavase fragment length polymorphism (CFLP) method engages endonuclease cleavase I recognizing structures formed by base-paired DNA at elevated temperatures and cutting DNA from the 5′ end to the first paired base at the junction of the single- and double-stranded regions. The resulting mixture contains a collection of randomly cleaved, labeled single-stranded DNA molecules extending from the 5′-labeled end to the terminal cleavage site. The discrimination power of CFLP is dependent on DNA sequence of the locus examined. The CFLP is also size limited; differences in the structural fingerprints of DNA fragments larger than 1 kbp cannot be reliably resolved by electrophoresis (Olive and Bean 1999).

Some typing methods are based on application of PCR (MLST, RAPD-PCR, REP-PCR, etc.). Multilocus sequence typing (MLST) reminds MLSA sequencing fragments of several housekeeping genes, but it is applied in population studies aimed at differentiation of strains within a species. For any gene fragment, different sequences are assigned as distinct alleles, and each isolate is defined by the alleles at each of the housekeeping loci (Lin et al. 2014; Ranjbar et al. 2014). In some variations of RFLP, restriction analysis is performed on PCR amplicons obtained using primers for specific sequences of interest (Ranjbar et al. 2014).

Random amplified polymorphic DNA (RAPD) technique employs primers of approximately ten bases for amplification of random segments from large template of genomic DNA. Amplified products are subsequently separated on agarose gel and stained with ethidium bromide. The genetic variation analysis based on RAPD ensures proper genetic diversity due to its capacity to generate random markers from the entire genome. This method doesn't require prior special knowledge of specific DNA target sequences (Lin et al. 2014; Ranjbar et al. 2014).

Repetitive extragenic palindromic (REP)-PCR amplifies spacer fragments lying between repeat motifs of the genome using two outwardly directed primers at high stringency. REP elements are 38-bp sequences consisting of six degenerate positions and a 5-bp variable loop between each side of a conserved palindromic stem (Stern et al. 1984). After amplification spacer fragments are analyzed by electrophoresis generating band patterns (Ranjbar et al. 2014). The enterobacterial repetitive intergenic consensus (ERIC)-PCR is similar to REP, but it is focused on 126-bp elements containing a highly conserved central inverted repeat and located in extragenic regions of the bacterial genome (Sharples and Lloyd 1990; Hulton et al. 1991). Other repetitive elements BOX are located within intergenic regions, but have no sequence relationship to either REP or ERIC. BOX are mosaic repetitive elements composed of various combinations of three subunit sequences referred to as boxA, boxB, and boxC with molecular lengths of 59, 45, and 50 nucleotides, respectively. They were first observed in *Streptococcus pneumonia*, but later BOX elements were found in a number of bacterial species (Olive and Bean 1999). The poly-trinucleotide $(GTG)_5$ motif represents a class of conserved repetitive sequences present in bacterial genomes, and it can also be used in corresponding PCR to type various bacterial groups (Ryberg et al. 2011).

Amplified fragment length polymorphism (AFLP) originally applied to characterization of plant genomes recently has been used in bacterial typing. The whole DNA is digested with one or more enzymes, and the resulting DNA fragments are ligated to linkers each containing restriction site and a sequence homologous to a PCR primer binding site. The differentiation power of AFLP appears to be higher than that of PCR-based ribotyping; however, it has not been compared to REP-PCR or PFGE. The AFLP procedure is more labor-intensive than REP-PCR, but the results are produced more rapidly than with PFGE (Olive and Bean 1999; Ranjbar et al. 2014).

The PCR melting profile (PCR MP) technique is based on low denaturation temperatures during ligation-mediated PCR. In the course of this process, genomic DNA is completely digested with restriction enzyme, and the restriction fragments are ligated with a synthetic adapter. Lower denaturation temperature in PCR decreases the number of amplified fragments because only single-stranded DNA may serve as a template for DNA synthesis. It allows gradual amplification of DNA fragments differing in thermal stability starting from the less stable DNA fragments amplified at lower denaturation temperature values to more stable ones amplified at higher values. The PCR MP fingerprinting method doesn't require prior knowledge of the examined sequence, its results can be readily analyzed on polyacrylamide gels stained with ethidium bromide, and the same adaptor and enzymes can be applied to analyze DNA from diverse species of bacteria. The technique shows similar power of discrimination to macrorestriction analysis of genomic DNA followed by PFGE (Krawczyk et al. 2006, 2009).

10.6 Bacteriophage Taxonomy and Classification

10.6.1 Significance of Bacteriophage Systematics

Bacteriophages are infectious agents selectively attacking bacterial cells. Phages are abundant microorganisms that can be found everywhere. Their population densities are estimated to be 10^9 g^{-1} of soil (Williamson et al. 2005) and 10^7 ml^{-1} of seawater (Bergh et al. 1989; Wommack and Colwell 2000). They take part in nutrient and energy transfer of global geochemical cycles and act as reservoirs of genetic diversity (Suttle 2005). Phages are components of microflora present at levels comparable to their bacterial hosts. It was shown that phages can even provide beneficial traits to their hosts helping to maintain community stability and afford resilience to invasion or disruption (Ogilvie and Jones 2015).

Ubiquitous distribution of bacteriophages in the environment and their enormous ecological impact and unique properties make them attractive objects for use in agriculture, medicine, and biotechnology. Preliminary generalization and categorization of bacterial viruses are essential to subdivide them into groups possessing common features. It allows to describe evolutionary links and horizontal gene transfer between different bacteriophages in order to get deeper insight into their biology.

Development of bacteriophage classification would allow to optimize search of efficient methods for identification of new phages isolated from natural sources. In turn, these methods would be engaged for detection of bacterial viruses that could be used in various industries and medicine or otherwise could cause economic losses in biotechnological sector. As a result, optimization of training programs in the area of microbiology and biotechnology, methods of maintaining bacteriophage collections, and database update may be achieved.

Thus viral systematics research contributes to progress in bacteriophage identification having significant implications in various areas of economic and scientific activities.

10.6.2 Historical Background of Bacteriophage Classification

In 1896 Hankin was the first to detect the ability of water from the Indian rivers to eliminate bacteria, especially causative agent of cholera. However, the author erroneously concluded that antimicrobial properties were shown by some volatile chemical substance (Hankin 1896). Frederick Twort and Felix d'Herelle independently in 1915 and 1917, respectively, observed action of biological agents causing bacterial lysis. The scientists guessed that the effect was induced by microorganism termed bacteriophage by d'Herelle (Twort 1915; d'Hérelle 1917). He supposed that only one bacteriophage named *Bacteriophagum intestinale* with many races existed (d'Hérelle 1918). However, subsequent studies have revealed a whole spectrum of viruses with distinct bacterial specificity, and attempts were made to classify bacteriophages. Burnet used serological tests to divide phages into several groups with the same plaque type and functional character. Nevertheless, viruses entering the same group may lyse different bacterial species (Burnet 1933a). Phages falling in a single serological group were approximately uniform in their response to biochemical tests evaluating rate of photodynamic inactivation by methylene blue, ability to lyse in the presence of citrate, and rate of inactivation by strong urea solutions (Burnet 1933b). In 1940 first pictures of bacteriophages taken by electron microscopy were published and proved the particulate nature of bacterial viruses (Pfankuch and Kausche 1940; Ruska 1940). Detailed investigation of viral structure followed. Ruska originally proposed morphological classification of viruses and reported at least four morphotypes of bacteriophages (Ruska 1943a, b). In 1948 all viruses were classified into three families based on host range and symptoms of disease. Family *Phagineae* included bacterial viruses (Holmes 1948). Phage systemization was also attempted using a mix of serological, morphological, and physiological criteria (Adams and

Wade 1954). In 1962 a new system (Lwoff-Horne-Tournier system or LHT) was proposed for classification of viruses. This system was founded on four features:

1. Nature of the nucleic acid: DNA or RNA
2. Symmetry of the viral particle: helical, cubical, or binal (virion with two different symmetrical structures—the head and the tail)
3. Presence or absence of nucleocapsid
4. Quantitative data such as diameter of the nucleocapsid for virion with helical symmetry or number of capsomers for the virion with cubical symmetry

Latinized nomenclature was proposed for viral classification. This system included three bacteriophage families: family *Phagoviridae* (order *Urovirales*) for tailed phages and families *Inoviridae* and *Microviridae* (order *Haplovirales*) for filamentous and ϕX-type phages, respectively (Lwoff et al. 1962). In 1965 Provisional Committee on Nomenclature of Viruses (PCNV) was founded. Later it was renamed as International Committee on Taxonomy of Viruses (ICTV). In 1967 Bradley examined phage particles under electron microscope and proposed classification based on type of nucleic acid and morphologic characters. He distinguished six basic types. The first group containing dsDNA included three basic types: phages with contractile tails, long and noncontractile tails, and short noncontractile tails. The second group with ssDNA embraced tailless phages with large capsomeres and filamentous viruses. The last group held tailless ssRNA phages with small capsomeres (Bradley 1967). Later Tikhonenko proposed a similar scheme of bacteriophage classification dividing bacterial viruses into five main morphological groups based largely on Bradley's classification: filamentous phages, phages with tail analogs, phages with a short tail, phages with a long and noncontractile tail, and phages carrying tail of complex structure with a contractile sheath (Tikhonenko 1970). In 1971 the ICTV issued its first report which included six phage genera: T-even phages, l, lipid phage PM2, the f X group, "filamentous phage," and the "ribophage group" (Wildy 1971). After 3 years, over 1150 phages have been categorized into 17 groups based on morphology and nucleic acid content. Some groups corresponded to Bradley morphological types such as tailed phages (A-C). However, they were additionally divided into three subtypes based on the shape of phage head: isometric (subtype 1) or prolate (subtypes 2 and 3). Other groups included phages with cubic symmetry (D, E), filamentous phages (F), and round/oval enveloped particles (G) (Ackermann and Eisenstark 1974). Later, the ICTV published nine reports concerning bacteriophage classification and taxonomy. Morphology of viral particle and composition of nucleic acid have remained main features for classification of viruses.

10.6.3 Current Classification of Bacteriophages

The first phage survey published in 1967 listed 111 negatively stained viruses, mostly tailed phages. The researchers also included phage ϕX174 (*Microviridae*), two ssRNA phages or *Leviviridae*, and nine filamentous phages of the family *Inoviridae* (Eisenstark 1967). Further on the number of studied bacteriophages steadily increased. By 1971 and 1996, the total 1150 and 4551 bacteriophages, respectively, were studied (Ackermann and Eisenstark 1974; Ackermann 1996). According to the latest update, nearly 6300 prokaryote viruses have been described morphologically, including 6196 bacterial and 88 archaeal viruses. The phages belong to 1 order and 10 families and infect members of 163 bacterial genera, mostly representatives of the *Firmicutes* and γ-proteobacteria. Phages of enterobacteria and the host genera *Streptomyces*, *Mycobacterium*, *Bacillus*, *Lactococcus*, *Pseudomonas*, and *Vibrio* predominate (Ackermann and Prangishvili 2012).

10.6.3.1 Tailed Phages: The Order *Caudovirales*

As previously mentioned, bacteriophages form only one order named *Caudovirales* (Latin *cauda*, tail). The order includes DNA tailed phages with icosahedral or prolate heads. Only 15% of capsids are prolate (e.g., φ29 and T4), and they are frequently found in enterobacterial and lactococcal phages. Tails are helical or consist of stacked disks and carry in most cases fixation structures such as baseplates, spikes, or terminal fibers. Tails have very different size and morphology with lengths ranging from ~100 to ~8000 Å. Phage particles are not enveloped. Despite general absence of lipids, about one third of tailed phages are chloroform-sensitive. The capsid contains linear double-stranded DNA packed with a high density of ~500 g/l and hence exerts internal pressure of dozens of atmospheres on the capsid walls. DNA may contain unusual bases such as 5-hydroxymethylcytosine or 5-hydroxymethyluracil. Phage genomes are large, complex, and organized as interchangeable building blocks or modules. Genes with related functions are assembled in clusters. Heads, tails, and tail fibers are synthesized separately and then assembled (Ackermann 2005, 2006; Ackermann and Prangishvili 2012; Fokine and Rossmann 2014).

Caudovirales are divided into three families depending on the length and contractility of tail. There are families *Myoviridae* (contractile tail), *Siphoviridae* (long and noncontractile tail), and *Podoviridae* (short tail). Tailed bacteriophages make the most numerous and widely distributed group of bacterial viruses. They represent over 96% of bacterial phages. *Siphoviridae* is the most profuse family both within the order and among other phages accounting for 57.3% share of the total population. Families *Myoviridae* and *Podoviridae* constitute 24.8 and 14.2% of all viruses, respectively (Ackermann and Prangishvili 2012).

Family *Myoviridae* represents phages with contractile tail consisting of a sheath and a central tube. Contraction enables the tail tube to penetrate through bacterial cell wall and serve as a channel for transport of the phage genome into cytoplasm. The family includes nearly 1600 species. To date *Myoviridae* are divided into 6 subfamilies with 39 unassigned genera: *Eucampyvirinae*, *Ounavirinae*, *Peduovirinae*, *Spounavirinae*, *Tevenvirinae*, and *Vequintavirinae*. Family *Siphoviridae* is the most abundant phage family with 3605 morphologically studied phages. The family contains 6 subfamilies and 94 unassigned genera: *Arquatrovirinae*, *Bclasvirinae*, *Guernseyvirinae*, *Mclasvirinae*, *Pclasvirinae*, and *Tunavirinae*. Family *Podoviridae* includes 891 tailed phages. Only 3 subfamilies and 20 unassigned genera were differentiated: *Autographivirinae*, *Picovirinae*, and *Sepvirinae* (Ackermann and Prangishvili 2012; ICTV 2016).

10.6.3.2 Non-tailed DNA-Containing Bacteriophages with Polyhedral Capsids

This group includes three bacteriophage families: *Microviridae*, *Corticoviridae*, and *Tectiviridae*.

Family *Microviridae* contains phages arranged as small non-enveloped virions (27 nm in diameter) with single piece of circular DNA (Ackermann 2006). The studies demonstrated that *Microviridae* could be divided into two groups depending on the number (one or two) of scaffolding proteins required for the assembly. The single-scaffolding systems contain an internal protein and have more complex coat protein fold. The two-scaffolding protein systems (φX174-like) encode internal and external proteins, as well as an additional structural protein: a spike on the icosahedral vertices (Doore and Fane 2016). Coliphage research revealed three distinct phylogenetic clades represented by φX174, G4, and α3 (Rokyta et al. 2006). Further studies described four subgroups based on genomic and structural characteristics: microviruses (genus *Microvirus*), gokushoviruses (subfamily *Gokushovirinae*), alphaviruses (subfamily *Alphavirinae*), and pichoviruses (subfamily *Pichovirinae*) (Roux et al. 2012). ICTV distinguishes two subfamilies: *Bullavirinae* (formerly *Microvirinae*) and *Gokushovirinae* (ICTV 2016).

Family *Corticoviridae* includes only one species, *Pseudoalteromonas* virus PM2, although related prophages are common in the genomes of aquatic bacteria (Krupovic and Bamford 2007). Phage capsid consists of two protein shells interlaid with a lipid membrane (Ackermann 2006). The capsid carries 200 major capsid protein P2 trimers located on a pseudo $T = 21$ lattice. The genome enclosed with the membrane is represented by supercoiled, circular double-stranded DNA. DNA comprises approximately 14% of the virion weight, and its G+C content is 42.2%. The genome has 21 putative genes, 17 of which have been shown to code for proteins. Proteins are responsible for nearly 72% of the virus particle weight, whereas lipids account for 14%. The membrane contains 34% phosphatidylethanolamine, 66% phosphatidylglycerol, and trace amounts of phosphatidic acid and acyl phosphatidylglycerol. The lipids are derived from the host plasma membrane, but their composition deviates from that of the host bacterium. Lipids form an internal membrane with virus-specific membrane-associated proteins (Oksanen and ICTV Report Consortium 2017).

Family *Tectiviridae* includes phages with a rigid outer protein capsid surrounding a thick, flexible lipoprotein vesicle. Upon adsorption to bacteria or chloroform treatment, this vesicle becomes a taillike tube about 60 nm in length, which acts as nucleic acid ejection device (Ackermann 2006). Genome is represented by linear double-stranded DNA. This family consists of five phages: four phages are included in genus *Tectivirus* and one is not affiliated (ICTV 2016).

10.6.3.3 Non-tailed RNA-Containing Bacteriophages with Polyhedral Capsids

This group consists of two families: *Leviviridae* and *Cystoviridae*.

Family *Leviviridae* are small non-enveloped poliovirus-like viruses with linear, positive-sense, single-stranded RNA. Their genome consists of four partially overlapping genes. Representatives of this family infect enterobacteria, pseudomonads, acinetobacters, and caulobacters. Many leviviruses adsorb to plasmid-dependent bacterial pili (Ackermann 2005, 2009). The family consists of two genera: *Allolevivirus* and *Levivirus* (ICTV 2016).

Family *Cystoviridae* includes a single member specific for the phytopathogenic bacterium *Pseudomonas syringae*. Viruses of this type bear icosahedral capsids surrounded by lipid-containing envelopes. Genome is represented by three molecules of linear segmented double-stranded RNA. Phages also contain a dodecahedral RNA polymerase complex (Ackermann 2005, 2006, 2009; ICTV 2016).

10.6.3.4 Filamentous Bacteriophages

This group consists of the sole family *Inoviridae.* The family is represented by non-enveloped phages with helical symmetry of virion. Single-stranded viruses replicate via the "rolling circle" model and generate double-stranded intermediate DNA. The particles are excreted from infected cells without killing the host. DNA of some family members is able to integrate into genome of host bacteria (Ackermann 2009). Until 2015 the family comprised two genera: *Inovirus* and *Plectrovirus*. Later *Inoviridae* was expanded to seven genera with new entries—*Fibrovirus*, *Habenivirus*, *Lineavirus*, *Saetivirus*, and *Vespertiliovirus* (ICTV 2016).

10.6.3.5 Pleomorphic Bacteriophages

Pleomorphic bacteriophages are exemplified by family *Plasmaviridae*. This family incorporates only one member—*Acholeplasma* virus L2. Virions don't have capsids, but they possess an envelope and a dense nucleoprotein granule. Virus infects mycoplasmas by membrane fusion and is excreted by budding. Thus, the propagation of *Plasmaviridae* reminds one of enveloped vertebrate viruses. The genome consists of a single molecule of circular, supercoiled double-stranded DNA (Ackermann 2005, 2009, 2011; ICTV 2016).

10.6.3.6 Archaea Viruses

Archaea viruses are less studied as compared to bacteriophages, but they are more morphologically diverse. Tailed viruses of archaea are highly similar in virion morphology to bacteriophages and divided into the same three families: *Myoviridae*, *Siphoviridae*, and *Podoviridae*. All tailed isolates infect euryarchaea, either halophilic or methanogenic (Pina et al. 2011; Pietilä et al. 2014).

Tailless icosahedral viruses are grouped in two families recently included in official classification: *Sphaerolipoviridae* and *Turriviridae*. Family *Turriviridae* unites two viruses STIV1 and STIV2 infecting hyperthermophilic *Sulfolobus*. Their icosahedral capsids contain an inner lipid membrane encasing the circular double-stranded DNA, and the capsid surface is decorated with turret-like structures at vertices. *Sphaerolipoviridae* are double-stranded DNA viruses with icosahedral capsid and an internal lipid membrane. All members of the family infect hosts from extreme environments, which are either halophilic archaea in case of genera *Alphasphaerolipovirus* and *Betasphaerolipovirus* or extreme thermophilic bacteria of genus *Thermus* in case of recently introduced genus *Gammasphaerolipovirus* (Pina et al. 2011; Pietilä et al. 2014; Pawlowski et al. 2014; ICTV 2016).

Filamentous archaea viruses are represented by the order *Ligamenvirales* divided into *Lipothrixviridae* and *Rudiviridae* and families *Clavaviridae* and *Spiraviridae*. *Lipothrixviridae* consists of three genera: *Beta-*, *Gamma-*, and *Deltalipothrixvirus*. Earlier the family also included genus *Alphalipothrixvirus*. Viral particles are characterized by combination of rodlike shape, lipoprotein envelope, and nucleosome-like core, although no lipids were found in the virions of deltalipothrixviruses. Genome is represented by linear double-stranded DNA. Viruses of this group infect hyperthermophilic archaea, like *Acidianus* and *Sulfolobus* (Ackermann 2009; Pietilä et al. 2014; ICTV 2016). *Rudiviridae* are rod-shaped, non-enveloped virions with short terminal fibers at each end. The particles bear close resemblance to the tobacco mosaic virus. All rudiviruses carry linear double-stranded DNA genomes with long inverted terminal repeats ending in covalently closed hairpin structures with 5′–3′ linkages. The family is represented by the only genus *Rudivirus* (Ackermann 2005; Pina et al. 2011; ICTV 2016). Family *Clavaviridae* includes one representative *Aeropyrum pernix* bacilliform virus 1, or APBV1. APBV1 has rigid bacilliform morphology, size about 140 × 20 nm, with one end pointed and the other rounded. It contains highly glycosylated single major protein and three minor proteins. The circular double-stranded DNA genome comprising 5278 bp encompasses 14 open reading frames, all localized on a single DNA strand (Mochizuki et al. 2010; ICTV 2016). *Spiraviridae* also includes one species: *Aeropyrum* coil-shaped virus, or ACV. ACV is non-enveloped, hollow, cylindrical virion formed from a coiling fiber, which consists of two intertwining halves of a single circular nucleoprotein. The virus is also exceptional for presence of single-stranded DNA (Mochizuki et al. 2012).

Families of pleomorphic archaea viruses are diverse. *Pleolipoviridae* contains three genera, *Alpha-*, *Beta-*, and *Gammapleolipovirus*. The virions of these viruses are composed of a pleomorphic membrane vesicle enclosing the genome. All pleolipoviruses have two major structural protein species, internal membrane and spike proteins. The genomes can be represented by single- or double-stranded, linear or circular DNA molecules (Pietilä et al. 2016). *Fuselloviridae* members are mostly spindle-shaped, although variations may range from a thin cigar to pear configuration. Fuselloviruses carry a set of short, thin fibers at one of the pointed ends. The genome is represented by circular double-stranded DNA. The family is divided into two genera *Alphafusellovirus* and *Betafusellovirus* (Pina et al. 2011; ICTV 2016). Spindle-shaped *Salterprovirus* is similar to *Fuselloviridae*, but it differs in sequence similarity, genome structure, and replication. So far, *Salterprovirus* has not been included in any family (ICTV 2011). *Guttaviridae* are droplet-shaped viruses having unique beehive-like structure with a pointed end densely covered with thin fibers. The genome is circular double-stranded DNA. The family includes two genera *Alphaguttavirus* and *Betaguttavirus* (Pina et al. 2011; ICTV 2016). Family *Ampullaviridae* consists of one species: *Acidianus* bottle-shaped virus (ABV). The bottle-shaped virion carries an envelope which encases a funnel-shaped core. The pointed end of the virion is likely to be involved in adsorption and channeling of viral DNA into host cells. The broad end exhibits about 20 thin filaments which appear to be inserted into a disk, or ring, and are interconnected at their bases. Virion contains 23.9-kb linear double-stranded DNA genome (Häring et al. 2005; ICTV 2016). *Bicaudaviridae* is represented by one species *Acidianus* two-tailed virus (ATV). This virus has a unique capacity to undergo major morphological changes that occur outside of its host. Virions are released from host cells as lemon-shaped tailless particles, and when incubated at temperatures close to the natural infection conditions, long tails develop at each pointed end. The circular double-stranded DNA genome encodes 72 predicted proteins, including 11 structural proteins. *Globuloviridae* comprises two species, PSV and TTSV1. The lipid-containing envelopes of their virions encase the linear double-stranded DNA genome, which is tightly packed into a helical nucleoprotein core (Pina et al. 2011).

Summarized data on bacteriophages and archaea viruses are presented in Table 10.1.

10.7 Problems of Bacteriophage Classification

As mentioned above, classification of bacteriophages provides an opportunity to understand better the biology of viruses and their application prospects in agriculture, medicine, and biotechnology. However, systemization attempts face some problems.

Historically bacteriophage classification was based on the following criteria: host spectrum and physical, chemical, and morphological characteristics. Systemizing approaches ultimately resulted in a comprehensive taxonomic system regularly updated by ICTV. A mandatory part of ICTV classification scheme is examination of bacteriophage virion morphology using electron microscopy technique.

Table 10.1 Overview of bacteriophages and archaea viruses

Family	Shape of virion	Genome	Example
Families common for bacteriophages and archaea viruses			
Myoviridae	Tailed, with a long contractile tail	dsDNA, L	T4 (bacteriophage), φH (archaea virus)
Siphoviridae	Tailed, with a long noncontractile tail	dsDNA, L	T1 (bacteriophage), ψM1 (archaea virus)
Podoviridae	Tailed, with a short tail	dsDNA, L	P22 (bacteriophage), HSTV-1 (archaea virus)
Families of bacteriophages			
Microviridae	Polyhedral	ssDNA, C	φX174
Corticoviridae	Polyhedral	dsDNA, C, S	PM2
Tectiviridae	Polyhedral	dsDNA, L	PRD1
Leviviridae	Polyhedral	ssRNA, L	BZ13
Cystoviridae	Polyhedral	dsRNA, L, seg	φ6
Inoviridae	Filamentous	ssDNA, C	MV-L51
Plasmaviridae	Pleomorphic	dsDNA, C, S	L2
Families of archaea viruses			
Sphaerolipoviridae	Polyhedral	dsDNA, L	SH1
Turriviridae	Polyhedral	dsDNA, C	STIV1
Lipothrixviridae	Filamentous	dsDNA, L	AFV9
Rudiviridae	Filamentous	dsDNA, L	SIRV-1
Clavaviridae	Filamentous	dsDNA, C	APBV1
Spiraviridae	Filamentous	ssDNA, C	ACV
Pleolipoviridae	Pleomorphic	ssDNA/ dsDNA, C/L, S	HRPV1
Fuselloviridae	Pleomorphic	dsDNA, C	SSV1
–	Pleomorphic	dsDNA, L	His1
Guttaviridae	Pleomorphic	dsDNA, C	SNDV
Ampullaviridae	Pleomorphic	dsDNA, L	ABV
Bicaudaviridae	Pleomorphic	dsDNA, C	ATV
Globuloviridae	Pleomorphic	dsDNA, L	TTSV1

C circular, *L* linear, *S* supercoiled, *seg* segmented, *ss* single-stranded, *ds* double-stranded

The nature of nucleic acid is also the essential classification parameter. They determine generally the order and family appurtenance of a virus. However, presence of double-stranded DNA common to all tailed phages doesn't serve as a reliable criterion for discrimination within this group, but it is very appropriate for identification in other families of bacteriophages. Added to family and order affiliation, electron microscopy allows to assume the presence of lipids, identify novel phages and attribute them to the described species, decide on the course of the following phage investigation, and estimate nucleic acid size from capsid diameter (Ackermann 2009). However, this method is difficult to apply in respect to prophages.

Nevertheless, accurate identification requires additional methods. Application of physical, chemical, and morphological tests may cost a lot of time and labor. Development of molecular-genetic methods opened frontiers for potential use of genome sequencing as alternative classification tool. Genome sequencing has brought enormous possibilities and novel problems. Analyses of the ribosomal DNA (rDNA) sequences revolutionized the taxonomic characterization of various forms of life. Phages do not contain a ribosomal sequence in contrast to other organisms. Structural proteins of capsid could hypothetically serve as a basis for phage taxonomy, but they are highly diverse and, unlike rDNAs, do not contain conserved regions facilitating identification. Moreover, investigation of bacteriophage genomes deposited at GenBank library did not reveal any gene common for all phages and potentially eligible for the role of the marker. As an alternative, taxonomic system could be based on the phage proteome. Detailed consideration of genomes allowed to chart a tree depicting relations between phage proteins and resolve a number of classification anomalies associated with the ICTV system (Rohwer and Edwards 2002).

Additionally, genome sequencing meets several challenges unique to phage genomes. Phages are not self-replicating and depend on cell host resources for their replication and growth. It requires sophisticated purification steps to dispose of host genetic material. Some phages, especially lytic, possess highly methylated genomes to ensure protection from bacterial restriction-modification systems. Such highly methylated sequences are recalcitrant to many routine genetic manipulations. Extreme GC content different from that of their host also may pose a problem for PCR and sequencing. Complex genomic structures such as extremely long direct or inverted repeats and terminal redundancies make problem for assembly of the whole-genome sequence from the reads (Klumpp et al. 2012).

Bacteriophage population is very diverse. Phages infecting the same host often display little or no sequence similarity. One of the distinctive features of bacteriophage genomes is their mosaic structure. Each genome can be considered as a unique combination of exchangeable modules transferred among the population. Sequence comparison and phylogenetic reconstruction show that different genes, groups of genes, or segments of genes reveal different ancestry and thus represent modules within a mosaic genome (Hatfull 2008). Similar genes can be observed in phages infecting different groups of bacteria: *Streptococcus* and *Bacillus* (Romero et al. 2004), *Acinetobacter* and *Escherichia* (Klovins et al. 2002), *Lactococcus* and *Bacillus* (Kotsonis et al. 2008), etc. In some cases, phages contain genes from other groups of organisms, even eukaryotes (Bordenstein and Bordenstein 2016). Mosaic structure of genomes is provided by various mechanisms: illegitimate, homologous, or site-specific recombination, transposition, and homing endonucleases (Hatfull and Hendrix 2011). Evolutionary mechanisms shaping the genomes of prokaryotic viruses vary between different families and depend on the type of the nucleic acid, characteristics of the virion structure, as well as the mode of the life cycle. The evolution of tailed virus genomes is driven mainly by illegitimate, rather than homologous, recombination occurring at essentially random positions within the genome, while genomes of *Tectiviridae* evolve mainly through the accumulation of point mutations and the seldom acquisition of new genes (Krupovic et al. 2011). Genome mosaic structure hinders the efforts to establish bacteriophage phylogeny.

Database operations also arouse certain problems. GenBank database is user-driven and accepts data from unpublished articles. Many journals insist that new deciphered sequences be deposited in recognized database before accepting the submitted papers for publication. Rejection of papers leads to accumulation of possibly worthless material. Furthermore, GenBank makes no difference between "phages" and "prophages" (Ackermann 2011). Now the NCBI phage genome database contains about 2200 and 80 complete genomes of bacteriophages and archaea viruses, respectively (NCBI 2017).

The presented data provide evidence of striking contradictions between different methods of bacteriophage classification. ICTV taxonomic scheme is not capable to accurately identify bacteriophages on species and strain level and investigate prophages, whereas molecular-genetic identification methods cannot always supply authentic information on bacteriophage affinity owing to absence of genetic marker common for all bacterial viruses and mosaic structure of phage genomes.

10.8 Conservation of Phages and Bacterial Populations

Microorganisms play an important role in ecosystem and biosphere balance. They are dominant living species on Earth, but only small part of the global microbial diversity has been characterized so far. Microorganisms are involved in nitrogen and carbon dioxide fixation, organic decomposition, reductive and oxidative transformations of a range of essential elements, etc. Microbial strains are exploited in industrial, medical, pharmaceutical, and agricultural sectors. The need to preserve a broad spectrum of microbial kingdoms spurred up development of various conservation methods (Cockell and Jones 2009; Sharma et al. 2016).

Microbial preservation strategies envisage "in situ" and "ex situ" technologies. "In situ" methods are applied for preservation of viable populations in original ecosystems and natural habitats. The biodiversity is managed and monitored within the native ecosystem. In situ management can either be targeted at single selected species or the whole populations of ecosystems. "In situ" approach allows to save microorganisms requiring special conditions and additionally protect organisms associated with certain bacteria (Krishnamurthy 2003; Sharma et al. 2016).

"Ex situ" conservation is carried out by microbial culture collections. In ex situ preservation methods, microbial activities are minimized or ceased artificially. These methods include subculturing, preservation on agar beads and in mineral oils, silica gel storage, spray-drying, fluidized bed drying, cryopreservation, lyophilization (freeze-drying), L-drying, and maintenance in sterile distilled water and gelatin disks. Cryopreservation and lyophilization are the most valuable and widespread methods to achieve long-term, stable storage of microorganisms (Sharma et al. 2016).

Lyophilization is a process of drying wherein water is sublimed from the product after its refrigeration. The sample to be dried is first frozen and then subjected to thermal treatment under deep vacuum so that frozen liquid sublimes, leaving only solid dried components. The process consists of four significant steps: culturing and preparing the cells, suspending the bacteria in a suitable freeze-drying medium, freeze-drying process itself, and post-lyophilization storage. Cell culturing and preparing manipulations are typical for standard cultivation methods. The medium composition must be conducive to maintain cell viability in the course of freezing, water removal, and subsequent storage. It usually includes mannitol, skim milk, and bovine serum albumin. Another vital component of the medium is lyoprotectant, ensuring preservation of the structure of biomolecules throughout lyophilization process. The classic lyoprotectant preferential for bacteria is sucrose. A basic lyophilization process can be divided into three stages: freezing, primary drying, and secondary drying. There are many complicated variations of this basic procedure, but most bacteria will dry well in a simple process. After freeze-drying, bacteria are stored at 4 °C during several years. Holding bacteria at temperatures above 4 °C for prolonged periods of time will dramatically decrease cell viability (OPS Diagnostics 2017).

Cryopreservation is the method of preserving microorganisms at low temperatures. Storage in the nitrogen vapor phase (−140 °C) or the liquid nitrogen phase (−196 °C) has attracted growing interest. At such low temperatures, cell viability is almost independent of the period of storage, and biological systems are believed to be genetically stable. Storage of cultures in the range of −60 to −80 °C also will often result in good viability and may be used when liquid nitrogen is not available or in noncritical applications where some loss of culture viability can be tolerated. Noteworthy, preservation at temperatures above −30 °C yields poor results (Perry 1995). Composition of the medium used to suspend the organisms for freezing is one of the most important factors for the efficiency of cryopreservation. In some cases, microorganisms are able to survive deep-freezing, but presence of suitable cryoprotective agents usually raises the survival rate considerably. There are many cryoprotective additives used in deep-freezing, but Me_2SO and glycerol are most popular. However, applied substances can be toxic to cells and should be removed by centrifugation or dilution after thawing (Hubálek 2003).

Subculturing is the recurrent transfer of microorganisms to fresh growth medium because with increasing time of cultivation the nutrients are depleted and toxic metabolites are accumulated. Nevertheless, this method isn't intended for long storage. In case of preservation on agar beads, the sections of agar measuring approximately 1 × 0.5 cm were cut from areas of dense bacterial growth, placed into individual sterile vials, and stored at temperature ranging from −40 to −70 °C (Winters and Winn 2010). The cultures can be grown on tube slants followed by covering every part of agar with mineral oil. This method is useful in preserving bacteria for periods over 12 months (Bagyaraj and Rangaswami 2007). Cell suspension may be mixed with silica gel particles, and storage time can reach several years at low temperatures (Sinclair and Dhingra 1995). Recently spray-drying method as alternative to lyophilization has been considered. Liquid in fine droplets is sprayed into a flow of hot and dry air (usually 150–250 °C). The subsequent increase in the air–liquid interface area dramatically promotes the drying kinetics, and the process takes only a few seconds. Spray-drying as compared to freeze-drying is distinguished by lower specific energy cost and higher productivity, but the problem of maintaining culture viability remains on the agenda. This method is mainly applied in food and pharmaceutical industry (Huang et al. 2017). Fluidized bed drying resembles spray-drying method, but lower temperatures lead to increased cell survival (Muller et al. 2009). Liquid drying (L-drying) is an alternative method of vacuum drying for preservation of bacteria that are particularly sensitive to the initial freezing stage of the normal lyophilization process. The intrinsic feature of this process is that cultures are prevented from freezing; drying occurs directly from the liquid phase (Annear 1956). Preservation of bacterial cultures can be carried out by transfer in sterile water. Bacteria can be stored in pure water for several years (Iacobellis and DeVay 1986; Liao and Shollenberger 2003). Gelatin disk method includes mixing bacterial suspension with melted gelatin followed by drying (Bagyaraj and Rangaswami 2007).

Cryopreservation and lyophilization are the most practiced techniques for phage preservation. Phages can be lyophilized using various conditions and media, but there is no similarity in survival rate of even closely related phages lyophilized under the same conditions (Cox et al. 1974; Merabishvili et al. 2013). Tailed phages were relatively stable when preserved at low temperatures. Some phages died in lysates stored at +4 °C and were recovered from frozen samples kept at −80 °C; other bacterial viruses were rapidly inactivated during freezing. Several phages were extremely unstable under any conditions (Ackermann et al. 2004). In contrast tailed phages could be stored inside the infected cells at −80 °C without significant loss of phage and host viability (Golec et al. 2011). Other methods such as L-drying, spray-drying at reduced temperatures, and emulsification can be used, although they aren't widespread (Annear 1957; Murthy and Engelhardt 2008; Matinkhoo et al. 2011).

None of the abovementioned methods appears universal due to the differences in sensitivity of individual phages to physical conditions and the content of storage media. Lysates of most tailed phages and polyhedral or filamentous phages without lipids can be stored over several years, but there are many individual peculiarities. The appropriate method of storage depends on phage context. Phage titers are expected to decline by 1 log per year. Lipid-containing phages tend to be more fastidious and unstable and should be controlled yearly. Chloroform may be added to individual phages if they are known to be resistant (Ackermann et al. 2004).

10.9 Applications of Phages and Bacteria in Biotechnology

Bacteria have been harnessed by humankind since ancient times. First bacteria were applied in food production, but studies and understanding of broad spectrum of their metabolic activity as well as opportunities to modify and select microorganisms with desirable characteristics led to widespread use of bacteria in numerous areas.

Food processing is one of the oldest applications of bacteria. Lactic acid bacteria (LAB) represent one of the most important groups of microorganisms used in food fermentations and include genera *Carnobacterium*, *Enterococcus*, *Lactobacillus*, *Lactococcus*, *Leuconostoc*, *Oenococcus*, *Pediococcus*, *Streptococcus*, *Tetragenococcus*, *Vagococcus*, and *Weissella*. These bacteria can degrade a variety of carbohydrates, with lactic acid being the predominant end product, and produce fermented foods of desirable appearance, body, texture, and flavor. LAB take part in manufacturing of wine, dairy (cheese, sour cream, yogurt), meat (sausages), and vegetable products (pickles, sauerkraut, olives). LAB supplements provide health benefits to the consumers by metabolizing indigestible food nutrients (e.g., lactose for lactose-intolerant consumers), stimulation of the immune system, and improvement of intestinal peristaltic activity. Besides, bacteriocins and other metabolites, like organic acids produced by LAB, display antimicrobial activity and hence can be applied in food preservation. In addition to LAB, *Propionibacterium* species are used in dairy fermentations (Swiss-type cheese brands) and contribute to natural fermentations of silage and olives; *Acetobacter* mediates production of vinegar; *Erwinia dissolvens* is essential for coffee bean production (Jay et al. 2006; Rosenberg et al. 2013).

Medicine is another sphere for bacterial use. On one hand, bacteria are able to cause infections responsible for severe consequences and even death. On the other hand, they produce a wide range of substances with potential medical applications. Bacteria synthesize antibiotics, antitumor agents, immunomodulators, and enzyme inhibitors. Since 1980s engineered strains have been introduced in manufacturing purified human proteins such as insulin and growth hormone. Prior to output of recombinant protein products, the substances were derived from cadavers or body fluids and could transmit diseases. Now products of *Escherichia coli* are applied in treatment of infections; diseases of blood, endocrine, and metabolic systems; and nutritional pathologies (Rosenberg et al. 2013).

Bacteria act as indicators of water quality. They may demonstrate the efficacy of detoxification process (e.g., chlorine disinfection), indicate the presence of fecal contamination, and detect pathogens by lack or presence of certain bacterial groups (Ashbolt et al. 2001). Apart from indication of contamination, microorganisms are able to degrade or transform environmental pollutants into relatively safe or less-hazardous substances (bioremediation). Bacteria even utilize spilled oil (Atlas and Hazen 2011). Nevertheless, bioremediation is a very slow process. Only certain species of bacteria proved their capacity of potent pollutant degraders on the industrial scale. Many strains were effective as bioremediation agents only under laboratory conditions (Karigar and Rao 2011).

Bioenergy is regarded as a sound alternative to fossil fuels and one of the ways to compensate for the escalating fuel demands and menacing global warming scenario. Bacteria possess flexible and diverse metabolic capability to convert/synthesize a variety of organics to different forms of bioenergy. Bacteria were shown to produce H_2, methane, ethanol and butanol, and bioelectricity in microbial fuel cell, although optimization of energy generation processes is indispensable for large-scale production of bioenergy (Mohan et al. 2013).

Generally bacteria produce a wide array of valuable substances. In addition to food processing, medicine, and biofuel sector, synthesized compounds can be used as detergents, pesticides, and insecticides; applied in laundry, textile, cosmetic, pulp, and paper industries; etc. (Rosenberg et al. 2013). Molecular-genetic studies make research outlet for microbial enzymes, whereas bacteria serve as model organisms in investigations of metabolic and genetic processes.

Since the discovery of bacteriophages, the latter were considered as promising antimicrobial agents. The initial phage applications were related to medicine. The first published paper on the clinical use of viruses appeared in 1921 and demonstrated potential of staphylococcal-specific phage in treatment of skin diseases (Bruynoghe and Maisin 1921). However, at that time, clinical trials were generally not consistent and contradictory. The discovery of antibiotics readily mass produced, administered, quite stable, and effective led to disappearance of phages from therapy, except USSR where phage treatment was developed, promoted, and widely employed (Summers 2012). Nevertheless, abuse of antibiotics in medicine, agriculture, and veterinary practice resulted in appearance of antibiotic-resistant bacteria. Some bacteria developed resistance to several antibiotics (multidrug-resistant pathogens) (Ventola 2015). Therefore, in recent decades, interest in phages as antimicrobial agents has revived.

The main distinguishing property of phages compared to antibiotics is their high specificity (Table 10.2). Phages eliminate only certain species and strains and don't affect beneficial bacteria. On the other hand, phage application makes no sense when pathogen is unknown. The limitation can be removed by phage cocktails containing various bacterial viruses. Antibiotics with their wide spectrum of action are able to cause multiple side effects, including intestinal disorders, allergies, and secondary infections. In turn, adverse impact of phages usually is induced by liberation of endotoxins from cells lysed by the bacterial viruses. However, the similar effect may be also observed when antibiotics are used. The problem can be solved by the construction of phage variants not affecting the integrity of cell envelope, although this method demands constant supply of phages (Hagens et al. 2004; Sulakvelidze and Kutter 2005; Loc-Carrillo and Abedon 2011).

Table 10.2 Comparison of antibiotics and bacteriophages

Feature	Antibiotics	Bacteriophages
Range of action	Wide	Narrow
Side effects	Multiple side effects (intestinal disorders, allergies, yeast infections)	Few side effects
"Auto-dosing"	No	Yes
A number of doses	Must be supplied repeatedly	One dose is enough
Site specificity	Not necessarily concentrate at the site of infection	Concentrate at host location area
Complexity of selection	Demands time and labor	Easily selected from the environment
Ability to overcome bacterial resistance	Cannot adapt to a bacterial mutation	Can mutate and evade host resistance

Phages in the body are capable to build up ("auto-dosing") specifically where hosts are located until complete elimination of bacteria. Therefore, one dose is enough to get the desired effect. Antibiotics do not necessarily concentrate at the site of infection, and they must be supplied repeatedly. Moreover, phages are easily screened from the environment, while development of new antibiotics demands meticulous investigations and time. Phages infect and kill via mechanisms distinguishing them from antibiotics, so specific antibiotic resistance mechanisms are not correlated and translated into mechanisms of phage resistance. It allows to couple phage and antibiotic treatment together (Loc-Carrillo and Abedon 2011).

Bacteria are able to develop resistance to certain bacteriophages, but it is always possible to select another phage or consortium of phages. In addition, bacteriophages in the process of coevolution developed mechanisms suppressing bacterial resistance (Samson et al. 2013).

Nevertheless, application of phages has several disadvantages. Obligate lytic bacteriophages should be used because temperate phages convert phage-sensitive bacteria into resistant ones, and sometimes they encode bacterial virulence factors, including toxins (Loc-Carrillo and Abedon 2011). Due to its viral nature, the phage eventually will be destroyed by the immune system, and therapeutic effect may be decreased. On the other hand, bacterial viruses can act as agents for treatment of patients with immunodeficiency (Borysowski and Górski 2008; Zimecki et al. 2009). Direct contact between phage and host cell followed by adsorption should be provided which is difficult to secure if bacteria are intracellular parasites. However, decreasing titer of pathogenic bacteria is observed when phage infiltrates with its host into eukaryotic cell (Kaur et al. 2014).

Phages as antibacterial agents have a number of properties that make them compelling alternatives to antibiotics. Clinical trials and experiments with animals showed a great viral potential in treatment of infections caused by *Pseudomonas aeruginosa* (Novik and Savich 2015), *Enterococcus faecium* (Biswas et al. 2002), *Escherichia coli* (Kwarcinski et al. 1994), *Klebsiella pneumonia* (Stroj et al. 1999), *Acinetobacter baumannii* (Shivaswamy et al. 2015), *Shigella* (Babalova et al. 1968) and *Staphylococcus* species (Meladze et al. 1982), etc. Phage therapy research and development are conducted by several companies, but only a few of them are involved primarily in phage product distribution (Abedon 2017).

Another trend in application of bacterial viruses is phage display. The DNA that encodes the polypeptide is fused with phage coat protein genes, and the desired protein is expressed on the surface of the phage particle (Smith 1985). This method allows to study protein–protein and protein–DNA interactions and relate proteins with the genetic basis that encodes them. Using this method, it is possible to synthesize or select proteins with desired characteristics (Lunder et al. 2005). Genetic engineering of bacteriophages can be applied to produce phages containing antigens on their surface. These viral particles bind with specific antibody and ensure output of large amounts of target proteins such as anti-Rh(D) antibodies required for blood group typing. The phage display technology has provided the ability to compile antibody libraries containing a great number of phage particles where each one encodes and displays different molecules (10^6–10^{11} different ligands in a population of >10^{12} phage molecules). Phage display offers the chance to produce diagnostic and therapeutic agents for autoimmune diseases. Many autoimmune disorders were studied and treated using modified phages. The technique is also applied to determine tumor antigens (Bazan et al. 2012a). The method allows to study immunization process, develop novel vaccines, and investigate allergen–antibody interactions (Bazan et al. 2012b). The researchers resort to display technology to promote phage adaptation to subsequent gene transfer into mammalian cells (Baird 2011).

Bacteriophages can be used for bacterial control. Treatment with specific phages in the food industry can prevent the decay of products and the spread of bacterial diseases. Several demands should be satisfied prior to phage application. They as well as phages for medicine should be strictly lytic to prevent appearance of resistant bacteria and change of their genome. Phages with broad host range capable of infecting many strains of the target species and/or genus are preferential for food biocontrol. Phages should be propagated on nonpathogenic host for safety reasons and show no adverse effects when consumed. Physicochemical food characteristics may affect the phage stability, and it is essential to maintain this vital parameter over a long time. Phages should be amenable to scale up for commercial production (Chibeu 2013). Phages reduce population of various food-borne pathogenic species. *Bacillus cereus* is a food-borne pathogen producing enterotoxins such as hemolysin BL, nonhemolytic enterotoxin, and cytotoxin K which can cause vomiting, diarrhea, and nausea. *B. cereus* is generally insusceptible to penicillin-related antibiotics due to production of β-lactamase and, in some cases, to erythromycin and tetracycline. Phage B4 effectively inhibited growth of members of the *B. cereus* group as well as *B. subtilis*, and growth inhibition persisted for over 20 h with burst size more than 200 plaque forming units (PFU) per infected host cell (Lee et al. 2013). *Salmonella* spp. are considered to be one of the major causes of zoonotic diseases worldwide. They can colonize a wide range of hosts, including all major livestock species generating contaminated meat and other food products. Bacteriophage FO1-E2 at 15 °C lowered *Salmonella* counts by 5 log units on turkey deli meat and in chocolate milk and by 3 logs on hot dogs and in seafood. In egg yolk, the effect was observed only after 2 days. At 8 °C no viable cells were detected (Guenther et al. 2012). Bacteriophage SE07 isolated from retail meat samples reduced *Salmonella* population by about 2 log cycles in fruit juice and fresh eggs and 2.1 and 2.0 log in treated beef and chicken meat samples (Thung et al. 2017). *Listeria monocytogenes* is another major food-borne pathogen concerned with a high mortality rate in high-risk group individuals such as pregnant women, neonates, immunocompromised patients, and the elderly (Farber and Peterkin 1991). Bacteriophage Listex P100 was able to control *L. monocytogenes* growth in raw salmon fillets, roast beef and cooked turkey, juices, and slices from melon and pear (Soni and Nannapaneni 2010; Chibeu et al. 2013; Oliveira et al. 2014). The combined treatment of phage P100 with potassium lactate-sodium diacetate mixture reduced the initial *L. monocytogenes* counts by 2–4 log CFU/cm^2 and also kept bacterial counts at that reduced level in queso fresco cheese for 28 days at 4 °C (Soni et al. 2012). Some strains of *Escherichia coli* are considered as food-borne pathogens because of serious clinical outcomes from infection. Experimental phage control of *E. coli* O157:H7 has been reported for lettuce, spinach, melon, tomato, broccoli, and ready-to-eat turkey and beef (Abuladze et al. 2008; Sharma et al. 2009; Anany et al. 2011; Hudson et al. 2013). Bacteriophages chemically labeled with horseradish peroxidase were able to detect *E. coli* O157:H7 in spinach consistently at levels of 1 colony-forming unit (CFU)/g and 1 CFU/100 cm^2 on swabbed meat samples and 10^2 CFU/100 ml in water samples (Willford et al. 2011). *Campylobacter* species have emerged over the last three decades as significant clinical pathogens causing gastrointestinal infections. The major source for infection with *Campylobacter* spp. is broiler meat (EFSA 2011). Phages can be used to control these pathogenic bacteria both at the farm level and in the processed carcass (Janež and Loc-Carrillo 2013).

Bacteriophages can be engaged in monitoring of environment. The fluorescent assay using 4′,6-diamidino-2-phenylindole (DAPI)-labeled T4 bacteriophages detects viable *E. coli* cells in canal water within 30 min, quickly delivering accurate findings for risk assessment of environmental contamination in water samples (Kenzaka et al. 2006). The bacteriophage specific toward *Bacillus anthracis* spores was immobilized onto all surfaces of magnetoelastic resonators. The biosensors tested in *B. anthracis* spore solutions have detection limit of 10^3 CFU/ml (Wan et al. 2007).

Bacteriophages are possible sources of enzymes. Lytic enzymes or lysins produced by bacterial viruses digest the bacterial cell wall for bacteriophage progeny release. Small quantities of purified recombinant lysin added to Gram-positive bacteria caused immediate lysis resulting in log-fold death of the target bacterium (Fischetti 2005). Purified lytic enzyme of pneumococcal bacteriophage was able to eliminate 15 common serotypes of *Streptococcus pneumoniae*, including highly penicillin-resistant strains (Loeffler et al. 2001). Lysin isolated from the γ phage of *Bacillus anthracis* specifically killed both vegetative cells and germinating spores of this bacterium (Schuch et al. 2002). Biofilms are communities of cells protected by an extracellular matrix. They allow cells to adhere to the surface and reduce negative influence of adverse factors. Biofilms are virulence factors preventing elimination of pathogenic bacteria. Specific enzymes as well as bacteriophages can be used for biofilm destruction. Purified bacteriophage-derived peptidase $CHAP_K$ completely removed the *Staphylococcus aureus* biofilms within 4 h and prevented further biofilm formation (Fenton et al. 2013). The ferments can be used in molecular-genetic methods. Bacteriophage ⍰29 DNA polymerase amplifies circular DNA by rolling circle amplification mechanism. Polymerase could be efficiently used for cloning of the complete genome of a *Begomovirus* from small amounts of viral DNA, avoiding errors in PCR mediated by non-proofreading enzymes and potential mutations incorporated by the primers (Inoue-Nagata et al. 2004).

Some phages induce detrimental effects in biotechnology, especially in food fermentation processes requiring lactic acid bacteria (LAB). Phages of LAB have short latent period, relatively large burst size, and/or resistance to pasteurization turning them into hardly disposable nuisance. Lysis leads to failed or slow fermentation, decrease in acid production, and deterioration of milk product quality (e.g., nutritive value, taste, texture, etc.) causing profound economic losses. Nevertheless,

LAB possess some defense systems against bacteriophages. These antiphage mechanisms include inhibition of phage adsorption, blocking of phage DNA injection, restriction modification systems, phage abortive infection systems, and CRISPR/Cas systems. Besides, genetic tools have been applied to engineer strains resistant to phages (Szczepankowska et al. 2013). Phages themselves can be used to select resistant LAB. Two methodologies are practiced to obtain mutants of *Lactobacillus delbrueckii*: the agar plate method and the secondary culture method. In the latter case, sensitive bacteria are inoculated into liquid medium and further infected with phage. Cultures exhibiting complete lysis and secondary growth after subsequent incubation are selected and streaked on MRS agar plates. The grown colonies are cultured again in liquid medium with the same selective phage during at least three rounds resulting in phage-resistant strains. All in all, 44 spontaneous mutants were isolated by two methods (Guglielmotti et al. 2006). After incubation with bacterial viruses, phage-resistant strains were selected for *Lactococcus lactis* subsp. *cremoris* (Weimer et al. 1993), *Streptococcus thermophiles* (Viscardi et al. 2003), and *Lactobacillus helveticus* (Quiberoni et al. 1998).

10.10 Conclusion

Microbial diversity encompasses a huge number of species. Systematization and full characterization of bacterial and bacteriophage species are the prerequisites for deeper insight into mechanisms underlying this diversity and elaboration of methods promoting their long-term preservation. New microbial cultures continue to be discovered and analyzed in morphological, physiological, and metabolic studies. Elucidation of properties inherent to new isolates facilitates development of optimal storage strategies and thus contributes to maintenance of global microbial diversity. Cryopreservation and lyophilization proved highly recognized and most popular strain preservation techniques, although phages require additional research.

Acknowledgments We thank Prof. A. Gamian, Dr. A. Grzhegorzhevich, and Dr. M. Pasciak from the Institute of Immunology and Experimental Therapy, Polish Academy of Sciences for their extensive consultations on the identification of bacterial glycolipids and phospholipids.

References

Abedon ST (2017) Phage companies. http://companies.phage.org. Accessed 31 May 2017

Abuladze T, Li M, Menetrez MY, Dean T, Senecal A, Sulakvelidze A (2008) Bacteriophages reduce experimental contamination of hard surfaces, tomato, spinach, broccoli and ground beef by *Escherichia coli* O157:H7. Appl Environ Microbiol 74:6230–6238. https://doi.org/10.1128/AEM.01465-08

Achtman M, Wagner M (2008) Microbial diversity and the genetic nature of microbial species. Nat Rev Microbiol 6(6):431–440. https://doi.org/10.1038/nrmicro1872

Ackermann HW (1996) Frequency of morphological phage descriptions in 1995. Arch Virol 141 (2):209–218. https://doi.org/10.1007/BF01718394

Ackermann HW (2005) Bacteriophage classification. In: Kutter E, Sulakvelidze A (eds) Bacteriophages: biology and applications. CRC Press, Boca Raton, pp 67–90

Ackermann HW (2006) Classification of bacteriophages. In: Calendar R (ed) The bacteriophages. Oxford University Press, Oxford, pp 8–16

Ackermann HW (2009) Phage classification and characterization. In: Clokie MRJ, Kropinski A (eds) Bacteriophages. Methods and protocols, vol 1: isolation, characterization, and interactions. Humana Press, New York, pp 127–140. https://doi.org/10.1007/978-1-60327-164-6

Ackermann HW (2011) Bacteriophage taxonomy. Microbiol Australia 32(2):90–94

Ackermann HW, Eisenstark A (1974) The present state of phage taxonomy. Intervirology 3 (4):201–219. https://doi.org/10.1159/000149758

Ackermann HW, Prangishvili D (2012) Prokaryote viruses studied by electron microscopy. Arch Virol 157(10):1843–1849. https://doi.org/10.1007/s00705-012-1383-y

Ackermann HW, Tremblay D, Moineau S (2004) Long-term bacteriophage preservation. World Fed Cult Collection Newslett 38:35–40

Adams MH, Wade E (1954) Classification of bacterial viruses: the relationship of two *Serratia* phages to coli-dysentery phages T3, T7, and D44. J Bacteriol 68:320–325

Anany H, Chen W, Pelton R, Griffiths MW (2011) Biocontrol of *Listeria monocytogenes* and *Escherichia coli* O157:H7 in meat by using phages immobilized on modified cellulose membranes. Appl Environ Microbiol 77:6379–6387. https://doi.org/10.1128/AEM.05493-11

Angert ER, Clements KD, Pace NR (1993) The largest bacterium. Nature 362:239–241. https://doi.org/10.1038/362239a0

Annear DI (1956) The preservation of bacteria by drying in peptone plugs. J Hyg 54(4):487–508

Annear DI (1957) The preservation of bacteriophage by drying. J Appl Microbiol 20(1):21–22. https://doi.org/10.1111/j.1365-2672.1957.tb04510.x

Ashbolt NJ, Grabow WO, Snozzi M (2001) Indicators of microbial water quality. In: Fewtrell L, Bartram J (eds) Water quality: guidelines, standards and health. IWA Publishing, London, pp 289–316

Atlas RM, Hazen TC (2011) Oil biodegradation and bioremediation: a tale of the two worst spills in U.S. history. Environ Sci Technol 45(16):6709–6715. https://doi.org/10.1021/es2013227

Babalova EG, Katsitadze KT, Sakvarelidze LA Imnaishvili NS, Sharashidze TG, Badashvili VA, Kiknadze GP, Meĭpariani AN, Gendzekhadze ND, Machavariani EV, Gogoberidze KL, Gozalov EI, Dekanosidze NG (1968) Preventive value of dried dysentery bacteriophage. Zh mikrobiol Epidemiol Immunobiol 45:143–145

Bagyaraj DJ, Rangaswami G (2007) Agricultural microbiology. Prentice Hall, New Delhi

Baird A (2011) Gene transfer into mammalian cells using targeted filamentous bacteriophage. Cold Spring Harb Protoc 2011:950–957. https://doi.org/10.1101/pdb.prot5653

Barry T, Colleran G, Glennon M, Dunican LK, Gannon F (1991) The 16S/23S ribosomal spacer region as a target for DNA probes to identify eubacteria. PCR Methods Appl 1:51–56. https://doi.org/10.1101/gr.1.1.51

Bavykin SG, Lysov YP, Zakhariev V, Kelly JJ, Jackman J, Stahl DA, Cherni A (2004) Use of 16S rRNA, 23S rRNA, and gyrB gene sequence analysis to determine phylogenetic relationships of *Bacillus cereus* group microorganisms. J Clin Microbiol 42:3711–3730. https://doi.org/10.1128/JCM.42.8.3711-3730.2004

Bazan J, Całkosiński I, Gamian A (2012a) Phage display – a powerful technique for immunotherapy: 1. Introduction and potential of therapeutic applications. Hum Vaccin Immunother 8 (12):1817–1828. https://doi.org/10.4161/hv.21703

Bazan J, Całkosiński I, Gamian A (2012b) Phage display – a powerful technique for immunotherapy: 2. Vaccine delivery. Hum Vaccin Immunother 8(12):1829–1835. https://doi.org/10.4161/hv.21704

Bergh O, Børsheim KY, Bratbak G, Heldal M (1989) High abundance of viruses found in aquatic environments. Nature 340:467–468. https://doi.org/10.1038/340467a0

Biswas B, Adhya S, Washart P, Paul B, Trostel AN, Powell B, Carlton R, Merril CR (2002) Bacteriophage therapy rescues mice bacteremic from a clinical isolate of vancomycin-resistant *Enterococcus faecium*. Infect Immun 70:204–210. https://doi.org/10.1128/IAI.70.1.204-210.2002

Bordenstein SR, Bordenstein SR (2016) Eukaryotic association module in phage WO genomes from *Wolbachia*. Nat Commun 7:13155. https://doi.org/10.1038/ncomms13155

Borysowski J, Górski A (2008) Is phage therapy acceptable in the immunocompromised host? Int J Infect Dis 12:466–471. https://doi.org/10.1016/j.ijid.2008.01.006

Bradley DE (1967) Ultrastructure of bacteriophage and bacteriocins. Bacteriol Rev 31:230–314

Bruynoghe R, Maisin J (1921) Essais de thérapeutique au moyen du bacteriophage du Staphylocoque. C R Soc Biol 85:1120–1121

Burnet FM (1933a) The classification of dysentery – coli bacteriophages. II. The serological classification of coli – dysentery phages. J Pathol Bacteriol 36:307–318. https://doi.org/10.1002/path.1700360212

Burnet FM (1933b) The classification of dysentery-Coli bacteriophages. III. A correlation of the serological classification with certain biochemical tests. J Pathol Bacteriol 37:179–184. https://doi.org/10.1002/path.1700370203

Case RJ, Boucher Y, Dahllöf I, Holmström C, Doolittle WF, Kjelleberg S (2007) Use of 16S rRNA and rpoB genes as molecular markers for microbial ecology studies. Appl Environ Microbiol 73:278–288. https://doi.org/10.1128/AEM.01177-06

Chibeu A (2013) Bacteriophages in food safety. In: Méndez-Vilas A (ed) Microbial pathogens and strategies for combating them: science, technology and education. Formatex Research Center, Badajoz, pp 1041–1052

Chibeu A, Agius L, Gao A, Sabour PM, Kropinski AM, Balamurugan S (2013) Efficacy of bacteriophage LISTEX™P100 combined with chemical antimicrobials in reducing *Listeria monocytogenes* in cooked turkey and roast beef. Int J Food Microbiol 167:208–214. https://doi.org/10.1016/j.ijfoodmicro.2013.08.018

Christensen H, Kuhnert P, Olsen JE, Bisgaard M (2004) Comparative phylogenies of the housekeeping genes atpD, infB and rpoB and the 16S rRNA gene within the *Pasteurellaceae*. Int J Syst Evol Microbiol 54:1601–1609. https://doi.org/10.1099/ijs.0.03018-0

Clarridge JE (2004) Impact of 16S rRNA gene sequence analysis for identification of bacteria on clinical microbiology and infectious diseases. Clin Microbiol Rev 17:840–862. https://doi.org/10.1128/CMR.17.4.840-862.2004

Cockell CS, Jones HL (2009) Advancing the case for microbial conservation. Oryx 43:520–526

Colwell RR (1997) Microbial diversity: the importance of exploration and conservation. J Ind Microbiol Biotechnol 18:302–307. https://doi.org/10.1038/sj.jim.2900390

Cox CS, Harris WJ, Lee J (1974) Viability and electron microscope studies of phages T3 and T7 subjected to freeze-drying, freeze-thawing and aerosolization. J Gen Microbiol 81(1):207–215. https://doi.org/10.1099/00221287-81-1-207

d'Hérelle F (1918) Technique de la recherche du microbe filtrant bactériophage (*Bacteriophagum intestinale*). C R Soc Biol 81:1160–1162

d'Hérelle F (1917) Sur un microbe invisible antagoniste des bacilles dysentériques. C R Acad Sci 165:373–375

Doore SM, Fane BA (2016) The *Microviridae*: diversity, assembly, and experimental evolution. Virology 491:45–55. https://doi.org/10.1016/j.virol.2016.01.020

EFSA Panel on Biological Hazards (BIOHAZ) (2011) Scientific opinion on *Campylobacter* in broiler meat production: control options and performance objectives and/or targets at different stages of the food chain. EFSA J9:2105. https://doi.org/10.2903/j.efsa.2011.2105

Eisenstark A (1967) Bacteriophage techniques. In: Maramorosch K, Koprowski H (eds) Methods in virology, vol 1. Academic, New York, pp 449–525

Farber JM, Peterkin PI (1991) *Listeria monocytogenes*, a foodborne pathogen. Microbiol Rev 55:476–511

Fenton M, Keary R, McAuliffe O et al (2013) Bacteriophage-derived peptidase CHAP (K) eliminates and prevents staphylococcal biofilms. Int J Microbiol 2013:625341. https://doi.org/10.1155/2013/625341

Fischetti VA (2005) Bacteriophage lytic enzymes: novel anti-infectives. Trends Microbiol 13 (10):491–496. https://doi.org/10.1016/j.tim.2005.08.007

Fokine A, Rossmann MG (2014) Molecular architecture of tailed double-stranded DNA phages. Bacteriophage 4:e28281. https://doi.org/10.4161/bact.28281

Glaeser SP, Kämpfer P (2015) Multilocus sequence analysis (MLSA) in prokaryotic taxonomy. Syst Appl Microbiol 38(4):237–245. https://doi.org/10.1016/j.syapm.2015.03.007

Golec P, Dąbrowski K, Hejnowicz MS et al (2011) A reliable method for storage of tailed phages. J Microbiol Methods 84:486–489. https://doi.org/10.1016/j.mimet.2011.01.007

Goodfellow M (2000) Microbial systematics: background and uses. In: Priest FG, Goodfellow M (eds) Applied microbial systematics. Springer, Dordrecht, pp 1–18. https://doi.org/10.1007/978-94-011-4020-1_1

Guenther S, Herzig O, Fieseler L et al (2012) Biocontrol of *Salmonella* Typhimurium in RTE foods with the virulent bacteriophage FO1-E2. Int J Food Microbiol 154:66–72. https://doi.org/10.1016/j.ijfoodmicro.2011.12.023

Guglielmotti DM, Reinheimer JA, Binetti AG, Giraffa G, Carminati D, Quiberoni A (2006) Characterization of spontaneous phage-resistant derivatives of *Lactobacillus delbrueckii* commercial strains. Int J Food Microbiol 111:126–133. https://doi.org/10.1016/j.ijfoodmicro.2006.04.035

Gürtler V, Stanisich VA (1996) New approaches to typing and identification of bacteria using the 16S-23S rDNA spacer region. Microbiology 142(Pt 1):3–16. https://doi.org/10.1099/13500872-142-1-3

Hagens S, Habel A, von Ahsen U, von Gabain A, Bläsi U (2004) Therapy of experimental *Pseudomonas* infections with a nonreplicating genetically modified phage. Antimicrob Agents Chemother 48:3817–3822. https://doi.org/10.1128/AAC.48.10.3817-3822.2004

Hamana K, Matsuzaki S (1992) Polyamines as a chemotaxonomic marker in bacterial systematics. Crit Rev Microbiol 18:261–283. https://doi.org/10.3109/10408419209113518

Hankin EH (1896) L'action bactericide des eaux de la Jumna et du Gange sur le vibrion du cholera. Ann Inst Pasteur (Paris) 10:511–523

Hansen DS, Skov R, Benedí JV, Sperling V, Kolmos HJ (2002) *Klebsiella* typing: pulsed-field gel electrophoresis (PFGE) in comparison with O:K-serotyping. Clin Microbiol Infect 8:397–404. https://doi.org/10.1046/j.1469-0691.2002.00411.x

Häring M, Rachel R, Peng X, Garrett RA, Prangishvili D (2005) Viral diversity in hot springs of Pozzuoli, Italy, and characterization of a unique archaeal virus, *Acidianus* bottle-shaped virus, from a new family, the *Ampullaviridae*. J Virol 79:9904–9911. https://doi.org/10.1128/JVI.79.15.9904-9911.2005

Hatfull GF (2008) Bacteriophage genomics. Curr Opin Microbiol 11:447–453. https://doi.org/10.1016/j.mib.2008.09.004

Hatfull GF, Hendrix RW (2011) Bacteriophages and their genomes. Curr Opin Virol 1:298–303. https://doi.org/10.1016/j.coviro.2011.06.009

He Y, Xie Y, Reed S (2014) Pulsed-field gel electrophoresis typing of *Staphylococcus aureus* isolates. Methods Mol Biol 1085:103–111. https://doi.org/10.1007/978-1-62703-664-1_6

Holmes FO (1948) Order Virales. In: Breed RS, Murray EGD, Hitchens AP (eds) Bergey's manual of determinative bacteriology, 6th edn. Williams & Wilkins, Baltimore, pp 1126–1286

Huang S, Vignolles ML, Chen XD, Le Loir Y, Jan G, Schuck P, Jeantet R (2017) Spray drying of probiotics and other food-grade bacteria: a review. Trends Food Sci Technol. 63:1–17. https://doi.org/10.1016/j.tifs.2017.02.007

Hubálek Z (2003) Protectants used in the cryopreservation of microorganisms. Cryobiology 46:205–229. https://doi.org/10.1016/S0011-2240(03)00046-4

Hudson JA, Billington C, Cornelius AJ et al (2013) Use of a bacteriophage to inactivate *Escherichia coli* O157:H7 on beef. Food Microbiol 36:14–21. https://doi.org/10.1016/j.fm.2013.03.006

Hulton CS, Higgins CF, Sharp PM (1991) ERIC sequences: a novel family of repetitive elements in the genomes of *Escherichia coli*, *Salmonella typhimurium* and other enterobacteria. Mol Microbiol 5(4):825–834. https://doi.org/10.1111/j.1365-2958.1991.tb00755.x

Iacobellis NS, DeVay JE (1986) Long-term storage of plant-pathogenic bacteria in sterile distilled water. Appl Environ Microbiol 52:388–389

ICTV (2011) ICTV ninth report; 2009 Taxonomy Release. https://talk.ictvonline.org/ictv-reports/ictv_9th_report/dsdna-viruses-2011/w/dsdna_viruses/103/bicaudaviridae. Accessed 31 May 2017

ICTV (2016) Virus taxonomy: 2016 release. https://talk.ictvonline.org/taxonomy/. Accessed 31 May 2017

Inoue-Nagata AK, Albuquerque LC, Rocha WB et al (2004) A simple method for cloning the complete begomovirus genome using the bacteriophage phi29 DNA polymerase. J Virol Methods 116:209–211. https://doi.org/10.1016/j.jviromet.2003.11.015

Jackman J (2012) The microbe: the basics of structure, morphology, and physiology as they relate to microbial characterization and attribution. In: Cliff JB, Kreuzer HW, Ehrhardt CJ, Wunschel DS (eds) Chemical and physical signatures for microbial forensics. Springer, New York, pp 13–34. https://doi.org/10.1007/978-1-60327-219-3_2

Janda JM, Abbott SL (2007) 16S rRNA Gene sequencing for bacterial identification in the diagnostic laboratory: pluses, perils, and pitfalls. J Clinl Microbiol 45(9):2761–2764. https://doi.org/10.1128/JCM.01228-07

Janež N, Loc-Carrillo C (2013) Use of phages to control *Campylobacter* spp. J Microbiol Methods 95:68–75. https://doi.org/10.1016/j.mimet.2013.06.024

Jay JM, Loessner MJ, Golden DA (2006) Modern food microbiology. Springer, New York. https://doi.org/10.1007/b100840

Karigar CS, Rao SS (2011) Role of microbial enzymes in the bioremediation of pollutants: a review. Enzyme Res 2011:11. https://doi.org/10.4061/2011/805187

Kaur S, Harjai K, Chhibber S (2014) Bacteriophage-aided intracellular killing of engulfed methicillin-resistant *Staphylococcus aureus* (MRSA) by murine macrophages. Appl Microbiol Biotechnol 98:4653–4661. https://doi.org/10.1007/s00253-014-5643-5

Kenzaka T, Utrarachkij F, Suthienkul O, Nasu M (2006) Rapid monitoring of *Escherichia coli* in Southeast Asian urban canals by fluorescent-bacteriophage assay. J Health Sci 52:666–671. https://doi.org/10.1248/jhs.52.666

KEYENCE (2017) Microscope types & principles. http://www.keyence.com/ss/products/microscope/bz-x700/study/principle/002/index.jsp. Accessed 31 May 2017

Klovins J, Overbeek GP, van den Worm SH, Ackermann HW, van Duin J (2002) Nucleotide sequence of a ssRNA phage from *Acinetobacter*: kinship to coliphages. J Gen Virol 83:1523–1533. https://doi.org/10.1099/0022-1317-83-6-1523

Klumpp J, Fouts DE, Sozhamannan S (2012) Next generation sequencing technologies and the changing landscape of phage genomics. Bacteriophage 2:190–199. https://doi.org/10.4161/bact.22111

Knirel YA, Valvano MA (eds) (2011) Bacterial lipopolysaccharides. Structure, chemical synthesis, biogenesis and interaction with host cells. Springer, Wien. https://doi.org/10.1007/978-3-7091-0733-1

Kotsonis SE, Powell IB, Pillidge CJ, Limsowtin GKY, Hillier AJ, Davidson BE (2008) Characterization and genomic analysis of phage asccφ28, a phage of the family *Podoviridae* infecting *Lactococcus lactis*. Appl Environ Microbiol 74:3453–3460. https://doi.org/10.1128/AEM.02379-07

Krawczyk B, Samet A, Leibner J, Sledzińska A, Kur J (2006) Evaluation of a PCR melting profile technique for bacterial strain differentiation. J Clin Microbiol 44:2327–2332. https://doi.org/10.1128/JCM.00052-06

Krawczyk B, Leibner-Ciszak J, Mielech A, Nowak M, Kur J (2009) PCR melting profile (PCR MP) – a new tool for differentiation of *Candida albicans* strains. BMC Infect Dis 9:177. https://doi.org/10.1186/1471-2334-9-177

Krishnamurthy KV (2003) Textbook of biodiversity. Science

Krupovic M, Bamford DH (2007) Putative prophages related to lytic tailless marine dsDNA phage PM2 are widespread in the genomes of aquatic bacteria. BMC Genomics 8:236. https://doi.org/10.1186/1471-2164-8-236

Krupovic M, Prangishvili D, Hendrix RW, Bamford DH (2011) Genomics of bacterial and archaeal viruses: dynamics within the prokaryotic virosphere. Microbiol Mol Biol Rev 75:610–635. https://doi.org/10.1128/MMBR.00011-11

Kwarcinski W, Lazarkiewicz B, Weber-Dabrowska B, Rudnicki J, Kamiński K, Sciebura M (1994) Bacteriophage therapy in the treatment of recurrent subphrenic and subhepatic abscess with jejunal fistula after stomach resection. Pol Tyg Lek 49:535

Lee JH, Shin H, Son B, Heu S, Ryu S (2013) Characterization and complete genome sequence of a virulent bacteriophage B4 infecting food-borne pathogenic *Bacillus cereus*. Arch Virol 158:2101–2108. https://doi.org/10.1007/s00705-013-1719-2

Liao CH, Shollenberger LM (2003) Survivability and long-term preservation of bacteria in water and in phosphate-buffered saline. Lett Appl Microbiol 37:45–50. https://doi.org/10.1046/j.1472-765X.2003.01345.x

Lin T, Lin L, Zhang F (2014) Review on molecular typing methods of pathogens. Open J Med Microbiol 4:147–152. https://doi.org/10.4236/ojmm.2014.43017

Loc-Carrillo C, Abedon ST (2011) Pros and cons of phage therapy. Bacteriophage 1(2):111–114. https://doi.org/10.4161/bact.1.2.14590

Loeffler JM, Nelson D, Fischetti VA (2001) Rapid killing of *Streptococcus pneumoniae* with a bacteriophage cell wall hydrolase. Science 294(5549):2170–2172. https://doi.org/10.1126/science.1066869

LPSN (2018) Classification of domains and phyla – hierarchical classification of prokaryotes (bacteria). http://www.bacterio.net/-classifphyla.html#bacteria. Accessed 30 July 2018

Lunder M, Bratkovic T, Kreft S, Štrukelj B (2005) Peptide inhibitor of pancreatic lipase selected by phage display using different elution strategies. J Lipid Res 46:1512–1516. https://doi.org/10.1194/jlr.M500048-JLR200

Lwoff A, Horne RW, Tournier P (1962) A system of viruses. Cold Spring Harb Symp Quant Biol 27:51–55

Mahilum-Tapy L (2009) The importance of microbial culture collection and gene banks in biotechnology. In: Doelle HW, Rokem JS, Berovic M (eds) Biotechnology, vol I: Fundamentals in biotechnology. EOLSS, Oxford

Matinkhoo S, Lynch KH, Dennis JJ, Finlay WH, Vehring R (2011) Spray-dried respirable powders containing bacteriophages for the treatment of pulmonary infections. J Pharm Sci 100:5197–5205. https://doi.org/10.1002/jps.22715

Meladze GD, Mebuke MG, Chkhetia NS, Kiknadze NI, Koguashvili GG (1982) The efficacy of staphylococcal bacteriophage in treatment of purulent diseases of lungs and pleura. Grudn Khir 1:53–56

Mendell JE, Clements KD, Choat JH, Angert ER (2008) Extreme polyploidy in a large bacterium. Proc Natl Acad Sci USA 105:6730–6734. https://doi.org/10.1073/pnas.0707522105

Merabishvili M, Vervaet C, Pirnay JP, De Vos D, Verbeken G, Mast J, Chanishvili N, Vaneechoutte M (2013) Stability of *Staphylococcus a*ureus phage ISP after freeze-drying (lyophilization). PLoS One 8:e68797. https://doi.org/10.1371/journal.pone.0068797

Mishra M, Chauhan P (2016) Applications of microscopy in bacteriology. Microscopy Res 4:1–9. https://doi.org/10.4236/mr.2016.41001

Mochizuki T, Yoshida T, Tanaka R, Forterre P, Sako Y, Prangishvili D (2010) Diversity of viruses of the hyperthermophilic archaeal genus *Aeropyrum*, and isolation of the *Aeropyrum pernix* bacilliform virus 1, APBV1, the first representative of the family *Clavaviridae*. Virology 402:347–354. https://doi.org/10.1016/j.virol.2010.03.046

Mochizuki T, Krupovic M, Pehau-Arnaudet G, Sako Y, Forterre P, Prangishvili D (2012) Archaeal virus with exceptional virion architecture and the largest single-stranded DNA genome. Proc Natl Acad Sci USA 109:13386–13391. https://doi.org/10.1073/pnas.1203668109

Mohan SV, Reddy MV, Chandra R, Subhash GV, Devi MP, Srikanth S (2013) Bacteria for bioenergy: a sustainable approach towards renewability. In: Gaspard S, Ncbi MC (eds) Biomass for sustainable applications: pollution, remediation and energy. RSC, London, pp 251–289. https://doi.org/10.1039/9781849737142-00251

Muller JA, Ross RP, Fitzgerald GF, Stanton C (2009) Manufacture of probiotic bacteria. In: Charalampopoulos D, Rastall RA (eds) Prebiotics and probiotics science and technology. Springer, New York, pp 725–759. https://doi.org/10.1007/978-0-387-79058-9_18

Murthy K, Engelhardt R (2008) Encapsulated bacteriophage formulation. US Patent US 13/528,215

NCBI (2017) Viruses – complete genomes. https://www.ncbi.nlm.nih.gov/genomes/GenomesGroup.cgi?taxid=10239&opt=Virus. Accessed 31 May 2017

Novik GI (1998) Structure-functional organization of bifidobacteria. Mikrobiologiia 67:376–383

Novik G, Savich V (2015) Bacteriophages of *Pseudomonas* bacteria: application in medicine and agriculture. In: Méndez-Vilas A (ed) The battle against microbial pathogens: basic science, technological advances and educational programs, vol 1. Formatex Research Center, Badajoz, pp 295–304

Novik GI, Vysotskiĭ VV (1995) Architectonics of Bifidobacteria populations: submicroscopic aspect of cell cohesion in *Bifidobacterium adolescentis* and *Bifidobacterium bifidum*. Mikrobiologiia 64(2):222–227

Novik GI, Vysotskiĭ VV, Bogdanovskaia ZN (1994) Cellular ultrastructure of various species of the genus *Bifidobacterium*. Mikrobiologiia 63:515–522

Novik GI, Astapovich NI, Samartsev AA (2001) Study of the physiological and biochemical characteristics of bifidobacteria at the late stages of population development. Mikrobiologiia 70:495–502

Novik GI, Astapovich NI, Kubler J, Gamian A (2002) Characterization of the cell-bound polysaccharides of *Bifidobacterium adolescentis* 94 BIM. Mikrobiologiia 71:205–210. https://doi.org/10.1023/A:1015194020327

Novik GI, Astapovich NI, Grzegorzewicz A, Gamian A (2005a) Isolation and comparative analysis of glycolipid fractions in bifidobacteria. Mikrobiologiia 74:772–780. https://doi.org/10.1007/s11021-005-0122-8

Novik GI, Astapovich NI, Pasciak M, Gamian A (2005b) Biological activity of polar lipids from bifidobacteria. Mikrobiologiia 74:781–787. https://doi.org/10.1007/s11021-005-0123-7

Novik GI, Astapovich NI, Grzhegorzhevich A, Gamian A (2006a) Analysis of phospholipids in bifidobacteria. Mikrobiologiia 75:40–45. https://doi.org/10.1134/S0026261706010061

Novik G, Gamian A, Francisco Jda C, Dey ES (2006b) A novel procedure for the isolation of glycolipids from *Bifidobacterium adolescentis* 94 BIM using supercritical carbon dioxide. J Biotechnol 121:555–562. https://doi.org/10.1016/j.jbiotec.2005.08.018

Ogilvie LA, Jones BV (2015) The human gut virome: a multifaceted majority. Front Microbiol 6:918. https://doi.org/10.3389/fmicb.2015.00918

Oksanen HM, ICTV Report Consortium (2017) ICTV virus taxonomy profiles: corticoviridae. J Gen Virol 98:888–889. https://doi.org/10.1099/jgv.0.000795

Olive DM, Bean P (1999) Principles and applications of methods for DNA-based typing of microbial organisms. J Clin Microbiol 37:1661–1669

Oliveira M, Viñas I, Colàs P, Anguera M, Usall J, Abadias M (2014) Effectiveness of a bacteriophage in reducing *Listeria monocytogenes* on fresh-cut fruits and fruit juices. Food Microbiol 38:137–142. https://doi.org/10.1016/j.fm.2013.08.018

OPS Diagnostics (2017) Bacteria lyophilization overview. http://opsdiagnostics.com/notes/ranpri/bacteria_lyophilization_overview.htm. Accessed 31 May 2017

Patel JB (2001) 16S rRNA gene sequencing for bacterial pathogen identification in the clinical laboratory. Mol Diagn 6:313–321. https://doi.org/10.1054/modi.2001.29158

Pawlowski A, Rissanen I, Bamford JK, Krupovic M, Jalasvuori M (2014) *Gammasphaerolipovirus*, a newly proposed bacteriophage genus, unifies viruses of halophilic archaea and thermophilic bacteria within the novel family *Sphaerolipoviridae*. Arch Virol 159:1541–1554. https://doi.org/10.1007/s00705-013-1970-6

Pei A, Nossa CW, Chokshi P, Blaser MJ, Yang L, Rosmarin DM, Pei Z (2009) Diversity of 23S rRNA genes within individual prokaryotic genomes. PLoS One 4:e5437. https://doi.org/10.1371/journal.pone.0005437

Perry SF (1995) Freeze-drying and cryopreservation of bacteria. Methods Mol Biol 38:21–30. https://doi.org/10.1385/0-89603-296-5:21

Pfankuch E, Kausche GA (1940) Isolierung und übermikroskopische Abbildung eines Bakteriophagen. Naturwissenschaften 28:46. https://doi.org/10.1007/BF01486932

Pietilä MK, Demina TA, Atanasova NS, Oksanen HM, Bamford DH (2014) Archaeal viruses and bacteriophages: comparisons and contrasts. Trends Microbiol 22:334–344. https://doi.org/10.1016/j.tim.2014.02.007

Pietilä MK, Roine E, Sencilo A, Bamford DH, Oksanen HM (2016) *Pleolipoviridae*, a newly proposed family comprising archaeal pleomorphic viruses with single-stranded or double-stranded DNA genomes. Arch Virol 161:249–256. https://doi.org/10.1007/s00705-015-2613-x

Pina M, Bize A, Forterre P, Prangishvili D (2011) The archeoviruses. FEMS Microbiol Rev 35:1035–1054. https://doi.org/10.1111/j.1574-6976.2011.00280.x

Quiberoni A, Reinheimer JA, Suárez VB (1998) Performance of *Lactobacillus helveticus* spontaneous phage-resistant mutants in hard cheese production. Int Dairy J 8:941–949. https://doi.org/10.1016/S0958-6946(98)00126-5

Rakhuba DV, Kolomiets EI, Dey ES, Novik GI (2010) Bacteriophage receptors, mechanisms of phage adsorption and penetration into host cell. Pol J Microbiol 59:145–155. https://doi.org/10.1016/j.micres.2015.01.008.1.94

Ranjbar R, Karami A, Farshad S, Giammanco GM, Mammina C (2014) Typing methods used in the molecular epidemiology of microbial pathogens: a how-to guide. New Microbiol 37:1–15

Razin S, Cosenza BJ (1966) Growth phases of *Mycoplasma* in liquid media observed with phase-contrast microscope. J Bacteriol 91:858–869

Rohwer F, Edwards R (2002) The phage proteomic tree: a genome-based taxonomy for phage. J Bacteriol 184:4529–4535. https://doi.org/10.1128/JB.184.16.4529-4535.2002

Rokyta DR, Burch CL, Caudle SB, Wichman HA (2006) Horizontal gene transfer and the evolution of microvirid coliphage genomes. J Bacteriol 188:1134–1142. https://doi.org/10.1128/JB.188.3.1134-1142.2006

Romero P, López R, García E (2004) Genomic organization and molecular analysis of the inducible prophage EJ-1, a mosaic myovirus from an atypical pneumococcus. Virology 322(2):239–252. https://doi.org/10.1016/j.virol.2004.01.029

Rosenberg E, DeLong EF, Lory S, Stackebrandt E, Thompson F (eds) (2013) The prokaryotes. Springer, Berlin. https://doi.org/10.1007/978-3-642-31331-8

Roux S, Krupovic M, Poulet A, Debroas D, Enault F (2012) Evolution and diversity of the *Microviridae* viral family through a collection of 81 new complete genomes assembled from virome reads. PLoS One 7:e40418. https://doi.org/10.1371/journal.pone.0040418

Ruska H (1940) Über die Sichtbarmachung der bakteriophagen Lyse im Übermikroskop. Naturwissenschaften 28:45–46. https://doi.org/10.1007/BF01486931

Ruska H (1943a) Versuch zu einer Ordnung der Virusarten. Archiv für die gesamte Virusforschung 2:480–498. https://doi.org/10.1007/BF01244584

Ruska H (1943b) Ergebnisse der Bakteriophagenforschung und ihre Deutung nach morphologischen Befunden. Ergeb Hyg Bakteriol Immunforsch Exp Ther 25:437–498

Ryberg A, Olsson C, Ahrné S, Monstein HJ (2011) Comparison of (GTG)5-oligonucleotide and ribosomal intergenic transcribed spacer (ITS)-PCR for molecular typing of *Klebsiella* isolates. J Microbiol Methods 84:183–188. https://doi.org/10.1016/j.mimet.2010.11.019

Samson JE, Magadán AH, Sabri M, Moineau S (2013) Revenge of the phages: defeating bacterial defences. Nat Rev Microbiol 11:675–687. https://doi.org/10.1038/nrmicro3096

Schleifer KH (2009) Classification of bacteria and archaea: past, present and future. Syst Appl Microbiol 32:533–542. https://doi.org/10.1016/j.syapm.2009.09.002

Schuch R, Nelson D, Fischetti VA (2002) A bacteriolytic agent that detects and kills *Bacillus anthracis*. Nature 418:884–889. https://doi.org/10.1038/nature01026

Schulz HN, Jorgensen BB (2001) Big bacteria. Annu Rev Microbiol 55:105–137. https://doi.org/10.1146/annurev.micro.55.1.105

Sentausa E, Fournier PE (2013) Advantages and limitations of genomics in prokaryotic taxonomy. Clin Microbiol Infect 19:790–795. https://doi.org/10.1111/1469-0691.12181

Sharma M, Patel JR, Conway WS, Ferguson S, Sulakvelidze A (2009) Effectiveness of bacteriophages in reducing *Escherichia coli* O157:H7 on fresh-cut cantaloupes and lettuce. J Food Prot 72:1481–1485. https://doi.org/10.4315/0362-028X-72.7.1481

Sharma SK, Gupta AK, Shukla AK, Ahmad E, Sharma MP, Ramesh A (2016) Microbial conservation strategies and methodologies: status and challenges. Indian J Plant Genet Resour 29:340–342. https://doi.org/10.5958/0976-1926.2016.00062.0

Sharples GJ, Lloyd RG (1990) A novel repeated DNA sequence located in the intergenic regions of bacterial chromosomes. Nucleic Acids Res 18:6503–6508

Shivaswamy VC, Kalasuramath SB, Sadanand CK, Basavaraju AK, Ginnavaram V, Bille S, Ukken SS, Pushparaj UN (2015) Ability of bacteriophage in resolving wound infection caused by multidrug-resistant *Acinetobacter baumannii* in uncontrolled diabetic rats. Microb Drug Resist 21:171–177. https://doi.org/10.1089/mdr.2014.0120

Sidorenko AV, Novik GI, Akimov VN (2008) Application of the methods of molecular systematics to classification and identification of bacteria of the genus *Bifidobacterium*. Mikrobiologiia 77:293–302. https://doi.org/10.1134/S0026261708030016

Sinclair JB, Dhingra OD (1995) Basic plant pathology methods. CRC Press, Boca Raton

Smith GP (1985) Filamentous fusion phage: novel expression vectors that display cloned antigens on the virion surface. Science 228:1315–1317. https://doi.org/10.1126/science.4001944

Sneath PHA, Brenner DJ (1992) "Official" nomenclature lists. ASM News 58:175

Soni KA, Nannapaneni R (2010) Bacteriophage significantly reduces *Listeria monocytogenes* on raw salmon fillet tissue. J Food Prot 73(1):32–38

Soni KA, Desai M, Oladunjoye A, Skrobot F, Nannapaneni R (2012) Reduction of *Listeria monocytogenes* in queso fresco cheese by a combination of listericidal and listeriostatic GRAS antimicrobials. Int J Food Microbiol 155:82–88. https://doi.org/10.1016/j.ijfoodmicro.2012.01.010

Stern MJ, Ames GF, Smith NH et al (1984) Repetitive extragenic palindromic sequences: a major component of the bacterial genome. Cell 37:1015-1026. doi: https://doi.org/10.1016/0092-8674(84)90436-7

Stroj L, Weber-Dabrowska B, Partyka K, Mulczyk M, Wójcik M (1999) Successful treatment with bacteriophage in purulent cerebrospinal meningitis in a newborn. Neurol Neurochir Pol 33:693–698

Sulakvelidze A, Kutter E (2005) Bacteriophage therapy in humans. In: Kutter E, Sulakvelidze A (eds) Bacteriophages: biology and applications. CRC Press, Boca Raton, pp 382–436

Summers WC (2012) The strange history of phage therapy. Bacteriophage 2:130–133. https://doi.org/10.4161/bact.20757

Suttle CA (2005) Viruses in the sea. Nature 437(7057):356–361. https://doi.org/10.1038/nature04160

Szczepankowska AK, Górecki RK, Kołakowski P, Bardowski JK (2013) Lactic acid bacteria resistance to bacteriophage and prevention techniques to lower phage contamination in dairy fermentation. In: Marcelino Kongo J (ed) Lactic acid bacteria – R&D for food, health and livestock purposes. InTech. https://doi.org/10.5772/51541. https://www.intechopen.com/books/lactic-acid-bacteria-r-d-for-food-health-and-livestock-purposes/lactic-acid-bacteria-resistance-to-bacteriophage-and-prevention-techniques-to-lower-phage-contaminat. Accessed 31 May 2017

Thung TY, Premarathne JMKJK, San Chang W, Loo YY, Chin YZ, Kuan CH, Tan CW, Basri DF, Radzi CWJWM, Radu S (2017) Use of a lytic bacteriophage to control *Salmonella* Enteritidis in retail food. LWT – Food Sci Technol 78:222–225. https://doi.org/10.1016/j.lwt.2016.12.044

Tikhonenko AS (1970) Ultrastructure of bacterial viruses. Springer, New York, pp 29–32. https://doi.org/10.1007/978-1-4684-1779-1

Twort FW (1915) An investigation on the nature of ultramicroscopic viruses. Lancet 186:1241–1243. https://doi.org/10.1016/S0140-6736(01)20383-3

Valueva OA, Rakhuba D, Shashkov AS, Zdorovenko EL, Kiseleva E, Novik G, Knirel YA (2011) Structure of the major O-specific polysaccharide from the lipopolysaccharide of *Pseudomonas fluorescens* BIM B-582: identification of 4-deoxy-D-xylo-hexose as a component of bacterial polysaccharides. J Nat Prod 74:2161–2167. https://doi.org/10.1021/np200472p

Valueva OA, Shashkov AS, Zdorovenko EL, Chizhov AO, Kiseleva E, Novik G, Knirel YA (2013) Structures of cell-wall phosphate-containing glycopolymers of *Bifidobacterium longum* BIM B-476-D. Carbohydr Res 373:22–27. https://doi.org/10.1016/j.carres.2013.03.006

Ventola CL (2015) The antibiotic resistance crisis: part 1: causes and threats. P T 40:277–283

Viscardi M, Capparelli R, Di Matteo R, Carminati D, Giraffa G, Iannelli D (2003) Selection of bacteriophage-resistant mutants of *Streptococcus thermophilus*. J Microbiol Methods 55:109–119. https://doi.org/10.1016/S0167-7012(03)00146-5

Wan J, Johnson ML, Guntupalli R, Petrenko VA, Chin BA (2007) Detection of *Bacillus anthracis* spores in liquid using phage-based magnetoelastic micro-resonators. Sens Actuators B 127:559–566. https://doi.org/10.1016/j.snb.2007.05.017

Wang LT, Lee FL, Tai CJ, Kasai H (2007) Comparison of gyrB gene sequences, 16S rRNA gene sequences and DNA-DNA hybridization in the *Bacillus subtilis* group. Int J Syst Evol Microbiol 57:1846–1850. https://doi.org/10.1099/ijs.0.64685-0

Wang M, Cao B, Yu Q, Liu L, Gao Q, Wang L, Feng L (2008) Analysis of the 16S–23S rRNA gene internal transcribed spacer region in *Klebsiella* species. J Clin Microbiol 46:3555–3563. https://doi.org/10.1128/JCM.00927-08

WDCM (2018) WDCM CCINFO cover. http://www.wfcc.info/ccinfo/. Accessed 30 July 2018

Weimer BC, Blake M, Hillier AJ, Davidson BE (1993) Studies on the isolation of phage-resistant derivatives of *Lactococcus lactis* subsp. *cremoris* FG2 with phage sk1. Aust J Dairy Technol 48:59–61

Whitman WB, Coleman DC, Wiebe WJ (1998) Prokaryotes: the unseen majority. Proc Natl Acad Sci USA 95:6578–6583. https://doi.org/10.1073/pnas.95.12.6578

Wildy P (1971) Classification and nomenclature of viruses. First report of the International Committee on Nomenclature of Viruses. Karger, Basel

Willford JD, Bisha B, Bolenbaugh KE, Goodridge LD (2011) Luminescence based enzyme-labeled phage (phazyme) assays for rapid detection of Shiga toxin producing *Escherichia* coli serogroups. Bacteriophage 1:101–110. https://doi.org/10.4161/bact.1.2.15666

Williamson KE, Radosevich M, Wommack KE (2005) Abundance and diversity of viruses in six Delaware soils. Appl Environ Microbiol 71:3119–3125. https://doi.org/10.1128/AEM.71.6.3119-3125.2005

Winters RD, Winn WC Jr (2010) A simple effective method for bacterial culture storage: a brief technical report. J Bacteriol Virol 40:99–101. https://doi.org/10.4167/jbv.2010.40.2.99

Wommack KE, Colwell RR (2000) Virioplankton: viruses in aquatic ecosystems. Microbiol Mol Biol Rev 64:69–114. https://doi.org/10.1128/MMBR.64.1.69-114.2000

Yamamoto S, Harayama S (1998) Phylogenetic relationships of *Pseudomonas putida* strains deduced from the nucleotide sequences of gyrB, rpoD and 16S rRNA genes. Int J Syst Bacteriol 48(Pt 3):813–819. https://doi.org/10.1099/00207713-48-3-813

Young KD (2006) The selective value of bacterial shape. Microbiol Mol Biol Rev 70:660–703. https://doi.org/10.1128/MMBR.00001-06

Zdorovenko EL, Kachala VV, Sidarenka AV, Izhik AV, Kisileva EP, Shashkov AS, Nobik GI, Knirel YA (2009) Structure of the cell wall polysaccharides of probiotic bifidobacteria *Bifidobacterium bifidum* BIM B-465. Carbohydr Res 344:2417–2420. https://doi.org/10.1016/j.carres.2009.08.039

Zimecki M, Artym J, Kocieba M, Weber-Dabrowska B, Borysowski J, Górski A (2009) Effects of prophylactic administration of bacteriophages to immunosuppressed mice infected with *Staphylococcus aureus*. BMC Microbiol 9:169. https://doi.org/10.1186/1471-2180-9-169

Zinder SH, Dworkin M (2006) Morphological and physiological diversity. In: Dworkin M, Falkow S, Rosenberg E, Schleifer KH, Stackebrandt E (eds) The prokaryotes, vol 1. Springer, New York, pp 185–220. https://doi.org/10.1007/0-387-30741-9_7

Chapter 11
Modern Immunochemical Approaches in Microbiology

Elena Kiseleva, Konstantin Mikhailopulo, and Galina Novik

Abstract Enzyme-linked immunosorbent assay (ELISA) is a method applicable for scientific and applied research in microbiology. In the study we reviewed the design, format, and main and auxiliary components of ELISA test systems and give a basic protocol of ELISA. Polyclonal antibodies (PAbs) to bacterial/yeast antigens are reagents suitable for ELISA. We presented six PAb preparations obtained as a result of rabbit immunization with whole cells of *Saccharomyces cerevisiae* BIM Y-195, *Debaryomyces hansenii* BIM Y-4, *Bacillus cereus* BIM B-491, *Lactobacillus plantarum* BIM B-495D, *Lactococcus lactis* subsp. *cremoris* BIM B-493D, and *Bifidobacterium bifidum* BIM B-733D. The protocols for rabbit immunization and determination of PAb primary characteristics, viz., antiserum working titer and cross-reactivity with bacteria/yeast related to strain immunogen are given. Two examples of PAb usage are shown: (1) test systems for quantification of bacterial cells in culture media or other fluids and (2) the methodology for studying the dynamics of synthesis and secretion of bacterial antigens. The data are useful for scientific researchers and biotechnologists, e.g., (1) to maintain microbial diversity through detection/comparison of microbial compounds and metabolites and (2) to optimize and control biotechnological processes.

11.1 Introduction

Microorganisms comprise the greatest numbers of individual organisms on Earth (Colwell 1997). They occupy different ecological niches, even extreme habitats, interact with related species and other taxa, and act as symbionts and pathogens of plants and animals (Whitman et al. 1998).

E. Kiseleva (✉) · K. Mikhailopulo
Institute of Bioorganic Chemistry, National Academy of Sciences of Belarus, Minsk, Republic of Belarus
e-mail: epkiseleva@yandex.by

G. Novik
Belarusian Collection of Non-pathogenic Microorganisms, Institute of Microbiology, National Academy of Sciences of Belarus, Minsk, Republic of Belarus

S. K. Sharma, A. Varma (eds.), *Microbial Resource Conservation*, Soil Biology 54,
https://doi.org/10.1007/978-3-319-96971-8_11

Bacteria/yeasts are widely used in biotechnology (Mohan et al. 2013; Vijayakumar and Saravanan 2015; Gupta et al. 2016). To optimize/control biotechnological processes, it is necessary to count bacterial/yeast cells and detect/quantify their components and secretory products. The rapid methods used for this are well reviewed and classified (Law et al. 2014; Sohier et al. 2014; Deshmukh et al. 2016; Priyanka et al. 2016). ELISA is an immunology-based method suitable for detection/quantification of microbial antigens and antibodies to microbial antigens and useful for scientific and applied research in microbiology; biotechnologies of natural bacterial proteins and bacterial-derived recombinant proteins, foods, and pharmaceuticals; food and feed safety; diagnosis of human, animal, and plant diseases; biodegradation and bioremediation; etc.

The systematization of microorganisms is an important part of a set of measures aimed at maintaining their diversity. One of the principles of systematics is the establishment of similarities and differences in metabolic pathways, structural and functional components of the cell, as well as secreted compounds. Since microbial biopolymers and many low-molecular compounds have antigen properties, immunochemical methods, including ELISA, can be used for maintaining microbial diversity.

11.2 ELISA in Microbiology: Main Principles of Designing Test Systems and Examples of Their Use in Scientific and Applied Research

11.2.1 ELISA as a Method for Scientific and Applied Research in Microbiology

ELISA is a sensitive, highly specific, and easily automated method suitable for screening since results are rapid, consistent, and relatively easily interpreted. Similar to other types of immunoassays, ELISA is based on highly specific binding between an antigen and an antibody where an epitope (immunodeterminant region) on the antigen surface is recognized by the antibody's binding site. The type of antibody and its affinity and avidity for the antigen determine ELISA sensitivity and specificity. Depending on the ELISA format, immunoassays can be qualitative or quantitative. In context of scientific and applied research in microbiology, ELISA is suitable for detection/quantification of microbial antigens and antibodies to microbial antigens.

11.2.1.1 ELISA for Detection/Quantification of Microbial Antigens

ELISA for detection/quantification of microbial antigens is used in:

- Biotechnologies of natural bacterial proteins and recombinant proteins for their application as biochemical reagents and components of pharmaceuticals/vaccines (Wu et al. 2009; Berger et al. 2011, Cheng et al. 2014; Feng et al. 2016)
- Production of bacteriocins for application in foods and pharmaceuticals (Keren et al. 2004; Richard et al. 2004; Criado et al. 2006; Aly et al. 2011; Yang et al. 2014)
- Food production (including dairy, bravery and bakery industry, winemaking) for monitoring of starting cultures (Garrote et al. 2005) and detection of spoilage species (March et al. 2005)
- Functional food industry and probiotic industry (Anonimous 2006)
- Detection of foodborne and waterborne pathogens (Aydin et al. 2014; Law et al. 2014; Tao and Guyer 2016) and their toxins (Klarić et al. 2009; Jain et al. 2011; Meulenberg 2012; Arya et al. 2013; Berthiller et al. 2013) for food and feed safety
- Plant disease detection (Fang and Ramasamy 2015)
- Technologies of biodegradation and bioremediation (O'Connor and Coates 2002; Shetty et al. 2010; Kriszt et al. 2012; Ferenczi et al. 2014; Vanhoutte et al. 2016) and so on

Whole cells of bacteria/yeast, their component, and secreted substances are suitable as analytes in ELISA test systems for microbial antigen detection/quantification.

11.2.1.2 ELISA for Detection/Quantification of Human/Animal Serum Antibodies to Microbial Antigens

ELISA for detection/quantification of human/animal serum antibodies to microbial antigens is widely used as a non-nucleic acid-based identification method for infections with viral, bacterial, and parasitic agents (Koivunen and Krogsrud 2006). A broad range of high-quality commercially available ELISA test systems constitutes indispensable and reliable means for accurate and individual diagnosis as well as prognosis.

Commercial ELISA kits for detection of antibodies to infection agents in human serum are used for:

- Blood bank screening (antihuman immunodeficiency virus (anti-HIV), anti-*Treponema pallidum* IgG/IgM, etc.)
- Prenatal care (anti-*Toxoplasma gondii* lgG/lgM, anti-rubella virus IgG/IgM; anti-cytomegalovirus lgG/IgM, anti-herpes simplex virus 1/2 IgG/IgM, etc.)
- Routine testing, including sexually transmitted diseases (anti-HIV, anti-hepatitis B surface antigen, anti-*Treponema pallidum* IgG/IgM), regional and endemic diseases (dengue virus IgG/IgM, anti-*Trypanosoma cruzi* IgG; anti-*Borrelia*

burgdorferi IgG/IgM, anti-*Brucella* IgG/IgM, etc.), and other diseases (anti-*Helicobacter pylori* IgG/IgA, etc.)

Commercial ELISA kits for detection of antibodies to nonpathogenic microorganisms in human serum are used for diagnosis of noninfectious diseases [e.g., anti-*Saccharomyces cerevisiae* antibodies (ASCA) IgG/IgA are recognized markers of Crohn's disease (Bertin et al. 2013)].

Indeed, it was shown recently that detection/quantification of human/animal serum antibodies to nonpathogenic microorganisms (including human commensals and probiotics in particular) is potentially useful for prognosis, timely prevention, and treatment of autoimmune diseases, e.g., type 1 diabetes (T1D) and autoimmune thyroid disease (Kiseleva et al. 2011; Talja et al. 2014; Gülden et al. 2015).

11.2.2 Design and Format of ELISA Test Systems

ELISA is a plate-based assay technique designed for detecting and quantifying high-molecular and low-molecular substances. In accordance with the term ELISA, using of immunosorbent and enzyme is a distinctive feature of the type of immunoassay. Immunosorbents in commercially available ELISA test systems are multi-well polystyrene plates. Magnetic particles are used relatively rare (Tudorache and Bala 2008; Shen et al. 2014; Tao and Guyer 2016).

An enzyme is used as a label necessary for detection of immune complexes as an alternative of radioactive, fluorescence, and chemiluminescence labels used in other types of immunoassay. Types of immobilization as well as most commonly used enzymes are discussed below, in Sects. 11.2.2.3 and 11.2.2.5.

11.2.2.1 Components of ELISA Test Systems

As stated above, ELISA tests are used for the detection of antigens or specific antibodies in a sample. A key event in ELISA is formation of immune complex between antibody and antigen. They are the main components of each test system. One of them plays a role of an analyte and is also used as a key component of calibration probes (standards). Another one is used for probing of analyte. Other components of ELISA test systems are auxiliary; they are used for immobilization either antibody or antigen and detection of immune complexes. Immobilization, probing, and detection are the main steps of each ELISA protocol. Affinity (K_a) of antigen-antibody interaction is in the range 10^5–10^{12} L/mol, which explains high sensitivity and specificity of ELISA as a method.

Design of most commonly used ELISA systems is shown in Fig. 11.1. There are two ELISA formats: noncompetitive (immunometric) and competitive (Fig. 11.1). Further in the text, we explain the terms used to designate main and auxiliary components of ELISA test systems (the legend of Fig. 11.1), give a basic

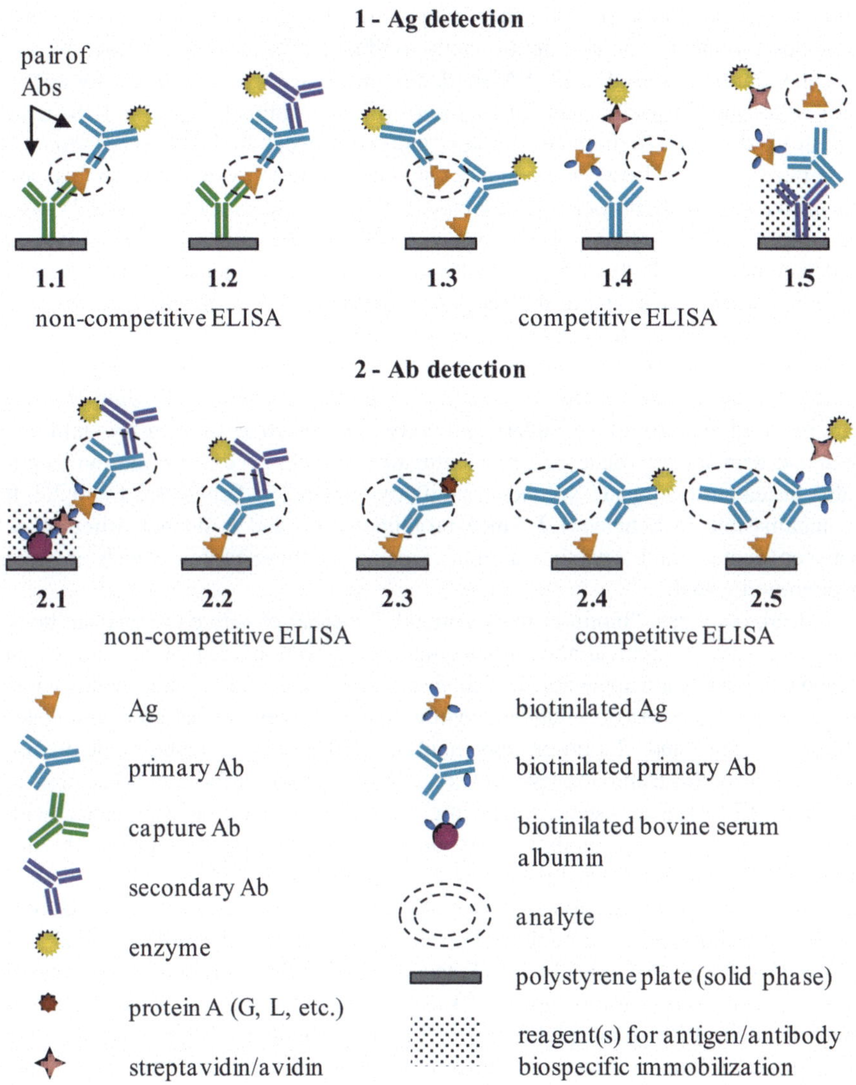

Fig. 11.1 Design of ELISA. Ag and Ab are antigen and antibody, respectively

immunological protocol for noncommercial (laboratory) ELISA test system, and discuss two types of detection strategies, direct *vs.* indirect.

In the case when an analyte is an antigen, specific antibody is denominated as primary. In the same ELISA tests, a matched pair of antigen-specific antibodies is used; they independently interact with two spatially isolated epitopes of the antigen that leads to formation of triplex complex (Fig. 11.1, Schemes 1.1, 1.2). The primary antibody used in solution is called the detection antibody; the antibody immobilized on solid phase is called the capture antibody or coating antibody (it coats immunosorbent and captures either antigen or immune complex via antigen). Schemes 1.1 and 1.2 are used traditionally to illustrate "sandwich" ELISA.

Other antibodies used in ELISA are denominated as secondary. They are necessary either for biospecific immobilization of primary antibody (Fig. 11.1, Scheme 1.5) or for detection of immune complex "antigen-primary antibody" (via formation of immune complex between secondary antibody and primary antibody; in this case the primary antibody plays a role of antigen) (Fig. 11.1, schemes 1.2, 2.1, 2.2). As a rule, the secondary antibody is anti-species PAb, e.g., the antibody of sheep is used to detect mouse antibody (sheep anti-mouse antibody); total mouse Igs were used as an immunogen for obtaining the secondary antibody. PAbs of rabbit, sheep, and goad are used as secondary antibody most frequently.

Protein A from *Staphylococcus aureus*, protein of group C and G streptococci, protein L from *Peptostreptococcus magnus*, and other bacterial Ig-binding proteins can be used instead of secondary antibody for detection of primary antibody bounded with antigen (Scheme 2.3). Conjugates "protein A/G/L-enzyme" are applicable instead of conjugate "secondary antibody-enzyme" in Schemes 2.1 and 2.2. It is inadmissible in Scheme 1.2, since proteins A, G, and L do not differentiate between capture and detecting antibody; using of these proteins leads to high background signal.

Indeed, conjugate "biotin-primary antibody" instead of native primary antibody and conjugate "streptavidin/avidin-enzyme" as detection reagent are shown in Scheme 2.5. They are applicable in Schemes 1.2, 1.3, 2.2, and 2.4. It is inadmissible in Scheme 2.2, since streptavidin and avidin do not differentiate between conjugates "biotin-antigen" and "biotin-primary antibody." Thus, it is inadmissible to use streptavidin-biotin technology simultaneously for immobilization and detection in the same ELISA test system. Interestingly, anti-biotin antibody conjugated with numerous enzymes is a modern development useful to detect conjugate "biotin-primary Ab"; the assay sensitivity may be amplified up to 200-fold over standard streptavidin-enzyme detection (Allan et al. 1989). Advantages of streptavidin/avidin-biotin technology for immobilization in ELISA are described in Sect. 11.2.2.5.

Competitive ELISA (Fig. 11.1, Schemes 1.3–1.5, 2.4, 2.5) is commonly considered "special" (i.e., exotic) type of ELISA used as the only way for detection/quantification of low-molecular antigens (Fig. 11.1, Schemes 1.4, 1.5). In the type of ELISA, probing is based on competition between conjugate "biotin-antigen" (added in equal quantity into all wells as a reference antigen) and non-labeled antigen (contained either in calibration probes (standards) or in analyzed samples) for antibody binding (Fig. 11.1, Schemes 1.4 and 1.5). Then conjugate "biotin-antigen" bounded with antibodies is detected using conjugate "streptavidin/avidin-enzyme."

In fact, the possibilities of using competitive ELISA format are much wider. It is suitable for detection/quantification of high-molecular antigens (Fig. 11.1, Scheme 1.3) and antibodies (Fig. 11.1, Schemes 2.4 and 2.5); examples of using Scheme 1.3 are in Sects. 11.2.3 and 11.2.4. In Scheme 1.3, antigen in liquid phase (contained either in calibration probes (standards) or in analyzed samples) competes with immobilized antigen for binding of antibodies. In Schemes 2.4 and 2.5, enzyme-labeled antibody

(Scheme 2.4) and biotin-labeled antibody (Scheme 2.5) compete with native antibody for binding with immobilized antigen. Note that ELISA format using the antigen attached directly to a plate is known as antigen-down assays. Besides competitive ELISA (Fig. 11.1, Schemes 1.3, 2.4, and 2.5 discussed above), the format is used in noncompetitive ELISA for detection of antibodies (Fig. 11.1, Schemes 2.2 and 2.3).

11.2.2.2 Basic Protocol of ELISA

Basic immunological protocol of noncommercial (laboratory) ELISA test is shown using Scheme 1.2 as an example. It includes 12 steps.

1. Immobilization of capture Ab by passive adsorption (in other words, plate coating to fix capture Ab on solid support).
2. Plate washing.
3. Plate blocking (blocking reagents are not shown in Scheme 1.2).
4. Binding of an antigen with capture Ab.
5. Plate washing.
6. Binding of an antigen with a primary detecting Ab.
7. Plate washing.
8. Binding of a primary Ab with an enzyme-labeled secondary Ab.
9. Plate washing.
10. Initiation of enzymatic reaction by adding of a substrate (it isn't shown in Scheme 1.2). The step is known as generation of a detectable signal.
11. Stopping of enzymatic reaction by adding of stop reagent. The step is used when a product of enzymatic reaction is colored/fluorescent substance.
12. Registration of results. An appropriate instrument is used depending of the type of substrate (spectrophotometer and fluorimeter for colored and fluorescent products of enzymatic reaction, respectively, and luminometer for detection of luminescence light emitted during the reaction) (Fig. 11.2).

Plate washing is necessary to remove molecules that were not immobilized/bounded at the previous step of ELISA and wash off nonspecific binding. Washing is accomplished by adding the detergents.

Plate blocking is performed by addition of irrelevant protein (bovine serum albumin (BSA), ovalbumin, etc.) or other molecules to cover all unsaturated surface sites and escape high background signal and false-positive/false-negative results.

Coated plates (contained immobilized substance and blocking reagents) can be used immediately or dried and stored at 4°C for later use, depending on stability of immobilized substance. When coated plates are stored, blocking solution contains preservatives effective against bacteria, yeasts, and molds. Plates of commercial ELISA kits are pre-coated, blocked, and ready to use.

As stated above, probing occurs via formation of primary immune complex either between analyzed antigen and immobilized primary antibody (Fig. 11.1, Ag detection, for exception of Schemes 1.1 and 1.2) or between analyzed antibody and immobilized antigen (Fig. 11.1, Ab detection). In Schemes 1.1 and 1.2, probing

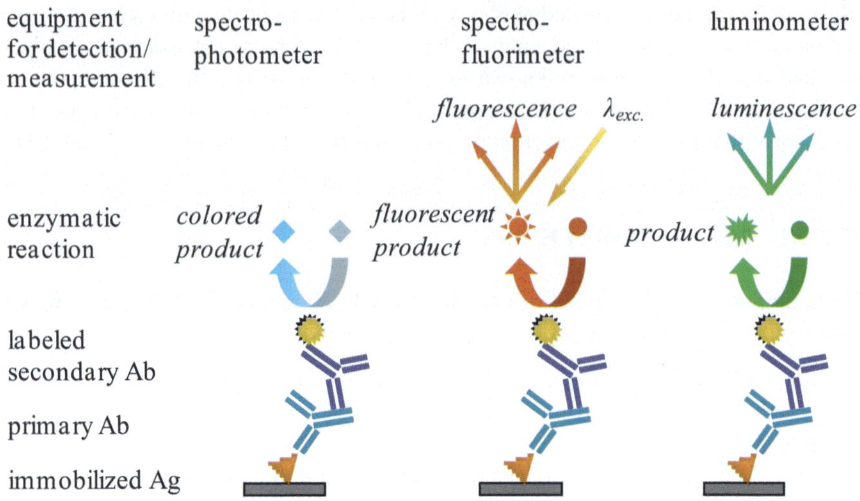

Fig. 11.2 Registration of ELISA results using of appropriate equipment in dependency of type of substrate. Luminometry is up to 100,000 times more sensitive than absorption spectroscopy and is at least 1000 times more sensitive than fluorometry

includes two steps corresponding to points 4 and 6 of the basic protocol of ELISA (Sect. 11.2.2.2), since two primary antibodies (capture and detecting) are used. In other schemes shown in Fig. 11.1, probing includes one step.

Detection in a restricted sense includes step 8 of the basic immunological protocol. To a wide extent, detection consists of step 8 and all subsequent steps including registration of results. One-step detection and two-step detection (in a restricted sense) correspond to two types of detection strategies, direct versus indirect. They are shown in Schemes 1.1, 1.3, and 2.4 (direct) and Schemes 1.2, 1.4, 1.5, 2.1, 2.2, and 2.3 (indirect). In direct ELISA, primary antibody is conjugated with an enzyme; in indirect ELISA primary antibody is detected using (1) secondary antibody conjugated with enzyme (Scheme 1.2, 2.1, and 2.2), (2) proteins A/G/L conjugated with enzyme (Scheme 2.3), and (3) streptavidin/avidin conjugated with enzyme in aggregate with pre-biotinylated primary antibody (Scheme 2.5). Other examples of indirect detection are shown in Schemes 1.4 and 1.5; in these schemes streptavidin/avidin conjugated with enzyme is used to detect pre-biotinylated antigen. Advantages of indirect detection are the following: (1) signal is amplificated, and (2) using of one conjugate "secondary antibody-enzyme" and numerous primary detecting antibodies instead of numerous conjugates "primary detecting antibody-enzyme" is beneficial in terms of effort and time.

Here is a note important for sandwich ELISA. As a rule, when direct strategy of detection is used, both antibodies in pair are MAbs (Scheme 1.1) since a variety of available MAbs facilitates the choice of a pair that satisfies the condition for the interaction of two antibodies with non-overlapping epitopes of the antigen. For indirect strategy of detection, capture antibody is MAb, and detecting antibody is

PAb from host other than mouse (rabbit, goad, or others) to exclude binding of labeled secondary antibody with capture antibody. Since cross-reaction of the enzyme-labeled anti-rabbit/goad/sheep antibody with the capture MAb is not excluded theoretically, it should be checked preliminary. PAb from animal 1 as capture antibody in combination with PAb from animal 2 as detecting antibody is problem design due to high probability of cross-reactions of the labeled secondary antibody with capture antibody; it leads to high background signal and false-positive results. PAb as capture antibody in combination with MAb as detecting antibody is inadmissible design due to heterogeneity of PAb (it means that PAbs are not specific to a single epitope); not all pairs of PAb-MAb are matched, which leads to underestimation of immune complexes "immobilized PAb-antigen" and false-negative results. In addition, it is necessary to carefully use proteins A, G, and L as the detection reagents in indirect sandwich ELISA since they interact with Ig from numerous species. The only variant of format that does not cause doubts is MAb IgM (capture antibody)-antigen–PAb from sheep/goad/rabbit/rat (detecting antibody)-enzyme-labeled protein G, since protein G does not interact with mouse IgM (Kruchen 2017).

11.2.2.3 Enzymes Used in ELISA

Enzymes commonly used in ELISA are horseradish peroxidase (HRP) and alkaline phosphatase (AP). Others enzymes used rarely are β-galactosidase, acetylcholinesterase, catalase, etc.

HRP is small a 40 kDa protein isolated from plant *Armoracia rusticana*. Catalytic function of the enzyme (1.11.1.7) is oxidation of organic (usually phenols or aromatic amines) and inorganic substrates by hydrogen peroxide; pH optimum is a near neutral; inhibitors are cyanides, sulfides, and azides. Due to its small size, HRP rarely causes steric hindrance problems; it means that the substance (antibody, streptavidin, protein A) labeled with an enzyme forms a specific complex (with antigen, biotin, and antibody, respectively) with the same probability as the native substance. Note that HRP is a glycoprotein with sugar about 18% that allows conjugation with an antibody through carbohydrate compound.

Commercial AP is a large protein isolated from either *Escherichia coli* (M ~80 kD (dimer) (Munson and Fall, 1978), optimum pH ~ 8.0) or calf intestine (M ~ 100 kD, optimal pH ~ 9.6) (Mössner et al. 1980). Catalytic function of the enzyme (3.1.3.1) is hydrolysis of phosphoric acid monoester; inhibitors are cyanides, arsenate, P_i, and divalent cation chelators (e.g., EDTA). Since AP is approximately double the size of HRP, a higher enzyme to antibody conjugation ratio (w/w) should be used. It also means that the larger molecular size of AP can cause steric hindrance issues due to the need for close packed antigen–antibody complexes. This also can result in lower enzymatic activity than expected for the estimated number of bound enzyme molecules. Indeed, AP is slightly more expensive than HRP but is considered to be more stable.

HRP and AP are probably the most popular in ELISA due to a wide range of commercially available substrates. Substrates for HRP and AP range from soluble to insoluble; soluble substrates giving soluble products are acceptable for ELISA. The substrates are chromogenic, fluorogenic, and chemiluminescent to be chosen according to required assay sensitivity and instrumentation available for signal detection (spectrophotometer, fluorometer, or luminometer) (Fig. 11.2). It is commonly accepted that sensitivity improves in a row of substrates: chromogenic < fluorogenic < chemiluminescent. Indeed, using of chemiluminescent substrates allows to accumulate signal, and in return, chromogenic/colorimetric substrates allow direct visualization.

Note that the maximum signal generated from HRP substrates is very fast, often within 5 min. The signal from AP substrates, however, gradually increases over time with a signal plateau around 60 min; also AP reaction rate is more linear in comparison with HRP. Thus, AP provides better reaction control, higher reproducibility, and lower scatter of the assay results. All the same, it is generally accepted that antibody-HRP conjugates are superior to antibody-AP conjugates due to their high activity (as a consequence of high turnover rate of the enzyme), good stability, and low cost. In addition synthesis of polyconjugates containing N molecules of HRP per molecule of antibody gives a possibility to increase the sensitivity of the analysis.

11.2.2.4 Calculation of Results

After registration of results using an appropriate instrument, average values for each set of duplicate standards and duplicate samples are calculated. Duplicates should be within 20% of the mean. Standard curve plot is created for quantification of an analyte by using a suitable computer program; it must be separated for each ELISA plate.

Standard curve plot for noncompetitive ELISA is the average absorbance for each calibration probe (OY axis) against the concentration of an analyte in calibration probes (standards) (OX axis) (Fig. 11.3a). In noncompetitive ELISA, the values of average absorbance are in direct dependency on an analyte concentration.

There are two standard curve plots for competitive ELISA: (1) coordinates which are identical to those for noncompetitive ELISA and (2) values of B_n/B_o (in percent) for each calibration probe (OY axis) against the concentration of an analyte in calibration probes (standards) (OX axis) (Fig. 11.3b, c). The values of average absorbance values B_n/B_o (in percent) are in inverse dependence on an analyte concentration (Fig. 11.4).

Figure 11.4 illustrates main stages of competitive ELISA for better understanding of the inverse dependence. Analyzed substance designated as yellow oval (Fig. 11.4) plays a role of antigen competitor in ELISA. The amount of immune complexes "primary antibody–immobilized antigen" (stage 1) and optical density in wells containing colored product of enzymatic reaction (stage "enzymatic reaction") are inversely proportional to the concentration of antigen competitor. The amount of

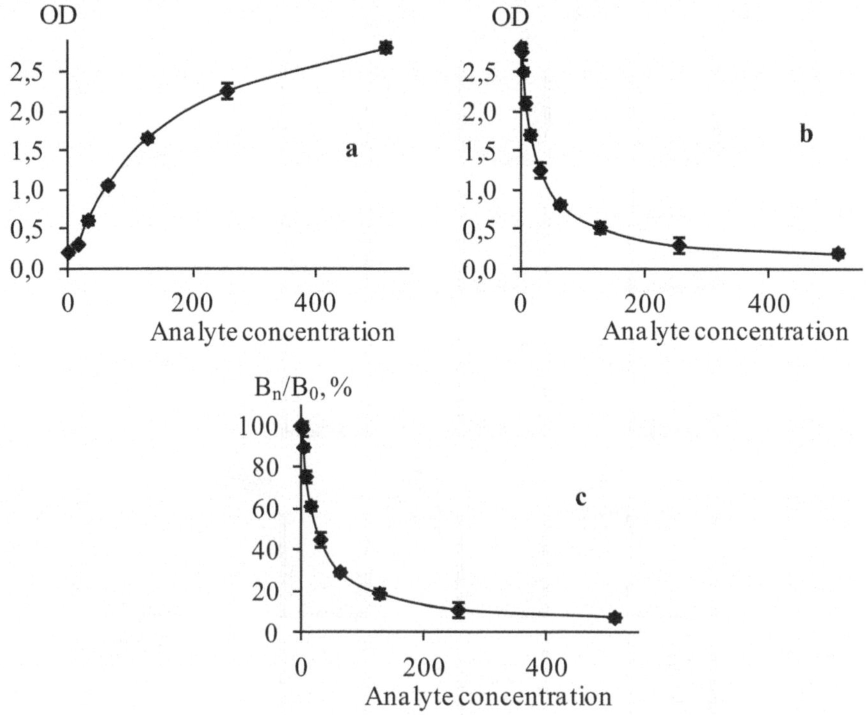

Fig. 11.3 Standard curve plots for non-competitive (**a**) and competitive (**b** and **c**) ELISA. OD is optical density

complexes "enzyme-labeled secondary antibody—primary antibody" (stage 2) is directly proportional to the amount of immune complexes "primary antibody–immobilized antigen."

11.2.2.5 Immobilization

ELISAs are typically performed in 96-well (386-well) polystyrene plates. The reactant immobilized on plate allows separating bound substances from unbound substances during the assay by the plate washing. It makes the ELISA a tool for measuring of a specific analyte within a crude preparation and other multicomponent media (culture media, bloods serum, and so on). As a rule, binding (adsorption) capacity of the plates is no less than 400 ng/cm^2, and value of coefficient of variation does not exceed 5% for limitation of deviations in assay results between wells and plates. The choosing of plate color depends upon the signal being detected. Clear polystyrene flat-bottom plates are used for colorimetric signals, while black or white opaque plates are used for fluorescence and chemiluminescence signals.

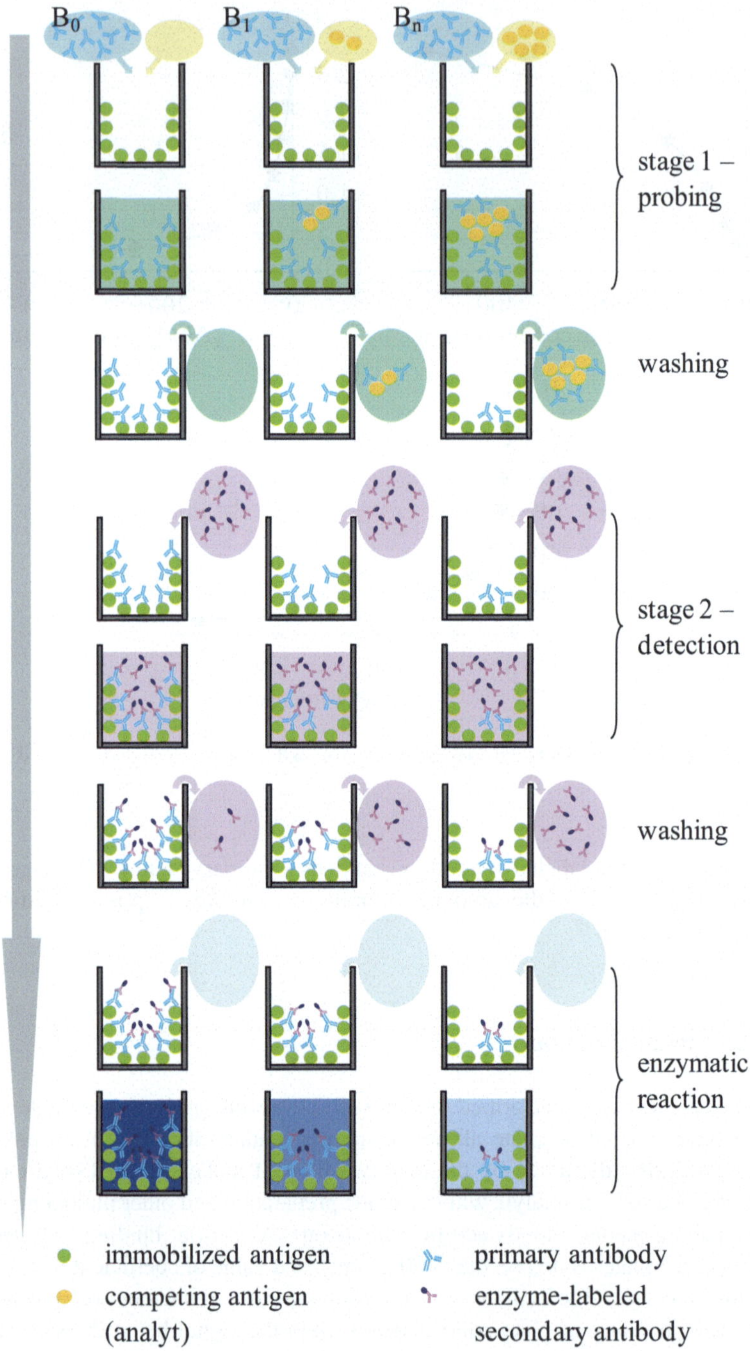

Fig. 11.4 Main stages of competitive ELISA. Steps of immobilization, and blocking are not shown

There are two traditionally used types of immobilization in ELISA: direct (passive adsorption) and indirect (biospecific). Indeed, covalent binding of biomolecules to the special microplate surface is used.

Polystyrene is the most commonly used base material for immobilization of biomolecules on the surface of ELISA microplates. Due to its chemical nature, polystyrene is a hydrophobic compound; however, its properties can be adjusted with a variety of physical surface treatments or coatings. If attachment to the solid surface is based upon passive adsorption, physiochemical forces like hydrophobic bonds, hydrophilic interactions, and H bonding are relevant. Therefore, ELISA microplates are most often physically treated to introduce a defined number of hydrophilic groups to the microplate surface.

Greiner Bio-One offers both a high and a medium binding surface for passive adsorption with high and relatively limited number of polar groups, respectively. Nunc plates are more manifold. MaxiSorp™ is a polystyrene surface with a high binding capacity for IgG and other molecules with both hydrophilic and hydrophobic regions. PolySorp™ is a polystyrene surface intended for non-antibody coatings; it binds antigens, especially large molecules with a predominance of hydrophobic moieties. MediSorp™ is a polystyrene surface treated to bind molecules with an intermediate hydrophobic/hydrophilic nature. This surface is less hydrophilic than MaxiSorp™ and is optimized for performing assays using human sera since it gives improved signal-to-noise ratio. MultiSorp™ is a very hydrophilic polystyrene surface suitable for the immobilization of highly polar molecules.

The selection of a surface that is best suited for a specific application should be evaluated empirically.

For some applications, adsorptive binding to a physically modified polystyrene surface is not feasible. Two most commonly used types of biospecific immobilization of primary components are shown in Fig. 11.1. In Scheme 1.5 a primary antibody is immobilized via secondary antibody. In Scheme 2.1 an antigen is immobilized using streptavidin-biotin technology; it includes adsorption of conjugate BSA-biotin (first layer) followed by a two-step biospecific immobilization of streptavidin/avidin (second layer) and conjugate "antigen-biotin" (third layer). Immobilization via secondary antibody can be applied instead of adsorption to Schemes 1.1, 1.2, and 1.4; streptavidin-biotin technology of antigen immobilization can be applied to Schemes 1.3 and 2.2–2.4. Indeed, in Scheme 1.5 protein A/G can be used for primary antibody immobilization instead of secondary antibody; this third type of biospecific immobilization is unacceptable for Schemes 1.1 and 1.2, since protein A/G is able to bind Ig used as soluble reagents (primary antibody at Scheme 1.1, primary and secondary antibody at Scheme 1.2).

Advantages of biospecific immobilization are high capacity of coated plates in aggregate with saving of immobilized antigen/primary antibody; high reproducibility and homogeneity of coating layer due to regular orientation of immobilized primary antibody (which is true when either secondary capture antibody or protein A/G is used for biospecific immobilization; streptavidin/biotin technology does not give regular orientation of immobilized pre-biotinylated substance), minimization of steric hindrances for primary immune complex formation due to increase of distance between plate surface and epitope of immobilized antigen/paratope of immobilized antibody.

Further, we outline in more detail the features of streptavidin/avidin-biotin technology for the immobilization of biomolecules.

Streptavidin is a 52.8 kDa protein purified from the bacterium *Streptomyces avidinii*. Streptavidin homo-tetramers have an extraordinarily high affinity for biotin (also known as vitamin B7 or vitamin H) ($K_a \sim 10^{14}$ L/mol) (Green 1975), and the binding is one of the strongest non-covalent interactions known in nature. Molar ratio of streptavidin tetramer to biotin is 4:1; each subunit binds biotin with equal affinity.

Avidin is the other biotin-binding protein isolated from the egg white (Piskarev et al. 1990). Its sequence identity with streptavidin is 30%, although avidin and streptavidin have almost identical secondary, tertiary, and quaternary structures. In contrast to streptavidin that lacks any carbohydrate modification and has a near-neutral pI (pI = 6.8–7.5), avidin is glycosylated and positively charged (pI = 10), which leads to nonspecific binding and high background signal. Indeed, avidin has pseudo-catalytic activity (it can enhance the alkaline hydrolysis of an ester linkage between biotin and a nitrophenyl group) and has a higher tendency for aggregation. Avidin has a higher affinity for free (unconjugated) biotin ($K_a \sim 10^{15}$ L/mol) than streptavidin but a lower binding affinity when biotin is conjugated to another molecule.

NeutrAvidin protein (Marttila et al. 2000) is a deglycosylated version of avidin with a mass of approximately 60 kDa. It offers the advantages of a near-neutral isoelectric point (pI = 6.3) to minimize nonspecific adsorption, along with lysine residues that remain available for derivatization or conjugation. NeutrAvidin yields the lowest nonspecific binding among the known biotin-binding proteins since it does not contain the RYD sequence of streptavidin mimicking RGD (Arg-Gly-Asp), the universal recognition domain presented in fibronectin and other adhesion-related molecules (Alon et al. 1990). For NeutrAvidin, the specific activity for biotin binding is approximately 14 µg per 1 mg of protein, which is near the theoretical maximum activity.

Streptavidin is used in ELISA for immobilization/detection more frequently than other biotin-binding proteins including chemically modified and genetically engineered avidins and streptavidins (Laitinen et al. 2006, 2007).

Polyvalence of streptavidin/avidin is a key point in their usage for immobilization and detection since it allows to significantly increase binding capacity (Fig. 11.1, Scheme 2.1) and multiply detection signal (Fig. 11.1, Schemes 1.4, 1.5, 2.5).

11.2.3 Rabbit Polyclonal Antibodies Against Whole Bacterial/Yeast Cells: Main Properties and Field of Application

Antibodies against microbial antigens obtained either by animal immunization (polyclonal antibodies, PAbs) or hybridoma technology (mouse monoclonal antibodies, MAbs) and recombinant antibodies against human proteins produced in bacterial cells are used for assaying the potency of vaccines (Schmeisser et al. 2014) and find therapeutic/diagnostic applications, e.g., for the diagnosis and treatment of cancer, immunotherapy of autoimmune diseases, inflammatory diseases, neutralization of toxins, and so on (Cheng et al. 2014; Groff et al. 2015; Hornsby et al. 2015; Sina et al. 2015; Zhong et al. 2015; Gupta and Shukla 2017). Indeed, they are key reagents in ELISA kits for detection/quantification of microbial antigens (Sect. 11.2.1.1) and widely used in research studies aimed at elucidating the localization and dynamics of the synthesis of bacterial/yeast substances with antigenic properties (Kiseleva et al. 2014, 2016, 2017).

To use PAb as reagents, it is necessary to determine their primary characteristics, including antiserum working titer (i.e., the functional dilution of antiserum sample necessary for a given assay, hereinafter referred to as PAb titer) and PAb specificity. The latter is defined as the cross-reactivity of the PAb with antigens of related microorganisms. The determination of the PAb titer allows assessing the economic feasibility of their usage; the determination of the PAb specificity is necessary to outline the range of objects and tasks for which the PAb is suitable.

It is generally accepted that the immunization of animals with whole bacterial/fungal cells of the selected strain leads to the production of antibodies to all antigens of this strain immunogen. In fact, two conditions are necessary for the production of antibodies. First, the antigen should be available to the cells of the immune system, and the availability of intracellular antigens depends on the composition of the cell wall and the possibility of its destruction by enzymes of antigen-presenting cells. Secondly, the antigen concentration should be within a well-defined range, which is determined empirically. Too high antigen concentration causes a state of tolerance; the antigen present in too low concentration is ignored by the immune system. In a whole cell, each antigen is in its inherent concentration, and the concentrations of some major/minor antigens do not fall within the range optimal for stimulating the immune response. Thus, when using whole bacterial/fungal cells as an immunogen, it is certainly impossible to obtain antibodies to all antigens.

The second generally accepted provision is that the use of whole bacterial/fungal cells as an immunogen allows to obtain antibodies cross-reacting with cells and cell components of other strains. In other words, antibodies are not strain-specific but species-specific or genera-specific. Specificity may be wider and extend to other members of the same family, which is explained by the similar component composition of cells of related microorganisms. Such antibodies are useful to biotechnologists and scientific researchers working with many representatives of one species, genus, or family, provided that they are cultivated separately.

11.2.3.1 Rabbit Immunization Protocol

All strains used for rabbit immunization were obtained from the Belarusian collection of nonpathogenic microorganisms in the Institute of Microbiology, National Academy of Sciences of Belarus. *Lactobacillus plantarum* BIM B-495D (*Lp* 495D), *Lactococcus lactis* subsp. *cremoris* BIM B-493D (*Ll* 493D), and *Bifidobacterium bifidum* BIM B-733D (*Bb* 733D) (cultures of third generation) were incubated 24 h in MRS media (Difco, USA) at 28–30 °C under aerobic conditions (*Lp* 495D and *Ll* 493D) or at 37 °C under anaerobic conditions (*Bb* 733D). *Bacillus cereus* BIM B-491 (*Bc* 491) (culture of third generation) was incubated 12 h under aerobic conditions at (28–30 °C) in meat-peptone broth (MPB) (HiMedia, India). *Saccharomyces cerevisiae* BIM Y-195 (*Sc* 195) and *Debaryomyces hansenii* BIM Y-4 (*Dh* 4) were cultivated for 3 days on wort agar at 26 °C. Wort agar: brewer's beer mixer 7B, agar-agar 2%, and pH 5.5–6.0. The yeast cells were washed off with 0.01 M sodium phosphate buffer and pH 7.5 (buffer 1).

The bacterial/yeast cells were precipitated by centrifugation at 2000 g for 20 min using a Beckman TJ-6 centrifuge (Beckman, USA) and washed three times with the same buffer. Yeast cells were suspended in 20% ethanol and held in a water bath at 90 °C for 1 h. The suspension of bacterial/yeast cells was transferred to ampoules (0.5 mL each) and lyophilized. For each strain, one of the ampoules was used to confirm loss of cell viability.

The scheme for obtaining PAb against antigens of *Lp* 495D, *Bb* 733D, and *Bc* 491 ($PAb_{anti\text{-}Lp\ 495D}$, $PAb_{anti\text{-}Bb\ 733D}$, $PAb_{anti\text{-}Bc\ 491}$) was as follows. Six-month-old male rabbits were immunized subcutaneously in the scapula area: injection is on the 1st, 14th, 28th, 60th, and 90th day and blood sampling on the 40th, 72th, and 100th day from the beginning of the procedure. Preparation for immunization of one animal: 2 mg of lyophilized nonviable cells (24 h of culture growth), 0.1 mL of sterile 0.15 M NaCl, and 2 mL of complete Freund's adjuvant.

The scheme for obtaining PAb against antigens of *Ll* 493D, *Sc* 195, and *Dh* 4 ($PAb_{anti\text{-}Ll\ 493D}$, $PAb_{anti\text{-}Sc\ 195}$, $PAb_{anti\text{-}Dh\ 4}$) resembled the above scheme, except for adjuvant. Preparation for immunization of one animal: 2 mg of lyophilized nonviable cells (24 h of culture growth), 250 μl of a sterile solution of 0.9% NaCl, 0.5 mL of Bayol F (Serva, Germany), and 12 μl of benzyl alcohol.

The blood was kept at 37 °C for 30 min and at 4 °C for 4 h and then centrifuged at 1000 g for 15 min. Each antiserum (supernatant) referred to below as $PAb_{anti\text{-}Lp\ 495D}$, $PAb_{anti\text{-}Bb\ 733D}$, etc. was divided into 100 mkL aliquots and stored at −70 °C.

11.2.3.2 Determination of PAb Titer and Cross-Reactivity

The design of ELISA shown in Fig. 11.1, Scheme 2.1, and cell wall-free fraction (CFF) of an appropriate strain as source of antigens was used to determine a working titer of PAbs. The graph for determination of $PAb_{anti\text{-}Ll\ 493D}$ working titer is shown as an example (Fig. 11.5). The data prove that $PAb_{anti\text{-}Ll\ 493D}$ has a relatively low

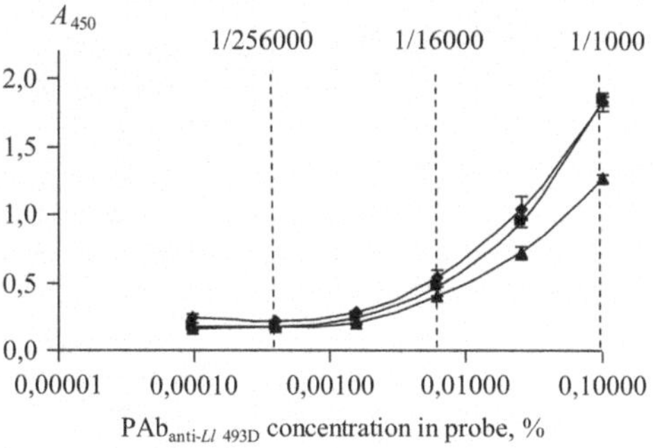

Fig. 11.5 $PAb_{anti\text{-}Ll493D}$ binding with antigens of *Ll* 493D. $PAb_{anti\text{-}Ll\ 493D}$.means antiserum of the rabbit immunized with whole cells of *Ll* 493D. CFF of *Ll* 493D immobilized from solutions with $A_{260\ nm,\ 1\ cm}$ values equal to (0.05–0.2) optical units was used as a source of antigens

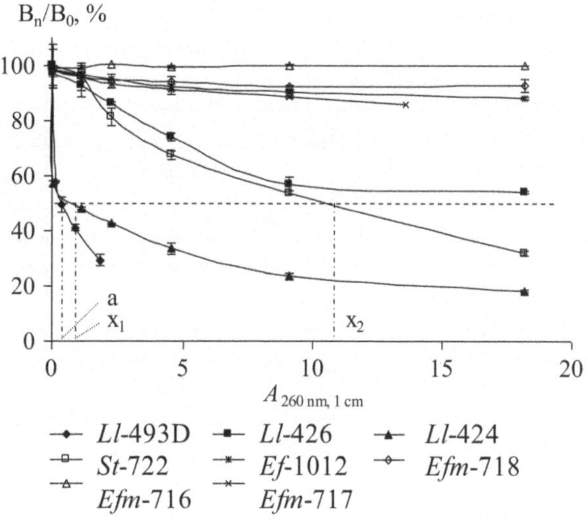

Fig. 11.6 The graph for determination of $PAb_{anti\text{-}Ll\ 493D}$ cross-reaction values with antigens of bacteria related to strain-immunogen

working titer in the range 1/1000–1/16,000 corresponding to the part of the curve with the maximum inclination angle to the OX axis.

ELISA performed in accordance with Scheme 1.3 (Fig. 11.1) was used for determination of each PAb cross-reactivity values. Briefly, antigens of each strain-immunogen were immobilized; PAb and antigens of related strains were in solution. PAb and antigens of strain immunogen in solution were used as a control probe. In each well, immobilized and contained in liquid phase antigens competed for PAb binding. Immune complexes “immobilized antigen-PAb” were detected with conjugate “sheep anti-rabbit-HRP.” CFF preparations of corresponding strains were used as sources of antigens. The value of the optical density at 260 nm in a cuvette with 1 cm optical path length (A_{260} nm, 1 cm) of each specimen CFF was used as the measure of CFF concentration. The value is proportional to the concentration of the total nucleic acid in the CFF composition and, consequently, proportional to the number of cells used to obtain CFF.

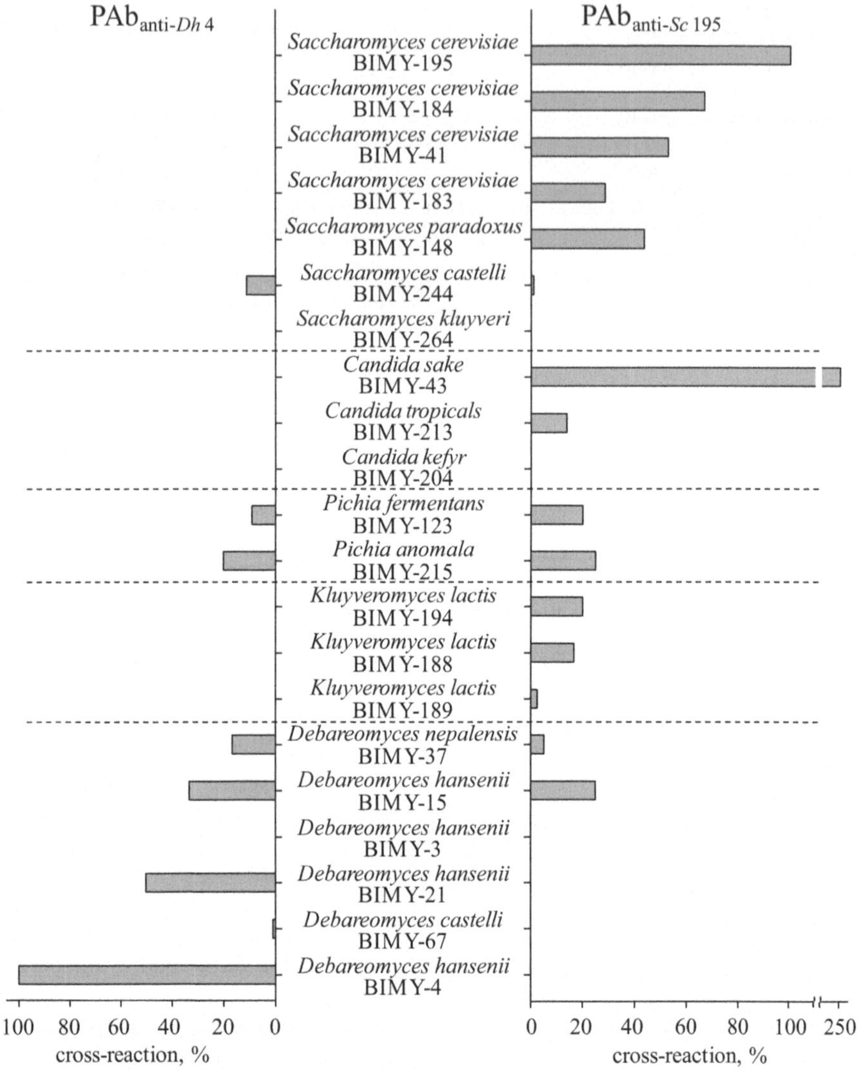

Fig. 11.7 Cross-reaction values calculated for $PAb_{anti\text{-}Sc195}$ and $PAb_{anti\text{-}Dh4}$

The data obtained for $PAb_{anti\text{-}Ll\ 493D}$ are shown in Fig. 11.6 as an example. Analogous graphs were obtained for other PAbs. To calculate cross-reactivity values, the following approach was used. The line *A* parallel to the OX axis was drawn through the OY axis point equal to 50%, and then the points of intersection of this line with each of the curves presented in the corresponding graph were determined (Fig. 11.2). If necessary, the curves were extended to their intersection with line *A*. Lines parallel to the OY axis were drawn through the obtained points, and their intersections with the OX axis were found. The resulting points were designated *a* (for strain immunogen) and x_1, x_2, etc. for other strains. Cross-reaction values (in percent) were calculated as ratios $(a/x_n) \times 100$.

The curves corresponding to some strains (e.g., strains *St* 722, *Efm* 716, and *Ll* 426 in Fig. 11.6) can't cross line *A* upon their extension. In this case, cross-reaction values were taken as less than 0.1%.

The values of cross-reactivity calculated for $PAb_{anti\text{-}Sc\ Y\text{-}195}$ and $PAb_{anti\text{-}Dh\ 4}$ are shown in Fig. 11.7 as an example. Interestingly, some cross-reaction values are above 100%, e.g., those that were calculated for $PAb_{anti\text{-}\ Sc\ 195}$ (antigens of *Candida sake* BIM Y-43, 250.0%) (Fig. 11.7) and $PAb_{anti\text{-}Bc\ 491}$ (antigens of *B. cereus* BIM B-205 and *B. mycoides* BIM B-179, 183.3% and 117.9%, respectively). The possible explanations are the following: (1) tested strain-competitor has antigens that are identical to antigens of strain immunogen, but their concentration per cell is higher and (2) tested strain-competitor has antigens that are similar but not identical to antigens of strain immunogen and interact with PAb with higher affinity than antigens of strain immunogen.

Similar data were obtained in other studies. For example, *Candida albicans* mannan was able to mimic *S. cerevisiae* mannan in its ability to detect ASCA, recognized markers of Crohn's disease. This overexpression of ASCA epitopes was achieved when *C. albicans* were grown in human tissues (Standaert-Vitse et al. 2006).

Working titers and cross-reaction values calculated for $PAb_{anti\text{-}Bb\ 733D}$, $PAb_{anti\text{-}\ Bc\ 491}$, $PAb_{anti\text{-}Lp\ 495D}$, $PAb_{anti\text{-}Ll\ 493D}$, $PAb_{anti\text{-}Sc\ 195}$, and $PAb_{anti\text{-}Dh\ 4}$ are discussed in Sect. 11.2.3.3.

11.2.3.3 PAb Against Whole Bacterial/Yeast Cells Have Unpredictable Titer and Vary from Species-Specific to Family-Specific

As shown above, we obtained six preparations of PAbs as a result of rabbit immunization with whole cells of yeasts *Saccharomyces cerevisiae* BIM Y-195 (*Sc* 195) and *Debaryomyces hansenii* BIM Y-4 (*Dh* 4) and bacteria *Bacillus cereus* BIM B-491 (*Bc* 491), *Lactobacillus plantarum* BIM B-495D (*Lp* 495D), *Lactococcus lactis subsp. cremoris* BIM B-493D (*Ll* 493D), and *Bifidobacterium bifidum* BIM B-733D (*Bb* 733D).

$PAb_{anti\text{-}Bb\ 733D}$ has a high working titer in the range 1/160,000–1/320,000, and other PAbs ($PAb_{anti\text{-}Sc\ 195}$, $PAb_{anti\text{-}Dh\ 4}$, $PAb_{anti\text{-}Bc\ 491}$, $PAb_{anti\text{-}Lp\ 495D}$, $PAb_{anti\text{-}Ll\ 493D}$) have a relatively low working titer in the range 1/1000–1/16,000 (each of 5 preparations).

The cross-reaction of $PAb_{anti\text{-}Sc\ 195}$, $PAb_{anti\text{-}Dh\ 4}$, $PAb_{anti\text{-}Bc\ 491}$, $PAb_{anti\text{-}Lp\ 423D}$, $PAb_{anti\text{-}Ll\ 493D}$, and $PAb_{anti\text{-}Bb\ 733D}$ was determined using 20, 20, 18, 23, 14, and 7 bacteria/yeast strains, respectively, related to respective strains used as immunogens.

$PAb_{anti\text{-}Sc\ Y\text{-}195}$ interacts with antigens shared by yeasts of genera *Saccharomyces*, *Kluyveromyces*, *Debaryomyces*, *Candida*, and *Pichia* and shows cross-reaction values no less than 10% with antigens of *Saccharomyces cerevisiae* BIM Y-184, BIM Y-41, and BIM Y-183, *Saccharomyces paradoxus* Y-148, *Debaryomyces hansenii* BIM Y-15, *Kluyveromyces lactis* BIM Y-194 and BIM Y-188, *Candida sake* BIM Y-43, *Candida tropicalis* BIM Y-213, *Pichia anomala* BIM Y-215, and *Pichia fermentans* BIM Y-123.

PAb$_{\text{anti-}Dh\ 4}$ shows cross-reaction values no less than 10% with antigens of *Debaryomyces hansenii* BIM Y-4, BIM Y-15, and BIM Y-21, *Debaryomyces nepalensis* BIM Y-37, *Saccharomyces castelli* BIM Y-244, and *Pichia anomala* BIM Y-215. Yeasts of genera *Kluyveromyces* and *Candida* do not contain antigens interacted with PAb$_{\text{anti-}Dh\ 4}$.

PAb$_{\text{anti-}Bc\ 491}$ shows cross-reaction values no less than 10% with antigens of bacilli of *Bacillus cereus* group (*Bacillus cereus* BIM B-205, BIM B-169, BIM B-206, *Bacillus mycoides* BIM B-179, *Bacillus thuringiensis* BIM B-180) and *Bacillus subtilis* group (*Bacillus subtilis* BIM B-25, *Bacillus amyloliquefaciens* BIM B-278, *Bacillus licheniformis* BIM B-175), as well as with antigens of geobacilli (*Geobacillus stearothermophilus* BIM B-295 and BIM B-203).

Contrary to expectations, PAb$_{\text{anti-}Lp\ 495\text{D}}$, PAb$_{\text{anti-}Ll\ 493\text{D}}$, and PAb$_{\text{anti-}Bb\ 733\text{D}}$ do not have as broad a specificity as other PAbs obtained using whole cells of yeast and bacilli as immunogens. PAb$_{\text{anti-}Lp\ 425\text{D}}$ shows a cross-reaction below 10% with antigens of other *Lactobacillus* strains (including strains of *L. plantarum*); the cross-reaction values are in the range 5–10% only for 2 out of 13 tested strains of lactobacilli. PAb$_{\text{anti-}Lp\ 425\text{D}}$ practically does not interact with antigens of bacteria of genera *Pediococcus*, *Leuconostoc*, *Lactococcus*, and *Enterococcus*. In turn, PAb$_{\text{anti-}Ll\ 493\text{D}}$ shows cross-reaction values not lower than 10% only in relation to antigens of lactococci and practically does not interact with antigens of bacteria of genera *Enterococcus*, *Streptococcus*, and *Lactobacillus*. PAb$_{\text{anti-}Bb\ 733\text{D}}$ is specific for antigens characteristic of bifidobacteria but does not interact with antigens of *B. bifidum* BIM B-465D, presumably because of the presence of unique polysaccharides in its cells (Zdorovenko et al. 2009).

In sum, PAbs against whole bacterial/yeast cells have unpredictable titer and vary from species-specific to family-specific.

The obtained PAbs are suitable for the quantitative determination of the cells of the microorganism-immunogen and related microorganisms under two conditions: (1) the values of the cross-reaction with the antigens of these microorganisms are not less than 10%; (2) other microorganisms with cross-reaction values exceeding 1% are not present in the analyzed material. In addition, PAbs can be used for affinity chromatography of bacterial/yeast antigens (cell components and secretory products) in the case of microorganisms satisfying the above conditions.

The field of application of PAbs is scientific researches, biotechnology for probiotic production, food industry, biosafety, and environmental protection.

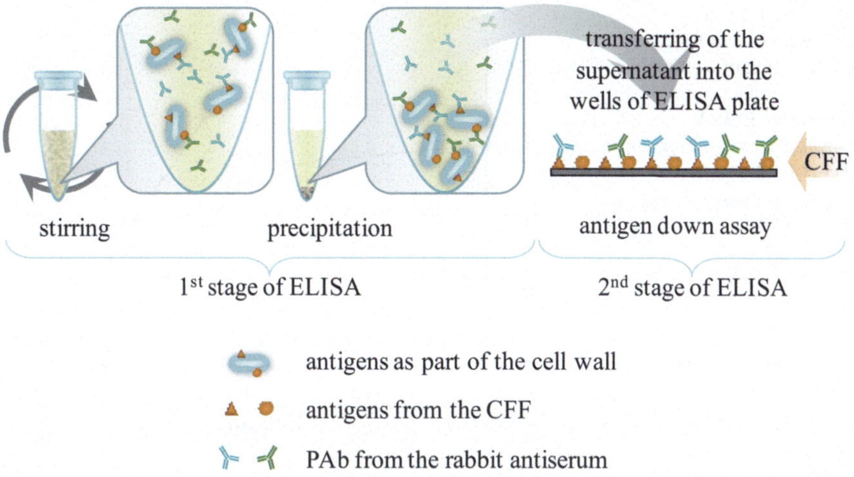

Fig. 11.8 The scheme of mediated competitive ELISA for the quantitative determination of cells of bifidobacteria and lactobacilli in culture media (or other fluids). The mediator is a solution containing the PAb, which was first in contact with bacterial cells (1st stage of ELISA), and then in contact with immobilized antigens of the same bacteria (2nd stage of ELISA)

11.2.4 PAb as a Component of Test System for Quantification of Bacterial Cells in Culture Media or Other Fluids

We give an example of the use of the genus-specific antibodies $\mathrm{PAb}_{\text{anti-}Bb\ 733\mathrm{D}}$ and $\mathrm{PAb}_{\text{anti-}Lp\ 425\mathrm{D}}$ as components of test systems for quantification of cells of bifidobacteria and lactobacilli, respectively, in culture media (or other fluids) by mediated competitive ELISA.

11.2.4.1 Components of the Test Systems and the Scheme of ELISA

The components of each of the two test systems are (1) plates with immobilized antigens of bacteria (*B. bifidum* BIM B-733D or *L. plantarum* BIM B-425D); (2) calibration probes (standards) containing either a known number of cells of the corresponding strain (probes B_1–B_n) or the buffer instead of cells (probe B_0); (3) appropriate PAb, viz., $\mathrm{PAb}_{\text{anti-}Bb\ 733\mathrm{D}}$ or $\mathrm{PAb}_{\text{anti-}Lp\ 425\mathrm{D}}$ (solution with a preselected titer); (3) ShaR-HRP conjugate (solution with a preselected titer); (4) solution containing HRP substrate (e.g., TMB) and hydrogen peroxide; (5) stop reactant; and (6) washing solution. The technology of preparing plates and calibration samples is presented below in Sects. 11.2.4.2 and 11.2.4.3. The analysis is carried out as follows (Fig. 11.8).

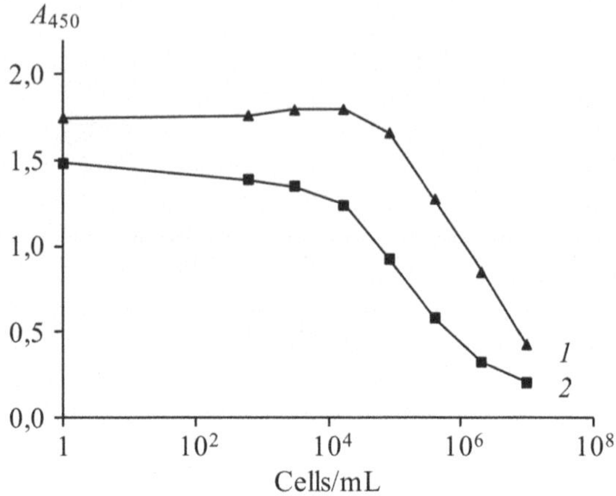

Fig. 11.9 Typical calibration curves for quantification of bacterial cells by mediated competitive ELISA. *1*—*Bifidobacterium bifidum* BIM B-733D, *2*—*Lactobacillus plantarum* BIM B-495D

First Stage of ELISA A homogeneous cell suspension of each tested culture is prepared. Equal volumes of suspensions (analyzed samples x_1–x_n) and pre-prepared calibration samples (standards) B_0; B_1–B_n are added into separate Eppendorf tubes. An equal volume of relevant PAb ($PAb_{anti\text{-}Bb\ 733D}$ or $PAb_{anti\text{-}Lp\ 425D}$) is added into each tube with samples x_1–x_n and standards B_0; B_1–B_n. Incubation is carried out with continuous stirring (Fig. 11.8). During the incubation, the formation of immune complexes of bacterial antigens (located on the surface of cells) with the PAb occurs.

After completion of the incubation process, an equal amount of 50% suspension of any available sorbent for gel filtration is added into each sample/standard to visualize the precipitate that is formed during Eppendorf tube centrifugation and contains immune complexes. It is inadmissible to use a sorbent capable of adsorbing cells, antibodies, or immune complexes, for example, a sorbent for ion-exchange or reverse-phase chromatography.

The amounts of free (non-bound with bacterial antigens) PAb in supernatants are inversely proportional to the number of bacterial cells that were in the analyzed samples and standards.

Second Stage of ELISA Supernatants containing free PAbs are added into corresponding plate wells prenumbered as x_1–x_n and B_0; B_1–B_n, each well in duplicates. During the incubation, the formation of immune complexes between the PAb and immobilized bacterial antigens occurs. The number of immune complexes is directly proportional to the amount of the free PAb added into corresponding wells.

The detection of immune complexes is carried out through their incubation with conjugate "ShaR-HRP" followed by the enzymatic reaction of HRP and its stopping. The results are recorded using the appropriate equipment, depending on the type of HRP substrate used. The cell concentration in samples x_1–x_n is determined using a calibration curve (OY axis: either A_{450} or B_n/B_0, %; OX axis—cell concentration in standards, cells/mL). Typical calibration curves are shown in Fig. 11.9.

The ELISA conditions used to obtain calibration curves (Fig. 11.9) are as follows. The corresponding series of Eppendorf tubes containing 0.06 mL of a homogeneous suspension of *B. bifidum* BIM B-733D cells (series 1) or *L. plantarum* BIM B-425D cells (series 2) with concentrations in the range 6×10^2–1×10^7 cells/mL (each series) were used as standards B_1–B_n. The Eppendorf tube containing 0.06 mL of culture medium was used as standard B_0. $PAb_{anti\text{-}Bb\ 733D}$ and $PAb_{anti\text{-}Lp\ 425D}$ (each 0.24 mL of a solution) were added in tubes of series 1 and 2, respectively; preselected PAb titers in the final volume were 1/180,000 ($PAb_{anti\text{-}Bb\ 733D}$) and 1/8000 ($PAb_{anti\text{-}Lp\ 423D}$). The Eppendorf tubes were incubated for 1.5 h at 37 °C (first stage of ELISA). 0.05 of a 50% suspension of Sephadex G-25 fine (Pharmacia, Sweden) was added to each tube and centrifuged for 10 min at 3000 rpm with centrifuge Eppendorf MiniSpin plus (Eppendorf AG, Germany).

Plates with immobilized CFF (from solutions with $A_{260\ nm,\ 1\ cm}$ values equal to 0.2, where $A_{260\ nm,\ 1\ cm}$ is absorption at a wavelength of 260 nm in a cuvette with an optical path length of 1 cm) of *B. bifidum* BIM B-733D (plate 1) and *L. plantarum* BIM B-425D (plate 2) were used. Supernatants from each tube of series 1 and 2 (0.1 mL) were added to the respective wells in duplicates and incubated for 1.5 h at 37 °C (2nd stage of ELISA). The wells were washed.

Conjugate "ShaR-HRP" (Xema Co. Ltd, Russia) was used to detect immune complexes. TMB liquid substrate system (Sigma, USA) and sulfuric acid (4.8%) were used to initiate and stop HRP enzymatic reaction, respectively. A_{450} in the wells was determined using automatic multichannel spectrophotometer (Uniplan, Russia).

It is obvious (Fig. 11.9) that in each of the two test systems, the range of detectable concentrations is 10^4–10^6 cells/mL. In accordance with user requirements, the range can be shifted to the area of lower or higher cell concentrations by changing the amounts of immobilized bacterial antigens and the PAb titer at the 1st stage of ELISA.

11.2.4.2 The Technology of Calibration Probe (Standard) Preparation

A homogeneous suspension of the bacterial culture is prepared using a vortex mixer or other available equipment of a similar purpose. A portion of the suspension is used to determine the cell number by any available method (e.g., manual cell counting using a counting chamber or automated cell counting). The concentration of cells (cells per mL of culture) is calculated.

A series of cell suspensions is prepared by the method of successive dilutions in the conditions of continuous shaking to obtain calibration probes (standards) B_1–B_n. Each sample is assigned a concentration of cells taking into account the dilution factor. The culture medium is used as standard B_0. Each standard is divided into aliquots (each with volume *a* mL) to obtain the required number of series of standards (B_0; B_1–B_n), a separate series for each analysis. During the preparation of aliquots, each suspension should be maintained in a homogeneous state by continuous shaking. All standards should be lyophilized and stored at −20 °C. Distilled water (*a* mL) is added in all standards just before the analysis and suspended until homogeneous.

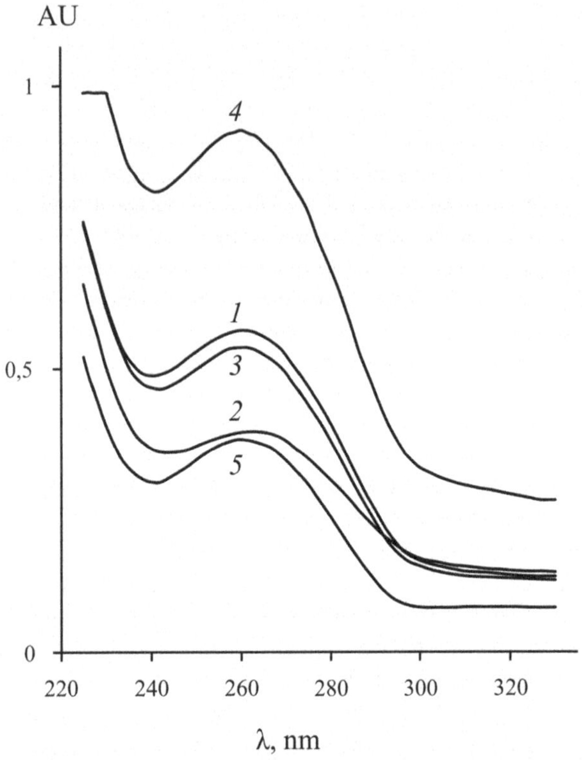

Fig. 11.10 Absorption spectra of cells of *B. bifidum* BIM B-733D, *Bifidobacterium adolescentis* 94 BIM, *Bifidobacterium longum* B379M, *L. plantarum* BIM B-425D and *Escherichia coli* M17 (strains *1–5*, respectively). Cells were solubilized by 10 M urea supplemented with 0.2 M NH_4OH, pH 11.5 for 2 h at room temperature. Spectra were detected in the compartment for turbid samples. Control was 10 M urea supplemented with 0.2 M NH_4OH, pH 11.5. Strains *2*, *3* and *5* (Institute of Microbiology, National Academy of Sciences of Belarus) were used to show that absorption maximum at a wavelength of 260 nm is the feature common for various bacteria

11.2.4.3 The Technology of Preparing Plates with Immobilized Bacterial Antigens

In the course of ELISA performed in accordance with the proposed scheme (Fig. 11.8), PAbs are distributed between the antigens of the suspended bacterial cells (first stage of ELISA) and identical antigens immobilized on the solid phase (second stage of ELISA). To ensure the identity of antigens interacting with PAb in two stages of ELISA, it is logical to immobilize bacterial cells rather than their fractions/components since intracellular antigens of whole cells are inaccessible to PAb during analysis. The problem is that cells are an inconvenient preparation for ELISA in general and for immobilization in wells in particular. We offer a simple, quick, and accurate method for standardizing the solid phase with immobilized bacterial cells. An additional advantage of the method is the possibility of its use

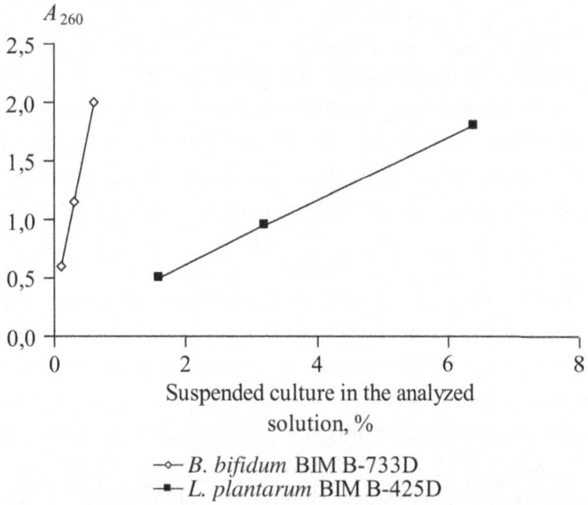

Fig. 11.11 Dependence of A_{260} values on the concentration of bacterial cells. The content of one ampoule (washed and lyophilized cells) was suspended in 500 μL of H_2O and the cell concentration was taken as 100%. Homogenous cell suspension of each strain (50, 100, 200 μL) was transferred into three tubes and H_2O was added to 200 μl. Cells were solubilized by 30% sulfuric acid (3 mL per tube) for 2 h at room temperature. Each sample containing *B. bifidum* BIM B-733D cells was further diluted 11 times with 30% sulfuric acid to obtain measurable A_{260} values

in a biochemical laboratory that is not equipped with standard microbiological equipment for cell counting.

The principle of the method consists in using the parameter $A_{260\ \mathrm{nm},\ 1\ \mathrm{cm}}$ of cell solution (in 30% sulfuric acid or 10 M urea supplemented with 0.2 M NH_4OH, pH 11.5; exposure with continuous shaking for 2 h at room temperature) as a measure of their concentration in the suspension. $A_{260\ \mathrm{nm},\ 1\ \mathrm{cm}}$ values are proportional to concentration of total nucleic acid in solution. Arguments in favor of the possibility of such an approach are the spectra of a cell solution with an absorption maximum at a wavelength of 260 nm (Fig. 11.10) and a linear dependence of $A_{260\ \mathrm{nm},\ 1\ \mathrm{cm}}$ values on the dilution factor (Fig. 11.11).

The $A_{260\ \mathrm{nm},\ 1\ \mathrm{cm}}$ values were recalculated to 1 mL of cell culture, taking into account the factor of its dilution during solubilization. Comparison of recalculated $A_{260\ \mathrm{nm},\ 1\ \mathrm{cm}}$ values shows that 30% sulfuric acid gives a more pronounced degree of cell dissolution compared to 12 M urea, pH 11.5, although the proportion of the $A_{260\ \mathrm{nm},\ 1\ \mathrm{cm}}$ values (strain 4 > strain 1 = strain 3 > strain 2 = strain 5; the numbers of strains are the same as in Fig. 11.10) does not depend on the method of cell solubilization.

The concentration of the cell suspension, which is optimal for cell immobilization on microplate wells, is selected empirically. For this, a series of cell suspensions is prepared by sequential dilution of pre-suspended cell culture. Two portions are taken from each suspension for (1) dissolving the cells and determining the values of $A_{260\ \mathrm{nm},\ 1\ \mathrm{cm}}$ and (2) cell immobilization followed by mediated competitive ELISA. The criterion for selecting the suspension is a calibration chart of the optimal shape (value B_0/B_n for standard with maximal cell concentration should not be higher than 20%) for the required range of detectable concentrations.

When preparing the next series of plates, the value of $A_{260\ \mathrm{nm},\ 1\ \mathrm{cm}}$ is determined only for the suspended cell culture, and then the suspension is diluted to the required number of times, which corresponds to the preselected optimal value of $A_{260\ \mathrm{nm},\ 1\ \mathrm{cm}}$.

The CFF is a more convenient preparation for immobilization on wells as compared to cells. To use CFF instead of cells, we proceed from the assumption that all bacterial antigens are synthesized inside the cell and then distributed between the cytosol, the cell membrane, the cell wall, and the external environment. Consequently, CFF contains antigens similar to those exposed on the cell wall surface (e.g., antigens A and B), plus other antigens (e.g., C and D antigens absent in the cell wall). C and D antigens could be inaccessible to the immune system when immunizing rabbits with whole bacterial cells; the availability of these antigens was dependent on the composition of the cell wall and its resistance to macrophage enzymes. In this case, PAb includes anti-A and anti-B antibodies but does not include anti-C and anti-D antibodies. This is a requirement for using CFF instead of bacterial cells. Otherwise, there will be a high background signal.

To test the possibility of using CFF instead of bacterial cells, we immobilized CFF and cells of each of the two strains in appropriate series of wells (series $1_{Bb\ 733\mathrm{D}}$ and series $2_{Bb\ 733\mathrm{D}}$, series $1_{Lp\ 425\mathrm{D}}$, and series $2_{Lp\ 425\mathrm{D}}$). Samples of human serum containing antibodies of *B. bifidum* BIM B-733D antigens (set$_{\mathrm{anti}\text{-}Bb\ 733\mathrm{D}}$; $n = 32$) were previously selected based on the results of the analysis performed in accordance with Scheme 2.2, Fig. 11.1. Samples of human serum containing antibodies of *L. plantarum* BIM B-425D antigens (set$_{\mathrm{anti}\text{-}Lp\ 425\mathrm{D}}$; $n = 32$) were selected in a similar way.

Selected serum samples (100-fold diluted) were added into the corresponding wells of two series (in duplicates), set$_{\mathrm{anti}\text{-}Bb\ 733\mathrm{D}}$ into wells of series $1_{Bb\ 733\mathrm{D}}$ and series $2_{Bb\ 733\mathrm{D}}$ and set$_{\mathrm{anti}\text{-}Lp\ 425\mathrm{D}}$ into wells of series $1_{Lp\ 425\mathrm{D}}$ and series $2_{Lp\ 425\mathrm{D}}$. The resulting immune complexes were detected using the "goat antihuman Igs antibody-HRP" conjugate (Xema Co. Ltd, Russia) and TMB liquid substrate system (Sigma, USA). The correlation coefficients (*r*) between the sets of A_{450} values obtained in the wells of series 1 and series 2 were calculated for each strain.

The *r* values obtained for *B. bifidum* BIM B-733D and for *L. plantarum* BIM B-425D were equal to 0.82 and 0.91, respectively. CFF can be used as an alternative to whole cells at *r* values greater than 0.85. Thus, only CFF of *L. plantarum* BIM B-425D can be used in place of the cells of this strain.

In sum, PAb$_{\mathrm{anti}\text{-}Bc\ 491}$ and PAb$_{\mathrm{anti}\text{-}Lp\ 425\mathrm{D}}$ are useful as components of test systems for quantification of cells of *B. bifidum* BIM B-733D or *L. plantarum* BIM B-425D, respectively, in culture media (or other fluids) by mediated competitive ELISA. We give the scheme of the ELISA, show typical calibration curves, and explain the main principles and methodological approaches underlying the preparation of standards and immunosorbents. Indeed, each PAb is suitable for quantification of cells of other species/strains of bifidobacteria and lactobacilli; the criterion for strain selection is the cross-reaction value of PAb$_{\mathrm{anti}\text{-}Bb\ 733\mathrm{D}}$ or PAb$_{\mathrm{anti}\text{-}Lp\ 425\mathrm{D}}$ at least 10%. To quantify the cells of any of the strains that satisfy the condition, it is necessary to prepare standards (Sect. 11.2.4.2) using cells of this strain in place of *B. bifidum* BIM B-733D cells or *L. plantarum* BIM B-425D cells.

The data are useful for scientific researchers and biotechnologists. These and similar test systems can be used to control biotechnological processes by quantifying bacterial cells.

11.2.5 A Methodology for Studying the Dynamics of Synthesis and Secretion of Bacterial Antigens of Unknown Nature and Structure

Using $PAb_{anti\text{-}Bb\ 733D}$, $PAb_{anti\text{-}Bc\ 491}$, and $PAb_{anti\text{-}Ll\ 493D}$ (Sect. 11.2.3) as key reagents, we elaborated the methodology for studying the dynamics of synthesis and secretion of bacterial antigens of unknown nature and structure, in other words, all antigens with which Pabs interact (Kiseleva et al. 2014, 2016, 2017). The methodology includes:

1. The method for selecting of bacteria suitable as research objects
2. Criteria for the quantification of cell fractions as well as liquid preparations used as a source of antigens
3. Scheme of ELISA used to study the dynamics (time dependence) of the specific concentration (C_{sp}) (calculated per cell) of intracellular antigens, secretory antigens and cell wall antigens
4. Methods for data processing and interpretation of the results.

It was proven using lactococci, bacilli, and bifidobacteria that C_{sp} of antigens contained in CFF, cell wall, and cultural liquid depends on the strain, stage of culture growth, and media composition. The dynamics of C_{sp} is unpredictable and for each object must be determined on a case-by-case basis (Kiseleva et al. 2014, 2016, 2017).

This methodology can be used to determine the dynamics of synthesis and secretion of any well-defined antigen, provided that MAb/PAb specific for this antigen is used instead of PAbs obtained with whole bacterial cells as immunogens.

The data will find an application in research studies and biotechnology to select growth stage for effective isolation of biopolymer with antigenic properties.

11.3 Conclusion

ELISA is a method applicable for scientific and applied research in microbiology. The highly specific interaction between the antigen and the antibody provides an opportunity to use ELISA for the conservation of biodiversity through identification of unique immunogenic biopolymers and low-molecular substances with antigenic properties. Indeed, ELISA is useful in biotechnologies of natural bacterial proteins and recombinant proteins, foods, and pharmaceuticals; food and feed safety; diagnosis of human, animal, and plant diseases; and biodegradation and bioremediation. The scientific and practical aspects of ELISA presented in this chapter and examples of using this method are a kind of advertising for specialists engaged in scientific research of microbiological objects and biotechnologists using microorganisms as effectors.

References

Allan GM, Todd D, Smyth JA, Mackie DP, Burns J, McNulty MS (1989) In situ hybridization: an optimised detection protocol for a biotinylated DNA probe renders it more sensitive than a comparable 35S-labelled probe. J Virol Methods 24:181–190

Alon R, Bayer EA, Wilchek M (1990) Streptavidin contains an RYD sequence which mimics the RGD receptor domain of fibronectin. Biochem Biophys Res Commun 170:1236–1241

Aly S, Floury J, Famelart MH et al (2011) Nisin quantification by ELISA allows the modeling of its apparent diffusion coefficient in model cheeses. J Agric Food Chem 59:9484–9490. https://doi.org/10.1021/jf2008474

Anonimous (2006) Chapter 7.2. Application of immunochemical techniques to potentially detect probiotic bacteria. Chapter 7.3 Methodology for immunochemical detection of probiotics. In: Goktepe I, Jineja VK, Ahmedna M (eds) Probiotics in food safety and human health. CRC Press, Taylor and Francis Group, Boca Raston, ISBN: 9781574445145 – CAT# DK3341, pp 167–171, 171–189

Arya P, Ponmariappan S, Singh L, Kumar O (2013) Development of ELISA based detection system for lethal toxin of *Clostridium sordellii*. Indian J Med Res 137:1180–1187

Aydin M, Herzig GP, Jeong KC, Dunigan S, Shah P, Ahn S (2014) Rapid and sensitive detection of *Escherichia coli* O157:H7 in milk and ground beef using magnetic bead-based immunoassay coupled with tyramide signal amplification. J Food Prot 77:100–105. https://doi.org/10.4315/0362-028X.JFP-13-274

Berger E, Crampton MC, Nxumalo NP, Louw ME (2011) Extracellular secretion of a recombinant therapeutic peptide by *Bacillus halodurans* utilizing a modified flagellin type III secretion system. Microb Cell Fact 10:62. https://doi.org/10.1186/1475-2859-10-62

Berthiller F, Crews C, Dall'Asta C, Saeger SD, Haesaert G, Karlovsky P, Oswald IP, Seefelder W, Speijers G, Stroka J (2013) Masked mycotoxins: a review. Mol Nutr Food Res 57:165–186. https://doi.org/10.1002/mnfr.201100764

Bertin D, Grimaud JC, Lesavre N, Benelmouloud C, Desjeux A, Garcia S, Desplat-Jégo S (2013) Targeting tissular immune response improves diagnostic performance of anti-*Saccharomyces cerevisiae* antibodies (ASCA) in Crohn's disease. PLoS One 8:e80433. https://doi.org/10.1371/journal.pone.0080433

Cheng CM, Tzou SC, Zhuang YH, Huang CC, Kao CH, Liao KW, Cheng TC, Chuang CH, Hsieh YC, Tai MH, Cheng TL et al (2014) Functional production of a soluble and secreted single-chain antibody by a bacterial secretion system. PLoS One 9:e97367. https://doi.org/10.1371/journal.pone.0097367

Colwell RR (1997) Microbial diversity: the importance of exploration and conservation. J Ind Microbiol Biotechnol 18:302–307. https://doi.org/10.1038/sj.jim.2900390

Criado R, Gutiérrez J, Martín M, Herranz C, Hernández PE, Cintas LM et al (2006) Immunochemical characterization of temperature-regulated production of enterocin L50 (EntL50A and EntL50B), enterocin P, and enterocin Q by *Enterococcus faecium* L50. Appl Environ Microbiol 72:7634–7643. https://doi.org/10.1128/AEM.00983-06

Deshmukh RA, Joshi K, Bhand S, Roy U (2016) Recent developments in detection and enumeration of waterborne bacteria: a retrospective minireview. MicrobiologyOpen 5:901–922. https://doi.org/10.1002/mbo3.383

Fang Y, Ramasamy RP (2015) Current and prospective methods for plant disease detection. Biosensors (Basel) 5:537–561. https://doi.org/10.3390/bios5030537

Feng X, Ma J-W, Sun S-Q, Guo HC, Yang YM, Jin Y, Zhou GQ, He JJ, Guo JH, Qi SY, Lin M, Cai H, Liu XT (2016) Quantitative detection of the foot-and-mouth disease virus serotype O 146s antigen for vaccine production using a double-antibody sandwich elisa and nonlinear standard curves. PLoS One 11:e0149569. https://doi.org/10.1371/journal.pone.0149569

Ferenczi S, Cserháti M, Krifaton C, Szoboszlay S, Kukolya J, Szőke Z, Kőszegi B, Albert M, Barna T, Mézes M, Kovács KJ, Kriszt B (2014) A new ochratoxin a biodegradation strategy using *Cupriavidus basilensis* Őr16 Strain. PLoS One 9:e109817. https://doi.org/10.1371/journal.pone.0109817

Garrote GL, Serradell MA, Abraham AG, Añon MC, Fossati CA, De Antoni GL (2005) Development of an immunochemical method to detect *Lactobacillus kefir*. Food Agric Immunol 16:221–233. https://doi.org/10.1080/09540100500244146

Green NM (1975) Avidin. Adv Protein Chem 29:85–133. https://doi.org/10.1016/s0065-3233(08)60411-8

Groff K, Brown J, Clippinger AJ (2015) Modern affinity reagents: recombinant antibodies and aptamers. Biotechnol Adv 33:1787–1798. https://doi.org/10.1016/j.biotechadv.2015.10.004

Gülden E, Wong FS, Wen L (2015) The Gut microbiota and type 1 diabetes. Clin Immunol 159:143–153. https://doi.org/10.1016/j.clim.2015.05.013

Gupta SK, Shukla P (2017) Microbial platform technology for recombinant antibody fragment production: a review. Crit Rev Microbiol 43:31–42. https://doi.org/10.3109/1040841X.2016.1150959

Gupta VK, Sharma GD, Tuohy MG, Gaur R (2016) The handbook of microbial bioresources. https://doi.org/10.1079/9781780645216

Hornsby M, Paduch M, Miersch S, Sääf A, Matsuguchi T, Lee B, Wypisniak K, Doak A, King D, Usatyuk S, Perry K, Lu V, Thomas W, Luke J, Goodman J, Hoey RJ, Lai D, Griffin C, Li Z, Vizeacoumar FJ, Dong D, Campbell E, Anderson S, Zhong N, Gräslund S, Koide S, Moffat J, Sidhu S, Kossiakoff A, Wells J (2015) A high through-put platform for recombinant antibodies to folded proteins. Mol Cell Proteomics 14:2833–2847. https://doi.org/10.1074/mcp.O115.052209

Jain S, Ponmariappan S, Kumar O (2011) Development of immunodetection system for botulinum neurotoxin type B using synthetic gene based recombinant protein. Indian J Med Res 134:33–39

Keren T, Yarmus M, Halevy G, Shapira R (2004) Immunodetection of the bacteriocin lacticin RM: analysis of the influence of temperature and Tween 80 on its expression and activity. Appl Environ Microbiol 70:2098–2104. https://doi.org/10.1128/AEM.70.4.2098-2104.2004

Kiseleva EP, Mikhailopulo KI, Sviridov OV, Novik GI, Knirel YA, Szwajcer Dey E (2011) The role of components of bifidobacterium and lactobacillus in pathogenesis and serologic diagnosis of autoimmune thyroid diseases. Benef Microbes 2:139–154. https://doi.org/10.3920/BM2010.0011

Kiseleva EP, Starovoitova TA, Mikhailopulo KI, Szwajcer Dey E, Novik GI (2014) Antigens of bifidobacteria: dynamics of synthesis and distribution in the system "cell–environment". In: Kolomiets EI, Lobanok AG (eds) Microbial biotechnology: fundamental and applied aspects, vol 6. Belarusian science, Minsk, Republic of Belarus, pp 286–300 (in Russian). http://mbio.bas-net.by/wp-content/uploads/2011/12/InMi_Proceedings_Vol-6_2014.pdf

Kiseleva E, Ezybets A, Mikhailopulo K, Novik G (2016) Methodological approach to the study of the dynamic of antigens synthesis and secretion by nonpathogenic organisms on the example of *Lactococcus lactis*. Proc Natl Acad Sci Belarus Ser Biol Sci 4:72–80 in Russian

Kiseleva EP, Mikhailopulo KI, Ladutska AI, Novik GI (2017) Methodological approach to the study of dynamics of specific concentration of cell wall antigens per cell of *Bacillus* species and examples of its application. EuroBiotech J 1:71–77. https://doi.org/10.24190/ISSN2564-615X/2017/01.10

Klarić MS, Cvetnić Z, Pepeljnjak S, Kosalec I (2009) Co-occurrence of aflatoxins, ochratoxin A, fumonisins, and zearalenone in cereals and feed, determined by competitive direct enzyme-linked immunosorbent assay and thin-layer chromatography. Arh Hig Rada Toksikol 60:427–434. https://doi.org/10.2478/10004-1254-60-2009-1975

Koivunen ME, Krogsrud RL (2006) Principles of immunochemical techniques used in clinical laboratories. Lab Med 37:490–497. https://doi.org/10.1309/MV9RM1FDLWAUWQ3F

Kriszt R, Krifaton C, Szoboszlay S, Cserháti M, Kriszt B, Kukolya J, Czéh A, Fehér-Tóth S, Török L, Szőke Z, Kovács KJ, Barna T, Ferenczi S (2012) A new zearalenone biodegradation strategy using non-pathogenic *Rhodococcus pyridinivorans* K408 strain. PLoS One 7:e43608. https://doi.org/10.1371/journal.pone.0043608

Kruchen B (2017) Lab Faqs manual from la Roche, 4th edn. ISBN: 3-88630-245-8, p 107

Laitinen OH, Hytönen VP, Nordlund HR, Kulomaa MS (2006) Genetically engineered avidins and streptavidins. Cell Mol Life Sci 63:2992–3017. https://doi.org/10.1007/s00018-006-6288-z

Laitinen OH, Nordlund HR, Hytönen VP, Kulomaa MS (2007) Brave new (strept)avidins in biotechnology. Trends Biotechnol 25:269–277. https://doi.org/10.1016/j.tibtech.2007.04.001

Law JW, Ab Mutalib NS, Chan KG, Lee LH (2014) Rapid methods for the detection of foodborne bacterial pathogens: principles, applications, advantages and limitations. Front Microbiol 5:770. https://doi.org/10.3389/fmicb.2014.00770

March C, Manclús JJ, Abad A, Navarro A, Montoya A (2005) Rapid detection and counting of viable beer-spoilage lactic acid bacteria using a monoclonal chemiluminescence enzyme immunoassay and a CCD camera. J Immunol Methods 303:92–104. https://doi.org/10.1016/j.jim.2005.06.002

Marttila AT, Laitinen OH, Airenne KJ, Kulik T, Bayer EA, Wilchek M, Kulomaa MS (2000) Recombinant NeutraLite avidin: a non-glycosylated, acidic mutant of chicken avidin that exhibits high affinity for biotin and low non-specific binding properties. FEBS Lett 467:31–36. http://polysciences.com/skin/frontend/default/polysciences/pdf/779.pdf

Meulenberg EP (2012) Immunochemical methods for ochratoxin a detection: a review. Toxins (Basel) 4:244–266. https://doi.org/10.3390/toxins4040244

Mohan SV, Reddy MV, Chandra R Subhash V, Devi MP, Srikanth S (2013) Bacteria for bioenergy: a sustainable approach towards renewability. In: Gaspard S, Ncibi MC (eds) Biomass for sustainable applications: pollution, remediation and energy. RSC, Cambridge, pp 251–289. https://doi.org/10.1039/9781849737142-00251

Mössner E, Boll M, Pfleiderer G (1980) Purification of human and bovine alkaline phosphatases by affinity chromatography. Hoppe Seylers Z Physiol Chem 361:543–549

Munson L, Fall RR (1978) Purification and characterisation of *Escherichia coli* alkaline phosphatase. a biochemical experiment. Biochem Educ 6:53–56. https://doi.org/10.1016/0307-4412(78)90059-6

O'Connor SM, Coates JD (2002) Universal immunoprobe for (per)chlorate-reducing bacteria. Appl Environ Microbiol 68:3108–3113. https://doi.org/10.1128/AEM.68.6.3108-3113.2002

Piskarev VE, Shuster AM, Gabibov AG, Rabinkov AG (1990) A novel preparative method for the isolation of avidin and riboflavin-binding glycoprotein from chicken egg-white by the use of high-performance liquid chromatography. Biochem J 265:301–304

Priyanka B, Patil RK, Dwarakanath S (2016) A review on detection methods used for foodborne pathogens. Indian J Med Res 144:327–338. https://doi.org/10.4103/0971-5916.198677

Richard C, Drider D, Fliss I, Denery S, Prevost H (2004) Generation and utilization of polyclonal antibodies to a synthetic c-terminal amino acid fragment of divercin V41, a class IIa bacteriocin. Appl Environ Microbiol 70:248–254. https://doi.org/10.1128/AEM.70.1.248-254.2004

Schmeisser F, Vasudevan A, Soto J, Kumar A, Williams O, Weir JP (2014) A monoclonal antibody-based immunoassay for measuring the potency of 2009 pandemic influenza H1N1 vaccines. Influenza Other Respir Viruses 8:587–595. https://doi.org/10.1111/irv.12272

Shen Z, Hou N, Jin M, Qiu Z, Wang J, Zhang B, Wang X, Wang J, Zhou D, Li J (2014) A novel enzyme-linked immunosorbent assay for detection of *Escherichia coli* O157:H7 using immunomagnetic and beacon gold nanoparticles. Gut Pathog 6:14. https://doi.org/10.1186/1757-4749-6-14

Shetty KG, Huntzicker JV, Rein KS, Jayachandran K (2010) Biodegradation of polyether algal toxins—isolation of potential marine bacteria. J Environ Sci Health A Tox Hazard Subst Environ Eng 45:1850–1857. https://doi.org/10.1080/10934529.2010.520510

Sina M, Farajzadeh D, Dastmalchi S (2015) Effects of environmental factors on soluble expression of a humanized anti-TNF-α scFv antibody in *Escherichia coli*. Adv Pharm Bull 5:455–461. https://doi.org/10.15171/apb.2015.062

Sohier D, Pavan S, Riou A, Combrisson J, Postollec F (2014) Evolution of microbiological analytical methods for dairy industry needs. Front Microbiol 5:16. https://doi.org/10.3389/fmicb.2014.00016

Standaert-Vitse A, Jouault T, Vandewalle P, Mille C, Seddik M, Sendid B, Mallet JM, Colombel JF, Poulain D (2006) *Candida albicans* is an immunogen for anti–*Saccharomyces cerevisiae* antibody markers of Crohn's disease. Gastroenterology 130:1764–1775. https://doi.org/10.1053/j.gastro.2006.02.009

Talja I, Kubo A-L, Veijola R, Knip M, Simell O., Ilonen J, Vähä-Mäkilä M, Sepp E, Mikelsaar M, Utt M, Uibo R (2014) Antibodies to lactobacilli and bifidobacteria in young children with different propensity to develop islet autoimmunity. J Immunol Res 325938, 6 p. https://doi.org/10.1155/2014/325938

Tao S, Guyer C (2016) Magnetic micro-particle immune separation combined with enzyme linked immunosorbent assay for quick detection of *Salmonella typhimurium* in raw eggs. Nano Res Appl 2:1. http://nanotechnology.imedpub.com/archive.php

Tudorache M, Bala C (2008) Sensitive aflatoxin B1 determination using a magnetic particles-based enzyme-linked immunosorbent assay. Sensors (Basel) 8:7571–7580. https://doi.org/10.3390/s8127571

Vanhoutte I, Audenaert K, De Gelder L (2016) Biodegradation of mycotoxins: tales from known and unexplored worlds. Front Microbiol 7:561. https://doi.org/10.3389/fmicb.2016.00561

Vijayakumar S, Saravanan V (2015) Biosurfactants – types, sources and applications. Res J Microbiol 10:181–192. https://doi.org/10.3923/jm.2015.181.192

Whitman WB, Coleman DC, Wiebe WJ (1998) Prokaryotes: the unseen majority. Proc Natl Acad Sci USA 95:6578–6583. https://doi.org/10.1073/pnas.95.12.6578

Wu H, Wang S, Qiao J, Liu J, Zhan J, Gao X (2009) Expression of $HpaG_{Xooc}$ protein in *Bacillus subtilis* and its biological functions. J Microbiol Biotechnol 19:194–203. https://doi.org/10.4014/jmb.0802.154

Yang SC, Lin CH, Sung CT, Fang JY (2014) Antibacterial activities of bacteriocins: application in foods and pharmaceuticals. Front Microbiol 5:241. https://doi.org/10.3389/fmicb.2014.00241

Zdorovenko EL, Kachala VV, Sidarenka AV, Kisileva EP, Kachala VV, Knirel YA (2009) Structure of the cell wall polysaccharides of probiotic bifidobacteria *Bifidobacterium bifidum* BIM B-465. Carbohydr Res 344:2417–2420. https://doi.org/10.1016/j.carres.2009.08.039

Zhong N, Loppnau P, Seitova A, Ravichandran M, Fenner M, Jain H, Bhattacharya A, Hutchinson A, Paduch M, Lu V, Olszewski M, Kossiakoff AA, Dowdell E, Koide A, Koide S, Huang H, Nadeem V, Sidhu SS, Greenblatt JF, Marcon E, Arrowsmith CH, Edwards AM, Gräslund S (2015) Optimizing production of antigens and Fabs in the context of generating recombinant antibodies to human proteins. PLoS One 10:e0139695. https://doi.org/10.1371/journal.pone.0139695

Chapter 12
Conservation and Application of Microalgae for Biofuel Production

Samadhan Yuvraj Bagul, Hillol Chakdar, K. Pandiyan, and Krishnashis Das

Abstract The world has been confronted with an energy crisis, depletion of fossil fuel and increasing concern of global climate change; therefore, there is a need to find an alternative biological solution which is green and free from pollution. Microalgae are highly diverse, single or multicellular, microscopic photosynthetic organisms and utilize sunlight and carbon dioxide for synthesis of food. Microalgal biomass has been considered to be an important and valuable resource due to appreciable growth rate and high levels of lipids, carbohydrates and other value-added products. Microalgae have been utilized for biofuel production such as biodiesel, bioethanol, biohydrogen and biomethane production over the years. Conservation of genetic resources is an important aspect for every nation. Various conservation techniques are available to conserve the microorganisms. This chapter is mainly focused on conservation of microalgae and application aspect of microalgae in the area of biofuel production.

12.1 Introduction

The environment, during the Cambrian period of about 4.5 billion years ago, was anaerobic when carbon dioxide (CO_2) concentration was 4500 ppm. During 2.5 billion years ago, cyanobacterial photosynthetic activity resulted in gradual rise in atmospheric oxygen level, referred as "Great Oxygenation Event" (Bekker et al. 2004). These organisms are primary producers and considered as ultimate source of atmospheric oxygen available during the early evolution of earth. Microalgae are reported to carry out about 40% of the total global photosynthesis (Falkowsky 1980).

S. Y. Bagul (✉) · H. Chakdar · K. Pandiyan
ICAR-National Bureau of Agriculturally Important Microorganisms, Maunath Bhanjan, Uttar Pradesh, India
e-mail: samadhanbagul@gmail.com

K. Das
ICAR-Indian Institute of Pulses Research, Kanpur, Uttar Pradesh, India

S. K. Sharma, A. Varma (eds.), *Microbial Resource Conservation*, Soil Biology 54,
https://doi.org/10.1007/978-3-319-96971-8_12

Many scientific reports have shown the existence of microalgae with diversity ranging from 200,000 to several million species as compared to 250,000 species of higher plants (Norton et al. 1996).

Microalgae are highly diverse, single or multicellular, microscopic photosynthetic organisms comprising eukaryotic photoautotrophs and prokaryotic cyanobacteria, also called as blue green algae due to its specialized blue pigment phycocyanin. They are found in freshwater as well as marine water and are part of food chain of aquatic environment having rapid growth rate and act as CO_2 sequestration agent under diverse situations. A microalga produces 1 kg biomass with the consumption of approximately 1.7–2 kg CO_2 (Clarens et al. 2010). It provides essential nutritional factors for aquatic animals, including omega-3 fatty acids and other lipids, proteins and carbohydrates (Belarbi and Molina 2000). Microalgae can be cultivated in waste water as these can utilize nutrients available in such waters and help in reducing pollution load (Bagul et al. 2017). Temperature requirement for microalgal growth is generally 20–30 °C. However, they have high requirement of water which is as high as 11–13 million L/ha/year for their growth (Amarasinghe et al. 2007). The size of microalgae varies depending upon the species ranging from micrometres (μm) to millimetres (mm).

Cyanobacteria are a diverse group of microorganisms present in different ecological niches which vary from aquatic to terrestrial and ultra-oligotrophic to hypereutrophic. They are also found in extreme habitats such as hot springs and hypersaline lakes (Bhatnagar and Bhatnagar 2005). They have the ability in photosynthesis as well as nitrogen fixation, and these properties with wider adaptability to various ecosystems make them ubiquitous. Cyanobacteria comprise about 150 genera with more than 2000 species, with a vast amount of diversity in their cell structure, unicellular and colonial to complex filamentous forms with or without branching. These cyanobacteria are vital species as they contribute towards carbon and nitrogen economy as they are nitrogen fixers and their biomass helps in increasing carbon in different ecological niches (Mishra and Pabbi 2004). Figure 12.1 depicts different microalgal genera.

12.2 Microalgae in Different Ecological Niches

12.2.1 Freshwater Microalgae

Chlamydomonas sp., *Chlorococcum* sp., *Chlorella* sp., *Cosmarium* sp., *Scenedesmus* sp., *Spirogyra* sp., *Kirchneriella* sp., *Navicula* sp., *Nitzschia* sp., *Anabaena* sp., *Nostoc* sp., *Microcystis* sp., *Chroococcus* sp., *Lyngbya* sp., *Gloeocapsa* sp., *Merismopedia* sp., *Oscillotoria* sp., *Plectonema* sp. and *Spirulina* sp.

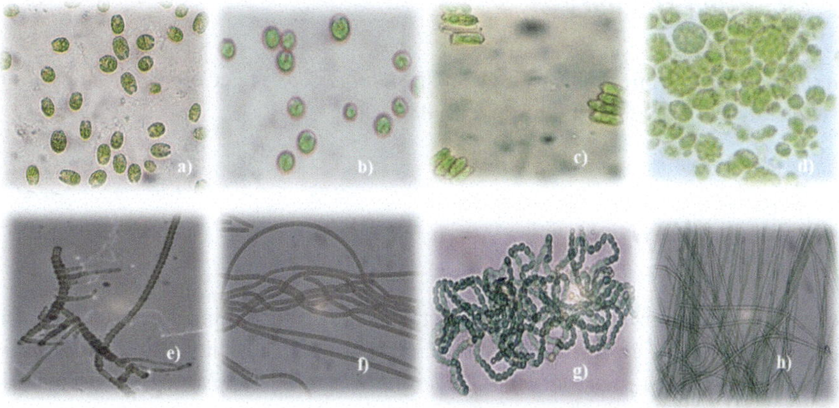

Fig. 12.1 Microphotographs of microalgal genera; (**a**) *Chlamydomonas* sp.; (**b**) *Chlorella* sp.; (**c**) *Scenedesmus* sp.; (**d**) *Chlorococcum* sp; (**e**) *Westielopsis* sp; (**f**) *Lyngbya* sp.; (**g**) *Nostoc* sp.; (**h**) *Phormidium* sp. (Source: Bagul et al. 2012; Sharma et al. 2016)

12.2.2 Marine Microalgae

Large numbers of marine algae are known to produce metabolites and pigments which have medical importance like neuroprotective traits, antiaging and antioxidant which make them attractive to use in cosmetics (Berthon et al. 2017). *Chlorella* sp., *Arthrospira* sp. and *Dunaliella* sp. were reported to erase vascular imperfection of epidermis, elevate collagen synthesis and prevent formation of wrinkle (Stolz and Obermayer 2005). *Navicula salinicola* and *Dunaliella* sp. have been reported from hypersaline lake of Turkey. *Aphanothece halophytica*, *Synechocystis* sp. and *Dactylococcopsis salina* are a few cyanobacteria described from the Great Salt Lake and the saline playas of the Thar Desert (Bhatnagar et al. 2003). In addition to the above mentioned microalgae, *Halomicronema* sp., *Spirulina* sp., *Aphanocapsa* sp., *Synechococcus* sp., *Nannochloropsis* sp., *Nodularia* sp., *Tetraselmis* sp. and *Asteromonas gracilis* have also been reported from marine habitats (Kacka and Donmez 2008; Kirkwood et al. 2008).

12.2.3 Acidic Environment

Euglena sp., *Stichococcus* sp., *Watanabea* sp., *Chlamydomonas sphagnophila* (<3 pH), *Chlamydomonas acidophila* (1.7 pH) and *Chlamydomonas pitschmannii* (1.5 pH) (Pollio et al. 2005).

12.2.4 Cold Habitat

Xanthonema sp. from alpine snowbanks, *Stichococcus* sp., *Chloromonas* sp. and *Trebouxia* sp. (Alfredo and Giuseppe 2008).

12.2.5 Desert Habitat

Deserts are characterized by moisture deficit (annual rain $\leq$254 mm) as a result of which biological activities are regulated by ephemeral water availability, e.g. *Chroococcus minutus*, *Oscillatoria pseudogeminata*, *Phormidium tenue* and *Nostoc* sp. (Bhatnagar and Bhatnagar 2005).

12.3 Conservation of Microalgae

Microalga is primarily maintained and conserved by sub-culturing and by cryopreservation for short- and long-term preservation, respectively. However, attempts have also been made by lyophilization, but the results were not appreciable. Many microalgal culture collections are working to conserve the microalgae which include the Culture Collection of Algae and Protozoa (CCAP), Scotland, where strains are primarily maintained by serial sub-culturing, although about 30% of the algal strains are cryopreserved (https://www.ccap.ac.uk/our-cultures). The Culture Collection of Algae at Gottingen University (SAG) has a collection of about 2250 strains from almost all evolutionary lineages of eukaryotic algae and prokaryotic cyanobacteria isolated mainly from freshwater or marine habitat and terrestrial habitats (https://www.uni-goettingen.de/en/184982.html). The University of Texas' (UTEX) algae collection has a staggering collection of over 2800 strains from 200 different genera (https://utex.org/). The Australian National Algae culture Collection (ANACC), Australia, holds 1000 strains of more than 300 microalgal species (https://www.csiro.au/en/Research/Collections/ANACC/About-our-collection). The Centre for Conservation and Utilisation of Blue Green Algae (CCUBGA), India, has a collection of over 550 cultures of microalgae (http://www.iari.res.in/ccubga/bga_catalog.php).

12.3.1 Cryopreservation

Presently microalgae are maintained by serial sub-culturing where microalgal cultures are transferred to a fresh medium after a certain period of time, which is a labour-intensive process, and chances of contamination are high. The cryopreservation of

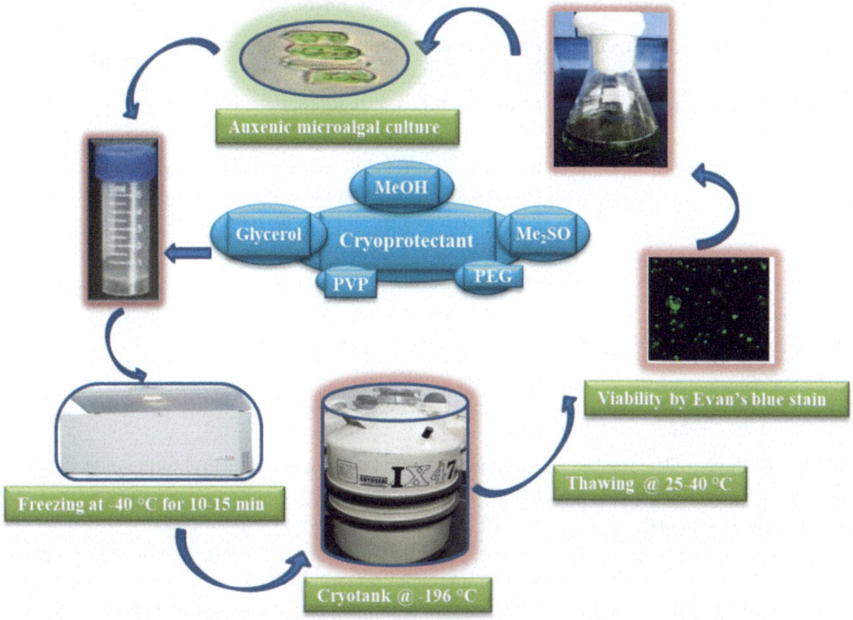

Fig. 12.2 Process of cryopreservation

microalgae can prevent genetic drift and minimize labour costs as compared to daily maintenance and sub-culturing (Day et al. 2000).

Cryopreservation process (Fig. 12.2) involves preparation of microalgal cell before freezing where a cryoprotective agent (CPA) is added into the cell suspension to avoid cellular damage due to ice formation (Taylor and Fletcher 1999). Microalgal cells are processed for cooling at −40 °C followed by direct transfer to the liquid nitrogen. Cooling is an important factor which could lead detrimental effects on cell. When the rate of cooling is faster, the chance of cell damage is higher due to the formation of ice crystals within the cell as the excess intracellular water is not sufficiently dehydrated and becomes super cool. However, slower cooling causes loss of intracellular liquid by excessive dehydration and is prone to chemical and physical damage by increase in toxicity of intracellular solutes and crushing of intracellular organelles. Finally these samples undergo quick thawing at 25–40 °C so that higher viability can be achieved. Concentration and duration of the cryoprotectant are important for successful cryopreservation. Generally, 5–10% concentration of cryoprotectant is used (Day et al. 1997; Rhodes et al. 2006). The cryoprotective agents applied for marine microalgae are methanol (MeOH), dimethyl sulfoxide (DMSO) and glycerol (penetrating type) which reduce the cytoplasm crystallization rate by decreasing the intracellular water-freezing temperature. Another cryoprotective agent (non-penetrating) such as polyvinylpyrrolidone (PVP) and polyethylene glycol (PEG) causes cell dehydration thus reducing the

Table 12.1 Cryoprotectants used for cryopreservation of microalgae (Hubalek 2003)

Group of cryoprotective agent	Representative compound of the group
Sulfoxides	Dimethylsulfoxide
Monohydric alcohols and derivatives	Methanol, Ethanol, Polyvinyl alcohol
Diols and derivatives	Ethylene glycol, Ethylene glycol, Trimethylene glycol, Diethylene glycol, Propylene glycol, Polyethylene glycol, Polypropylene glycol, Polyethylene oxide
Triols	Glycerol
Polyalcohols	Mannitol, Sorbitol, Dulcitol
Monosaccharides	Glucose, Xylose
Disaccharides	Sucrose, Lactose, Maltose, Trehalose
Trisaccharides	Raffinose
Polysaccharides	Dextran, Mannan, Dextrin, Hydroxyethyl starch, Ficoll, Gum arabic
Amides, *N*-alkylamides, imides	Acetamide, Methyl acetamide, Dimethyl formamide, Dimethyl acetamide
Heterocyclic compounds	Methyl pyrrolidone, Polyvinyl pyrrolidone
Amino acids and carbonic acids	Proline, Glycine, Glutamic acid, Aminobutyric acid, Glutaric acid, Ammonium acetate, EDTA
Proteins, peptides, polypeptides, and glycoproteins	Blood serum, albumins, Gelatin, peptones, Shell extract, Glycoproteins, Mucin, Valinomycin, Gramicidin
Complex substrates	Yeast extract, Malt extract, Skimmed milk, Honey
Nonionic surfactants	Tween 80, Triton, Macrocyclon

amount of available water to form intracellular ice (Hubalek 2003). Table 12.1 represents various cryoprotectant agents commonly used for cryopreservation of microalgae. Many studies have been reported for successful cryopreservation of microalgae such as *Amphidinium carterae*, *Chaetoceros muelleri*, *Chaetoceros calcitrans*, *Chaetoceros* sp., *Isochrysis galbana* and *Nitzschia ovalis* using 10% and 15% DMSO (Day et al. 1997; Rhodes et al. 2006). Nakanishi et al. (2012) have reported the preservation of *Chlorella vulgaris*, *Nannochloropsis oculata* and *Tetraselmis tetrathele* with 50% viability after 15 years in a mixture of 5% cryoprotectant (DMSO, ethylene glycol and proline). Park (2006) has preserved many bloom-forming cyanobacteria for more than 2.5 years.

12.3.2 *Immobilization*

Sodium alginate beads are used for immobilization of microalgae. Generally the concentration of 1–4% of sodium alginate is being used for the preparation of beads. This solution is then kept overnight and filtered to remove contaminants like proteins and polyphenols. Sodium alginate and algal suspension are mixed thoroughly and added dropwise into calcium chloride (2M) solution using micropipette which forms the transparent beads (Hirata et al. 1995). This method is nontoxic and allows different types of microalgae to grow inside the matrix with additional advantages of being a cheap, easy and feasible method of preservation. Chen (2001) reported successful preservation of *Scenedesmus quadricauda* by alginate beads for more than 3 years. *Euglena gracilis* also has been preserved for more than 2 years by Tamponnet et al. (1985). *Chlorella* sp. was successfully preserved by Day et al. (1987) by air-dried alginate beads for 1 year.

12.3.3 Lyophilization

The principal involved in the method is sublimation of frozen microbial cultures under vacuum. Protective agents such as different carbohydrates, proteins, polymers, polyols, amino acids and oxygen scavengers are mixed with microbial cultures in an ampoule. These ampoules were then frozen at −20 °C for 2 days. Primary drying takes place at 0.05 mbar for 4–6 h. The ampoules are then transferred in soft glass tubes containing silica gel with cotton plug. Outer tubes are then fitted into manifold and mounted on lyophilization. Secondary drying then takes place about14–16 h at 0.001 mBar; finally ampoules are sealed under vacuum. Malik (1993) successfully preserved the microalgal cultures up to 1 year by freeze-drying which includes *Chlorella pyrenoidosa*, *Chlorella vulgaris*, *Selenastrum capricornutum*, *Scenedesmus subspicatus* and *Euglena gracilis.*

Another method of conservation of microalgae has been reported by Prasad et al. (2016); they have used cotton cloth and soaked it in cetrimide and washed with double-distilled water to remove cetrimide. After drying, the cloth was cut into small pieces, autoclaved thereafter incubated with 100 mL of BG11 medium containing 5-day-old culture in a culture room. The viability of microalgal cells of *Chlorella minutissima* and *Chlamydomonas* sp. is achieved even after 18 months.

Syiem and Bhattacharjee (2010) used the entrapment method for preservation of cyanobacteria in which 10 mL of concentrated cyanobacterial culture is mixed with 2.4% nutrient agar. This mixture is then solidified on petri plates for 30 min under aseptic conditions. After solidification small cubes were prepared and dried under lamina for 24 h. They successfully used this method for the preservation of *Nostoc* sp., *Anabaena* sp., *Cylindrospermum* sp., *Plectonema* sp., *Fischerella* sp. and *Mastigocladus* sp. for 3 years with good viability.

12.4 Application of Microalgae for Biofuel Production

The rising level of carbon dioxide, i.e. 36.2 billion tonnes in the atmosphere, increasing demand for fossil fuel and its depleting reserves are some of the serious concerns of the environment. Fossil fuels are available reasonably at low cost, but this scenario will change in the coming years, and there is an urgent need to identify alternative renewable sources of fossil fuel that are potentially carbon neutral (Hill et al. 2006; Demirbas 2009). At present 80–90% of fuel is being produced from non-renewable resources, while only 10% are from renewable resources (Demirbas 2010; Maity et al. 2014). Global biofuel production has reached a level of 130.7 billion litres in 2015 with a share of 75.2% of ethanol, 23% biodiesel and 3.7% hydrotreated vegetable oil (HVO). This is 3.6% increase in biofuel production compared to last year REN 21 (2015). Microalgal biomass has been considered to be promising feedstock due to its appreciable growth rate and high levels of lipids, carbohydrates and other value-added products. Therefore, it can be exploited for production of various types of renewable, clean and green fuels such as biomethane production through anaerobic digestion, biodiesel by transesterification, bioethanol through fermentation and photobiologically produced biohydrogen (Spolaore et al. 2006). Figure 12.3 depicts integrated model for biofuel production from microalgae.

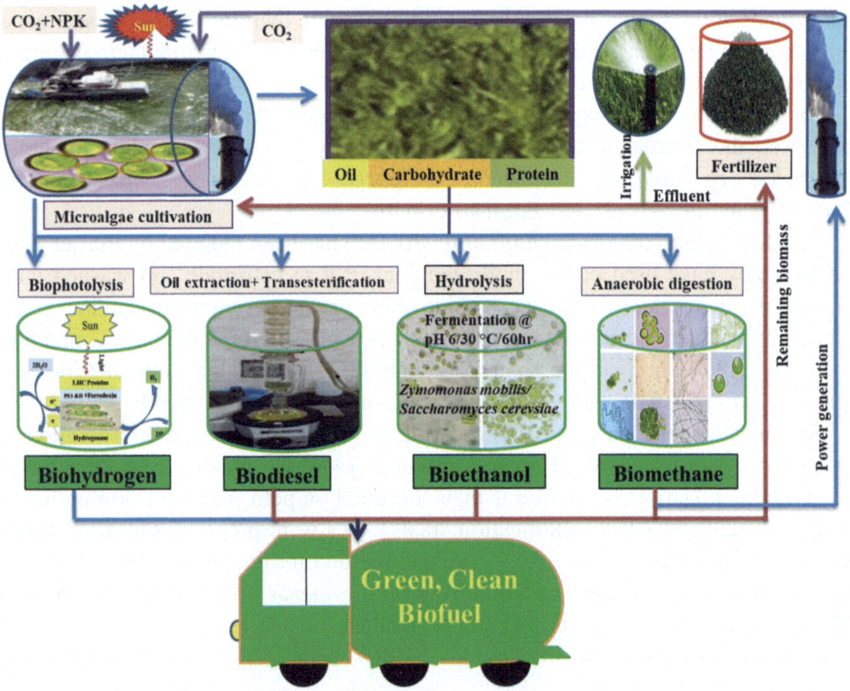

Fig. 12.3 Integrated model for biofuel production from microalgae

12.4.1 Bioethanol

Currently, most of the biofuels are produced from sugar crops, agricultural feedstocks or starch of edible grains by bioconversion. First-generation feedstock for bioethanol production includes cereals and legumes which are actually meant for human consumption. The diversion of these food crops to biofuel production resulted in increase in food price and at the cost of agricultural land. However, these problems are solved to some extent by the use of second-generation feedstocks including lignocellulosic materials such as agro- or forest residues and woody biomass. These feedstocks have some advantages over first generation feedstocks as they require minimal land and do not compete with food (Mohr and Raman 2013). However, the problem associated with this feedstock is complex cell structure as lignocellulosic residues are recalcitrant in nature which requires more expensive pretreatment processes to release sugars, and their harvesting, purification and various pretreatment process makes these feedstock more challenging and uneconomical. Microalgae, considered as third-generation feedstock for biofuel, are an alternative for the first- and second-generation feedstocks due to their high productivity, appreciable growth rate and easy and round-the-year cultivation (Nigam and Singh 2011; Daroch et al. 2013). Bioethanol is an alternative biofuel to gasoline since they share similar physical and chemical properties. The energy content of bioethanol is 21.2 MJ/dm^3 while that of gasoline is 31.3 MJ/dm^3. The energy content of a single unit of carbohydrate is 467 kJ/mol. Photosynthetic active radiation (PAR), having wavelength range of 400–700 nm, can generalize 217 kJ/mol of energy. To produce one carbohydrate unit, eight photons of PAR are required; therefore, energy storage efficiency in the form of carbohydrate is 11.6% (Wegeberg 2010; Kumar et al. 2016). Microalgae contain 21–64% carbohydrate on dry weight basis which is higher than lignocellulosic biomass (Lam and Lee 2015). Additional benefit is that lignin, a recalcitrant substance, is not present; therefore it requires mild pretreatment before converting it into bioethanol. However, generally microbial biomass goes through a hydrolysis, a mild pretreatment which releases the carbohydrate (glucose) which is a preferred carbon source for *Saccharomyces cerevisiae* or *Zymomonas mobilis* for fermentation. Fermentation process is carried out by two methods such as separate hydrolysis fermentation (SHF) in which reactions occur in two different reactors, while another method is simultaneous saccharification fermentation (SSF) in which hydrolysis and fermentation occur simultaneously (Hill et al. 2006). High-starch-containing algae *Chlorella*, *Dunaliella*, *Scenedesmus* and *Chlamydomonas* are useful for bioethanol production. Ho et al. (2013) reported 11.66 g/L of ethanol yield by SHF process with 1% dilute acid hydrolysis which produced 87.6% theoretical yield in 12 h. Algae can produce 46,760–140,290 L/ha of ethanol yield (Mussatto et al. 2010).

12.4.2 Biodiesel

Microalga is a well-known candidate to use solar energy efficiency and has a huge potential for biomass production with minimal land requirement. They appear to be the only source of biodiesel that has the potential to completely displace fossil diesel. For biofuel production, algae need to have a lipid content in the range of 20–80% by weight of dry biomass, but oil levels of 20–50% are quite common (Illman et al. 2000; Chisti 2007; Maity et al. 2014). Table 12.2 shows the oil content in different microalgae. B100 biodiesel is considered as transesterified product of long-chain fatty acids derived from vegetable oils or animal fats, which fulfil the criteria of ASTM D 6751 standards. However, biodiesel blend (BXX) is a mixture of biodiesel fuel with petroleum-based diesel fuel; here, XX is the percent volume of biodiesel (Howell 2012). Transesterification, the process in which triglycerides are reacted with methanol in the presence of a catalyst to produce methyl esters of fatty acids and glycerol, is the most fundamental step in commercial production of biodiesel (Harun et al. 2006). This process can run at atmospheric pressure and temperatures of approximately 60 °C (140 °F) (Hussain 2010). The addition of 10% methanol and 90% oil can produce 90% biodiesel and 10% glycerine as a by-product. The reaction during transesterification occurs in a stepwise manner in which triglycerides are first converted to diglycerides and then to monoglycerides and finally to glycerol. Normally, when transesterification is performed with algal oil, the biodiesel yield is around 80% of the volume of the used algal oil (El-Shimi et al. 2013).

Table 12.2 Oil content of certain microalgae suitable for biodiesel production (Illman et al. 2000; Chisti 2007)

Microalga	Oil content (% dry weight)
Botryococcus braunii	25–75
Chlorella emersonii	63
Chlorella minutissima	57
Chlorella vulgaris	40
Chlorella protothecoides	23
Chlorella sorokiniana	22
Crypthecodinium cohnii	20
Cylindrotheca sp.	16–37
Dunaliella primolecta	23
Isochrysis sp.	25–33
Monallanthus salina	>20
Nannochloris sp.	20–35
Nannochloropsis sp.	31–68
Neochloris oleoabundans	35–54
Nitzschia sp.	45–47
Phaeodactylum tricornutum	20–30
Schizochytrium sp.	50–77
Tetraselmis sueica	15–23
Chlorella sp.	28–32

Transesterification process is catalysed by acids, alkalis and lipase enzymes, and the alkali-catalysed transesterification is about 4000 times faster than the acid catalysed reaction. Commercially, alkali such as sodium and potassium hydroxide are commonly used as catalysts at a concentration of about 1% by weight of oil. However, alkoxides such as sodium methoxide are considered to be better catalysts and are increasingly being used. The use of lipases offers important advantages but is not currently feasible because of the relatively high cost of the catalyst. Alkali-catalysed transesterification is carried out at about 60°C under atmospheric pressure, as methanol boils at 65 °C at atmospheric pressure, and under these conditions, reaction takes about 90 min to complete (Fukuda et al. 2001). A higher temperature can be used in combination with higher pressure, but this becomes an expensive process. To prevent yield loss due to saponification reactions, the oil and alcohol must be dry, and the oil should have a minimum of free fatty acids. Biodiesel can be recovered by repeated washing with water to remove glycerol and methanol (Chisti 2007).

12.4.3 Biohydrogen

German plant physiologist Hans Gaffron discovered the hydrogen metabolism in 1939 in green alga *Scenedesmus obliquus* and demonstrated that in the anaerobic dark condition H_2 evolution was at low rate, while H_2 production simulated largely in the light for brief period of time (Gaffron 1939). Hydrogen (H_2) is a potential source of a clean and renewable energy. Process of hydrogen takes place at ambient temperature and pressure which makes it more sustainable and attractive. Combustion of some hydrogen-rich fuels releases water that is subsequently evaporated in the combustion chamber known as the heating value. Its high energy of 2.2 pounds of hydrogen contains an equal amount of 1 gallon or 6.2 pounds of gasoline which makes it a more attractive biofuel. Higher heating value (HHV) of hydrogen is 140 kJ/g and 120 kJ/G (LHV) (considering water as a product). In combustion, water is the main product; thus, hydrogen is regarded as a clean non-polluting fuel. Apart from being used as fuel, hydrogen (H_2) is also used for the manufacture of ammonia, removal of impurities in oil refineries, in methanol production and as a fuel in rocket engines. Hydrogen production processes include direct biophotolysis which involves the splitting of water molecules to (H^+), (e^-) and (O_2) under sunlight in the presence of eukaryotic microalgae and recombining them by hydrogenase to produce H_2 (Ayhan 2009). Figure 12.4 shows a schematic presentation of biophotolysis of water. Microalgae have the high photon conversion efficiency of >80%. H_2-producing microalgae include *Scenedesmus obliquus*, *Chlorococcum littorale*, *Platymonas subcordiformis* and *Chlorella fusca* (Oncel 2013). However, *Chlamydomonas reinhardtii* has been more exploited for H_2 production which showed 5.94 mmol/g Chl/h of H_2 yield.

Cyanobacteria evolves H_2 by indirect biophotolysis; in this process light-harvesting complex (LHC) proteins channel the photons to photosystem II which utilizes the radiant energy in photosynthesis to split water molecules. Extracted

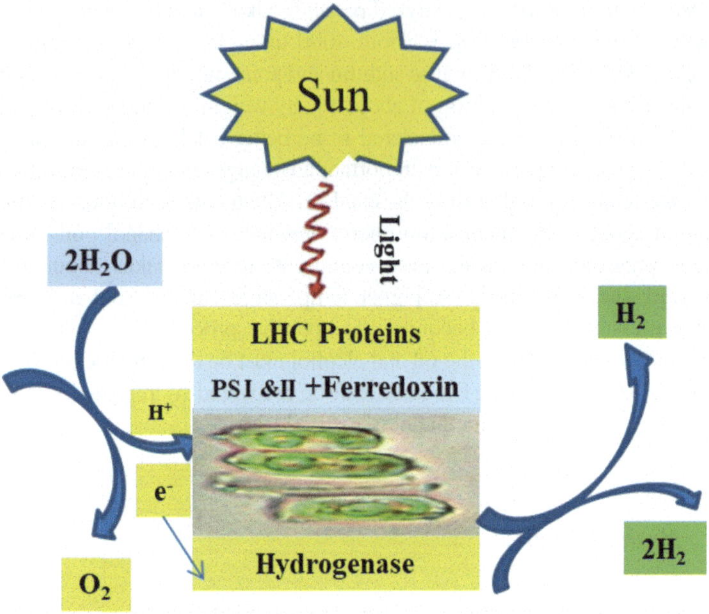

Fig. 12.4 Mechanism of direct biophotolysis

electrons after oxidation are transported to the Fe–S protein in ferredoxin on the reducing side of photosystem I. The hydrogenase enzyme present in the stroma of the microalgal chloroplast accepts electrons from reduced ferredoxin and donates them to two protons to generate one H_2 molecule. Many cyanobacteria are reported to produce a large quantity of hydrogen gas, e.g. *Gloeocapsa alpicola*, *A. cylindrica*, *Cyanothece*, *A. variabilis*, *Anabaena azollae*, *Arthrospira (Spirulina) platensis* and *Nostoc muscorum* (Dutta et al. 2005). A report also showed that nitrogen-starved cells of *A. cylindrica* produced highest amount of hydrogen (30 mL H_2/lit/h) (Tiwari and Pandey 2012). Hydrogen (H_2) production is the by-product of fixation of nitrogen into ammonia. A total of 16 ATPs are required for the fixation of 1 mol of N_2 which results into the formation of 1 mol of hydrogen gas. Depending on species, cyanobacteria can use both nitrogenases (unidirectional) and hydrogenases (bidirectional) for hydrogen production, while eukaryotic microalgae depend solely on hydrogenases. Two different types of hydrogenase enzymes which catalyse the process in various cyanobacteria are uptake hydrogenase (*hup*SL), responsible for the 'Knallgas' reaction, and reversible or bidirectional hydrogenases (*hox*FUYH) that can produce or take up hydrogen. Ni–Fe and Fe–Fe hydrogenases based on metal as cofactor are prevalent among microbes (Margheri et al. 1990). In cyanobacteria H_2 evolution occurs temporally during light and dark periods or spatially in heterocyst cells. Microalgae have high H_2 production efficiency due to its Fe–Fe hydrogenase iron-containing protein of 48 kDa which is 100 times higher than other hydrogenases (Happe and Naber 1993). On the basis of optimal assumptions, green algae could produce maximum of 10 mol (20 g) H_2/m^2per day (Melis and Happe 2001).

12.4.4 Biomethane

Eutrophication is the phenomenon where microalgae growth is robust due to the presence of nitrogen and phosphorous. Another way to produce biomass is utilization of waste water as substrate as they are rich in nutrients which ultimately reduce the cost of cultivation with the advantage of waste water treatment. Various microalgae such as *Scenedesmus* sp., *Spirulina* sp., *Euglena* sp., *Chlorella vulgaris*, *Melosira* sp. and *Oscillatoria* sp. have been used as a feedstock for anaerobic digestion by many researchers (Uziel 1978; Samson and Leduy 1986; Yen and Brune 2007; Ras et al. 2011). This microalgal biomass could be utilized to produce biomethane by anaerobic digestion which is a biochemical process where decomposition of organic matter takes place by anaerobic microorganisms. The process involves different steps such as hydrolysis, fermentation, acetogenesis and methanogenesis (Vavilin et al. 2008). During hydrolysis complex cell materials such as lipids, polysaccharides, proteins and nucleic acids are converted into simple compounds like fatty acids, monosaccharides, amino acids, purines and pyrimidines. These substances then undergo fermentation to produce acetate, hydrogen, carbon dioxide, methanol, formate, propionate, butyrate, methylamines, etc., and in the final stages, methane is produced by acetoclastic (acetate consumer) and hydrogen-utilizing (H_2 and CO_2 consumers) methanogens (Clark et al. 2009).

The methane production is about 55–75% and carbon dioxide 25–45% (Harun et al. 2010). Estimates show that heating value of biogas is 600 BTU/ft^3 (Noorollahi et al. 2015). Scientific literature shows that from 1 tonne of algae, 9.4 GJ biomethane could be produced. According to Harun and co-workers, microalgae produce more energy in the form of biomethane (14.04 MJ/kg) as compared to biodiesel (6.6 MJ/kg) and bioethanol (1.79 MJ/kg) on dry weight basis. Microalgae are favourable substrate for biomethane production due to its biodegradable components such as carbohydrate, proteins and lipids for anaerobic digestion. The yield of methane depends on species of microalgae, pretreatment of microalgal biomass, presence or absence of methanogenesis inhibitors, pH, C/N ratio and temperature. Carbohydrates are part of the microfibrillar polysaccharides embedded in matrix of polysaccharides and proteoglycans and are difficult to extract from microalgae. This requires pretreatment to release those fermentable sugars (Gonzalez-Fernandez et al. 2012). Mendeza et al. (2015) reported an increase of 50% methane production from *Chlorella vulgaris* by application of thermal pretreatment. Microalgae have been extensively studied by many researchers as a substrate for methane production and reported in the range of 143–400 L-Ch_4 (kg/VS) (Perazzoli et al. 2016).

Two-step processes have been found suitable for production of methane which also has high energetic gain from microalgal biomass. In the first step, the carbohydrates are fermented to produce hydrogen and volatile fatty acids (VFA) and in the

Table 12.3 Comparative methane yield in mono and co-digestion of feedstock

Co-digestion feedstock	Methane yield		References
	Co-digestion	Single feedstock	
Scenedesmus and *Chlorella* consortia + food waste	639.8 ± 1.3 mL/g VS	106.9 ± 3.2 mL/g VS (consortia)	Zhen et al. (2016)
Algae sludge (AS)+ waste paper	1170 ± 75 mL/l/day	573 ± 28 mL/l/day (AS)	Yen and Brune (2007)
Water Hyacinth + Cow manure + Sewage sludge	812 mL	335 mL (kitchen waste and cow manure	Tasnim et al. (2017)
Cattle manure (CM) +Organic Kitchen Waste (OKW)	14653.5 mL	7971.19 mL (OKW) and 1988.69 mL (CM)	Aragaw et al. (2013)
Dairy Manure (DM) + Tomato Residue (TR)	415 L/kg VS	275 L/kg/VS (DM), 45 L/kg VS (TR)	Li et al. (2016)
Corn Stover (CS) + tomato residue	315 L/kg VS	210 L/kg VS (CS)	Li et al. (2016)
Fruit and vegetable waste (FVW) + meat residue	0.9 m^3/kg VS	0.07 m^3/kg VS (FVW)	Garcia-Pena et al. (2011)
Taihu blue algae (TBA) and corn straw	325 mL/g/VS	201 mL/g/VS (TBA)	Zhong et al. (2012)

second step, the VFA are digested under methanogenic conditions to generate methane. This two-step strategy has been used in microalgae residues after lipid extraction which increased methane yield by 22 and up to 67% as compared to methanogenesis by using a single step (Reyes and Buitron 2016). Microalgal biomass has low C/N ratio (6–9) due to their relative high protein content which causes high total ammonia nitrogen release and high volatile fatty acid accumulation (Gonzalez-Fernandez et al. 2012). They are the inhibitors of the anaerobic digestion process which affects the methanogenic activity. Therefore, many researchers use carbon-rich feedstock to get rid of toxic ammonia. Ideal C/N ratio suggested by different researchers for anaerobic digestion is 20–30 (Zhong et al. 2012). Many investigators suggested co-digestion by adding 50% waste paper (C/N ratio 25:1) with microalgal biomass resulted in doubling of the methane production rate to 1170 ± 75 mL/l/day as compared to 573 ± 28 mL/l/day of algal sludge digestion alone because of balanced C/N ratio (Yen and Brune 2007). Co-digestion of feedstock, sisal pulp with fish waste showed 59–94% increase in methane production (Mshandete et al. 2004). Switchgrass with dairy manure co-digestion showed improved buffering capacity of the feedstock and further improved fermentation efficiency which resulted in 39% increase in methane yield (Zheng et al. 2015). A mixture of *Dunaliella salina* and Olive mill solid waste showed 48% increased methane yield when both the substrates were used in the ratio of 50:50 (Fernandez-Rodriguez et al. 2014). Addition of soybean oil, maize silage, corn straw and olive mill solid waste also reported appreciable methane yield. Table 12.3 depicts comparative methane yield between mono- and co-digestion of substrates.

12.5 Conclusion and Future Prospectus

Microalga is an important feedstock for biofuel production which is a biodegradable, nontoxic and sustainable option. Microalgae for energy production can be seen in the future as an alternative for fossil fuels. Microalgae conservation could be achieved by lyophilization and cryopreservation for long-term storage by optimizing conditions such as different cryoprotectants, ratio of cryoprotectants and manipulation of cultural conditions. Still there is a strenuous research needed in the area of biofuel production. Future study needs to find promising microalgal strain with novel properties. The study also needs to find efficient cultivation and harvesting technology for algal biomass production. Continuous efforts are also required to improve preservation methods for long-term conservation of microalgae.

Acknowledgements Authors are grateful to the Indian Council of Agricultural Research-National Bureau of Agriculturally Important Microorganisms for providing financial support.

References

Alfredo T, Giuseppe F (2008) Osmotic adjustments in a psychrophilic alga, *Xanthonema* sp. (Xanthophyceae). Environ Exp Bot 63:342–350

Amarasinghe UA, Shah T, Turral H, Anand BK (2007) India's water future to 2025–2050: business-as-usual scenario and deviations. (IWMI Research Report 123). International Water Management Institute/American Public Health Association, Colombo/New York, 47 p

Aragaw T, Andargie M, Gessesse A (2013) Co-digestion of cattle manure with organic kitchen waste to increase biogas production using rumen fluid as inoculums. Int J Phy Sci 8:443–450

Ayhan D (2009) Progress and recent trends in biodiesel fuels. Energ Conver Manage 50:14–34

Bagul SY, Prasanna R, Dhar DW (2012) Identification of efficient microalgal strains for growth potential and total lipids. Indian J Plant Physiol 17(3&4) (N.S.):203–206

Bagul SY, Prasanna R, Dhar DW (2017) Biomass productivity and nutrient removal study with wastewater grown microalgae for biodiesel production. Green Farm 8:672–675

Bekker A, Holland HD, Wang PL, Rumble D, Stein HJ, Hannah JL, Coetzee LL, Beukes NJ (2004) Dating the rise of atmospheric oxygen. Nature 427:117–120

Belarbi EH, Molina EY (2000) A process for high yield and scaleable recovery of high purity eicosapentaenoic acid esters from microalgae and fish oil. Process Biochem 35:951–969

Berthon JY, Kappes RN, Bey M, Cadoret JP, Renimel I, Filair E (2017) Marine algae as attractive source to skin care. Free Radic Res 6:555–567

Bhatnagar A, Bhatnagar M (2005) Microbial diversity: microbial diversity in desert ecosystems. Curr Sci 89:91–100

Bhatnagar A, Bhatnagar M, Makandar MB, Garg MK (2003) Satellite centre for microalgal biodiversity in arid zones of Rajasthan. Project completion Report, Funded by Department of Biotechnology, New Delhi

Chen YC (2001) Immobilized microalga *Scenedesmus quadricauda* (Chlorophyta, Chlorococcales) for long-term storage and for application for water quality control in fish culture. Aquaculture 195:71–80

Chisti Y (2007) Biodiesel from microalgae. Biotechnol Adv 25:294–306

Clarens AF, Resurreccion EP, White MA, Colosi LM (2010) Environmental life cycle comparison of algae to other bioenergy feedstocks. Environ Sci Technol 44:1813–1819

Clark DP, Dunlap P, Madigan M, Martinko J (2009) Brock biology of microorganisms. Scientific Publisher, Beijing

Daroch M, Geng S, Wang G (2013) Recent advances in liquid biofuel production from algal feedstocks. Appl Energy 102:1371–1381

Day JG, Priestley IM, Codd GA (1987) Storage, recovery and photosynthetic activities of immobolised algae. In: Webb C, Mavituna F (eds) Plant and animal cells, process possibilities. Ellis Harwood Ltd., Chichester, pp 257–261

Day JG, Watanabe MM, Morris GJ, Fleck RA, McLellan MR (1997) Long-term viability of preserved eukaryotic algae. J Appl Phycol 9:121–127

Day JG, Fleck RA, Benson EE (2000) Cryopreservation-recalcitrance in microalgae: novel approaches to identify and avoid cryo-injury. J Appl Phycol 12:369–377

Demirbas A (2009) Biofuels securing the planets future energy needs. Energy Convser Manage 50:2239–2249

Demirbas A (2010) Social, economic, environmental and policy aspects of biofuels. Energy Educ Sci Technol Part B Soc Edu Stud 2:75–109

Dutta D, De D, Chaudhuri S, Bhattacharya SK (2005) Hydrogen production by cyanobacteria. Microb Cell Fact. https://doi.org/10.1186/1475-2859-4-36

El-Shimi HI, Attia NK, El-Sheltawyi ST, El-Diwani GI (2013) Biodiesel production from *Spirulina Platensis* microalgae by in-situ transesterification process. J Sustain Bioenergy Sys 3:224–233

Falkowsky PG (1980) Primary productivity in the sea. In: Falkowski PG (ed) Environmental science research, vol 19. Plenum Press, New York

Fernandez-Rodriguez MJ, Rincon B, Fermoso FG, Jimenez AM, Borja R (2014) Assessment of two-phase olive mill solid waste and microalgae co-digestion to improve methane production and process kinetics. Bioresour Technol 157:263–269

Fukuda H, Kondo A, Noda H (2001) Biodiesel fuel production by transesterification of oils. J Biosci Bioeng 92:405–416

Gaffron H (1939) Reduction of CO_2 with H_2 in green plants. Nature 143:204–205

Garcia-Pena EI, Parameswaran P, Kang DW, Canul-Chan M, Krajmalnik-Brown R (2011) Anaerobic digestion and co-digestion processes of vegetable and fruit residues: process and microbial ecology. Bioresour Technol 102:9447–9455

Gonzalez-Fernandez C, Sialve B, Bernet N, Steyer JP (2012) Impact of microalgae characteristics on their conversion to biofuel Part II: focus on biomethane production. Biofuels Bioprod Biorefin 6:205–2018

Happe T, Naber JD (1993) Isolation, characterization and N-terminal amino acid sequence of hydrogenase from the green alga *Chlamydomonas reinhardtii*. Eur J Biochem 214:475–481

Harun M, Ali M, Mustafa I (2006) Environmentally friendly diesel for Pakistan. Environ Monit 6:14–18

Harun R, Singh M, Forde GM, Danquah MK (2010) Bioprocess engineering of microalgae to produce a variety of consumer products. Renew Sustain Energy Rev 14:1037–1047

Hill J, Nelson E, Tilman D, Polasky S, Tiffany D (2006) Environmental, economic, and energetic costs and benefits of biodiesel and ethanol biofuels. PNAS 103:11206–11210

Hirata K, Phunchindawan M, Kanemoto M, Miyamoto K, Sakai A (1995) Cryopreservation of shoot primordia induced from horseradish hairy root cultures by encapsulation and two-step dehydration. Cryo-Letters 16:122–127

Ho SH, Huang SW, Chen CY, Hasunuma T, Kondo A, Chang JH (2013) Bioethanol production using carbohydrate-rich microalgae biomass as feedstock. Bioresour Technol 135:191–198

Howell S (2012) Biodiesel Industry and ASTM Standards for: Washington State DOT March 22, 2012. National Biodiesel Board

Hubalek Z (2003) Protectants used in the cryopreservation of microorganisms. Cryobiology 46:205–229

Hussain K (2010) Economically effective potential of algae for biofuel production. World Appl Sci J:1313–1323

Illman AM, Scragg AH, Shales SW (2000) Increase in *Chlorella* strains calorific values when grown in low nitrogen medium. Enzyme Microb Technol 27:631–635

Kacka A, Donmez G (2008) Isolation of *Dunaliella* spp. from a hypersaline lake and their ability to accumulate glycerol. Bioresour Technol 99:8348–8352

Kirkwood AE, Buchheim JA, Buchheim MA, Henley WJ (2008) Cyanobacterial diversity and halotolerance in a variable hypersaline environment. Microb Ecol 55:453–465

Kumar K, Ghosh S, Angelidaki I, Holdt SL, Karakashev DB, Morales MA, Das D (2016) Recent developments on biofuels production from microalgae and macroalgae. Renew Sustain Energy Rev 65:235–249

Lam MK, Lee KT (2015) Bioethanol production from microalgae. Handbook of marine microbiology. https://doi.org/10.1016/B978-0-12-800776-1.00012-1

Li Y, Li Y, Zhang D, Li G, Lu J and Li S (2016). Solid state anaerobic co-digestion of tomato residues with dairy manure and corn stover for biogas production. Bioresour Technol. https://doi.org/10.1016/j.biortech.2016.01.111

Maity JP, Bundschuh J, Chen CY, Bhattacharya P (2014) Microalgae for third generation biofuel production, mitigation of greenhouse gas emissions and wastewater treatment: present and future perspectives e a mini review. Energy 78:104–113

Malik KA (1993) Preservation of unicellular green algae by a liquid-drying method by a liquid. J Microbiol Meth 18:41–49

Margheri MC, Treddici MR, Allotta G, Vagnoli L (1990) Heterotrophic metabolism and regulation of uptake hydrogenase activity in symbiotic cyanobacteria. In: Polsinelli M, Matterassi R, Vincenzini M (eds) Developments in plant and soil sciences bio nitrogen fixation. Kluwer Academic, Dordrecht, pp 481–486

Melis A, Happe T (2001) Hydrogen production. Green algae as a source of energy. Plant Physiol 127:740–748

Mendeza L, Mahdya A, Ballesterosa M, Fernándeza CG (2015) Biomethane production using fresh and thermally pretreated *Chlorella vulgaris* biomass: a comparison of batch and semi-continuous feeding mode. Ecol Eng 84:273–277

Mishra U, Pabbi S (2004) Cyanobacteria: a potential biofertilizer for rice. Resonance 9:6–10

Mohr A, Raman S (2013) Lessons from first generation biofuels and implications for the sustainability appraisal of second generation biofuels. Energy Policy 63:114–122

Mshandete A, Kivaisi A, Rubindamayugi M, Mattiasson B (2004) Anaerobic batch co-digestion of sisal pulp and fish wastes. Bioresour Technol 95:19–24

Mussatto SI, Dragone G, Pedro MR, Guimaraes JPA, Silva LM, Carneiro IC, Roberto AV, Lucilia D, Jose A (2010) Technological trends, global market, and challenges of bioethanol production. Biotechnol Adv 28:817–830

Nakanishi K, Deuchi K, Kuwano K (2012) Cryopreservation of four valuable strains of microalgae, including viability and characteristics during 15 years of cryostorage. J Appl Phycol 24:1381–1385

Nigam PS, Singh A (2011) Production of liquid biofuels from renewable resources. Prog Energy Combust Sci 37:52–68

Noorollahi Y, Kheirrouz M, Asl HF, Yousefi H, Hajinezhad A (2015) Biogas production potential from livestock manure in Iran. Renew Sustain Energy Rev 50:748–754

Norton TA, Melkonian M, Andersen RA (1996) Algal biodiversity. Phycologia 35:308–326

Oncel S (2013) Microalgae for a macro energy world. Renew Sustain Energy Rev 26:241–264

Park HK (2006) Long-term preservation of bloom-forming cyanobacteria by cryopreservation. Algae 21:125–131

Perazzoli S, Bruchez BM, Michelon W, Steinmetz RLR, Mezzari MP, Nunes EO, Silva MLB (2016) Optimizing biomethane production from anaerobic degradation of *Scenedesmus* spp. biomass harvested from algae-based swine digestate Treatment. Int Biodeter Biodegr 109:23–28

Pollio A, Cennamo P, Ciniglia C, Stefano MD, Pinto G, Huss VAR (2005) *Chlamydomonas pitschmannii* Ettl, a little known species from thermoacidic environments. Protist 156:287–302

Prasad R, Shabnam N, Pardha-Saradhi P (2016) Immobilization on cotton cloth pieces is ideal for storage and conservation of microalgae. Algal Res 20:172–179

Ras M, Lardon L, Bruno S, Bernet N, Steyer JP (2011) Experimental study on a coupled process of production and anaerobic digestion of *Chlorella vulgaris*. Bioresour Technol 102:200–206

REN 21 (2015) Renewable energy policy network for the 21st century, 2015 Global status report, Paris

Reyes JC, Buitron G (2016) Biohydrogen and methane production via a two-step process using an acid pretreated native microalgae consortium. Bioresour Technol 221:324–330

Rhodes L, Smith J, Tervit R, Roberts R, Adamson J, Adams S, Decker M (2006) Cryopreservation of economically valuable marine micro-algae in the classes Bacillariophyceae, Chlorophyceae, Cyanophyceae, Dinophyceae, Haptophyceae, Prasinophyceae, and Rhodophyceae. Cryobiology 52:152–156

Samson R, Leduy A (1986) Detailed study of anaerobic digestion of *Spirulina maxima* algal biomass. Biotechnol Bioeng 28:1014–1023

Sharma SK, Singh UB, Karthikeyan N, Sahu PK, Bagul SY (2016) Techniques of handling cyanobacteria: microbial culture handling and maintenance. ICAR-NBAIM, Maunath Bhanjan, pp 34–39

Source: https://utex.org/

Source: https://www.uni-goettingen.de/en/184982.html

Source: http://www.iari.res.in/ccubga/bga_catalog.php

Source: https://www.ccap.ac.uk/our-cultures.htm

Source: https://www.csiro.au/en/Research/Collections/ANACC/About-our-collection

Spolaore P, Joannis-Cassan C, Duran E, Isambert A (2006) Commercial applications of microalgae. J Biosci Bioeng 101:87–96

Stolz P, Obermayer B (2005) Manufacturing microalgae for skin care. Cosm Toiletries 120:99–106

Syiem M, Bhattacharjee A (2010) An efficient protocol for long-term preservation of cyanobacteria. J Adv Lab Res Biol 1:41–45

Tamponnet C, Costantino F, Barbotin JN, Clavayrac R (1985) Cytological and physiological behaviour of *Euglena gracilis* cells entrapped in a calcium alginate gel. Physiol Plant 63:277–283

Tasnim F, Iqbal SA, Chowdhury AR (2017) Biogas production from anaerobic co-digestion of cow manure with kitchen waste and Water Hyacinth. Renew Energy 109:434–439

Taylor R, Fletcher R (1999) Cryopreservation of eukaryotic algae a review of methodologies. J Appl Phycol 10:481–501

Tiwari A, Pandey A (2012) Cyanobacterial hydrogen production – a step towards clean environment. Int J Hydrogen Energy 37:139–150

Uziel M (1978) Solar energy fixation and conversion with algal bacterial systems, Ph.D. Thesis, University of California, Berkeley

Vavilin V, Fernandez B, Palatsi J, Flotats X (2008) Hydrolysis kinetics in anaerobic degradation of particulate organic material: an overview. Waste Manage 28:939–951

Wegeberg S (2010) Algae biomass for bioenergy in Denmark. Biological/technical challenges and opportunities

Yen HW, Brune DE (2007) Anaerobic co-digestion of algal sludge and waste paper to produce methane. Bioresour Technol 98:130–134

Zhen G, Lu X, Kobayashi T, Kumar G, Xu K (2016) Anaerobic co-digestion on improving methane production from mixed microalgae (*Scenedesmus* sp., *Chlorella* sp.) and food waste: kinetic modelling and synergistic impact evaluation. Chem Eng J. https://doi.org/10.1016/j.cej.2016.04.118

Zheng Z, Liu J, Yuan X, Wang X, Zhu W, Yang F, Cui Z (2015) Effect of dairy manure to switchgrass co-digestion ratio on methane production and the bacterial community in batch anaerobic digestion. Appl Energy 151:249–257

Zhong W, Zhang Z, Luo Y, Qiao W, Xiao M, Zhang M (2012) Biogas productivity by co-digesting Taihu blue algae with corn straw as an external carbon source. Bioresour Technol 114:281–286

Chapter 13
Desert Truffles in Saudi Arabia: Diversity, Ecology, and Conservation

Abeer Hashem, Abdulaziz A. Alqarawi, Manzoor Ahmad Shah, Stephan Wirth, Dilfuza Egamberdieva, B. Tabassum, and Elsayed Fathi Abd_Allah

Abstract Desert truffles have been an important source of food and medicine in many countries across the world. These fungi are a rich source of amino acids, fiber, protein, and carbohydrates. In the peninsula of Saudi Arabia, desert truffles comprise an important seasonal commercial crop. To improve the ecological and agricultural activities in the desert and semiarid areas of Saudi Arabia, mycorrhizal technology promises to be quite useful for the cultivation of desert plants at a commercial scale. They are considered as a source of a valuable antibiotic alternative to fight antibiotic-resistant pathogens. With regard to achieving better growth, yield, and quality of desert truffles, in vitro mycorrhization is a powerful technique with equally good implications for the conservation of truffle biodiversity in Saudi Arabia. The present review briefly discusses some aspects of truffles regarding their biochemical aspects and medicinal values, with a special reference to their association with other mycorrhizal fungi.

A. Hashem (✉)
Faculty of Science, Botany and Microbiology Department, King Saud University, Riyadh, Saudi Arabia

Mycology and Plant Disease Survey, Plant Pathology Research Institute, Agriculture Research Center, Giza, Egypt
e-mail: habeer@ksu.edu.sa

A. A. Alqarawi · E. F. Abd_Allah
Faculty of Food and Agricultural Sciences, Department of Plant Production, King Saud University, Riyadh, Saudi Arabia

M. A. Shah
Department of Botany, University of Kashmir, Srinagar, Jammu and Kashmir, India

S. Wirth · D. Egamberdieva
Institute of Landscape Biogeochemistry, Leibniz Centre for Agricultural Landscape Research (ZALF), Müncheberg, Germany

B. Tabassum
Department of Zoology, Toxicology Laboratory, Government Raza PG College, Rampur, Uttar Pradesh, India

S. K. Sharma, A. Varma (eds.), *Microbial Resource Conservation*, Soil Biology 54,
https://doi.org/10.1007/978-3-319-96971-8_13

13.1 Introduction

Desert truffles are widely used throughout the world, and they have special economic and medicinal importance. The medicinal mushrooms have been reported as a source of polysaccharides and showed antibacterial and anticancer activities (Schillaci et al. 2017). These fungi are extensively utilized in many countries in the world including several Arabian countries (El-Enshasy et al. 2013), for several purposes, including as food and as medicine. The medicinal properties of truffle extracts have been reasonably well studied. For instance, the antibacterial properties of truffle extract for treating a wide range of diseases, such as trachoma, are well documented (Al-Qarawi and Mridha 2012; El-Enshasy et al. 2013). Schillaci et al. (2017) studied the edible desert truffle mushrooms *Tirmania pinoyi*, *Terfezia claveryi*, and *Picoa juniperi* and found that they have antibacterial properties against the strains *Staphylococcus aureus* ATCC 29213 and *Pseudomonas aeruginosa* ATCC 15442. Paul et al. (2006) observed antifungal activity of truffles of *Terfezia* against plant pathogenic fungi. In other studies, the extract of *Tirmania pinoyi* showed antimicrobial activities against *B. subtilis* and *S. aureus* (Dib-Bellahouel and Fortas 2011). In addition, truffles contain a significant amount of quality proteins, fiber, amino acids, carbohydrates, and fatty acids; thus they are used as an edible food resource (El-Enshasy et al. 2013; Kagan-Zur et al. 2014). In desert truffles, proteins comprise approximately 20% of the dry weight, which is substantially higher than the protein content of many other vegetables and fungi.

Many areas in Saudi Arabia grow truffles, and for a long time, they also constituted an essential food for daily usage and income. The people of Saudi Arabia are traditionally well experienced in the collection and cultivation techniques of desert truffles under existing natural conditions. However, continued overexploitation of this valuable natural resource has posed a great threat to this crop, thereby underscoring the need for its conservation (Fig. 13.1).

The arbuscular mycorrhizas have been found to have a close association with desert truffles. Despite the importance of mycorrhizal association to desert truffles, there has been a paucity of studies in this area of research. Hence, the primary objective of the present review is to document the status of information available regarding arbuscular mycorrhiza-desert truffle interactions, identify its key gaps, and suggest strategies to overcome these information gaps (Fig. 13.2).

13.2 Distribution and Ecology: A Brief Overview

Desert truffles commonly comprise different hypogeous members of ascomycetes and include genera such as *Balsamia*, *Delastria*, *Delastreopsis*, *Leucangium*, *Mattirolomyces*, *Phalangium*, *Terfezia*, and *Tirmania* (El-Enshasy et al. 2013; Mandeel and Al-Laith 2007; Trappe et al. 2010). In addition to these examples, a few tuber species are also included within this economically important group of

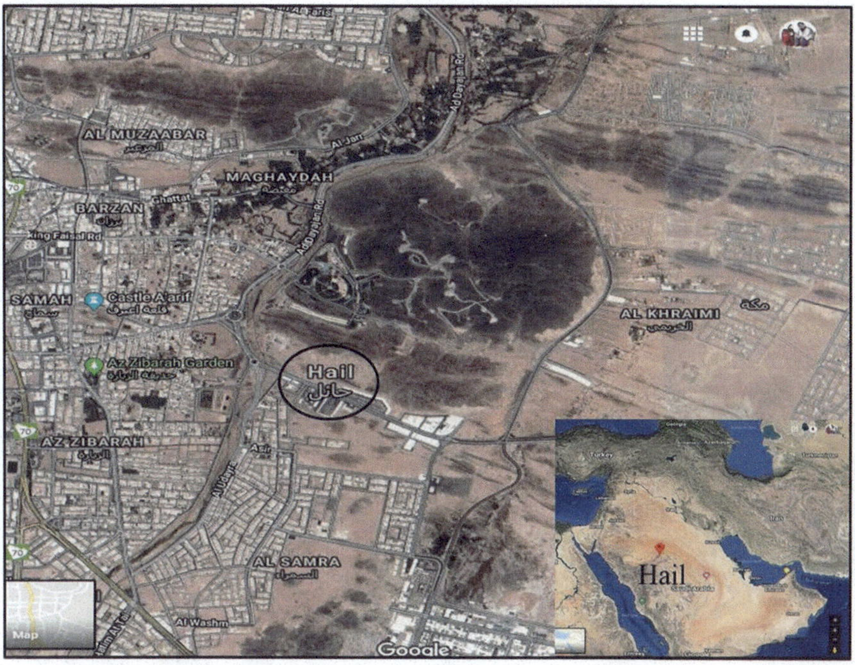

Fig. 13.1 Google map location showing the area of occurrence of desert truffles (*Tirmania nivea*) Zabide in Ha'il region, Saudi Arabia

fungi. For instance, in several countries, *Tuber magnatum* Pico and *Tuber melanosporum* Vittad., also called the white truffle and the black truffle, respectively, are mostly appreciated for their special taste and smell (Kagan-Zur et al. 2014). Their indigenous flavor is usually the blend of several hundred volatile ingredients. The development of truffles is believed to be influenced by several factors, such as adequate and timely rainfall, the physicochemical properties of soil, and the meteorology of the particular region. The production of truffles is also affected by air temperature, because minimum temperature has negative effect on host plant physiology (Morte et al. 2010). This may explain the negative effect of minimum temperatures on *Terfezia claveryi* productivity following the GLM.

Among these important factors, rainfall is the most important factor for determining the truffle fruiting and, hence, the distribution. In desert environments, the average rainfall required as an indicator of the possible timing of truffle emergence is 256 mm (Ceruti et al. 2003; Claridge et al. 2013). It is believed that the geographical distribution of desert truffles is usually confined to arid and semiarid regions, mostly in and around the Mediterranean basin, such as the southern region of Spain, Saudi Arabia, Iran, Iraq, Libya, Syria, Kuwait, Turkey, Egypt, Israel, Portugal, France, and Hungary. In addition to these famous places, a few desert truffles are also documented as coming from South Africa, Australia, North America, and Japan

Fig. 13.2 Showing the native growth of desert truffles *Tirmania nivea* (Zabide) along with the dominant host species (*Helianthemum lippii*) in Hail region, Saudi Arabia. [(**a**&**b**) The native habitat of *Helianthemum lippii* showing swelling of the ground (arrow) caused by Truffle *Tirmania nivea* (Zabide). (**c**&**d**) Initial growth of ascocarp (*Tirmania nivea*) over the soil surface (arrow) surrounded by young seedlings of their host plant (*Helianthemum lippii*). (**e**) Showing complete growth of ascocarp over soil surface adhering *Helianthemum lippii* (host plant). (**f**) Desert truffles *Tirmania nivea* (Zabide) in the last marketing stage.]

(Claridge et al. 2000). Regions with annual rainfalls of 50–380 mm usually support the emergence of desert truffles. In North African countries and some parts of Europe, good yields of truffles were reported to have been achieved from rainfall in the ranges of 70–120 mm and 100–350 mm (Morte et al. 2010). Bradai et al. (2015) reported that desert truffles in hyper-arid condition of Sahara desert of Algeria require adequate autumnal rainfall during October–December for better production of truffle. The total production for all three harvested species such as *Terfezia arenaria*, *Terfezia claveryi*, and *Tirmania nivea* was 785.43 ± 743.39 g/ha. In addition to this report, Morte et al. (2010) also observed that the timing of rainfall is as important as the quantity. For example, in North African and Middle Eastern countries, truffle emergence needs rainfall to occur exactly at the beginning of December and no later. According to Diez et al. (2002), the host specialization and soil pH can play an important role in the distribution pattern of the desert truffle species. For example, *H. salicifolium* and *H. ledifolium* species are found in basic soils and *H. guttatum* only in acid soils. In addition, host plant fitness and fungi spore germination and growth are also influence production of truffles (Sitrit et al. 2014). For example, Le Tacon et al. (2014) observed a decreased production of truffles (*Tuber melanosporum* Vittad.) due to low precipitations.

Ammarellou et al. (2007) studied desert truffles in Zanjan, Qazvin, Bandar-e Abbas, and Tabriz provinces of Iran and found *Terfezia boudieri.* Trappe et al. (2010) reported seven truffle species distributed in Australia, namely, *Elderia arenivaga, Mattirolomyces mulpu* sp. nov., *Mycoclelandia arenacea, M. bulundari, Reddelomyces westraliensis, Ulurua nonparaphysata* gen. and sp. nov., and *Horakiella watarrkana* sp. nov. Several other truffles were observed in Southern Africa which include *Eremioniyces echinulatus, Kalaharituber pfeilii,* and *Mattirolomyces austroafricanus.* Three species of the family Pezizaceae, namely, *Terfezia arenaria* (Moris) Trappe, *Terfezia claveryi* Chatin, and *Tirmania nivea* (Desf.) were found in North Algerian Sahara (Bradai et al. 2014). Owaid (2016) found two genera of desert truffles, namely, *Terfezia* sp. and *Tirmania* sp., in Iraq. Al-Rawi and Taha (2010) reported *Tirmania* sp. found in Iraq, Anbar desert. Bouzadi et al. (2017) reported *Tirmania pinoyi* (Maire) Malencon, *Tirmania nivea* (Desf.) Trappe, and *Terfezia boudieri* Chatin found in Libya. Table 13.1 shows some examples of different types of truffles found in various regions.

13.3 Biochemical and Physiological Aspects

Very few reports are available regarding the composition and chemistry of truffles; these qualities often reflect their economic importance. In fact, truffles are a good and rich source of amino acids, fiber, quality protein, and carbohydrates (Murcia et al. 2003; Mandeel and Al-Laith 2007). It is believed that proteins constitute 20% of the total dry mass of desert truffles, and the protein content is considered to be higher than that present in most vegetables and fungi. Consumption of desert truffles is also widely recommended for their medicinal properties against several ailments (El-Enshasy et al. 2013; Murcia et al. 2003). In addition to this fact, truffles act as a unique source of several medicinally important therapeutic compounds possessing anti-inflammatory, antimutagenic, antibacterial, and anticarcinogenic properties (Murcia et al. 2002; Al-Laith 2010; Dib-Bellahouel and Fortas 2011). In addition, truffles have been reported to act as immunosuppressors and antioxidants (Murcia et al. 2002; Kagan-Zur et al. 2014).

The bioactive organic compounds such as ergosterol, phenolic acids, carotenoids, flavonoids, and phenolic derivatives were found in truffles which attributed their antioxidant capacity (Villares et al. 2012; Al-Laith 2010). Bokhary et al. (1989) reported saturated and unsaturated fatty acids found in *Terfezia boudieri*, such as pentadecanoic, margaric, stearic, arachidic, behenic, palmitic, palmitoleic, stearic, oleic, linoleic, and linolenic. In another study Bouzadi et al. (2017) reported protein (16.3 and 18.5%), lipid (6.2 and 5.9%), and carbohydrates (67.2 and 65%) of Libyan wild truffle (Terfezia and Tirmania). Vahdani et al. (2017) also reported occurrence of high quantity of proteins, phenolics, potassium and iron in *Terfezia claveryi*

Table 13.1 Biodiversity of desert truffles in Saudi Arabia and some Arab countries

Country	Taxonomy of desert truffles		Host plant	References
Saudi Arabia	*Terfezia*	*Terfezia claveryi* Chatin	*Helianthemum vesicarium* Boiss. *Helianthemum kahiricum* Del. *Helianthemum lippii* (L.) Dum. Cours	Sawaya et al. (1985), Bokhary (1987), Hashem and Al-Obaid (1996), Janakat et al. (2005), Bawadekji et al. (2012), Schillaci et al. (2017)
		Terfezia boudieri Chatin		
	Tirmania	*Tirmania nivea* (Desf.) Trappe		
		Tirmania pinoyi (Maire)		
	Picoa	*Picoa juniperi* Vittad.		
		Picoa lefebvrei (Pat.) Maire		
	Phaeangium	*Phaeangium lefebvrei* (Pat.) Maire		
Egypt	*Terfezia*	*Terfezia arenaria* (Moris) Trappe	*Helianthemum lippii* (L.) Dum.-Courset	El-Kholy (1989)
		Terfezia claveryi Chatin		
	Tirmania	*Tirmania nivea* (Desf.) Trappe		
Bahrain	*Tirmania*	*Tirmania nivea* (Desf.) Trappe	*Helianthemum kahiricum* Del.	Mandeel and Al-Laith (2007)
	Phaeangium	*Phaeangium lefebvrei* Pat.		
Kuwait	*Tirmania*	*Tirmania nivea* (Desf.) Trappe	*Helianthemum kahiricum* Dei.	Al-Sheikh and Trappe (1983)
Tunisia	*Terfezia*	*Terfezia boudieri* Chatin	*Helianthemum sessiliflorum* Desf.	Hamza et al. (2016)
Jordan	*Terfezia*	*Terfezia claveryi* Chatin	*Helianthemum* Spp.	Janakat et al. (2004)
Algeria	*Tirmania*	*Tirmania pinoyi* (Maire)	*Helianthemum lippii* (L.) Dum.-Courset	Dib-Bellahouel and Fortas (2011), Bradai et al. (2014)
Iraq	*Terfezia*	*Terfezia claveryi* Chatin	*Helianthemum* Spp.	Dahham et al. (2016)

(black truffle) and *Tirmania nivea* (white truffle). The medicinal properties of desert truffles, such as their antibiotic and antimicrobial activities, have been demonstrated by several researchers, such as Chellal and Lukasova (1995) and Janakat et al. (2004, 2005). Splivallo (2008) found about 200 secondary metabolites in truffles, which include essential oils, fatty acids, aromatic compounds, phenolics, flavonoids, and terpenoids.

Hussain and Al-Ruqaie (1999) reported 16.6% and 24.8% carbohydrates, 8.1–13.8% proteins, and 9.7–25.5% phosphorus in dry biomass of Iraqi truffles (*Terfezia claveryi*, *Tirmania nivea*, and *Tirmania pinoyi*). The truffles also contain essential minerals including Na, K, Ca, Al, Mn, Zn, and Mg (Wang and Marcone 2011). In addition, insoluble polysaccharides, chitin, pectin, and hemicelluloses contribute to up to 80% of truffle biomass (Cheung 1997). The biological active compounds including phenols, carotenoids, and anthocyanin were also found in desert truffles (Gouzi et al. 2013). Slama et al. (2009) have observed that the fruiting bodies of desert truffles contain significant sugar (15.4%) content, and they also observed that, among the different individual sugars reported, 2.02% constitute soluble sugar content and 10.5% constitute insoluble sugar content. Desert truffles contain good quantities of potassium, calcium, phosphorus, iron, and magnesium. For example, *Terfezia boudieri* contains potassium, calcium, phosphorous, and magnesium in sufficient amounts. Normally, desert truffles have more calcium than *Tuber melanosporum* and *T. magnatum*. A comprehensive study will be required to achieve complete knowledge of the desert truffles available in Saudi Arabia. However, all these findings make desert truffles a remarkable subject for future researchers of these valuable fungi. Al-Sheikh and Trappe (1990) have analyzed the general features and spore-related characteristics, such as germination and their number, in the brown Kame truffle, *Terfezia claveryi*. Mycelia of desert truffles grown on modified Melin-Norkrans agar media and potato dextrose media showed their best growth at neutral pH, a clue to its use for the efficient and good generation and culturing of desert truffles for economic use (Morte et al. 1994; Morte and Honrubia 1995, 1997). Good desert growth and production need several intense laboratory and physio-biochemical requirements. The characterization and biochemical properties of the latent desert truffle *Terfezia claveryi* have been reported by Pervez-Gilabert et al. (2001). In two desert truffles, polyethylene glycol-induced osmotic stress effects have been studied by several groups because desert truffles have the ability to grow under water stress conditions (Mexal and Reid 1973; Coleman et al. 1989; Navarro-Rodenas et al. 2011). Navarro-Rodenas et al. (2011) reported that alkaline phosphatase (ALP) activity is a key indicator of the metabolic activity of these fungi and it was higher under drought condition. They observed that two species of desert truffle, *Terfezia claveryi* strain TcS2 and *Picoa lefebvrei* strain OL2, were able to tolerate water stress, below −1.07 MPa, and water stress even improved mycelial inoculum production of these desert truffles. Akyuz et al. (2012) reported medicinal use of desert truffles *Terfezia boudieri* that contains antioxidants, phenolic content, and other minerals.

13.4 Taxonomy of Desert Truffles

Truffles belong to ascomycetes and form symbiotic associations with host plants for the successful completion of their life cycle (El-Enshasy et al. 2013; Kagan-zur et al. 2014). Based on the morphometric studies of their spore ornamentation, these desert truffles cannot be well distinguished into individual species (Al-Qarawi and Mridha 2012). Morphological identification keys that have been proposed from time to time for the identification of these truffles serve as important basic tools for their clear identification (Trappe 1979; Castellano et al. 1989). However, convincing delimitation between different species and taxa has been achieved based on molecular studies and gene sequencing. In Saudi Arabia, desert truffles were recorded as early as quarter of a century ago (Bokhary 1987; Al-Qarawi and Mridha 2012; Bokhary and Parvez 1987, 1988). However, wide gaps exist in the literature regarding these truffles in Saudi Arabia. In the Arabian region, different species of desert truffles have been recorded from several countries including Egypt, Qatar, Algeria, Iraq, Kuwait, Bahrain, and Libya, and they are also found in the Iberian Peninsula (Kagan-zur et al. 2014). Trappe and Sundberg (1977) and Al-Sheikh and Trappe (1983) reported and described two different desert truffle species belonging to *Terfezia* and *Tirmania*. It is worth mentioning that the *Terfezia* have been reported in the Mediterranean and Middle East regions (Kovacs et al. 2009, 2011). Awameh and Al-Sheikh (1978) and Bokhary and Parvez (1988) have also identified certain desert truffles in Saudi Arabia. For the initial and first-hand identification and characterization, traditional criteria, such as macro- and micromorphological attributes of the fungal fruiting body, were used to fulfill the basic requirement. On the other hand, morphological identification of truffle fungi within their mycorrhizal structures is somewhat difficult. Thus, the practice of molecular approaches has become important for handling such taxonomically challenging issues (Gutierrez et al. 1995). In this direction, the analysis of DNA using polymerase chain reaction is one of the most powerful techniques for easy identification, and it has been accepted as a good approach for studying the intraspecific variability of genes (Gandeboenuf et al. 1994). At the molecular level, employing sequencing mediated by amplification of random amplified polymorphic DNA (RAPD) helps in detecting and differentiating the species and individuals (Cenis et al. 1993).

Desert truffles are found in many different countries of the world; for example, Morte et al. (2000) have given a short description of the characteristics of the main desert truffles collected in Spain. A spatial difference between truffles of the *Terfezia* genus has been studied by Kovacs et al. (2008) through molecular methods, and they have reported that most members of this genus probably belong to species reported from the Mediterranean and the Middle East regions. More extensive molecular studies have resulted in the inclusion of *Terfezia gigantea* Imai from North America and Japan and *T. pfeilii* Henn from South Africa, which are within the genera *Imaia* (Kovacs et al. 2008) and *Kalaharituber*, respectively (Ferdman et al. 2005). Studies have revealed that truffles commonly known as American *T. longii* Gilkey and *T. spinosa* Harkn belong to separate genera (Kovacs et al. 2011), while

T. austroafricana Marasas and Trappe of South Africa are included within *Mattirolomyces* (Trappe et al. 2010). Montecchi and Sarasini (2000) reported five frequently occurring Middle Eastern and Mediterranean species of *Terfezia*, including *Terfezia arenaria*, *T. boudieri*, *T. claveryi*, *T. leptoderma*, and *T. olbiensis*. Nevertheless, additional species of *Terfezia* have been described from the same zone from time to time but have been totally or more often treated as similar to the five species mentioned above. For molecular taxonomy purposes, *T. claveryi* and *T. boudieri* have been widely used as model species (Zaretsky et al. 2006; Navarro-rodenas et al. 2009; Zitouni-Haouar et al. 2014). It has been well reported that species such as *T. arenaria* and *T. claveryi*, as well as *T. boudieri*, are more easily identified and separated on the basis of morphological characteristics than *T. olbiensis* and *T. leptoderma*. Depending on the keys utilized to differentiate species of the genus, *T. olbiensis*, *T. leptoderma*, and other associated and relevant species are considered spiny-spored *Terfezia* species, reflecting their different developmental stages and therefore are treated as the same group by most authors. Taxonomic studies on *Terfezia* based on morphology, as well as molecular studies, have been well studied in several species. Diez et al. (2002) have studied four species by employing sequencing of ITS regions and have observed considerable differences in the sequences of all of the studied species. They further proposed that variations in the ITS sequences of *T. leptoderma* may possibly be due to the combined effects of several edaphic as well as biotic circumstances. However, studying a smaller number of species creates a major problem in such observations, and a similar problem was counteracted by Kovacs et al. (2011); several such species identification-related questions are still unresolved. Therefore, the hypothesis based on the intraspecific variation at the ITS region can be better established by the observation of more species, so that solid and widely acceptable identification can be done. Moreover, molecular differences can also result in identification of several different lineages. The results have indicated that there might possibly be more species of *Terfezia* that are included within the desert truffles belonging to Mediterranean and the Middle East regions, as compared with the approximately four to five *Terfezia* species that are normally discussed and admitted in the available literature. For example, Kovacs et al. (2011) observed a great degree of intraspecific and intrasporocarpic variation in morphological attributes, as well as the ITS sequences of their DNA. These differences may be even greater when the changes in edaphic and other associated factors are considered.

13.5 Cultivation and Mycorrhizal Association

Desert truffles also live symbiotically with plants and also form mutualistic interaction with mycorrhizal fungi. Several studies (Diez et al. 2002; Mandeel and Al-Laith 2007) reported that desert truffles in the Sahara of Algeria establish symbioses with *Helianthemum lippii*. Kagan-Zur et al. (2014) and Mandeel and Al-Laith (2007) reported that the hypogeous ascomycetes form mycorrhizal associations sharing mutualistic associations with several species of *Helianthemum* roots. Reports are available that indicate that they also form good mycorrhizal associations with *Helianthemum* species including *H. ledifolium* and *H. salicifolium* and have been reported to have mycorrhizal associations with various *Terfezia* species including *T. boudieri* and *T. claveryi*, as well as *Tirmania* species including *T. nivea* and *T. pinoyi* (Awameh and Al-Sheikh 1980; Kovacs and Trappe 2014). Species of the genus *Helianthemum* form ectomycorrhizal, as well as ecto-endomycorrhizal associations (Claridge et al. 2013). In addition to this fact, under in vitro conditions, during associations, desert truffles are able to form sheathing ectomycorrhiza in association with several species, such as *Terfezia claveryi*, *Picoa lefebvrei* (=*Phalangium lefebvrei*), etc. (Gutierrez et al. 2004). In the roots of *Helianthemum* species that colonized or formed association with the endomycorrhizal desert truffles, these species have been reported to lack a Hartig net as well as their mantle, nevertheless displaying undifferentiated hyphae, for example, in association between *Helianthemum* spp. and various species of both *Terfezia* and *Tirmania* (Awameh 1981; Dexheimer et al. 1985). Similar observations have been reported for the association between *Citrullus vulgaris* roots and *Kalaharituber pfeilii*, which also presents a good example of the point mentioned above (Kagan-Zur et al. 1999). Hence, desert truffle could be regarded as some sort of transitional link between true ecto- and endomycorrhizal associations.

However, the gap between these two widely occurring mycorrhizal associations is something crystalline and fluid. It should be noted here that the characteristics and attributes of mycorrhizal associations are usually and often determined by the environmental situations, as well as internal factors. Mycorrhizal associations can tolerate certain levels of harsh environmental conditions well, compared with the plants in arid and semiarid regions of the globe (Morte et al. 2009). Some truffle species, e.g., *Terfezia claveryi* and *Picoa lefebvrei* (=*Phalangium lefebvrei*), are hypogeous ascomycetes growing in the soils of semiarid regions that contain marl and gypsum; these species also establish mycorrhizal associations with most annual, as well as perennial, *Helianthemum* species (Honrubia et al. 1992). An ectomycorrhizal association of *Helianthemirhiza hirsuta* and *Helianthemum ovatum* has been described by Kovacs et al. (2011). This particular association is usually cottony, ranges from ochre to brown in color, and possesses a simple mycorrhizal system that has straight and slightly curved, bent, or tortuous margins. A comprehensive study about desert truffles and their association with plants has been made by Kovacs et al. (2002). Working with *Mattirolomyces terfezioides*, the association with plants and the subsequent behavior of both partners were investigated under aseptic conditions using Modified Melin-Norkrans (MMN) substrates supplemented with various phosphate concentrations. This task was performed to assess the changes that occur during the interaction of *Terfezia terfezioides* with *Robinia pseudoacacia* and *Helianthemum ovatum*. Colonization of the root system of black locust has often remained sensitive and weaker compared with that of *Helianthemum*. Among the few key characteristics, the presence of the intracellular coiled, branched, and septate hyphae in dead root cells was prominent. In addition, the intercellular hyphae usually show the formation of a Hartig net having such fingerlike structures only in *Helianthemum*; furthermore, these types of interactions could not be considered as mycorrhizae without any prior unambiguous proof. Further, they also observed that no lucid difference was obvious between the RFLP profiles of the nuclear DNA ITS of 19 fungal fruiting bodies that have been collected from the same habitat at the same time. They further demonstrated that the ITS of the randomly chosen three specimens showed similarity at the sequence level. Invariability between such types highlights the need for the development and design of PCR primers that are species-specific and would have a possibly important role in unambiguously identifying the host plants. Gutierrez et al. (2004) studied the morphological attributes of the mycorrhizal associations established between *Helianthemum almeriense* Pau and *Terfezia claveryi* and *Picoa lefebvrei* and provided an important useful and comprehensive review of mycorrhizal associations of desert truffles in relation to genus *Helianthemum*. In addition to this result, they also described the detailed anatomical features of the mycorrhizal associations in which *Helianthemum almeriense* serves as one partner and the structure, as well as ultrastructure, of the established mycorrhiza of this plant species with *Terfezia claveryi* and *Picoa lefebvrei*. An understanding of the use of biotechnological tools, as well as management techniques, in further studying the fungal inocula and mycorrhizal associations with plants and subsequent plantation, has been developed and introduced for some selected species of *Terfezia* (Honrubia et al. 2001). To mediate efficient and successful mycorrhization in *Helianthemum* spp. under in vitro conditions, Morte and Honrubia (1997) and Zamora et al. (2006) designed, developed, and introduced micropropagation protocols. Morte et al. (1994) and Morte and Honrubia (1997) mentioned that inocula that have been developed in modified Melin-Norkrans (MMN) or agar media can be well used from the plates as inocula directly for the production and synthesis of in vitro mycorrhization. True *Terfezia* species are mycorrhizal and have been used for the mycorrhizal bioassay experiments with plants mainly belonging to genus *Helianthemum* (Kovacs et al. 2002). They investigated the in vitro interaction of the *Terfezia terfezioides* with *Robinia pseudoacacia* and *Helianthemum ovatum* and made a detailed anatomical and ultrastructural study of the truffle. A similar study has been carried out by Morte et al. (2008) regarding the methods involved in the efficient synthesis of mycorrhiza between desert truffles and *Helianthemum* species based on the source of the fungal inocula (whether it is from spores or mycelia), as well as the plant source (whether it is seedlings or micropropagated plantlets), in addition to the culture conditions (in vitro or in vivo).

In another study, Khanaqa (2006) suggested to produce desert truffles in plantation with *Olea europaea* in which dual benefits could be achieved. The tree will improve water retention, soil erosion, and soil productivity. The compound β-cyclodextrin was able to stimulate the mycelial growth of *T. claveryi* (Lopez-Nicolas et al. 2013). The plants, *Helianthemum kahiricum* and *Helianthemum lippii*, were the dominant plants in Libya and found to form a mycorrhiza with desert truffles (Bouzadi et al. 2017). Most desert truffles are good options for the establishment of mycorrhizal associations, but *Terfezia claveryi* and *Picoa lefebvrei* are the two most widely accepted candidates for this particular attribute. To improve the ecological, biotechnological, and agricultural activities in the desert and semiarid areas of Saudi Arabia, implementing the mycorrhizal technology could prove to be an important component for the improvement of many desert plants. Considerable attention is needed to investigate this corner thoroughly.

13.6 Methods of Preserving Desert Truffles

Desert truffles have short shelf life when kept as fresh products, as its taste is completely altered when stored. Storage for longer durations brings changes in smell, aroma, taste, and, hence, the marketability, which needs special attention. Information on their preservation and increasing their shelf life by applying different preservation methods are limited. The most often and commonly employed methods for storing desert truffles involve chilling, followed by drying and subsequently freezing. Arab desert inhabitants, also called Bedouins, store and preserve pure truffles after pickling them in vinegar (4–6%) and brine (Kovacs et al. 2009). Treatment with gamma rays is believed to enhance the shelf life and safeguard the inherent characteristics of the important truffles. Nazzaro et al. (2007) studied the effect of 1.5 kGy gamma radiation on the biochemical and morphological properties, as well as the shelf life, of black desert truffles, and after 30 days of storage at 4 °C, they observed this treatment to be suitable for preservation. In addition, evidence in favor of treatment with brine has been reported by Al-Ruqaie (2006), who observed that pretreatment with sodium chloride (4%) helps to retain the color, texture, and flavor of the stored truffles. However, comparing these two widely employed techniques, freezing is more widely accepted when compared with the brine treatment. Furthermore, Al-Ruqaie (2006) also observed that treatment with acetic acid and sodium oxalate before radiation treatment significantly enhances the shelf life of desert truffles, which showed further improvement after being stored at lower temperatures. Separate or synergistic effects of applications of heat, fungicide, and radiation in the prevention of fungal infection were demonstrated by Al-Rawi and Aldin (1979), who, in an experiment, observed that the combined effect of 2000 ppm propionic acid, 150 krad ionizing radiation, and heating at 56 °C for a period of at least 5 min completely inhibited growth of microbes, thereby preventing the truffles from possible effects of microbial spoilage. Increased shelf life due to the synergistic treatment of gamma radiation and vacuum packaging under low temperature has been reported by Adamo et al. (2004). They further reported that enhanced shelf life is directly correlated with the efficient techniques involved in the removal of microbes and other related spoiling agents. An optimal treatment with radiation, chemicals, or heat is always required; however, in cases in which the optimal ranges are not available, truffles undergo unwanted chemical modification. Widely accepted by most workers, the recommended temperature for the safe storage of truffles has been considered and has been confirmed to be 4 °C. At this temperature, both biochemical and microbiological attributes of fresh desert truffles are preserved (Saltarelli et al. 2008). Hajjar et al. (2010) demonstrated that exposure of fresh desert truffles to elevated levels of CO_2 and hypoxia results in reductions in the polyphenol content, anaerobic metabolic pathways, and polyamine biosynthetic pathways, hence slowing senescence. Thus, in order to increase the shelf life and preserve the texture, as well as aroma of the truffles, it is very important to have a thorough understanding of the preservation techniques that can lead to the efficient preservation of truffles. Adopting efficient preservation techniques can be beneficial in maintaining the marketability of the truffles.

13.7 Conclusion and Future Prospects

Desert truffles are economically important, with great medicinal value. Significant progress is, however, needed for better taxonomic delimitation of desert truffles and detailed analysis of their medicinal values and biochemical properties, through both morphological and molecular approaches. With regard to achieving better growth, yield, and quality of desert truffles, in vitro mycorrhization is a powerful technique with equally good implications for the conservation of truffle biodiversity in Saudi Arabia. In view of its huge market value and economic benefits, the precise taxonomic delimitation of truffles through both conventional and molecular tools and their bioprospecting through biotechnological interventions attains special priority. The benefits accrued to truffles through mycorrhizal associations need to be properly tapped and scaled up. In a nutshell, better agricultural techniques, selection of proper species/strains, use of efficient bioinoculants, and good preservation techniques promise to boost the desert truffle industry in Saudi Arabia.

Acknowledgment The authors would like to extend their sincere appreciation to the Deanship of Scientific Research at King Saud University for its funding to the Research Group number RGP-271.

References

Adamo M, Capitani D, Mannina L, Cristinzio M, Rangni P, Tata A, Coppola R (2004) Truffles decontamination treated by ionizing radiation. Radiat Phys Chem 71:167–170

Akyuz M, Kirbag S, Kursat M (2012) Ecological aspects of the arid and semi-arid truffle in Turkey: evaluation of soil characteristics, morphology, distribution, and mycorrhizal relationships. Turk J Bot 36:386–391

Al-Laith AAA (2010) Antioxidant components and antioxidant, antiradical activities of desert truffle (*Tirmania nivea*) from various Middle Eastern origins. J Food Compos Anal 23:15–22

Al-Qarawi AA, Mridha MAU (2012) Status and need of research on desert truffles in Saudi Arabia. J Pure Appl Microbiol 6:1051–1062

Al-Rawi AMA, Aldin MM (1979) New mycorrhizal identification, truffle cultivation and truffles radiation preservation. Radiat Phys Chem 14:759–767

Al-Rawi AA, Taha AM (2010) Chemical study for three Iraqi truffles types. Al-Anbar J Agr Sci 8:33–41

Al-Ruqaie MA (2006) Effect of different treatment processes and preservation methods on the quality of truffles. 1. Conventional methods (drying/freezing). J Food Process Preserv 30:335–351

Al-Sheikh AM, Trappe JM (1983) Desert truffles: the genus *Tirmania*. Trans Br Mycol Soc 81:83–90

Al-Sheikh AM, Trappe JM (1990) Features and analysis of spore germination in the brown Kame *Terfezia claveryi*. Mycologia 72:495–499

Ammarellou A, Saremi H, Gucin F (2007) Evaluation of morphology, cytology and mycorrhizal relationships of desert truffles (*Terfezia boudieri*) in Iran. Pak J Biol Sci 10:1486–1490

Awameh MS (1981) The response of *Helianthemum salicifolium* and *H. ledifolium* to infection by the desert truffle *Terfezia boudieri*. Mushroom Sci 11:843–853

Awameh MS, Al-Sheikh AM (1978) Laboratory and field study of four kinds of truffles *Terfezia* and *Tirmania* species for cultivation. Mushroom Sci 10:507–517

Awameh MS, Al-Sheikh AM (1980) Ascospore germination of black kame (*Terfezia boudieri*). Mycologia 72:50–54

Bawadekji A, Gargano ML, Saitta A, Venturella G (2012) A new record of the desert truffle Picoa lefebvrei in Saudi Arabia. Mycotaxon 122:243–247

Bokhary HA (1987) Desert Truffle “Al-Kamah” of the Kingdom of Saudi Arabia. 1. Occurrence, identification and distribution. Arab Gulf Sci Res Agric Biol Sci 5:245–255

Bokhary HA, Parvez S (1987) Occurrence of desert truffle in Harrat Al-Harra (Northern region), Saudi Arabia. A survey report submitted to National Commission for wildlife conservation and development (NCWCD), Riyadh

Bokhary HA, Parvez S (1988) Desert Truffle “Al-Kamah” of the Kingdom of Saudi Arabia. additional contribution. Arab Gulf J Sci Res Agric Biol Sci 6:103–112

Bokhary HA, Suleiman AAA, Basalah MO (1989) The fatty acid components of the desert truffle “AI Kamah” of Saudi Arabia. J Food Prot 52:668–669

Bouzadi M, Grebenc T, Turunen O, Kraigher H, Taib H, Alafai A, Sbissi I, Assad MEH, Bedade D, Shamekh S (2017) Characterization of natural habitats and diversity of Libyan desert truffles. 3 Biotech 7:328. https://doi.org/10.1007/s13205-017-0949-5

Bradai L, Bissati S, Chenchouni H (2014) Desert truffles of the North Algerian Sahara: diversity and bioecology. Emir J Food Agric 26:425–435

Bradai L, Bissati S, Chenchouni H, Amrani K (2015) Effects of climate on the productivity of desert truffles beneath hyper-arid conditions. Int J Biometeorol 59:907–915

Castellano MA, Trappe JM, Maser Z, Maser C (1989) Key to spores of the genera of hypogenous fungi of north temperate forests. Mad River Press, Eureka, p 184

Cenis JL, Perez P, Fereres A (1993) Identification of aphid (*Homoptera*: *Aphididae*) species and clones by random amplified polymorphic DNA. Ann Entomol Soc Am 86:545–550

Ceruti A, Fontana A, Nosenzo C (2003) Le specie europee del genere Tuber, Una revisione storica. Regione Piemonte, Torino

Chellal A, Lukasova E (1995) Evidence for antibiotics in the two Algerian truffles *Terfezia* and *Tirmania*. Pharmaceut 50:228–229

Cheung PCK (1997) Dietary fibre content and composition of some edible fungi determined by two methods of analysis. J Sci Food Agric 72:255–260

Claridge AW, Cork SJ, Trappe JM (2000) Diversity and habitat relationships of hypogeous fungi. I. Study design, sampling techniques and general survey results. Biodivers Conserv 9:151–172

Claridge AW, Trappe JM, Paull DJ (2013) Ecology and distribution of desert truffles in the Australian outback. In: Zur et al (eds) Desert truffles, soil biology. Springer, Berlin

Coleman MD, Bledsoe CS, Lopushinsk W (1989) Pure culture response of ectomycorrhizal fungi to imposed water stress. Can J Bot 67:29–39

Dahham SS, Al-Rawi SS, Ibrahim AH, Abdul Majid AS, Abdul Majid AMS (2016) Antioxidant, anticancer, apoptosis properties and chemical composition of black truffle *Terfezia claveryi*. Saudi J Biol Sci. https://doi.org/10.1016/j.sjbs.2016.01.031

Dexheimer J, Gererd J, Leduc LP, Chevalier G (1985) Etude ultrastucturale compare des associations symbiotiques mycorrhiziennes *Helianthemum salicifolium- Terfezia claveryi* et *Helianthemum salicifolium-Terfezia leptoderma*. Can J Bot 63:582–591

Dib-Bellahouel S, Fortas Z (2011) Antibacterial activity of various fractions of ethyl acetate extract from the desert truffle, *Tirmania pinoyi*, preliminarily analyzed by gas chromatography mass spectrometry (GC-MS). Afr J Biotechnol 10:9694–9699

Diez J, Manjon JL, Martin F (2002) Molecular phylogeny of the mycorrhizal desert truffles (*Terfezia* and *Tirmania*), host specificity and edaphic tolerance. Mycologia 94:247–259

El-Enshasy H, Elsayed EA, Aziz R, Wadaan MA (2013) Mushrooms and truffles: historical biofactories for complementary medicine in Africa and in the Middle East. Evid-Based Complement Alternat Med 620451, 10 p. https://doi.org/10.1155/2013/620451

El-Kholy H (1989) Genetical and physiological studies on truffles. Ph.D. Thesis, Cairo University, Egypt

Ferdman Y, Aviram S, Roth-Bejerano N, Trappe JM, Kagan-Zur V (2005) Phylogenetic studies of *Terfezia pfeilii* and *Choiromyces echinulatus* (Pezizales) support new genera for southern African truffles: Kalaharituber and Eremiomyces. Mycol Res 109:237–245

Gandeboenuf D, Dupre C, Henrion B, Martin F, Chevalier G (1994) Characterization and identification of Tuber species using biochemical and molecular criteria. IV European Symposium on Mycorrhizae, p 17

Gouzi H, Leboukh M, Bouchouka E (2013) Antioxidant and antiradical properties of methanolic extracts from algerian wild edible desert truffles (Terfezia and Tirmania, Ascomycetes). Int J Med Mushrooms 15:471–486

Gutierrez A, Garre V, Honrubia M, Torres-Martinez S, Morte A (1995) Caracterizacione identificacion de hongos hipogeos micorricicos coneluso de metodos moleculares. XI Simposio Nacional de Botanica Criptogamica, September, Santiago de Compostela, pp 18–21

Gutierrez A, Moreno G, Morte A, Honrubia M (2004) *Terfezia olbiensis* Ness in south-east of Spain: A different species from *T. leptoderma* Tul. Actes du Premier Symposium sur les Champignons hypoges du Basin Mediterraneen. Rabat, Morocco, p 102

Hajjar SE, Massantini R, Botondi R, Kefalas P, Mencarell F (2010) Influence of high carbon dioxide and low oxygen on the post harvest physiology of fresh truffles. Post Harvest Biol Technol 58:36–41

Hamza A, Zouari N, Zouari S, Jdir H, Zaidi S, Gtari M, Neffati M (2016) Nutraceutical potential, antioxidant and antibacterial activities of Terfezia boudieri Chatin, a wild edible desert truffle from Tunisia arid zone. Arab J Chem 9:383–389

Hashem AR, Al-Obaid AM (1996) Mineral composition of soil and wild desert truffles in Saudi Arabia. J King Saud Univ 8:5–10

Honrubia M, Cano A, Molina-Niñirola C (1992) Hypogeous fungi from Southern Spanish semiarid lands. Persoonia 14:647–653

Honrubia M, Gutierrez A, Morte A (2001) Desert truffles plantations from South–east Spain, 3rd International Workshop on Mycorrhiza, Adelaide

Hussain G, Al-Ruqaie JM (1999) Occurrence, chemical composition and nutritional value of truffles: an overview. Pak J Biol Sci 2:510–514

Janakat SM, Al-Fakhiri SM, Sallal AKJ (2004) A promising peptide antibiotic from *Terfezia claveryi* aqueous extract against *Staphylococcus aureus in vitro*. Phytother Res 18:810–813

Janakat SM, Al-Fakhiri SM, Sallal AKJ (2005) Evaluation of antibacterial activity of aqueous and methanolic extracts of the truffle *Terfezia claveryi* against *Pseudomonas aeruginosa*. J Saud Med 26:952–955

Kagan-Zur V, Kuang J, Tabak S, Taylor FW, Roth-Bejerano N (1999) Potential verification of a host plant for the desert truffle *Terfezia pfeilii* by molecular methods. Mycol Res 103:1270–1274

Kagan-Zur V, Roth-Bejerano N, Sitrit Y, Morte A (2014) Desert truffles phylogeny, physiology, distribution and domestication. Springer, Heidelberg

Khanaqa A (2006) Truffle production in the Kingdom of Saudi Arabia – potential and limitation. J Appl Bot Food Qual 80:14–18

Kovacs GM, Trappe JM (2014) Nomenclatural history and genealogies of desert truffles. In: Kagan-Zur et al (eds) Desert truffles, soil biology. Springer, Berlin, pp 21–37

Kovacs GM, Bagi I, Vagvolgyi C, Kottke I, Oberwinkler F (2002) Studies on the root associations of the truffle *Terfezia terfezioides*. Acta Microbiol Immunol Hung 49:207–213

Kovacs GM, Trappe JM, Al-Sheikh AM, Boka K, Elliott TF (2008) Imaia, a new truffle genus to accommodate *Terfezia gigantea*. Mycologia 100:930–939

Kovacs GM, Martyn MP, Calonge FD (2009) First record of *Mattirolomyces terfezioides* from the Iberian Peninsula: its southern– and westernmost locality. Mycotaxon 110:235–330

Kovacs GM, Balazs TK, Calonge FD, Martin MP (2011) The diversity of *Terfezia* desert truffles: new species and a highly variable species complex with intrasporocarpic nrDNA ITS heterogeneity. Mycologia 103:841–853

Le Tacon F, Marçais B, Courvoisier M, Murat C, Montpied P, Becker M (2014) Climatic variations explain annual fluctuations in French Périgord black truffle wholesale markets but do not explain the decrease in black truffle production over the last 48 years. Mycorrhiza 24:115–125

Lopez-Nicolas JM, Perez-Gilabert M, Garcia-Carmona F, Lozano-Carrillo MC, Morte A (2013) Mycelium growth stimulation of the desert truffle *Terfezia claveryi* chatin by β-cyclodextrin. Biotechnol Prog 29:1558–1564

Mandeel QA, Al-Laith AAA (2007) Ethnomycological aspects of the desert truffle among native Bahraini and non-Bahraini peoples of the Kingdom of Bahrain. J Ethnopharmacol 110:118–129

Mexal J, Reid CPP (1973) The growth of selected mycorrhizal fungi in response to induced water stress. Can J Bot 51:1579–1588

Montecchi A, Sarasini M (2000) Funghi Ipogei d Europa. Associazione Micologica Bresadola, Fondazione Centro

Morte A, Honrubia M (1995) Improvement of mycorrhizal synthesis between micro propagated *Helianthemum almeriense* plantlets with Terfezia claveryi (desert truffle). In: Elliot TJ (ed) Science and cultivation of edible fungi. BaIkerna, Rotterdarn, pp 863–868

Morte A, Honrubia M (1997) Micropropagation of *Helianthemum almeriense*. In: YPS B (ed) Biotechnology in agriculture and forestry: high-tech and micropropagation. Springer, Berlin, pp 163–177

Morte A, Cano A, Honrubia M, Torres P (1994) *In vitro* mycorrhization of micro propagated *Helianthemum almeriense* plantlets with *Terfezia claveryi*. Agric Sci Finland 3:309–314

Morte A, Lovisolo C, Schubert A (2000) Effect of drought stress on growth and water relations of the mycorrhizal associations *Helianthemum almeriense-Terfezia claveryi*. Mycorrhiza 10:115–119

Morte A, Gutierrez A, Honrubia M (2008) Biotechnology and cultivation of desert truffles. In: Varma A (ed) Mycorrhiza: biology, genetics, novel endophytes and biotechnology, 3rd edn. Springer, Berlin, pp 467–483

Morte A, Zamora M, Gutierrez A, Honrubia M (2009) Desert truffle cultivation in semiarid Mediterranean areas. In: Gianinazzi-Pearson V, Azcón C (eds) Mycorrhizas: functional processes and ecological impact. Springer, Heidelberg, pp 221–233

Morte A, Navarro-Ródenas A, Nicolás E (2010) Physiological parameters of desert truffle mycorrhizal *Helianthemum almeriense* plants cultivated in orchards under water deficit conditions. Symbiosis 52:133–139

Murcia MA, Martinez-Tome M, Jimenez AM, Vera AM, Honrubia M, Parras P (2002) Antioxidant activity of edible fungi (truffles and mushrooms): losses during industrial processing. J Food Prot 65:1614–1622

Murcia MA, Martinez-Tome M, Vera A, Morte A, Gutierrez A, Honrubia M, Jimenez AM (2003) Effect of industrial processing on desert truffles *Terfezia claveryi* and *Picoa juniperi* Vitt: proximate composition and fatty acids. J Sci Food Agric 83:535–541

Navarro-Rodenas A, Morte A, Perez-Gilabert A (2009) Partial purification, characterisation and histochemical localisation of alkaline phosphatase from ascocarps of the edible desert truffle *Terfezia claveryi* Chatin. Plant Biol 11:678–685

Navarro-Rodenas A, Lozano-Carrillo MC, Perez-Gilabert A, Morte A (2011) Effect of water stress on in vitro mycelium cultures of two mycorrhizal desert truffles. Mycorrhiza 21:247–253

Nazzaro F, Fratianni F, Picariello G, Coppola R, Reale A, Luccia AD (2007) Evaluation of gamma rays influence on some biochemical and microbiological aspects in black truffles. Food Chem 103:344–354

Owaid MN (2016) Biodiversity and bioecology of Iraqi desert truffles (Pezizaceae) during season 2014. J Arid Land Agric 2:22–25. https://doi.org/10.19071/jaa.2016.v2.3046

Paul NC, Kim WK, Woo SK, Park MS, Yu SH (2006) Diversity of endophytic fungi associated with *Taraxacum coreanum* and their antifungal activity. Mycobiology 34:185–190

Pervez-Gilabert M, Morte A, Honrubia M, Garcia-Carmona F (2001) Partial purification, characterization and histochemical localization of fully latent desert truffle (*Terfezia claveryi* Chatin) polyphenol oxidase. J Agric Food Chem 49:1922–1927

Saltarelli R, Ceccaroli PP, Cesari P, Barbieri E, Stocchi V (2008) Effect of storage on biochemical and microbiological parameters of edible truffle species. Food Chem 109:8–16

Sawaya WN, Al-Shalhat A, Al-Sogair A, Mohammad M (1985) Chemical composition and nutritive value of truffles of Saudi Arabia. J Food Sci 50:450–453

Schillaci D, Cusimano MG, Cascioferro SM, Di Stefano V, Arizza V, Chiaramonte M, Inguglia L, Bawadekji A, Davino S, Gargano ML, Venturella G (2017) Antibacterial activity of desert truffles from Saudi Arabia against *Staphylococcus aureus* and *Pseudomonas aeruginosa*. Int J Med Mushrooms 19:121–125

Sitrit Y, Roth-Bejerano N, Kagan-Zur V, Turgeman T (2014) Pre-symbiotic interactions between the desert truffle Terfezia boudieri and its host plant *Helianthemum sessiliflorum*. In: Zur et al (eds) Desert truffles, soil biology. Springer, Berlin, pp 81–92. https://doi.org/10.1007/978-3-642-40096-4_6

Slama A, Neffati M, Boudabous A (2009) Biochemical composition of desert truffles *Terfezia boudieri* Chatin. In: ISHS Acta Horticulturae: international symposium on medicinal and aromatic plants-SIPAM, p 853

Splivallo R (2008) Biological significance of truffle secondary metabolites. In: Karlovsky P (ed) Secondary metabolites in soil ecology, of soil biology, vol 14. Springer, Berlin, pp 141–165

Trappe JM (1979) The orders, families, and genera of hypogeous *Ascomycotina* (truffles and their relatives). Mycotaxon 9:297–340

Trappe JM, Sundberg WJ (1977) *Terfezia gigantea* (Tuberales) in North America. Mycologia 69:433–437

Trappe JM, Kovacs GM, Claridge AW (2010) Comparative taxonomy of desert truffles of the Australian Outback and African Kalahari. Mycol Prog 9:131–143

Vahdani M, Rastegar S, Rahimzadeh M, Ahmadi M, Karmostaji A (2017) Physicochemical characteristics, phenolic profile, mineral and carbohydrate contents of two truffle species. J Agric Sci Technol 19:1091–1101

Villares A, García-Lafuente A, Guillamón E, Ramos A (2012) Identification and quantification of ergosterol and phenolic compounds occurring in *Tuber* spp. truffles. J Food Compos Anal 26:177–182

Wang S, Marcone MF (2011) The biochemistry and biological properties of the world's most expensive underground edible mushroom: truffles. Food Res Int 44:2567–2581

Zamora M, Morte A, Gutiérrez A, Honrubia M (2006) *Helianthemum violaceum* Pers., a new hostplant for mycorrhizal desert truffle plant production. 5th Inter Conf Mycorrhiza, Granada, Spain, pp 223

Zaretsky M, Kagan-Zur V, Mills D, Roth-bejerano N (2006) Analysis of mycorrhizal associations formed by *Cistus incanus* transformed root clones with *Terfezia boudieri* isolates. Plant Cell Rep 25:62–70

Zitouni-Haouar FH, Fortas Z, Chevalier G (2014) Morphological characterization of mycorrhizae formed between three *Terfezia* species (desert truffles) and several *Cistaceae* and Aleppo pine. Mycorrhiza 24:397–403

Chapter 14
Maintenance, Conservation and Regulation of Microbial Resources for Defense Applications

Pritam Sangwan, Kirty Sharma, Mary Celin, and Bala Kiran

Abstract Microbes have a comprehensive assortment of utility in various fields and industries. This chapter deals with the maintenance, conservation, and regulation of microbial resources for defense applications. In the defense sector, microbes have been utilized in microbial warfare, as biosensors and in bioremediations of energetic materials. They have been also used as a source of food and medicines and to produce biofuel or propellants for missiles, rockets, jets, etc. Proper handling and conservation of microbial resources are required to maintain the balance of nature and the ecosystem and to avoid the loss of biodiversity. There are many organizations in India and abroad which keep a collection of these microbes. Since these microbes need to be handled carefully, there are many rules and regulations on handling, distribution, transport, and packaging set by different countries. Technical research has been carried out for employment of microorganisms in real-time monitoring of soldier's health, and new advance technologies are under progress exploring their vast potential for use in defense applications.

P. Sangwan · M. Celin
Centre for Fire, Explosive and Environment Safety, Defence Research and Development Organization, New Delhi, India

K. Sharma
Centre for Fire, Explosive and Environment Safety, Defence Research and Development Organization, New Delhi, India

Department of Environmental Science and Engineering, Guru Jambheshwar University of Science and Technology, Hisar, Haryana, India

B. Kiran (✉)
Discipline of Biosciences and Biomedical Engineering, Indian Institute of Technology Indore, Simrol, Madhya Pradesh, India
e-mail: kiranb@iiti.ac.in

S. K. Sharma, A. Varma (eds.), *Microbial Resource Conservation*, Soil Biology 54,
https://doi.org/10.1007/978-3-319-96971-8_14

14.1 Introduction

Microbes play a significant role in sustenance and working of various ecosystems. Due to their huge abundance and diversity, microbial resources have been utilized in many fields like healthcare, biotechnology, agriculture, and food technology including defense and offer unexplored solutions (Stackebrandt et al. 2015). Microbial resource conservation is a collection and preservation of different microbial species for their utilization in varied fields.

Microorganisms have been employed as explosive degrading agents to remediate explosive contamination from polluted sites (Sangwan et al. 2015). The manufacturing and explosive usage for military operations have led to the contamination of water and soil resources which is a major environmental concern. These contaminants may likely to enter in the environment through the formation of leachates at waste disposal sites and from the effluents coming out from the manufacturing plants (Adamia et al. 2006). The water, soils, and sediments at many army industrial sites have been contaminated by production and handling of explosives and propellants (Boopathy 2014). The US army has approximated that there is 1.2 million tons of soil polluted with explosives (Hampton and Sisk 1997). More than 1000 sites with explosive contamination have been identified by the US Department of Defense (DoD) (Walsh et al. 1993). There are many treatment technologies for remediation of nitro-organic compounds. Bioremediation is a technique which utilizes biological organisms to aid in the removal of hazardous substances from contaminated sites (Head et al. 2006). Most of the times, microorganisms naturally exist at contaminated sites and have fostered the capacity of deteriorating the contaminants. Sometimes due to lack of environmental conditions, competent microbes required for remediation of the contaminants are absent. Therefore stimulated bioremediation is required to increase the growth of microorganisms. Microbial remediation along with other methods has been recognized as a successful remediation technology with increasing numbers of remediated sites. Microorganisms have also been used as biological weapons (Purver 1995; Stuart 2005). Microbial resources are also used in making biosensors which can detect the presence of explosives and buried landmines (Shemer et al. 2017). New applications of microbes have also been worked out for new medicines and in organ regeneration technology for injured soldiers. Research on biosensors to monitor the health and body response of soldiers is also going on using microbial resources. Microbes can also be used in making missile propellants which can reduce the environmental pollution caused by conventional propellants. Conservationists had understood very well the value of microbes in biogeochemical processes and industries (Hawksworth 1993; Colwell 1997). Their further applications comprise of waste detoxification and bioremediation. They have a considerable role in plant and animal health and in soil fertility. They are used in medical diagnostics for testing performance of biocides, drugs, and disinfectants or as reference strains. Still there is hidden enormous number of microbes which require exploring, identifying, conserving, and utilizing for the welfare of mankind and the environment. In many countries around the world, microbial culture collections are established for various purposes. Isolation and conservation of microorganisms may have a variety of interests. Therefore they must be conserved so that they can be available to researchers for research work and for teaching purpose. This chapter focuses on the conservation of microbial resources for defense applications and their need for maintenance due to a vast scope of appliance in military-related areas.

14.2 Microbial Resources for Defense Applications

14.2.1 Microbial Remediation of Explosive/Ammunition-Contaminated Defense Sites

Energetic materials and their by-products generated during military exercises, their erroneous management, and improper disposal methods have resulted in the contamination of environment which has endangered the living beings and ecosystems (Pichtel 2012). High explosives-contaminated soil and wastewater are one of the major environmental concerns. 2,4,6-Trinitrotoluene, royal demolition explosive or hexahydro-1,3,5-trinitro-1,3,5-triazine, high melting explosive or octahydro-1,3,5,7-tetranitro-1,3,5,7-tetrazocine, and dynamite are some examples of high explosives. TNT and RDX have been identified as carcinogen. Abnormal functioning of the liver and anemia in humans can be induced by TNT, and RDX can induce seizures if considerable quantities are inhaled or consumed. Health issues caused by contamination of HMX are less known (www.atsdr.cdc.gov). Pichtel (2012) has reported that iodine uptake by the thyroid gland is affected by perchlorate inhalation in humans. Microorganisms play an important role in the decontamination of environmental pollutants and reclamation of polluted ecosystems. They can be utilized for the degradation of recalcitrant compounds used for defense purposes like explosives, propellants, etc. The utilization of biological organisms in the removal of hazardous substances from contaminated sites is called bioremediation (Head et al. 2006). Explosives like TNT, RDX, and HMX have been tried for degradation by an extensive diversity of microorganisms in aerobic or anaerobic settings using TNT as a carbon and/or nitrogen source. Microbial degradation of TNT has been observed by many researchers (Anuradha et al. 2015; Mulla et al. 2014) and found that native flora is more efficient for the degradation of TNT (Rahal and Lobna 2011; Gumuscu and Tekinay 2013; Boopathy 2014). In India, research has been started and subsequently advanced using microbes for reclamation of contaminated defense sites using bioremediation technology. Sangwan et al. (2015) evaluated bioremediation of 2,4,6-trinitrotoluene (TNT) using *Acinetobacter nosocomialis* isolated from actual polluted sites in aqueous phase using free and immobilized cultures. Under aerobic conditions, RDX mineralization using bacteria has been reported by utilizing the compound as a nitrogen source. TNT and RDX degradation by fungi has also been carried out by many researchers (Kalderis et al. 2011). Tope et al. (1999)

reported the biotransformation of TNT using immobilized *Arthrobacter* sp. The Bhabha Atomic Research Centre (BARC) facility Kalpakkam, India, has developed technology for the removal of strategic wastes using microorganisms. Mixed-culture microbial granules were developed by Nancharaiah et al. (2006) in sequencing batch reactor (SBR) technology and efficiently degraded recalcitrant synthetic chelating agents having potential application in heavy metals or radionuclide disposal. Nancharaiah et al. (2006) developed aerobic self-immobilizing microbial consortia and studied the potential for biosorption of uranium from aqueous solutions. It was found that aerobic granular biomass can be used as an efficient biosorbent substance for removing and recovering uranium or other radionuclides from dilute nuclear wastes. The Longhorn Army Ammunition Plant (LHAAP), Karnack, Texas, used fluidized bed reactor (FBR) with the objective of removing perchlorate from contaminated groundwater being discharged by treatment plant (AFCEE/ERT Fact Sheet August 2002). Microbes have evolved novel and diverse pathways for the mineralization of explosives and other compounds and can prove as a potential biodegraders in this field. Different microbes are suitable to degrade different contaminants depending upon the nature and concentration of explosive contaminant and the metabolic needs of the microbes (Sangwan and Celin 2016a). Some of the contaminants related to defense and microorganisms used for their remediation are given in Table 14.1.

14.2.2 Bioreporters/Biosensors

Microbial biosensors are being used for the detection of trace explosives and monitoring other contaminants. Some bacterial strains used for detecting trace explosives are given in Table 14.2. Shemer et al. (2017) analyzed the new development in design and building of microbial sensor. Such bioreporter strains may be used for monitoring environmental pollution of explosive contaminants and can also be used for buried landmines in remote areas. These molecularly engineered strains in the presence of minute quantities of explosives are used to produce dose-dependent signal. Altamirano et al. (2004) developed a microalgal biosensor for the detection of TNT by using *Dictyosphaerium chlorelloides* (Chlorophyceae). A genetically engineered yeast strain *Saccharomyces cerevisiae* was used to detect 2,4-DNT (2,4-dinitrotoluene) by Radhika et al. (2007). Miniscule biosensors for monitoring the health of soldiers are being designed by the Defense Advanced Research Project Agency (DARPA; which would be implanted on each soldier for constantly monitoring different health parameters). These sensors could also be helpful to measure the effect of adverse environmental conditions, stress, or any changes in food habits on soldier body functions. Understanding of biochemical and hormonal changes in the soldier body, like cortisol and histamine levels, could help the military to prepare and modify individual's diet chart and sleep patterns for best performance. Although these types of biosensors are not at the advance stage of development, research is being carried out in this direction. In the near future, real-time information to doctors may be provided by these sensors

Table 14.1 Degradation of contaminants produced through military activities using microbes

S. no.	Contaminants	Microorganisms	References
1	TNT	*Enterobacter cloacae PB2* *Pseudomonas savastanoi* *Stenotrophomonas maltophilia* *Irpex lacteus* *Escherichia coli* *Acinetobacter noscomialis*	French et al. (1998) Martin et al. (1997) Oh and Kim (1998) Kim and Song (2000) Yin et al. (2005) Sangwan and Celin (2016a)
2	RDX	*Shewanella sediminis* *Klebsiella pneumonia* *Clostridium bifermentans* *Rhodococcus rhodochrous* *Acetobacterium paludosum* *Enterobacter cloacae ATCC 43560* *Acetobacterium malicum* *Gordonia and Williamsia* spp.	Zhao et al. (2005) Zhao et al. (2002) Zhao et al. (2003) Smith et al. (2002) Sherburne et al. 2005 Pudge et al. (2003) Adrian and Arnett (2004) Thompson et al. (2005)
3	HMX	*Phanerochaete chrysosporium*	Fournier et al. (2004)
4	Uranium	*Clostridium* sp.	Francis et al. (1994)
5	Nitroglycerine (NG)	*Penicillium corylophilum Dierckx* *Bacillus thuringiensis* *Bacillus cereus*	Marshall and White (2001) Meng et al. (1995)
6	Perchlorate	Strain GR-1 (DSM 11199) GR-1 *Proteobacteria*	Kengen et al. (1999) Rikken et al. (1996)
7	Unsymmetrical dimethyl-hydrazine (UDMH)	*Stenotrophomonas* sp. M12 (M12)	Liao et al. (2016)

helping in better care and diagnosis for military as well as civilian population. The Defence Research & Development Establishment (DRDE), Defence Research and Development Organisation (DRDO), has also developed sensors, detection systems, and kits for biological and chemical agents including their subsystems and materials for defense applications (www.drdo.gov.in).

Table 14.2 Microbial strains used to detect trace explosives

S. no.	Organisms	Reporting element	Target analytes	References
1	*Escherichia coli*	Green Fluorescent Protein (GFP)	TNT, DNT, DNB	Looger et al. (2003), Lönneborg et al. (2012), Davidson et al. (2012), Tan et al. (2015)
2	*Dictyosphaerium chlorelloides*	Intrinsic chlorophyll A fluorescence	TNT	Altamirano et al. (2004)
3	*Saccharomyces cerevisiae*	Green Fluorescent Protein (GFP)	DNT	Radhika et al. (2007)
4	*Pseudomonas putida*	lux AB, GFP (Green Fluorescent Protein)	DNT	Garmendia et al. (2008)

14.2.3 Microbial Warfare

In the scenario of numerous international wars and advancements in microbiology, biological weapons were notably developed. The 1925 Geneva Protocol was drafted after the World War I for the prevention of the usage of biological weapons and poisonous and asphyxiating gases due to the increased threat of the biological and chemical weapons. Many countries were a signatory in this protocol with the clause that it did not restrict research and development in this field (www.state.gov). The refining and advancement programs for biological weapons were started by countries like the USA, Japan, USSR, and Germany (Moon 1991; Christopher et al. 1997). The Convention on the Prohibition of the Development, Production, and Stockpiling of Bacteriological and Toxin Weapons and their Destruction was held in Geneva in 1972 and signed by many countries (www.state.gov). In the USA, most of the research and development related to biothreat agents were of a defensive nature, but many countries like Iraq and the Union of Soviet Socialist Republics carried on offensive research and development also. Some regional terrorist and radical groups are also working independently on obnoxious implementation of biological weapons (Lim et al. 2005).

A quick and accurate diagnosis of biological agents involved and appropriate steps to be taken during bioterrorist attacks should be ensured to protect the public. So there is an urgent need to analyze and detect the threat of biological weapons in the current scenario of bioterrorism. Laboratory Response Network is a systematic system in the USA prepared by the Centers for Disease Control and Prevention (CDC) to give instructions for the identification and reporting of bioagents (www.bt.cdc.gov) in which carrying many tests is fast, precise, and reliable, and few tests have been assessed under field trials.

14.2.4 Source of Nutrition

Microbial resources can be used to develop the source of nutrition for the army working in adverse climatic conditions. The Defence Food Research Laboratory (DFRL), Defence Research and Development Organisation (DRDO), is working for food availability in remote areas and many other food-related problems faced by the Indian Army, Navy, Air Force, and other paramilitary forces. DRDO, India, has successfully tried to make a tomato variety, Pusa Ruby, more tolerant to cold stress with the help of osmotin gene introgressed through *Agrobacterium*-mediated genetic transformation for fulfilling the food demands of soldiers at high altitude (www.drdo.gov). At the time of the World War I, Germany used single-cell protein technology to substitute yeast protein sources. The term single-cell protein (SCP) was coined in the 1960s for fermented microbial cultured products, and using this technology protein scarcity problem can be solved worldwide. In this technology, low-value by-products, like wastes, are biotransformed into enhanced nutritional and market value products (Ugaldea and Castrillo 2002).

A method called "Zulaufverfahren," in which sugar solution was put into an aerated suspension of yeast, was developed in Denmark and Germany in 1919 (Ezejiofor et al. 2014). By the beginning of the World War II, yeasts had been made an integral part of food in army and after that added as civilian food (Ugaldea and Castrillo 2002).

14.2.5 Regenerative Medicine Technology

Microbial resources can be used to develop new medicines and gene source for organ regeneration technology. This technology can be used to treat soldiers who are wounded or have lost their organs during warfare. Biotechnological advancement has made it possible to potentially use adult stem cells in producing regenerative medicine for both military and civilians (Christopherson and Nesti 2011).

Regenerative medicine technology for wounded US military personnel has been developed by The Armed Forces Institute of Regenerative Medicine (AFRIM) which includes a network of many leading universities and hospitals doing efforts in this direction (www.army-technology.com). Due to latest developments in biotechnology and genetic engineering, the desired genes are modified and incorporated with the help of microbes and enzymes secreted by them. For example, bacterial nanocellulose is produced by a bacteria *Gluconacetobacter xylinus* and is a natural polymer which can be used in restoration of damaged tissues, and after some specific modifications, the same can be used to make cartilage-like material which can help in the restoration of damaged peripheral nerves (Ludwicka et al. 2013).

14.2.6 Scientific Devices

A type of cellulose is produced from some bacteria which is called microbial cellulose. It is a natural substitute for a large number of medical and tissue-engineered applications because of its unique nanostructure and properties. This biomaterial has wide applications in the field of biomedical scientific devices, electronics, products for injuries and the regeneration of damaged or infected organs, acoustics, and other related scientific R&D programs. If it can be successfully produced in sufficient quantity, it has the potential to become a promising tool for the production of numerous medical devices and consumer products. Membranes developed from this type of cellulose have been successfully utilized for injured skin healing and blood vessel replacement. It can also be used in periodontal treatments, guided tissue regeneration (GTR), or as a substitute for dura mater (Czaja et al. 2007). Microbial cellulose is generated by acetic acid bacteria in synthetic as well as non-synthetic medium (Esa et al. 2014). One of the most investigated and the most efficient microbial cellulose producers is *Acetobacter xylinum* (El-Saied et al. 2004).

14.2.7 Chemical Propellants for Missile

Microbial resources can be used to produce chemical propellants for missiles. The genetically modified bacteria like *Escherichia coli* and *Pseudomonas* could be used as mini-factories for producing enough chemicals like butanetriol to make propellants for weapons like missiles (www.genomenewsnetwork.org). A project financed by the Office of Naval Research is underway at the Michigan State University, and research and development is going on for utilizing microbes to produce chemical propellants for missiles although these propellants are still at testing stage. The green propellants produced from microbial resources would be much cheaper and cleaner than the current practices of producing propellants saving thousands of dollars and the environment.

14.2.8 Microbial Forensics and Bioterrorism

A biological weapon is easy to produce and could lead to numerous deaths similar to nuclear weapons and is a big terrorism threat (Purver 1995). Many researchers have estimated the probability of the serious bioterrorist attacks in the last few decades (Wein et al. 2003; WHO 1970; Office of Technology Assessment, USA 1993). There are examples of such type of attacks like when sarin gas was released in a city and Tokyo subway in Japan in the 1990s (Stuart 2005; Kaplan and Marshall 1996). A widespread fear, number of deaths, and hospitalizations were caused by the attack of anthrax spores intentionally sent hidden in letters in 2001 in the USA (Jernigan et al. 2001; Olson and Shinrikyo 1999). These types of attacks can cause serious health concerns and can affect the food supply, global economy, and the environment (Pattnaik and Jana 2005).

Advanced molecular techniques and practices like whole-genome sequencing, microarray analysis, molecular phylogeny, and DNA fingerprinting have been proven to be very useful in the field of microbial forensics. It can help in determining the origin, type, and health effects in case of bioterrorism and the accidental release of any obnoxious microbes (Pattnaik and Jana 2005). Scientific information related to microbial genetics, ecology, physiology, and evolution is also required for better results in this field.

There are many success stories of microbial forensics in defense, and investigations have resulted in several high-profile findings. It has proved the link of microbes with the crime. Anthrax spores were recovered by an Arizona researcher from an ampoule of slime scratched 10 years ago from walls of Tokyo headquarters of a Japanese doomsday cult and investigated the particular strain held by the terrorists (Pattnaik and Jana 2005). A weapon attack was suspected during the unexpected outbreak of West Nile virus in 1999 in the Northern USA, and it was very difficult to find out the cause. Later in the investigation, it was found a natural outbreak due to a virus from dead goose (Jia et al. 1999; Lanciotti et al. 1999). The case of the Florida mail anthrax attack was compared with a portion isolate of the Ames strain (by CDC in 2001). Bioterrorism-related practice of the microbial forensic also requires the geographic origin details of the isolated and identified microbes further making it more challenging.

14.2.9 Production of Biofuel in the Defense Sector

Biofuels are getting a lot of attention globally as mixing components or direct substitute for conventional fuels in automobiles. Biofuels produced from microalgae are environment-friendly, renewable, nontoxic, and biodegradable. Some of the examples of microalgae used in the production of biofuels are *Chlorella* sp., *Arthrospira* sp., *D. salina*, and *H. pluvialis* (Greenwell et al. 2010). Oil can be extracted from microalgae by the process of chemical, mechanical, and enzymatic extraction which is later converted into biofuel by the process of biochemical and thermochemical transesterification or hydrogenation. The main constituents of this oil are oleic (18:1), palmitoleic (16:1), linolenic (18:3), and linoleic (18:2) acids which are a mixture of unsaturated fatty acids. Some saturated fatty acids like stearic (18:0) and palmitic (16:0) are also present in very less amount. Biofuels have been tested for their applications in the defense by some researchers. The defense sector worldwide is exploring a wide range of alternative energy sources in an effort to reduce military's reliance on fossil fuels. Department of Defense (DOD), USA, is the largest consumer of energy and aviation fuel in the world, and it is exploring significant resources to address its dependence on fossil fuels in the near future (Sangwan and Celin 2016b).

The kerosene-based high-performance fuel is less hazardous and flammable and allows more safety and combats survivability. Nearly 4.5 billion gallons per year of JP-8 fuel is consumed by the US Army, US Air Force, and NATO. Along with jets, JP-8 is also utilized as fuel in heaters, tanks, stoves, and other automobiles in military service. Jet A and Jet A-1 are also kerosene based which are used by commercial airliners. UOP, a Honeywell company, is a well-known leader in refining process developing technology and petrochemical industries. It had created a renewable energy and chemical unit for business in 2006 with a purpose to provide commercial solutions for the production of biofuels. According to the news released by the company [DES PLAINES, Ill., June 28, 2007—UOP LLC, a Honeywell (NYSE: HON)], UOP was given a project for military jets in the USA to develop biofuel technology, and the project was funded by the Defense Advanced Research Projects Agency, USA. This project was focused on showing efficient transformation of algal oils and vegetable extracts to jet propellant-8. The goal of the project was to improve and commercialize a process to make jet propellant 8 which has to be used by NATO and US militaries. Defense Advanced Research Projects Agency provided a sum of $6.7 million for the development of this technology. Biofuels have been tested for their applications in defense by some researchers. The defense sector worldwide is exploring a wide range of alternative energy sources in an effort to reduce military's reliance on fossil fuels. Department of Defense (DOD), USA, is the largest consumer of energy and aviation fuel in the world, and it is exploring significant resources to address its dependence on fossil fuels in the near future.

14.3 Microbial Preservation Methods

Microbial preservation methods can be broadly categorized as in situ and ex situ conservation. In the first type of conservation, the microbes are preserved in their native environment. Bacterial diversity of endemic nature can be preserved in the tropical forest soil, desert soil crust, hypersaline areas, glaciers, hot springs, etc., preserving microflora of hindgut of critically endangered animals like Somali wild ass and red kangaroo (Sharma et al. 2016). In the second type, i.e., ex situ conservation, microbes are preserved in laboratories using specialized methods. These microorganisms can be used in varied fields later. There are various organizations involved in the preservation of microbes and cell cultures on both international and national level.

Microorganisms are preserved as per the guidelines of WFCC and Organization for Economic Cooperation and Development (OECD) mainly by techniques like freeze-drying, liquid-drying, or cryopreservation. Microorganisms can also be preserved under mineral oil. Recalcitrant fungi and algae may be preserved by means of subcultivation (Kirsop and Doyle 1991). Appropriate preservation method is chosen depending on past experience or depositor suggestions. Biological Resource Centers (BRCs) preserve diverse biological materials including bacteria, viruses, fungi, and cell lines. The following quality control criteria should be maintained for a particular method:

1. Preserved culture should be authentic.
2. Preserved culture should be genetically stable.
3. High fraction of the preserved culture should be viable.
4. Preserved culture should be contamination-free.

Microbes can be maintained and preserved in both metabolically active and inactive forms in laboratory. Metabolically active microbes can be preserved for shorter duration of time in comparison to metabolically inactive microbe. Metabolically active microbes are generally maintained in agar or mineral oil, whereas metabolically inactive microbes are generally maintained and preserved by cryopreservation or freeze-drying.

For short-term storage of microbial cultures in laboratory, microbes are streaked on agar plates in aseptic conditions and stored at 4 °C by keeping agar side up to reduce the probability of contamination after sealing the culture dish with paraffin. This method of storage is suitable for daily use of microbial cultures and also keeps the culture and agar hydrated.

Some bacterial strains can also be successfully stored at 4 °C in agar stab cultures for up to 1 year. Stab cultures are prepared by transferring the sterilized strain-compatible agar into the screw-cap vials using the proper aseptic techniques. After the solidification of agar, the colonies of bacteria are picked up with the help of sterile straight wire and plunged deep into the soft agar several times. These inoculated vials are incubated at 37 °C for 8–12 h with slightly loose caps. After incubation the vials are sealed tightly and stored in the dark at 4 °C (www.thermofisher.com).

Cryopreservation is the rapidly freezing of microbes in liquid nitrogen in the presence of stabilizing agent, such as glycerol, which prevents the formation of ice crystals and promotes survival of the cells. Freeze-drying and cryopreservation are the most commonly employed methods for conservation of microorganisms (Day and Stacey 2007).

A. Cryopreservation can be achieved by the following freezing and low-temperature storage methods:

    (a) In or above liquid nitrogen storage
    (b) Below −70 °C storage

B. Drying can be done by the following preservation methods:

    (a) Liquid-drying (L-drying)
    (b) Shelf freeze-drying
    (c) Vacuum drying
    (d) Spin freeze-drying

Details of these methods can be found on Common Access to Biological Resources and Information website (www.cabri.org). Microorganisms in sealed ampoules can be simply transported and stored at normal temperature after freeze-drying (Morgan et al. 2006). BRCs guide their users on the prescribed procedure for revival of microorganisms after drying (Janssens et al. 2010). A series of processes are included in freeze-drying in which metabolism is brought down to zero without affecting viability. Well-documented preservation techniques are used at BRCs to ensure reproducibility, and replicate collection is maintained at a different place to escape accidental loss. Freeze-drying is also known as lyophilization in which culture is rapidly frozen at a very low temperature (−70 °C) and then dehydrated by vacuum. In this way microbes become dormant and remain viable for years. Classification of laboratories can be done into different biosafety levels on the basis of microbial risk group. Risk Group 1 laboratories work on well-identified nonpathogenic microbes and are called Biosafety Level 1 laboratory, whereas Biosafety Level 2 laboratory works on moderate hazardous microbes and Biosafety Level 3 includes laboratories working on potentially hazardous and pathogenic microbes for diagnostic and clinical production and research purpose. BSL-1 and BSL-2 have autoclaves and biosafety cabinets. BSL-3 laboratories have specifically trained personnel working in controlled conditions and are well equipped with all scientific instruments required for conservation and maintenance of microbial resources. These laboratories follow strict protocols for disposal of microbial waste and have HEPA (high-efficiency particulate air)-filtered exhaust having unidirectional airflow created by room pressure gradients of negative pressure. BSL-4 laboratories handle microbes of Risk Group 4 (highly dangerous and lethal microbes) and are equipped with positive pressure BSL-4 suits and advanced technologies for the safety of workers and to avoid any other risk associated with these microbes (Mourya et al. 2014).

14.4 Microbial Culture Organization for Defense

For the sustainable use of microbial populations in defense, culture collections are required for their conservation. Culture collections serve as the reliable microbial resources for research and teaching in the form of reagents for quality control and reference strains and have an important role in the conservation and sustainable utilization (Smith et al. 2002; Daniel and Prasad 2010).

In the USA, the US Army Medical Research Institute of Infectious Diseases (USAMRIID) under the Department of Defense (DoD) is the primary institution and facility for defensive research against bioterrorism and biological warfare. Highly hazardous viruses at Biosafety Level 4 are also investigated at USAMRIID within positive pressure personnel suits. It also conducts research to protect military personnel from biological threats and to provide medical solution in the form of information, drugs, vaccines, and diagnostics.

Similarly in India Defence Research and Development Establishment (DRDE), an establishment of DRDO, works for the defense against hazardous chemical and biological agents as well as associated toxicological problems. This laboratory also works in the fields of toxicology, microbiology, biotechnology, virology, etc. for defense purposes (www.drdo.gov.in). According to Arora et al. (2005), Defence Material and Stores Research and Development Establishment Culture Collection (DMSRDECC), Defence Research and Development Organization, New Delhi has 1100 fungal strain collection preserved for different defense applications.

14.5 Rules and Regulations Related to Microbial Resources

14.5.1 Laws and Regulation for Biosafety and Biosecurity

On matters of biosafety and biosecurity, regional, local, and global exchanges of biological materials are controlled for their access and use by numerous national and international rules and regulations (Bussas 2017). The Convention on Biological Diversity has been in action and had 191 countries as members including India and was adopted during the Earth Summit in Rio de Janeiro in 1992. It states the sovereign right of the country on the biological resource which is originated from the country, and it has made it difficult to share biological materials for international collaboration. All signatory countries of CBD have framed their own legislations and national agenda to reaffirm the conservation of their national microbial resources in view of CBD.

Initially the World Federation for Culture Collections (WFCC 1990) published guidelines on all aspects of culture collection which were accepted internationally. New guidelines on culture collection operation are provided by OECD (Organization for Economic Cooperation and Development) which were approved by OECD member nations in 2007 (www.oecd.org). The Bonn Guidelines (2002) stress on "Access to Genetic Resources and the Fair and Equitable Sharing of Benefits Arising from their Utilization to the CBD" (Convention on Biological Diversity 2002; www.cbd.int). In this regard, a code of conduct is provided by Microorganisms Sustainable Use and Access Regulation International Code of Conduct (MOSAICC), by the help of the Directorate General XII for Science, Research and Development of the European Commission (Janssens et al. 2010).

The Nagoya Protocol which was adopted on October 29, 2010 in Nagoya, Japan, on Access to Genetic Resources and the Fair and Equitable Sharing of Benefits Arising from their Utilization to the CBD made a clear legal framework for providers and users of microbial resources (www.cbd.int).

14.5.2 Handling and Distribution of Microbial Resources

At the local, national, and international levels, extensive legislation has been made regarding the safe handling and distribution of biological materials. Microbes are classified into various risk groups depending on the levels of hazards caused by them, and different guidelines are there for handling each category according to the risk involved. As per WHO, Risk Group 1 has the least harmful microorganisms like *Lactobacillus helveticus*, *Bacillus subtilis*, *Aspergillus oryzae*, etc., and Risk Group 4 has the most hazardous microbes causing severe diseases to common population, like herpes B virus, poxvirus, Lassa fever virus, Ebola virus, etc. The World Health Organization's *Laboratory Safety Manual* (WHO 2004) provides very good compilation of the relationship between the biosafety levels and risk groups. The manual also provides practical microbiological techniques and procedures that when applied in laboratory will prevent the contamination of the laboratory workers, environment, and the experiment setup.

14.5.3 Packaging of Microbial Resources

Three-layer packaging system is adopted for the safe packaging of microbial material which includes an outer packaging, watertight secondary packaging, and an innermost watertight primary container. An absorbent material should be kept between primary and secondary packaging. Outer packaging has all the labels and documentation (WHO 2015). Hazardous biological substances are regulated by dangerous goods transport regulations and nonhazardous (Risk Group 1) by postal regulations and requirements as set by the Universal Postal Union (UPU) (International Bureau of the Universal Postal Union 2013) (www.upu.int). Fast and undamaged reach of packets should be ensured to their destination. Risk Group 2 and upper biological material fall within "UN Model Regulations on the Transport of Dangerous Goods" affecting all modes of transport (United Nations 2015).

14.5.4 Transport of Microorganisms

Microbes of interest can be sent if needed to other laboratories for further research operations. Therefore there is a requirement for safe transportation of these microbes. Regulations pertaining transport have been formulated to ensure that microorganism will reach its terminus in the original status maintaining the viability and uncontaminated. Proper packaging, classification, labeling, and marking of the material should be ensured before transportation. The dispatch of a living microorganism should be legitimate and ensured not to go into wrong hands. As a common rule, biological material should not be handed over to private/untrained persons and should be received in well-equipped laboratories by trained staff. Another very crucial restriction is that before every shipment of a microorganism, the addressee has to be verified against applicable sanction lists following antiterrorism laws (e.g., European Union EEC/881/2002) (European Parliament 2016).

14.5.4.1 Export

Biological material export is generally regulated by national laws. International or local legislation also occurs targeting at the harmonization on the lists of microorganisms affected. Biological materials classified in Risk Group 1 are accepted for export without limitation with few exceptions. United Nations sanctions may be in there and have to be taken cared. The exports of materials with a dual use purpose for civil or military and biological materials classified as biological weapons require more restrictions regarding laws and directives and might be regulated by a signed end-user certificate if necessary. The EU list of dual-use goods (European Community 2000) has restrictions on certain crop plants, genetically modified microorganisms, and pathogenic species. Transport of biological substance in other nations might be subject to customs inspection and must carry the green customs declaration CN22. Shipping of infectious substances by airmail is regulated by the International Air Transport Association and Dangerous Goods Regulations (IATA 2017).

14.5.4.2 Import

The European Community, 2000 (European Community Commission Directive 2000/29/EC), limits import of biological materials harmful to organisms and the environment. Similar laws exist in other countries also. The German Federal Law (2012) (German Plant Protection Act), Germany, regulates *Synchytrium endobioticum*, the agent responsible for potato cancer from import. Similarly, the German Federal Law (German Infectious Diseases of Animals Imports Enactment 1982, 2015) restricts the import of *Chlamydia psittaci* causative agent for lymphatic vessel inflammation of one-hoofed animals. The restrictions on import of biological material are generally based on the climatic or ecological conditions of a particular country and subjected to regulations of recipient country. Some countries like Brazil, New Zealand, Australia, and Canada need an import allowance for transport of microorganisms.

14.5.4.3 Shipment

A system called "Orange Book" for the classification of all types of dangerous materials has been settled by United Nations Committee of Experts for the Transport of Dangerous Goods (United Nations 2015). The purpose of this committee was to ensure safe shipping of dangerous materials during transport in all modes. Dangerous goods were divided into nine different classes including one miscellaneous. For shipping procedures, proper shipping name, the UN number, class, hazard label, packing group, and packing instructions are a requisite.

Guidelines for the transport of dangerous materials by air are also established by International Civil Aviation Organization (ICAO) and had been implemented by the

IATA. In transboundary exchange of biological material, air transport is playing a significant role. In this regard, the IATA Infectious Substances Shipping Guidelines (IATA 2017) are particularly concerned with the transport of infectious biological goods. A declaration from the shipper has to be filled for dangerous goods while transporting infectious substances of category A by air which is a legal document and needs to be signed by a well-trained technical person involved in the packing and shipping of the biological material.

14.5.4.4 Transport of Genetically Modified Organisms

The Cartagena Protocol on Biosafety (Secretariat of the Convention on Biological Diversity 2000) has to be followed for the transport of genetically modified organisms. CBD (United Nations 1992) in its framework has an international agreement to ensure the safe handling, use, and transport of living modified organisms (LMOs). Its aim is to make microbial diversity sufficiently protected from any adverse effects of LMOs, and it was implemented on September 2003. It focuses on transboundary movements of LMOs and also takes care of the risks related to human health. Genetically modified organism transportation is also regulated by IATA, DGR.

14.5.4.5 Dry Ice Use

Dry ice or solid carbon dioxide is a hazardous material and used as a refrigerant in the packaging of biological materials. It should be used as per packing guidelines of UN 1845 (www.iata.org).

14.6 Future Prospects and Challenges

Microbial applications to decontaminate the explosives contaminated soil and water are being widely used in countries like the UK and USA. Work has also been initiated in India, and research is going on using microbes for restoration of polluted defense sites. Many researchers have evaluated biodegradation of TNT, RDX, and HMX by native microorganisms isolated from actual contaminated sites in aqueous phase employing free and immobilized cultures. The US Department of Defense is initiating new programs related to microbial research and expanding its horizons aiming protection against possibility of biological warfare and health of American forces. An inventory of microorganisms and their distributions should be prepared which can help in protecting forces against infectious diseases as microorganisms can cause infectious disease and could be used as biological weapons. The one more reason for interest in microbial research is cleanup and restoration of polluted military sites in which bioremediation technology has emerged as an eco-friendly and economical technology for achieving cleanup. Microbial remediation for cleanup and destruction of munitions can save several million dollars and can help in environmental restoration naturally (ASM). The roles of biological resource centers are becoming important in strengthening international network of microbial resources, and it will enhance the use and applications for scientific community and society. These BRCs are using modern tools like bioinformatics for the identification, characterization, and implication of a huge variety of microorganisms.

14.7 Conclusion

Microbial repositories are sources of beneficial metabolic products and genetic information that can be used for various applications. Proper handling and conservation of microbial resources are required to maintain the balance of nature and the ecosystem to avoid the loss of biodiversity. They can be both beneficial and hazardous depending on their application. There are microbial culture collection centers that provide information about identity and characters of the microbes. They also provide live cultures and other preserved forms of microbes and related data to promote research and development across the globe. Some international and national initiatives have also been taken in the form of various treaties which provide guidelines for microbial resource conservation, handling, and transportation. Application of microbes has been extended to defense related area which has resulted in the initiation of research on various microorganisms for their potential utility for military use. Microbes have been used in biological warfare, biodegradation of explosives, and biosensors for detection of explosives. Further research has been carried out for employment of microbes in real-time monitoring of soldier's health, and new advance technologies are under progress exploring their vast potential for use in defense applications.

References

Adamia G, Ghoghoberidze M, Graves D, Khatisashvili G, Kvesitadze G, Lomidze E, Ugrekhelidze D, Zaalishvili G (2006) Adsorption, distribution and transformation of TNT in higher plants. Ecotoxicol Environ Safety 64:136–145

Adrian NR, Arnett CM (2004) Anaerobic biodegradation of hexahydro-1,3,5-trinitro-1,3,5-triazine (RDX) by *Acetobacterium malicum* strain HAAP-1 isolated from a methanogenic mixed culture. Curr Microbiol 48:332–340

Altamirano M, García-Villada L, Agrelo M, Sánchez-Martın L, Martın-Otero L, Flores-Moya A, Rico M, López-Rodas V, Costas E (2004) A novel approach to improve specificity of algal biosensors using wild-type and resistant mutants: an application to detect TNT. Biosens Bioelectron 19:1319–1323

Anuradha, Paul B, Jagdish (2015) Microbial degradation of expired slurry explosives in mines: a review. Int J Environ Stud 72:117–131

Arora DK, Saikia R, Dwievdi R, Smith D (2005) Current status, strategy and future prospects of microbial resource collections. Curr Sci 89:488–495

Boopathy R (2014) Biodegradation of 2,4,6-trinitrotoluene (TNT) under sulfate and nitrate reducing conditions. Biologia 69:1264–1270

Bussas V (2017) Biosafety, transport and related legislation concerning microbial resources—an overview. In: Microbial resources. Academic, London, pp 293–315

Christopher G, Cieslak T, Pavlin J, Eitzen E (1997) Biological warfare: a historical perspective. JAMA 278:412–417

Christopherson GT, Nesti LJ (2011) Stem cell applications in military medicine. Stem Cell Res Ther 2(5):40. https://doi.org/10.1186/scrt81

Colwell RR (1997) Microbial diversity: the importance of exploration and conservation. J Ind Microbiol Biotechnol 18:302–307

Czaja WK, Young DJ, Kawecki M, Brown JRM (2007) The future prospects of microbial cellulose in biomedical applications. Biomacromolecules 8:1–12

Daniel HM, Prasad GS (2010) The role of culture collections as an interface between providers and users: the example of yeasts. Res Microbiol 161:488–496

Davidson ME, Harbaugh SV, Chushak YG, Stone MO, Kelley- Loughnane N (2012) Development of a 2,4-dinitrotoluene-responsive synthetic riboswitch in *E. coli* cells. ACS Chem Biol 8:234–241

Day JG, Stacey GN (2007) Cryopreservation and freeze-drying protocols, 2nd edn. Humana Press, Totowa, pp 1–14

El-Saied H, Basta AH, Gobran RH (2004) Research progress in friendly environmental technology for the production of cellulose products (bacterial cellulose and its application). Polym Plast Technol Eng 43:797–820

Esa F, Tasirin SM, Rahman NA (2014) Overview of bacterial cellulose production and application. Agric Agric Sci Proc 2:113–119

European Community (2000) List of dual-use items and technology. Article 3 of regulation (EC) no 1334/2000

European Parliament (2016) Council regulation (EU) 2016/363 of 14 March 2016, amending regulation (EC) no 881/2002 imposing certain specific restrictive measures directed against certain persons and entities associated with the Al-Qaida network

Ezejiofor TIN, Enebaku UE, Ogueke C (2014) Waste to wealth-value recovery from agrofood processing wastes using biotechnology: a review. Br Biotechnol J 4:418–481

Fournier D, Halasz A, Thiboutot S, Ampleman G, Manno D, Hawari J (2004) Biodegradation of octahydro-1,3,5,7-tetranitro-1,3,5,7-tetrazocine (HMX) by *Phanerochaete chrysosporium*: new insight into the degradation pathway. Environ Sci Technol 38:4130–4133

Francis AJ, Dodge CJ, Lu F, Halada GP, Clayton CR (1994) XPS and XANES studies of uranium reduction by *Clostridium* sp. Environ Sci Technol 28:636–639

French CE, Nicklin S, Bruce NC (1998) Aerobic degradation of 2,4,6-trinitrotoluene by *Enterobacter cloacae* PB2 and by pentaerythritol tetranitrate reductase. Appl Environ Microbiol 64:2864–2868

Garmendia J, De Las Heras A, Galvão TC, De Lorenzo V (2008) Tracing explosives in soil with transcriptional regulators of *Pseudomonas putida* evolved for responding to nitrotoluenes. Microb Biotechnol 1:236–246

Greenwell HC, Laurens LML, Shields RJ, Lovitt RW, Flynn KJ (2010) Placing microalgae on the biofuels priority list: a review of the technological challenges. J R Soc Interface 7:703–726

Gumuscu B, Tekinay T (2013) Effective biodegradation of 2,4,6-trinitrotoluene using a novel bacterial strain isolated from TNT-contaminated soil. Int Biodeterior Biodegrad 85:35–41

Hampton ML, Sisk WE (1997) Environmental stability of windrow composting of explosives-contaminated soil. In: Tedder DW (ed) Emerging technologies in hazardous waste management IX. Division of Industrial and Engineering Chemistry, American Chemical Society, Washington, DC, pp 252–257

Hawksworth DL (1993) The tropical fungal biota: census, pertinence, prophylaxis, and prognosis. In: Isaac S, Frankland JC, Watling R, Whalley AJS (eds) Aspects of tropical mycology. Cambridge University Press, Cambridge, pp 265–293

Head IM, Jones DM, Roling WFM (2006) Marine microorganisms make a meal of oil. Nat Rev Microbiol 4:173–182

IATA (2017) Infectious substances shipping guidelines, 15th edn

Janssens D, Arahal DR, Bizet C, Garay E (2010) The role of public biological resource centers in providing a basic infrastructure for microbial research. Res Microbiol 161:422–429

Jernigan DB, Raghunathan PL, Bell BP, Brechner R, Bresnitz EA, Butler JC (2001) Investigation of bioterrorism related anthrax, United States, 2001: epidemiological findings. Emerg Infect Dis 8:1019–1028

Jia XY, Briese T, Jordan I, Rambaut A, Chi HC, Mackenzie JS, Hall RA, Scherret J, Lipkin WI (1999) Genetic analysis of West Nile New York 1999 encephalitis virus. Lancet 354:1971–1972

Kalderis D, Juhasz AL, Boopathy R, Comfort S (2011) Soils contaminated with explosives: environmental fate and evaluation of state-of the-art remediation processes (IUPAC Technical Report). Pure Appl Chem 83:1407–1484

Kaplan DE, Marshall A (1996) The cult at the end of the world. Crown, New York, pp 1–310

Kengen SWM, Rikken GB, Hagen WR, Ginkel CGV, Stams AJM (1999) Purification and characterization of (Per) Chlorate reductase from the chlorate-respiring strain GR-1. J Bacteriol 181:6706–6711

Kim HY, Song HG (2000) Comparison of 2,4,6-trinitrotoluene degradation by seven strains of white rot fungi. Curr Microbiol 4:317–320

Kirsop BE, Doyle A (1991) Maintenance of microorganisms and cell cultures, 2nd edn. Academic, London, pp 1–288

Lanciotti RS, Roehrig JT, Deubel V, Smith J, Parker M, Steele K, Crise B, Volpe KE, Crabtree MB, Scherret JH, Hall RA, MacKenzie JS, Cropp CB, Panigrahy B, Ostlund E, Schmitt B, Malkinson M, Banet C, Weissman J, Komar N, Savage HM, Stone W, McNamara T, Gubler DJ (1999) Origin of the West Nile virus responsible for an outbreak of encephalitis in the northeastern United States. Science 286:2333–2337

Liao Q, Feng C, Wang L (2016) Biodegradation of unsymmetrical dimethylhydrazine in solution and soil by bacteria isolated from activated sludge. Appl Sci 6:95. https://doi.org/10.3390/app6040095

Lim DV, Simpson JM, Kearns EA, Kramer MF (2005) Current and developing technologies for monitoring agents of bioterrorism and biowarfare. Clin Microbiol Rev 18:583–607

Lönneborg R, Varga E, Brzezinski P (2012) Directed evolution of the transcriptional regulator DntR: isolation of mutants with improved DNT-response. https://doi.org/10.1371/journal.pone.0029994

Looger LL, Dwyer MA, Smith JJ, Hellinga HW (2003) Computational design of receptor and sensor proteins with novel functions. Nature 423:185–190

Ludwicka KK, Cala J, Grobelski B, Sygut D, Kupnicka DJ, Kolodziejczyk M, Bielecki S, Pasieka Z (2013) Modified bacterial cellulose tubes for regeneration of damaged peripheral nerves. Arch Med Sci 9:527–534

Marshall SJ, White GF (2001) Complete denitration of nitroglycerin by bacteria isolated from a washwater soak away. Appl Environ Microbiol 67:2622–2626

Martin JL, Comfort SD, Shea PJ, Kokjohn TA, Drijber RA (1997) Denitration of 2,4,6-trinitrotoluene by *Pseudomonas savastanoi*. Can J Microbiol 43:447–455

Meng M, Sun WQ, Geelhaar LA, Kumar G, Patel AR, Payne GF, Speedie MK, Stacy JR (1995) Denitration of glycerol trinitrate by resting cells and cell extracts of *Bacillus thuringiensis/cereus* and Enterobacter agglomerans. Appl Environ Microbiol 61:2548–2553

Moon JEVC (ed) (1991) Biological and toxin weapons: research, development, and use from the middle ages to 1945, vol 1. Stockholm International Peace Research Institute, Stockholm, Sweden, pp 8–34

Morgan CA, Herman N, White PA, Vesey G (2006) Preservation of micro-organisms by drying. A review. J Microbiol Methods 66:183–193

Mourya DT, Yadav PD, Majumdar TD, Chauhan DS, Katoch VM (2014) Establishment of Biosafety Level-3 (BSL-3) laboratory: Important criteria to consider while designing, constructing, commissioning & operating the facility in Indian setting. Indian J Med Res 140:171–183

Mulla SI, Talwar MP, Ninnekar HZ (2014) Bioremediation of 2, 4, 6-trinitrotoluene explosive residues. In: Singh SN (ed) Biological remediation of explosive residues. Springer, Switzerland, pp 201–233. https://doi.org/10.1007/978-3-319-01083-0

Nancharaiah YV, Joshi HM, Mohan TVK, Venugopalan VP, Narasimhan SV (2006) Aerobic granular biomass: a novel biomaterial for efficient uranium removal. Curr Sci 91:503–509

Office of Technology Assessment (1993) Proliferation of weapons of mass destruction: assessing the risks. OTA-ISC-559. U.S. Government Printing Office, Washington, DC. http://www.wws.princeton.edu/_ota/disk1/1993/9341_n.html

Oh K-H, Kim Y-J (1998) Degradation of explosive 2,4,6-trinitrotoluene by s-triazine degrading bacterium isolated from contaminated soil. Bull Environ Contam Toxicol 61:702–708

Olson KB, Shinrikyo A (1999) Once and future threat? Emerg Infect Dis 5:513–516

Pattnaik P, Jana AM (2005) Microbial forensics: applications in bioterrorism. J Environ Forensics 6:197–204. https://doi.org/10.1080/15275920590952874

Pichtel J (2012) Distribution and fate of military explosives and propellants in soil: a review. Appl Environ Soil Sci 617236:1–33. https://doi.org/10.1155/2012/617236

Pudge IB, Daugulis AJ, Dubois C (2003) The use of Enterobacter cloacae ATCC 43560 in the development of a two-phase partitioning bioreactor for the destruction of hexahydro-1,3,5-trinitro-1,3,5-s-triazine (RDX). J Biotechnol 100:65–75

Purver R (1995) Chemical and biological terrorism: the threat according to the Open Literature. Canadian Security and Intelligence Service. http://www.csis-scrs.gc.ca/eng/miscdocs/tabintr_e.html

Radhika V, Proikas-Cezanne T, Jayaraman M, Onesime D, Ha JH, Dhanasekaran DN (2007) Chemical sensing of DNT by engineered olfactory yeast strain. Nat Chem Biol 3:325–330

Rahal AG, Lobna AM (2011) Degradation of 2,4,6-Trinitrotoluene (TNT) by soil bacteria isolated from TNT contaminated soil. Aust J Basic Appl Sci 5:8–17

Rikken GB, Kroon AGM, Ginkel CG (1996) Transformation of (per) chlorate into chloride by a newly isolated bacterium: reduction and dismutation. Appl Microbiol Biotechnol 45:420–426

Sangwan P, Celin SM (2016a) Microbial biotechnology – an efficient tool for pollution control in explosive industry. Microorg Tools Sustainab 3:45–74

Sangwan P, Celin SM (2016b) Microalgae as a potential biofuel for defence applications. In: Sharma M, Bansal D, Chauhan A (eds) Microalgae: windows of opportunity. SPW Publishers, New Delhi, pp 64–83

Sangwan P, Celin SM, Hooda L (2015) Response surface methodological approach for optimizing process variables for biodegradation of 2, 4, 6-trinitrotoluene using *Acinetobacter noscomialis*. EJAET 2:1–6

Sharma SK, Gupta AK, Shukla AK, Ahmad E, Sharma MP, Ramesh A (2016) Microbial conservation strategies and methodologies: status and challenges. Indian J Plant Genet Resour 29:340–342

Shemer B, Koshet O, Yagur-Kroll S, Belkin S (2017) Microbial bioreporters of trace explosives. Curr Opin Biotechnol 45:113–119

Sherburne LA, Shrout JD, Alvarez PJ (2005) Hexahydro-1,3,5-trinitro-1,3,5-triazine (RDX) degradation by *Acetobacterium paludosum*. Biodegradation 16:539–547

Smith SHMB, Rosser SJ, Basran A, Travis ER, Dabbs ER, Nicklin S, Bruce NC (2002) Cloning, sequencing, and characterization of the hexahydro-1,3,5-trinitro-1,3,5-triazine degradation gene cluster from *Rhodococcus rhodochrous*. Appl Environ Microbiol 68:4764–4771

Stackebrandt E, Schüngel M, Martin D, Smith D (2015) The microbial resource research infrastructure MIRRI: strength through coordination. Microorganisms 3:890–902

Stuart A (2005) Degradation of biological weapons agents in the environment: implications for terrorism response. Environ Sci Technol 39:2736–2743

Tan J, Kan N, Wang W, Ling J, Qu G, Jin J, Shao Y, Liu G, Chen H (2015) Construction of 2,4,6-trinitrotoluene biosensors with novel sensing elements from *Escherichia coli* K-12 MG1655. Cell Biochem Biophys 72:417–428

Thompson KT, Crocker FH, Fredrickson HL (2005) Mineralization of the cyclic nitramine explosive hexahydro-1,3,5-trinitro-1,3,5-triazine by *Gordonia* and *Williamsia* spp. Appl Environ Microbiol 71:8265–8272

Tope AM, Jamil K, Baggi TR (1999) Transformation of 2,4,6-trinitrotoluene (TNT) by immobilized and resting cells of *Arthrobacter* sp. J Hazard Subst Res 2:3–9

Ugaldea UO, Castrillo JI (2002) Single cell proteins from fungi and yeasts. Appl Mycol Biotechnol 2:123–149

United Nations (1992) Convention on biological diversity. UN, Rio de Janeiro. 5 June

United Nations (2015) Committee of experts on the transport of dangerous goods and on the globally harmonized system of classification and labelling of chemicals, 47th edn. UN, Geneva

Walsh ME, Jenkins TF, Schnitker PS, Elwell JW, Stutz MH (1993) USA Cold Regions Research and Engineering Laboratory CRREL Special Report 93–5. Hanover N.H. 1–17

Wein LM, Craft DL, Kaplan EH (2003) Emergency response to an anthrax attack. Proc Natl Acad Sci 100:4346–4351

WFCC (World Fedration for culuture collections) (1990) Guidelines for the establishment and operation of collections of cultures of microorganisms. WFCC chaminas, Brazil

WHO (2004) Laboratory biosafety manual. World Health Organisation, Geneva. ISBN: 92 4 154650 6

WHO (2015) Guidance on regulations for the transport of infectious substances 2015–2016. World Health Organisation, Geneva. WHO/HSE/GCR/2015.2

World Health Organization: Geneva (1970) Health aspects of chemical and biological weapons. http://www.who.int/csr/delibepidemics/biochem1stenglish/en/index.html

Yin H, Wood TK, Smet BF (2005) Reductive transformation of TNT by *Escherichia coli*: pathway description. Appl Microbiol Biotechnol 67:397–404

Zhao JS, Halasz A, Paquet L, Beaulieu C, Hawari J (2002) Biodegradation of Hexahydro-1,3,5-Trinitro-1,3,5-Triazine and Its Mononitroso Derivative Hexahydro-1-Nitroso-3,5-Dinitro-1,3,5-Triazine by *Klebsiella pneumoniae* strain SCZ-1 isolated from an anaerobic sludge. Appl Environ Microbiol 68:5336–5341

Zhao JS, Paquet L, Halasz A, Hawari J (2003) Metabolism of hexahydro-1,3,5-trinitro-1,3,5-triazine through initial reduction to hexahydro-1-nitroso-3,5-dinitro-1,3,5-triazine followed by denitration in *Clostridium bifermentans* HAW-1. Appl Microbiol Biotechnol 63:187–193

Zhao JS, Manno D, Beaulieu C, Paquet L, Hawari J (2005) *Shewanella sediminis* sp. nov., a novel Na+-requiring and hexahydro-1,3,5-trinitro-1,3,5-triazine-degrading bacterium from marine sediment. Int J Syst Evol Microbiol 55:1511–1520

Chapter 15
Diversity, Ecology, and Conservation of Fungal and Bacterial Endophytes

Pramod Kumar Pandey, Siddhartha Singh, Mayanglambam Chandrakumar Singh, Amit Kumar Singh, Sudheer Kumar Yadav, Ajai Kumar Pandey, and Punabati Heisnam

Abstract Every plant present on the earth is accompanied by internal association of microbes generally pronounced as endophytes. Bacterial and fungal endophytic microbes are among one of them. Their presence was noted in various niches ranging from mountain to sea, from forest to desert, and, moreover, in cold and in hot spring, copper mine wasteland, agronomic crops, prairie plants, deepwater rice, and grass ecosystem. These endophytic microbes are agriculturally important because of their various plant growth-promoting traits. They are found to inhabit the seeds, roots, stems, and leaves and even the periderm. These endophytic microbes generally enter the plant tissues via several "hotspots" like root system and mitigate with biotic and abiotic stresses, help to cure human diseases by producing several secondary metabolites, help in the induction and expression of plant immunity, exclude plant pathogens by niche competition, as well as actively participate in phenylpropanoid metabolism and antioxidant activities. The discoveries of novel bioactive compound and defense activator like antifungal, antibacterial, antiviral, and antitumor compounds, antibiotics, secondary metabolites, and volatile insecticides attributed to these endophytes are utilized as therapeutic agents in the field of pharmaceutical, medicine, agriculture, and industries. The conservation of endophytic microbes and their gene pools is an emerging and vital issue, even though the development is scary.

P. K. Pandey (✉)
Centre for Biotechnology and Bioinformatics, Dibrugarh University, Dibrugarh, Assam, India

College of Horticulture & Forestry, Central Agricultural University, Pasighat, Arunachal Pradesh, India
e-mail: pramod.pandey84@gmail.com

S. Singh · M. C. Singh · A. K. Singh · A. K. Pandey · P. Heisnam
College of Horticulture & Forestry, Central Agricultural University, Pasighat, Arunachal Pradesh, India

S. K. Yadav
Department of Botany, Faculty of Science, Banaras Hindu University, Varanasi, Uttar Pradesh, India

S. K. Sharma, A. Varma (eds.), *Microbial Resource Conservation*, Soil Biology 54,
https://doi.org/10.1007/978-3-319-96971-8_15

15.1 Introduction

The plant's intercellular or intracellular spaces harbor a huge sort of unnoticed group of microorganisms known as endophytes. Fungal and bacterial microbes are among one of them. De Bary conceptualized the term "endophyte" in (1866). With the advancement of research in the field, definitions of endophytes have been modified several times. According to Tervet and Hollis (1948), endophytes are those microorganisms which live inside the plant without any disease symptoms. Petrini (1991) explained endophytes as an internal association of microorganism with plant at any stage of its life cycle without giving any disease symptoms which are further modified by Sikora et al. (2007) who termed this internal association of microorganism with plants may be beneficial, neutral, or detrimental to its host plant. The literature meaning of endophyte is "inside the plant," and the microorganisms which inhabit the inner tissue of the plant are known as endophytic microorganism (Fig. 15.1). It may be further extended to endophytic fungus, bacteria, or actinomycetes (Pandey et al. 2012, 2014, 2016). Almost all plants are the dwellers of these indigenous microbes which can reside intercellularly or intracellularly and responsible for nutrient assimilation, their processing, induction of the defense system, and synthesis of secondary metabolites (Pandey et al. 2017).

History of exploitation of fungal and bacterial endophytes for their plentiful novel bioactive compounds (Strobel et al. 2002; Harper et al. 2003) and their use in agriculture (James 2000), biotechnology (Berger 2009), and pharmaceutical (Strobel et al. 2004; Gangadevi and Muthumary 2008) is a century long. The fungal and bacterial endophytes are reported to produce novel bioactive compounds which are previously untapped such as insecticides (Ryan et al. 2008), antimicrobials (Liu et al. 2010a; De Melo et al. 2009; Wang et al. 2009; Bacilio-JimeÂnez et al. 2001), etc. responsible for induction of immune response and restriction of pathogen entry inside the plant tissues (Sturz et al. 2000; Lodewyckx et al. 2002). Production of non-food crops for biomass and biofuel production is also reported from bacterial endophytes (Rogers et al. 2012). Fungal and bacterial endophytes have been known to play a significant role in soil fertility, plant growth, and biocontrol activities through various mechanisms. Stable microbial endophytic communities with defense arsenals could able to control plant pathogens by producing antifungal metabolites and antibiotics, de novo synthesis of structural compounds, induction and expression of molecular-based plant immunity, exclusion of other organisms from niche competition, or induced systemic resistance (ISR). Fungal and bacterial endophytes are sometimes much more bioactive by inducing rapid and stronger responses against stress in the host than the rhizospheric microbes or any other microbes (Pandey et al. 2012). Understanding of the plant endophytes interaction should provide a range of application.

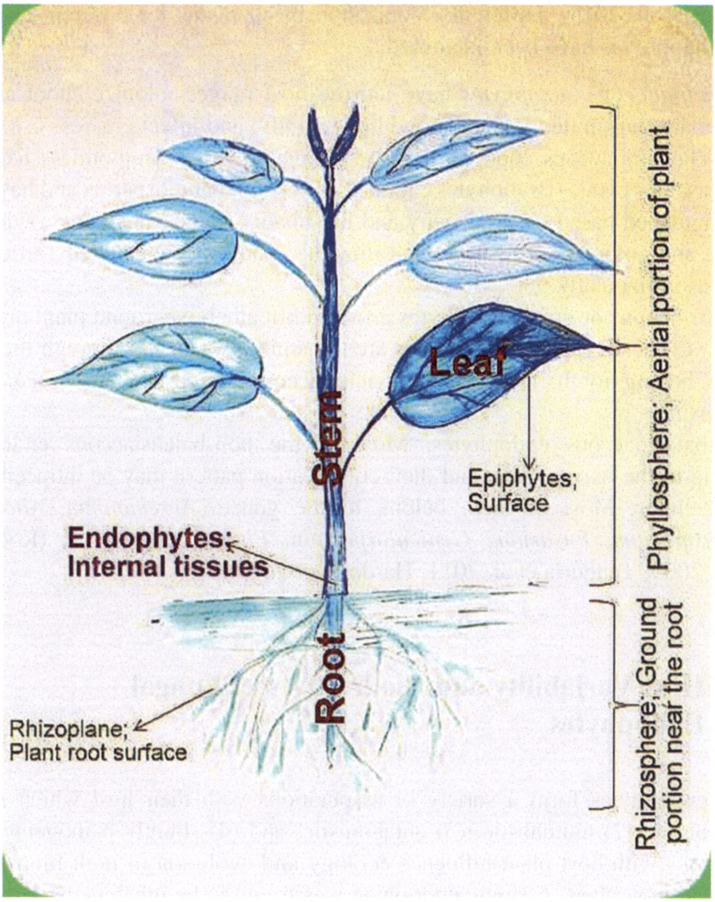

Fig. 15.1 Plant part showing localization of endophytic and epiphytic microbes

15.2 Evolutionary Origins and Ecological Consequences of Fungal Endophytes

Vogl in 1898 reported the presence of an endophytic mycelium inside the grass seed *Lolium temulentum.* Freeman in 1904 reported an endophytic fungus in annual grass *Persian darnel.* It is found that certain microorganisms able to penetrate the plant tissues, either with the help of cell wall-hydrolyzing enzymes like pectinase and cellulase or certain other mechanisms, and reside inside the plant tissue and coevolved. During the course of coevolution, they may be adapted toward the internal milieu of the host plant by the mechanisms of cross talk (Pathak 2011).

Colonization behavior and degree of colonization of fungal endophytes are influenced by numerous factors such as host tissue type, host genotype, the microbial taxon type, and strain type and several biotic and abiotic environmental factors

(Hardoim et al. 2015). Different colonization mechanisms for different classes of fungal endophytes have been identified:

1. *Clavicipitaceous endophytes* have narrow host range; colonize shoot and rhizome, are transmitted vertically and horizontally; and infect grasses.
2. Non-clavicipitaceous endophytes were recovered from symptomless tissues of nonvascular plants (Bryophytes), ferns, conifers, and angiosperms and have been distinguished based on phylogeny and life history traits. They have a wide host range and colonize root, shoot, and rhizome. They are transmitted vertically as well as horizontally.
3. *Balansiaceous* or grass endophytes grow within all aboveground plant organs of grasses. These types of endophytes are transmitted vertically through the seeds. They belong to the genera of clavicipitaceous which include *Epichloe* and *Balansia*.
4. Non-balansiaceous endophytes: Most of the non-balansiaceous endophytes belong to the *Ascomycota*, and their colonization pattern may be intracellular or intercellular. Most of them belong to the genera *Acremonium*, *Alternaria*, *Cladosporium*, *Fusarium*, *Geniculosporium*, *Phoma*, and so on (Rodriguez et al. 2009; Tenguria et al. 2011; Hardoim et al. 2015).

15.3 Host Variability and Biodiversity of Fungal Endophytes

Fungal endophytes form a variety of associations with their host which include (1) symbiotic, (2) mutualistic, (3) antagonistic, and (4) slightly pathogenic. Their associations with host plant influence ecology and evolution of both fungal endophytes and host plant. A single endophyte was found to be inhabitants of different hosts rather than in a specific host. Their distribution in the different plant tissues affected by their ability to utilize nutritional substances which could be synthesized in different plant part of the same host and their relationship in the different host varies with the nature of ecological and geographical conditions. The interaction between fungal endophytes and the host is controlled at the gene level, involving the genes of both the partners which are inflected by the environment. The encounter of endophyte by the host plant modulates the level and pattern of gene expression in the host plant. The diversity of fungal endophytes has multidisciplinary effects on the ecosystem such as the enhancement of primary productivity, retention, and flow of nutrient and along with the development of resistance to the pathogen entry. Fungal endophytes have been isolated from almost all plant groups ranging from palms, grasses, sea grasses, large trees, lichens, and medicinal plants. The diversity of fungal endophytes may vary in different plant parts with the location of host plant in different geographic sites such as temperate or tropical. Most of the fungal endophytes isolated belong to ascomycetes, and several may also belong to basidiomycetes. Endophytic fungi from aquatic, temperate, tropic, and xerophytic conditions are also reported (Pandey et al. 2014).

Endophytic fungi, *Alternaria alternata*, *Cladosporium cladosporioides*, *Chaetomium globosum*, *C. herbarum*, *Gliocladium roseum*, *Curvularia lunata*, *Nigrospora sphaerica*, and *Phyllosticta* spp. were reported as frequently occurring fungi in a leaf of medicinal climber and grasses (Shankar and Shashikala 2010). A semiarid region of Pakistan, where climatic conditions are exceedingly harsh with low rainfall, was helpful in understanding the frequency of colonization of endophytic mycoflora and species richness of medicinal plant *Withania somnifera*. Among the endophytic fungal isolates, 4 isolates belong to the class *Ascomycetes*, and 20 isolates belong to the class *Deuteromycetes*. The highest species richness was noted from the stem, and *Alternaria alternata* was found to be the most dominant endophyte. In the isolated class of fungi, *Deuteromycetes* were found to be the most prevalent (Khan et al. 2010). In a survey, it was found that sapwood of *Hevea brasiliensis* had greater endophytic fungal diversity in comparison to the leaves, while their colonization frequency was more in the leaves as compared to the sapwood (Gazis and Chaverri 2010). Endophytic fungal diversity has been investigated based on its relative frequency, isolation, and colonization rates in medicinal plants (*Adhatoda vasica*, *Ocimum sanctum*, *Withania somnifera*, *Cannabis sativa*, and *Viola odorata*) of Himachal Pradesh, India, which covers 15 fungal genera and 18 species (*A. clavatus*, *A. flavus*, *A. variecolor*, *Penicillium chrysogenum*, *Aspergillus niger*, *Alternaria alternata*, *Curvularia lunata*, *Haplosporidium* sp., *Phoma* sp., *Nigrospora* sp., *Colletotrichum* sp., *Cladosporium* sp., *Stemphylium* sp., *Fusarium* sp., *Geotrichum* sp., *Phomopsis* sp., *Trichoderma* sp., and *Rhizopus* sp.) (Gautam 2013). An effort was made to examine the diversity, distribution, and phylogeny among the endophytic fungi from the different plant sources on the basis of ITS1–5.8S-ITS2 sequence of Western Himalayas which showed diverse taxonomic affinities among isolated fungal endophytes. In the sum of 72 endophytic fungal strains isolated, only two belong to *Basidiomycetes*, whereas the rest belongs to *Ascomycetes*. *Alternaria* spp. and *Fusarium* spp. consist more than half (54.2%) of the strains isolated. Most of the genera of *Alternaria* strains were reported from *Artemisia annua* and *Rauwolfia serpentina*, whereas most of the *Fusarium* spp. were reported to be inhabitants of *Artemisia annua*, *Withania somnifera*, and *Platanus orientalis*. The fungal endophytes obtained from the conifers such as from *Pinus roxburghii*, *Cedrus deodara*, and *Abies pindrow* harbored the most diverse endophytic fungi of 13 different genera and offer significant taxonomic variants (Qadri et al. 2013).

15.4 Evolutionary Origins and Ecological Consequences of Bacterial Endophytes

Bacterial endophytes have been studied for more than 100 years (Anu Ranjan 2012). The first endophytic bacterial presence within the plant tissues was reported in 1926 (Hallman et al. 1997). Their relationship may have been intended to develop during the hundred million years ago along with the higher plants appeared on the Earth

(Anu Ranjan 2012). The evolution of endophytic bacteria is believed to be initiated from the symbiosis of rhizobia with leguminous plants. The interaction of rhizobia and leguminous plants is mediated by the signal exchange between them through the following steps: (1) curling of root hairs, (2) the appearance of infection threads, and, finally, (3) the appearance of nodule primordia in the inner root layers. During the course of evolution, initially the bacterial cell becomes surrounded by the plant membrane and forms the structure called symbiosome resembling to the mycorrhizal arbuscules, in which atmospheric nitrogen is fixed and transferred in exchange for the carbohydrates (Hardoim et al. 2015).

The entry of bacterial endophytes into the host involves recognition and chemotaxis for the attachment to the root surface of the host plant in the rhizosphere (Lugtenberg and Kamilova 2009). Type IV pili encoded by *pilAB* (Dorr et al. 1998), LPS (lipopolysaccharide) (Balsanelli et al. 2010), and EPS (exopolysaccharide) (Meneses et al. 2011) along with the various other components of bacterial surface are thought to be of essence for the endosphere colonization. It is believed that the endophytic bacteria in the form of several microcolonies may enter into the cells (1) through thin-walled apical root zone, (2) from the root hair zone, and (3) from the basal root zone of the emergence site of the lateral roots (Zachow et al. 2010) either as active or passive penetration and sometimes a combination of both. Active penetration takes place with the aid of cell wall-degrading enzymes (cellulolytic) like endoglucanase which hydrolyze the exodermal cell wall of plants (hydrolysis of β(1–4) linkage of the cellulose) (Reinhold-Hurek et al. 2006). The passive penetration takes place via cracks produced by the emergence of lateral roots in the basal root zone (Compant et al. 2005).

Bacterial endophytes provide a defense to the host plant by colonizing the tissues and the other parts earlier than the pathogenic microorganism. The colonization extent of the endophytic bacteria in the host's tissues makes them suitable for the adaptation in a specific location (Hallman et al. 1997). They have been reported to successively cross the endodermis and flourish in different plant parts as the endophytes such as in leaves, roots, stem, tubers, etc. (Yonebayashi et al. 2014; Upreti and Thomas 2015; Perez-Rosaleset et al. 2017; Compant et al. 2005). The strain-dependent mechanism of O-antigen chain of bacterial lipopolysaccharides (Dekkers et al. 1998; De Weger et al. 1989), vitamin B1, NADH dehydrogenases from the bacteria having a high growth rate (Dekkers et al. 1998; Simons et al. 1996), and type IV pili (Dorr et al. 1998) was found to be involved in the root colonization of the bacterial endophytes.

Bacterial endophytes from the endodermis and the root cortex of the plants have been employed to advocate that the bacteria from the rhizosphere are able to penetrate and colonize the root tissues (Sturz et al. 2000). During colonization process root cortex becomes the part of the soil–root microbial environment, which provides an apoplastic pathway of entry of microorganism (from the root epidermis to the shoot) (Algam et al. 2005). Secondary roots and the intercellular space colonization were observed after the disruption of the mucilage layer with the aid of scanning electron microscopy (Bacilio-Jimenez et al. 2001). Studies of the bacterial endophytic colonization patterns in the vegetative tissues and effects of bacterial endophytes on the plant growth offer a better understanding of the biological interaction of the plant-bacterial endophytes (Algam et al. 2005). These studies have been performed by inoculating the endophytic bacteria into the host plants and comparing the inhibition of disease symptoms (Hastuti et al. 2012; Kumar et al. 2015; Zhao et al. 2017). Bacterial endophytes are able to penetrate deep within the plant and spread systemically to reach the aerial tissues and have been found to be an efficient biological nitrogen fixer (Verma et al. 2001). According to a report, the entry of the bacterial endophytes in the plant tissue may be accompanied (1) through the root zone, (2) by the aerial parts of the plant, (3) through the stems, or (4) by the cotyledons (Compant et al. 2010). They may either be localized at the point of entry or are able to spread throughout the plant, and such isolate can live (1) in the intercellular spaces, (2) inside cells, and up to some extent (3) in the vascular system (Sharma et al. 2005). Autofluorescent protein (AFP) and green fluorescent proteins (GFP) were employed to study the colonization and localization patterns of endophytic bacterial microbes and plant-endophytic bacterial interaction (Ryan et al. 2008; Akbari and Akbari 2017). GFP was used for the chromosomal integration and the expression of *gfp* gene in endophytic bacteria (Montalbán et al. 2017) and can be identified by the epifluorescence microscopy or confocal laser scanning microscopy (Villacieros et al. 2003; Germaine et al. 2004). The use of the β-glucuronidase (GUS) reporter system has been developed to monitor the bacterial colonization within the root intercellular spaces, stele, xylem vessels, stems, and other several parts of rice seedlings (James et al. 2002). Nitrogenase (nif) gene expression in the rice root and the responses of rice roots toward the endophytic bacterial colonization were reported (Miche et al. 2006). Investigation of the endophytic bacterial gene expression in the rhizosphere and phyllosphere by in vivo expression technology (IVET) and the recombination in vivo expression technology is very useful to study and understand the essential endophytic bacterial genes used to enter inside, to compete, to colonize, and to suppress the phytopathogens (Zhang et al. 2006; Ryan et al. 2008).

The bacterial endophytes may be beneficial, neutral, or detrimental to its host (Sikora et al. 2007). This association may be some time pronounced as endophytism, and probably all plants in their life cycle may harbor bacterial endophytes (Mercado-Blanco and Bakker 2007). They are reported to be mostly positive symbiotic to its host (Li et al. 2012; Egamberdieva et al. 2017), in which host plants provide nutrients for the growth and multiplication of bacterial endophytes and endophytes pay off in the form of releasing various active and novel metabolites which are used by the plants for its the development, regulation of physiology, and induction of defense mechanisms (Lodewyckx et al. 2002; Li et al. 2008). It is reported that the symbiotic relationship between the bacterial endophytes and plants can act as a switch between mutualistic and parasitic form, where a change in a particular gene segment may convert the endophytes into a detrimental parasitic form (Kogel et al. 2006). Hence, the bacterial endophytes from one host can obey different lifestyles in a different host. On the basis of life strategies, Hardoim et al. (2008) have described the endophytic bacteria in the following groups: (1) facultative bacterial endophytes, organisms that can (optionally or without restriction) live inside plants and in other habitats also; (2) obligate bacterial endophytes, organisms that are (restricted) strictly bound to life inside a plant; (3) opportunistic bacterial endophytes, organisms that occasionally enter plants for their own needs; and (4) passenger bacterial endophytes, organisms that enter the plant by chance.

15.5 Host Variability and Biodiversity of the Bacterial Endophytes

The bacterial endophytes are found to inhabit the majority of healthy and symptomless plants. They are found to inhabit the seeds, roots, stems, and leaves and even the periderm also (Johri 2006; Mastretta et al. 2009). The presence of endophytic bacteria in the copper mine wasteland, agronomic crops and prairie plants, deepwater rice, grass ecosystem, pea cultivars in field condition, tropical grasses of the *Brachiaria* genus, and prairie plants has been reported (Zinniel et al. 2002; Verma et al. 2001; Ryan et al. 2008).These endophytic bacteria live in a variety of tissue types, inside many plant species, and thereby suggested their ubiquitous existence. Endophytic bacteria of a plant host are not confined to a single species, but they may go to several genera and species (Ryan et al. 2008). A number of endophytes from rice (Biswas et al. 2000; Ji et al. 2014; Sev et al. 2016), maize (Rosenblueth and Martinez Romero 2004; Mousa et al. 2015; de Abreu et al. 2017; Bodhankar et al. 2017), wheat (Zinniel et al. 2002; Pan et al. 2015; Robinson et al. 2015; Herrera et al. 2016), cotton (Reva et al. 2002), sorghum (Nascimento et al. 2014; Li et al. 2016), cacti (Puente et al. 2009), apple tree (Miliūtė and Buzaitė 2011; Miliute et al. 2016), bananas (*Musa* spp.) (Ngamau et al. 2012; Souza et al. 2014; Karthik et al. 2017), Antarctic moss (*Sanionia uncinata*) (Park et al. 2013), chili pepper (Paul et al. 2013), sphagnum (Shcherbakov et al. 2013), tomato (*Solanum lycopersicum* L.) (Romero et al. 2014), ginger (Chen et al. 2014), *Helianthus tuberosus* (Montalbán et al. 2017), potato, and *Arabidopsis* (Hallmann et al. 2001; Sessitsch et al. 2004; Hong et al. 2015) were reported. Beneficial bacterial endophytes have been isolated from the various plant species (Mastretta et al. 2009; Forchetti et al. 2007) along with the different plant parts like tubers, seeds, stems, buds, roots, seedlings, root bark, and plant tissues such as the intercellular space, the vascular tissue, and the periderm (Fig. 15.2), which make them suitable candidate for the plant growth promotion as they are less exposed to the external biotic and the abiotic stresses. According to an estimate, approximately 129 kinds of the bacterial endophyte consisting of both Gram-negative and Gram-positive species representing more than 54 genera were already identified from the various host crops (Zhenhua et al. 2012). Endophytic bacterial isolates from the leaves of neem (*Azadirachta indica*) have hydrocarbon degradation potential (Singh and Padmavathy 2015). Endophytic bacterial strains from the roots and crude seed extracts of a Cu-tolerant population of *Agrostis capillaris* were assessed for their influence on Cu tolerance and phytoextraction (Kolbas et al. 2015). A total of 536 bacterial and fungal endophytes from the root,

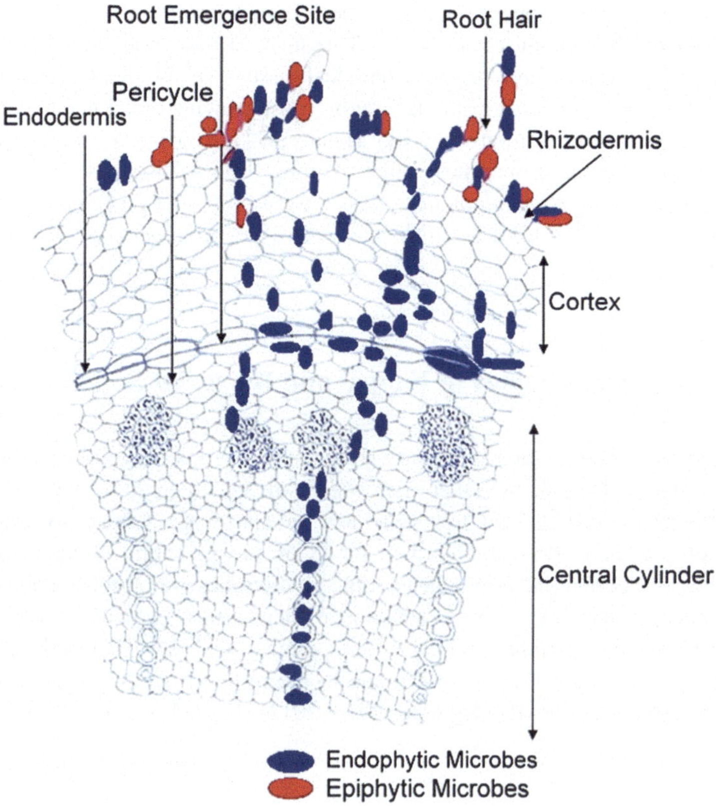

Fig. 15.2 Sites of colonization by endophytic microbes in plant's root

stem, and leaves of *Amaranthus spinosus* were recovered successfully. It was found that the bacterial endophytes are more prominent in the root than in either stem or leaves, whereas, fungal endophytes are more prominent in stem than in either roots or leaves. Furthermore, Gram-negative endophytic bacteria were found to be more in comparison to Gram-positive endophytic bacteria in total isolate isolated (Sharma and Roy 2015). Among the 62 endophytic bacterial isolates, isolated from the leaves, roots, and stems of a healthy *Lepidium perfoliatum* L. plant, a sum of 34 isolates was isolated from the leaves, 27 were isolated from roots, and 1 isolate was isolated from stems. Among them, 66% were similar to *B. subtilis*, 14% were similar to *B. flexus*, 6% were similar to *B. licheniformis*, 5% were similar to *B. mojavensis*, 5% were similar to *B. sonorensis*, 2% were similar to *B. cereus*, and 2% were similar to *B. safensis* (Li et al. 2017). In a study, 125 strains of bacterial endophytes were identified based on 16S rRNA gene sequencing from the endangered traditional medicinal plants *Ferula* spp. in arid regions of Xinjiang belonging to 3 different phyla and 29 genera. Among the isolates, 3 strains isolated from the roots were identified as a novel species of the genera *Porphyrobacter*, *Paracoccus*, and *Amycolatopsis* (Yonghong et al. 2017). A total of 133 bacterial endophytes were isolated from the root crown, stem, and leaf tissues of 34 plants consisting of 8 different species, collected from the Oregon coast growing under harsh conditions from 4 different sites. Among the 133 isolates, 94 are identified as unique and represent 36 different taxonomic groups. Among the total isolates, 20.3% of the isolates belong to *Pseudomonads*, 8.2% were *Curtobacterium*, 8.2% were *Microbacterium*, 7.5% were *Bacillus*, and 6% were *Xanthomonas* (Dombrowski et al. 2017).

15.6 Chemical Ecology of Fungal and Bacterial Endophytes

Nowadays world population is facing various health problems caused by bacteria, virus, fungus, protozoan, and other microorganisms which could be due to the drug resistance or parasitism, and this generates an alarming situation to the world population for their survival. Research based on the invention of medicinal and novel bioactive compounds from the endophytic microbes is a promising task. There is a substantial demand for the production of new drugs, particularly antibiotics, anticancer agents, immunomodulator compounds, and agrochemicals that are eco-friendly and cause less or no damage to the milieu and replace the synthetic fertilizers and pesticides (Pandey et al. 2013, 2014) (Fig. 15.3).

15.6.1 Chemical Ecology of Fungal Endophytes

A fungal endophyte produces a large number of diverse bioactive molecules which are previously unknown. A number of bioactive compounds from the endophytic fungus have been isolated which provides protection by producing antimicrobial compounds like antifungal, antibacterial, and antiviral (Fig. 15.3). Researches are mostly concentrated on the investigation and discovery of natural bioactive compounds and improving the productivity by taking advantage of genetic engineering and other measures (Pandey et al. 2014). The endophytic fungus is also reported to produce the bioactive compound that acts as an anticancer agent (Strobel et al. 1996, 1997). Paclitaxel (Taxol) is the world's first billion dollar anticancer drug mainly obtained from the bark of the Pacific yew tree which is a tetracyclic diterpenoid (Strobel et al. 1996, 1997). The *Taxus* trees are very rare and unable to meet the market need. Hence, an alternative strategy should be needed to develop. The production of Taxol (paclitaxel) by endophytic fungus, such as *Phyllosticta spinarum* (Kumaran et al. 2008), *Bartalinia robillardoides* (Gangadevi and Muthumary 2008), *Pestalotiopsis terminaliae* (Gangadevi and Muthumary 2009), and *Botryodiplodia theobromae* (Pandi et al. 2010), is an attractive invention and generated more attention in the treatment of various cancers because of its unique

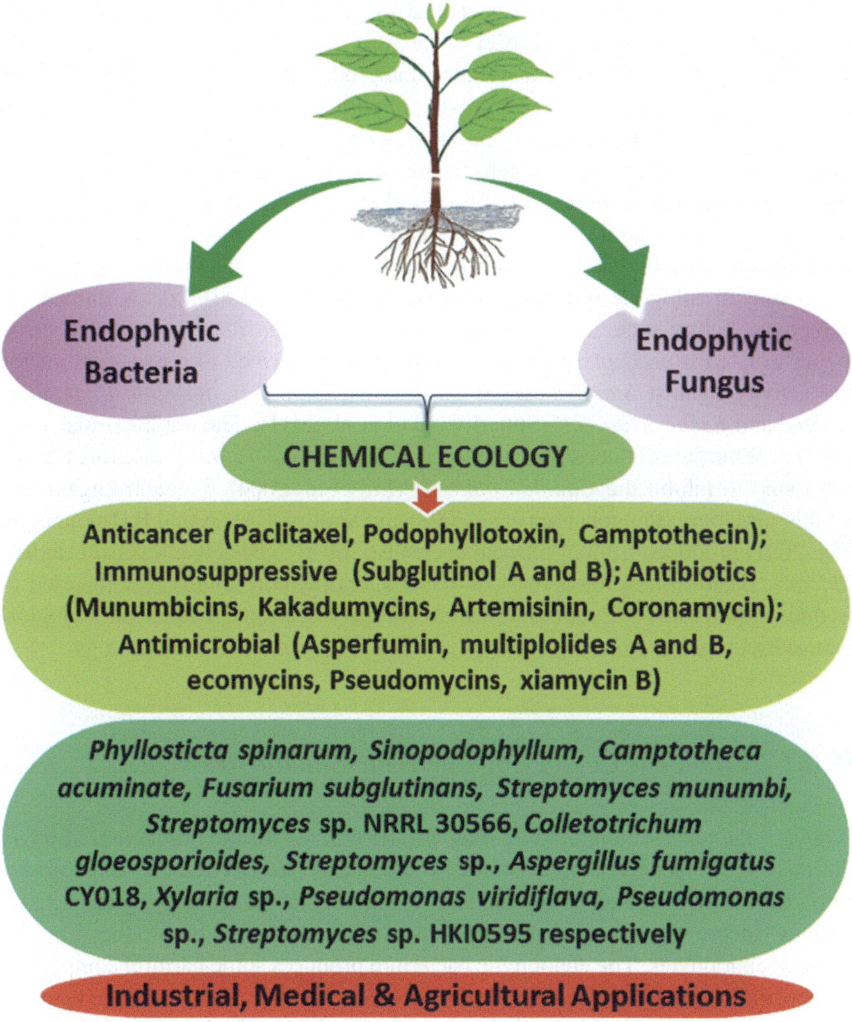

Fig. 15.3 Overview of chemical ecology of fungal and bacterial endophytic microbes

mode of action as compared to the other anticancer agents. Podophyllotoxin has been used as a precursor for synthesis of anticancer drug and having the antimicrobial and antioxidative properties which are mainly produced by the genera of an endangered fungal species *Sinopodophyllum* (or *Podophyllum*) and *Aspergillus fumigatus*, *Fusarium oxysporum*, and *Phialocephala fortinii* (Eyberger et al. 2006; Kusari et al. 2008, 2009; Kour et al. 2008). Similarly, camptothecin, a pentacyclic quinoline alkaloid, was found to be working as an antineoplastic agent and important precursor of anticancer drugs topotecan and irinotecan (Shaanker et al. 2008), which are reported to inhibit the intranuclear enzyme topoisomerase-I required for the DNA replication and transcription (Pandey et al. 2014), and is predominantly isolated from the *Camptotheca acuminate* (Liu et al. 2010b). The anticancer terpenoid indole alkaloids known as vinblastine and vincristine produced from an endophytic fungus interfere with microtubule and mitotic spindle (Zhao et al. 2010). Pestacin (1,3-dihydroisobenzofuran) and isopestacin (isobenzofuranone) having antioxidant activity are obtained from an endophytic fungus *P. microspora* and able to scavenge superoxide and hydroxyl free radicals (Strobel et al. 2002; Harper et al. 2003). Immunosuppressive nontoxic bioactive compound subglutinols A and B are identified from the endophytic fungus *Fusarium subglutinans* and propose to have an active role in allograft rejection in transplant (Lee et al. 1995). A number of endophytic fungi such as *Phyllosticta* spp., *Nodulisporium* spp., and *Xylaria* sp. are isolated from *Dipterocarpus* trees and were described to possess antimicrobial activity against *Staphylococcus aureus*, *Pseudomonas aeruginosa*, *Bacillus subtilis*, and *Escherichia coli* (Sutjaritvorakul et al. 2011). The antimicrobial compound asperfumin produced by an endophytic fungus *Aspergillus fumigatus* CY018 was found to inhibit the *Candida albicans* (Liu et al. 2004). The antifungal compounds sordaricin (Pongcharoen et al. 2008) and multiplolides A and B (Boonphong et al. 2001) produced from *Xylaria* sp. were also found to be active against *Candida albicans*. Besides these bioactive compounds, endophytic fungus was reported to produce various sorts of antibiotic, which is briefly dealt under the head biological control by fungal and bacterial endophytes in the section functional ecology.

15.6.2 Chemical Ecology of Bacterial Endophytes

Likewise, the bacterial endophytes may also be utilized for the production of novel pharmaceutical agents and agrochemical compounds. There is tremendous scope for the isolation of novel bioactive medicinal compounds from endophytic bacteria. Various categories of natural products have been reported to be produced by the bacterial endophytes. The endophytic bacterium produces some harmful compounds to keep out the predators from the host plants (Suto et al. 2002). An ample range of low molecular weight compounds has been isolated from the bacterial endophytes which are active at their low concentrations against a range of animal and plant pathogens. Many bacterial endophytes such as *Pseudomonas* and *Burkholderia* are well recognized for producing secondary metabolites that include antibiotics, anticancer compounds, and insecticidal, antifungal, antiviral, and immunosuppressant agents (Lodewyckx et al. 2002). The grass endophytic bacteria *Pseudomonas viridiflava* are reported to produce ecomycins, a family of novel lipopeptides, which was found to be active against the pathogenic fungi *Cryptococcus neoformans* and *Candida albicans*. The antifungal compound pseudomycins was also described to be produced by plant-associated pseudomonads (Strobel and Daisy 2003). The previously unknown novel compounds xiamycin B, indosespene, and sespenine together with the previously known compound xiamycin A are also reported from the culture broth of endophytic *Streptomyces* sp. HKI0595. These compounds are used as a biocontrol agent and found to have strong antimicrobial activities against different pathogenic bacteria such as *Staphylococcus aureus* and *Enterococcus faecalis* (Ding et al. 2011). Several compounds and antibiotics, namely, ammonia, butyrolactones, 2,4-diacetylphloroglucinol, kanosamine, oligomycin A, oomycin A, phenazine-1-carboxylic acid, pyoluteorin, pyrrolnitrin, viscosinamide, xanthobaccin, and zwittermicin A, were reported to be produced by antagonistic endophytic bacteria (Whipps 2001). Several endophytic bacteria were known to produce hydrolytic enzymes that cause cell wall lysis and utilized to control fungal pathogens (Backman and Sikora 2008). The endophytic bacteria were found to produce biosurfactants and used as an antimicrobial compound. The endophytic bacteria *Pseudomonas fluorescens* was found to produce cyclic lipopeptide surfactants such as viscosinamide and tensin and act as antifungal against *Rhizoctonia solani* and *Pythium ultimum* (Nielson et al. 1999, 2000). The protein fractions 42 and 75 from the *Bacillus amyloliquefaciens* and *Bacillus pumilus* act as elicitor molecules in the induced resistance against the pathogen *Xanthomonas vesicatoria* in the tomato plant with increasing peroxidase (POX) and polyphenol oxidase (PPO) activities (Lanna-Filho et al. 2013). The antimicrobial compounds surfactin and fengycin A homologs were identified from the endophytic *Bacillus amyloliquefaciens* and are found to be active against the peanut bacterial wilt (BW) caused by *Ralstonia solanacearum* (Wang and Liang 2014). The presence of tannins, alkaloids, saponins, glycosides, amino acids, proteins, and phenolic compounds was described from the endophytic bacteria *Lactobacillus* sp. isolated from the leaf tissues of *Adhatoda beddomei* (Swarnalatha et al. 2015).

15.7 Functional Ecology of Fungal and Bacterial Endophytes

15.7.1 Biofertilization

The plant growth promotion by increasing the availability or supply of major and minor nutrients is designated as biofertilization (Gaiero et al. 2013) (Fig. 15.4).

15.7.1.1 Production of Phytohormones

It is well known that the plant hormones are organic compounds effective in very minute concentration and subsequently on their synthesis transported to the other location where they can interact with specific target tissue and regulate the physiological function of the host plant and hence referred to as plant growth regulators or phytohormones (auxins, gibberellins, ethylene, cytokinins, and abscisic acid) (Pandey et al. 2017). Phytohormones producing endophytic microbes under extreme conditions also involve the output of numerous other secondary metabolites such as

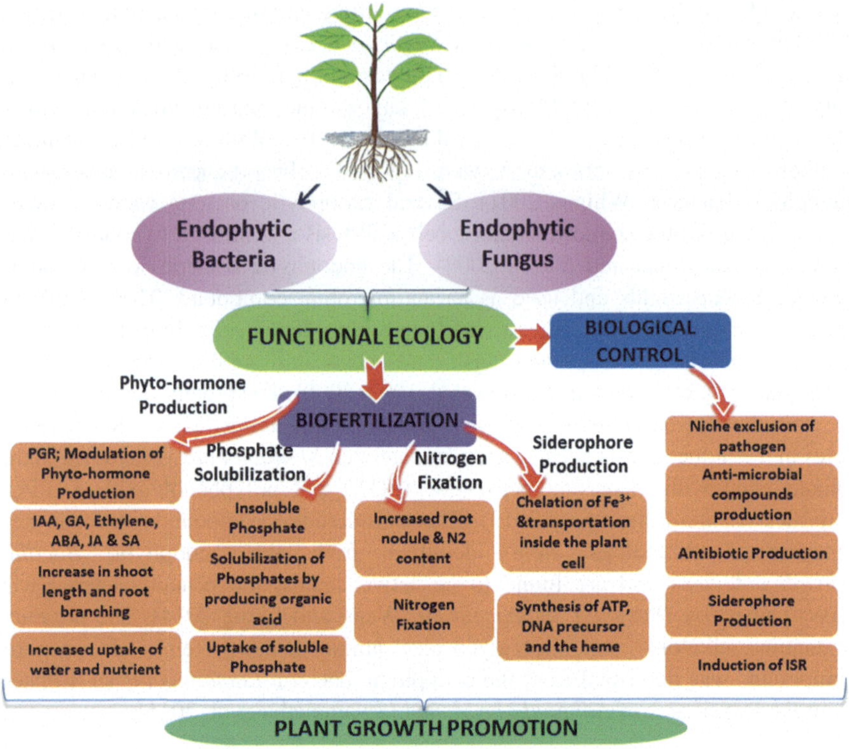

Fig. 15.4 Overview of functional ecology of fungal and bacterial endophytic microbes

flavonoids to avoid the unfavorable conditions (Khan et al. 2011). Different microbial endophytic groups are reported to produce the IAA (indole-3-acetic acid), the most important auxin which regulates plant development such as the cell expansion, division, and differentiation, gene regulation, and the other tropic response (Ratul et al. 2013; Pandey et al. 2017). The endophytic bacterial isolates from the root of *Hyptis marrubioides* Epling, a *Lamiaceae*, were studied to synthesize IAA in DYGS medium supplemented with tryptophan. In the total endophytic bacterial isolates, 52% isolates were able to synthesize IAA, and the isolates RF18 (95.13 μg/ml), RG9 (39.28 μg/ml), RF13 (16.21 μg/ml), and RG24 (11.96 μg/ml) were identified as to produce significant amounts of IAA under test conditions (Vitorino et al. 2012). Endophytic bacterial isolates from the duckweed (*Landoltia punctata*) were examined for their plant growth promotion ability by the production of phytohormone indole-3-acetic acid (IAA). Out of 71 isolates, 27 were identified as positive for IAA production in which *Deinococcus* sp. L2–88 was reported to produce the highest quantity of IAA, i.e., 713.2 + 11.6 μg/ml (Kittiwongwattana 2015). Endophytic bacterial isolates from sugar beet roots were reported to produce indole-3-acetic acid (IAA) in vitro and are reported to positively affect the plant height, their fresh and dry weights, and leaves per plant, along with levels of phytohormones when compared to the control plant (Shi et al. 2009). The fungal endophytes such as *Fusarium oxysporum* CSE15, *Trichoderma harzianum* CSE10, *Cadophora malorum* CSE14, and *Phialophora mustea* CS7E2 recovered from the *Crocus sativus* Linn were reported to produce 109.5, 95.12, 92.94, and 73.52 mg L^{-1} of IAA, respectively (Wani et al. 2016). Fungal endophytes like *Gibberella fujikuroi* (Bomke et al. 2008), *Penicillium citrinum* (Khan et al. 2008), *Chrysosporium pseudomerdarium* (Hamayun et al. 2009), and *Aspergillus fumigatus* (Khan et al. 2011) were reported to produce GAs in their culture medium and have an active GA biosynthesis pathway. The culture filtrates of endophytic fungi *Paecilomyces formosus* LHL10 isolated from the roots of cucumber plant showed the presence of different forms of GAs such as GA1, GA3, GA4, GA8, GA9, GA12, GA20, and GA24. When the endophytic *Paecilomyces formosus* LHL10 is evaluated under salinity stress by inoculation, it was found that the endophyte significantly enhanced vegetative parameters of cucumber as compared to uninoculated control plant and elevated production of GA3, GA4, GA12, and GA20 contents modulates the salinity stress (Khan et al. 2012). Another most highly studied example of production of plant growth regulators is the production of the enzyme 1-aminocyclopropane-1-carboxylate (ACC) deaminase, which brings down the level of plant hormone ethylene and reduces the abiotic stresses. It is reported that the elevated ethylene levels inhibit cell division, DNA synthesis, and root/shoot growth. A number of microbial endophytes have the ability to produce ACC deaminase, and thereby increased plant growth was reported (Gaiero et al. 2013). Besides IAA, GA, and ethylene, there are several other phytohormones such as abscisic acid (ABA), salicylic acid (SA), and jasmonic acid (JA) which are also responding in several abiotic stresses as a stimulus and hence act as a defense signaling substances (Shinozaki and Yamaguchi-Shinozaki 2007). Jasmonic acid (JA) induces the biosynthesis of defense proteins and defense-related secondary metabolites along with modulation of various physiological events associated with the biotic and abiotic stresses (Brodersen et al. 2006; Balbi and Devoto 2008; Lorenzo et al. 2004). Abscisic acid (ABA) modulates the stomatal closure and regulates the plant growth and development during stress conditions (Wasilewska et al. 2008; Shinozaki and Yamaguchi-Shinozaki 2007). Salicylic acid (SA) was likewise found to modulate the induction of flowering, biosynthesis of ethylene, the behavior of stomata, and respiration in several plants (Wildermuth et al. 2001; Rivas-San Vicente and Plasencia 2011; Rabe et al. 2013; Ambrose et al. 2015). Role of salicylic acid in defense is well studied; its application induces pathogenesis-related (PR) genes (Rabe et al. 2013). A novel gibberellin (GA)-producing basidiomycetous endophytic fungus *Porostereum spadiceum* AGH786 was found to be capable of producing six types of GAs besides modulating the JA and ABA production in salt-stressed soybean seedlings. It was found that endophytic fungus *Porostereum spadiceum* AGH786 inoculated seedlings were able to synthesize high levels of GAs and low ABA and JA. The finding indicates that the endophytic fungus ameliorates the effect of salinity by modulating endogenous phytohormones of the seedlings. Interestingly, in GA production when compared with the most efficient GA-producing fungus *Gibberella fujikuroi*, the endophytic fungus *Porostereum spadiceum* AGH786 was found to produce higher amounts of bioactive GA3 significantly (Hamayun et al. 2017).

15.7.1.2 Nitrogen Fixation

Nitrogen fixation by the conversion of atmospheric nitrogen to ammonia is a well-studied form of biofertilization employed by endophytic microorganism in plant growth promotion (Gaiero et al. 2013). Nitrogen (N) is the most common nutrient required for the optimal growth of the plant, and in spite of its 78% of presence in the atmosphere, its availability is limited to the plants. The availability of nitrogen to the plant could be enhanced by the application of endophytic microbes. Endophytic microbes have the capacity to convert free nitrogen into the ammonia by biological nitrogen fixation using a complex enzyme system of nitrogenase. A full 60% of the earth's total nitrogen fixation is escorted by biological nitrogen fixation (BNF) (Pandey et al. 2017). In 1986, the discovery of nitrogen-fixing endophyte *Gluconacetobacter diazotrophicus* in the sugarcane stem by Brazilian scientists draws attention of several scientists for investigations of more nitrogen-fixing endophytes such as *Herbaspirillum seropedicae*, *H. rubrisubalbicans*, *Azospirillum lipoferum*, *Klebsiella pneumoniae*, *Azorhizobium caulinodans*, *Novosphingobium sediminicola*, and *Ochrobactrum intermedium* (Cavalcante and Dobereiner 1988; Schloter et al. 1994; Muangthong et al. 2015; Moyes et al. 2016). Further, the presence of *Gluconacetobacter diazotrophicus* was also noticed in sugarcane, sweet potato, and pineapple (Silva-Froufe et al. 2009). The study of DNA sequence of nitrogenase reductase gene (nifH) from the endophytic bacterial strain isolated from the Brachiaria forage grasses was demonstrated that the nifH gene sequences were highly similar to the N2-fixing organisms (Kelemu et al. 2011). Although the sufficient work was held by the researcher to investigate the nitrogen-fixing endophytes from the sugarcane, the mechanism dealing with the proportion of fixed nitrogen contributed to plant and the level of nitrogen fixed by endophytes still remains unclear (Giller and Merckx 2003). The cocultivation of *Arabidopsis* and tobacco seedlings with endophytic fungus *Piriformospora indica* demonstrates a considerable transfer of nitrogen from the agar plates into the aerial part of the seedlings and helps in growth promotion of *Arabidopsis* and tobacco seedlings by stimulation of the NADH-dependent nitrate reductase enzyme and expression of a nitrate reductase gene (Sherameti et al. 2005).

15.7.1.3 Phosphate Solubilization

Phosphorus (P) is the second most important major nutrient after nitrogen required for the optimal growth of plants. Their role in the metabolic processes, signal transduction, macromolecular biosynthesis, photosynthesis, and respiration is well established (Khan et al. 2009). The immobilization limits the availability of phosphorous to the plants as the phosphorous rapidly forms complex with the other elements in the field (Miller et al. 2010; Gurikar et al. 2016; Edwards et al. 2016). Hence, phosphate fertilizers are needed to meet the demands of phosphorus to the developing plants (Yazdani et al. 2009; Oliveira et al. 2009). But, the addition of phosphatic fertilizers is very costly and ecologically unfair, and this problem is overcome by the application of endophytic microorganisms in environmentally friendly manner. The endophytic microorganisms provide the available phosphorous to the plants by various mechanisms (Pandey et al. 2017; Senthilkumar et al. 2009). These endophytic microorganisms are capable to secrete different types of organic acids which help in the lowering of pH in the soil delimited by the plant's root and consequently release the bound forms of phosphate in the soils for uptake of plant (Pandey et al. 2017). Among the organic acids, the gluconic acid (GA) and to a lesser extent α-ketogluconic acid are found to be the most frequent agents of inorganic phosphate solubilization (Oteino et al. 2015). The endophytic bacterial strains when inoculated into *P. sativum* L. are able to produce 14–169 mM gluconic acid (GA) which helps in the solubilization phosphate in the range of 400–1300 mg L $^{-1}$ (Oteino et al. 2015). The capacity of endophytes from the root of *Hyptismar rubioides* Epling, a *Lamiaceae*, to solubilize calcium phosphate in GELP medium and iron phosphate in the modified Reyes basal medium indicates that of the 42 endophytic isolates, 20% isolates were able to solubilize the inorganic phosphate. It is observed that endophytic strains of root zone make available the soil phosphorus to the plant, thereby increasing vegetation and improving plant growth (Vitorino et al. 2012). The phosphate-solubilizing bacterial endophytes from the meristematic tissues of strawberry promote plant growth, which is signified by the biomass accumulation and improvement in the development of other plant parts (Armando et al. 2009). The endophytic isolates from the ginseng plants were reported to solubilize phosphate, detected by extracellular solubilization of precipitated tricalcium phosphate with glucose as a sole source of carbon (Thamizhvendan et al. 2010). The *Parthenium hysterophorus* L. root endophytic fungus *Curvularia geniculata* (Tracy & Earle) Boedijn was investigated for their ability to solubilize different sources of phosphorus (P) [$AlPO_4$, $FePO_4$, and $Ca_3(PO_4)_2$] by inoculation into pigeon pea (*Cajanus cajan*) plants. It is found that the plants exhibited superior growth over uninoculated control plants and the fungus solubilized different sources of P in the order of $FePO_4 > AlPO_4 > Ca_3(PO_4)_2$ (Priyadharsini and Muthukumar 2017).

15.7.1.4 Siderophore Production

Siderophore is an iron-chelating molecule of about 1000 molecular weight and reported to be produced by many microorganisms (Logeshwaran et al. 2009). Their biosynthesis is influenced by the presence of iron, and their synthesis is suppressed if the iron is abundant in the environment (Verma et al. 2011; Loaces et al. 2011; Rungin et al. 2012). The production of siderophore is considered as very important due to the leading role of iron in the nitrogen fixation and assimilation processes (Sauvêtre and Schröder 2015; Mukherjee et al. 2017). Many endophytic microbes were reported to secrete siderophores under iron-limiting conditions (Senthilkumar et al. 2009; Logeshwaran et al. 2009). The siderophore-secreting endophytic microbes inside the plant tissues help in the transport of Fe^{3+} inside the plant cell and intensify the plant growth and productivity through the synthesis of ATP, DNA precursor, and the heme (Stintzi et al. 2000; Chincholkar et al. 2000; Logeshwaran et al. 2009; Loaces et al. 2011; Kumar et al. 2016). Its deficiency can alter the DNA synthesis and growth of cell and hamper with the several metabolic processes such as photosynthesis and the mitochondrial reactions (Chincholkar et al. 2000). The functions of siderophore in the inhibition of plant pathogens were identified (Compant et al. 2005). Siderophore production ability of endophytic bacteria *Methylobacterium* spp. was tested by chromeazurol agar assay test (CAS), Csáky test (hydroxamate type), and Arnow test (catechol type), and it was found that a total of 37 strains of *Methylobacterium* spp. were positive for chromeazurol agar assay test. *Methylobacterium* spp. was found to be producing a hydroxamate type of siderophore in the Csáky assay (Lacava et al. 2008). The production of siderophore provides competitive advantages to endophytic microbes for their successful colonization into the plant tissues and exclusion of other harmful microorganisms from the same ecological niche (Loaces et al. 2011). The endophytic bacterial isolates HKA-72 and HKA-113 were reported to synthesize Fe (III)-chelating siderophores, indicated by the formation of a yellowish-gold halo around the endophytic bacterial colonies grown on CAS agar plates. It was proposed that the siderophore production was very important for the iron supply to the growing plant as well as associated endophytes as endophytic bacteria have to also compete with the plant cell for its growth (Senthilkumar et al. 2009).

15.7.2 Biological Control

The use of chemical pesticides may be pessimistic because it develops pathogen resistance (Vinale et al. 2008), and thus, scientist focuses on the alternative means by utilizing nonchemical agents such as microorganisms in the agricultural fields to control phytopathogen (Barakat and Al-Masri 2005; Pandey et al. 2012). The promotion of plant growth and protection from phytopathogens by taking advantage of biological agents is considered as biological control or biocontrol (Fig. 15.4). Among these biological agents, the scientist gave special attention toward endophytes by taking account of their beneficial role on host plant (Firáková et al. 2007) because they are systemically distributed inside the plant and more competent to colonize the plant tissues than to phytopathogens (Rai et al. 2007). Numerous endophytes have been reported with biocontrol potential against various plant pathogens such as fungus, bacterial, viral, insect, and nematodes (Hallman et al. 1997; Ryan et al. 2008; Monteiro et al. 2017). The endophytes have been proven to prevent disease development through endophyte-mediated de novo synthesis of novel compounds (Ryan et al. 2008). Studies of the endophytic strains of novel metabolites should identify new drugs for effective treatment of disease in humans, plants, and animals (Strobel et al. 2004). The endophytic inoculant trades for the biological control in the past few decades witnessed an approximately annual growth rate of 10% worldwide (Lahlali and Hijri 2010). Host specificity and pattern of host colonization, population dynamics, potentiality to enter within host tissues, and the ability to induce systemic resistance are certain factors which accompanied the effectiveness of endophytes as biocontrol agents (Gonzalez et al. 2016; Yadav and Yadav 2017; Hu et al. 2017). A number of different mechanisms that are used by the endophytic microbes to enhance the plant growth by suppression of phytopathogens are explained by researchers (Backman and Sikora 2008; Bacilio-JimeÂnez et al. 2001; Weyens et al. 2009) which include (1) ability to synthesize chelators that effectively chelate elements necessary for growth and making it unavailable to phytopathogens; (2) capacity to synthesize antimicrobial compounds, antibiotics, cell wall-lysing enzymes, and hydrogen cyanide, which effectively suppresses the growth of phytopathogens; (3) potentiality to enhance growth of plant by effectively competing with phytopathogens for nutrition and space; (4) synthesis of phytohormones to colonize specific niches inside the plant; and (5) ability to induce systemic resistance.

15.7.2.1 Biological Control Through Fungal Endophytes

The endophytic fungi were isolated from an aromatic and medicinal plant *Monarda citriodora* Cerv. ex Lag. which showed variable degrees of antagonism in different assays against plant pathogenic fungi *Sclerotinia* sp., *Colletotrichum capsici*, *Aspergillus flavus*, and *Aspergillus fumigatus*, and this suggests that the endophytes from the plants with medicinal values could be explored as promising biocontrol agents and employed as a defensive metabolite resource (Katoch and Pull 2017). The volatile organic compounds (VOCs) of endophytic fungi *Muscodor* spp. and *Simplicillium* sp. from the *Coffea arabica* were found to be effective against the phytopathogens *Rhizoctonia solani*, *Fusarium oxysporum*, *Phoma* sp., *Botrytis cinerea*, *Fusarium solani*, *Fusarium verticillioides*, *Cercospora coffeicola*, and *Pestalotia longisetula* (Monteiro et al. 2017). Several mechanisms may be involved by endophytic fungi in biological control, in which antibiotic production is one of the chief mechanisms (Lu et al. 2000; Pandey et al. 2014). An endophytic fungus *Colletotrichum gloeosporioides* isolated from Chinese traditional herb *Artemisia Annua* was found to produce artemisinin with their fungistatic nature to the plant pathogen (Lu et al. 2000). *Phomopsis* sp. isolated from *Erythrina crista*, a medicinal plant, was reported to produce phomal, a polyketide lactone identified as anti-inflammatory metabolite (Weber et al. 2004). Endophytic fungus was also reported to have anti-insect property to the host plant against herbivorous insects through the production of toxic repellent (Pandey et al. 2014). The antimicrobial activity of endophytic fungus isolates isolated from the leaves and branches of *Ocimum* sp. (Tulsi) was assessed and found effective against pathogenic microorganisms such as *Candida albicans*, *Penicillium chrysogenum*, *Pseudomonas aeruginosa*, *Salmonella typhimurium*, and *Mycobacterium smegmatis* (Pavithra et al. 2012).

A diverse number of endophytic fungus isolates belong to the species of *Phyllosticta* spp., *Nodulisporium* spp., and *Xylaria* sp.1 isolated from *Dipterocarpus* trees which were reported to produce antimicrobial compounds with antimicrobial activities against *Staphylococcus aureus*, *Pseudomonas aeruginosa*, *Bacillus subtilis*, and *Escherichia coli* (Sutjaritvorakul et al. 2011). The endophytic fungi *Aspergillus fumigatus* CY018 and *Xylaria* were reported to inhibit *Candida albicans* (Liu et al. 2004; Boonphong et al. 2001; Pongcharoen et al. 2008). *Xylaria* sp.1 was found to have broad spectrum antimicrobial activity (Liu et al. 2008). Entomopathogenic endophytic fungi *B. bassiana* and *Clonostachys rosea* were active against the coffee berry borer (Vega et al. 2008).

15.7.2.2 Biological Control Through Bacterial Endophytes

Bacterial endophytes attribute to antagonistic activity due to their faster colonization than phytopathogen, induction of defense mechanisms, and synthesis of antimicrobial secondary metabolites. To understand the interaction and colonization of bacterial endophytes with plants and their proposed role in plant growth promotion and biological control, many studies were performed (Algam et al. 2005; Bacilio-JimeÂnez et al. 2001). The production and secretion of antibiotic, a molecule that either kills the target pathogens or inhibits their growth, is a well-known mechanism by which bacterial endophytes control the plant diseases (Compant et al. 2005). Bacterial endophytes were reported to synthesize a wide variety of antibiotics having antagonistic activity against several pathogens and commercially utilized in pharmaceutical, medical, and agricultural field (Castillo et al. 2002; Castillo et al. 2003; Ezra et al. 2004; Park et al. 2005; Wang et al. 2007). A wide spectrum antibiotic munumbicins by an endophytic *Streptomyces munumbi* isolated from the *Kennedia nigricans* and a novel antibiotic kakadumycins from the endophytic *Streptomyces* sp. NRRL 30566 of *Grevillea pteridifolia* were reported from the same author (Castillo et al. 2002; Castillo et al. 2003). The endophytic *Streptomyces* sp. isolated from *Monstera* sp. was found to produce a novel antibiotic coronamycin active against the pathogenic fungus *Cryptococcus neoformans* (Ezra et al. 2004). Endophytic bacterial strains *Corynebacterium flavescens* and *Bacillus pumilus* from rice were evaluated for their ability to enhance plant growth by successfully excluding rice pathogen *Azospirillum brasilense* (Bacilio-JimeÂnez et al. 2001). Endophytic bacterial strains of *Pseudomonas* sp., *Bacillus* sp., and *Methylobacterium* sp. isolated from different agroecosystems were isolated and screened for their nematicidal action against *Meloidogyne incognita* in host *Abelmoschus esculentus* L. (Vetrivelkalai et al. 2010). Possible mechanism of nematode control by bacterial endophytes might be due to production of secondary metabolites, e.g., 2,4-diacetylphloroglucinol and other lytic enzymes, antibiotics, hydrogen cyanide (Ahl et al. 1986; Vetrivelkalai et al. 2010), toxic metabolites such as bacillopeptidase, subtillin E, and a lactamase (Vetrivelkalai et al. 2010). Plant growth-promoting endophytic bacteria isolated from the corm and roots of banana were tested for their ability to induce systemic resistance against BBTV (Harish et al.

2008). The endophytic bacterial strains of genus *Phyllobacterium* isolated from *Epimedium brevicornum* Maxim displayed a wide spectrum antimicrobial activity against fungal and bacterial phytopathogens such as *Alternaria alternata, Sclerotinia sclerotiorum, Verticillium dahliae, Botrytis cinerea*, and *Botrytis fabae* (He et al. 2009). Endophytic bacteria trigger a phenomenon known as induced systemic resistance (ISR) which is phenotypically similar to systemic-acquired resistance (SAR), but ISR differs from SAR in that the inducing bacterium does not have any visible symptoms on the host plant (Van Loon et al. 1998; Ryan et al. 2008).

15.8 Conservation of Fungal and Bacterial Endophytic Microbial Diversity

As discussed in the previous section, the importance of endophytic microorganisms has been ever realized from the ancient period of time in the dairy industry, fermentation industry, biocontrol, production of antibiotics, and novel secondary metabolites. Each individual plant on earth is host to one or several endophytes (Strobel et al. 2004; Ryan et al. 2008). Endophytes have been isolated from surface-sterilized plant tissues or extracted from internal plant tissues (Pandey et al. 2012). Of the total plant species reported (approximately 300,000 present on the earth), each plant species carry one or more endophytes in their lifetime, and nearly all vascular plant species investigated to date were reported to harbor endophytic microorganisms (Pathak 2011). The presence of endophyte in the copper mine wasteland, agronomic crops and prairie plants, deepwater rice, grass ecosystem, pea cultivars in field condition, tropical grasses of the *Brachiaria* genus, and prairie plants has been reported earlier (Zinniel et al. 2002; Verma et al. 2001; Ryan et al. 2008). Endophytes have been reported to benefit the human welfare directly and serve as a major resource in the development of biotechnology. The conservation of endophytic microbial gene pools is an emerging and vital issue, even though development is scary. In the previous studies, conservation dealt with endophytic bacteria almost entirely addressed in context of ex situ conservation. However, obviously this strategy is quite inadequate for ensuring conservation of huge endophytic microbial diversity.

15.8.1 In Situ Conservation of Fungal and Bacterial Endophytic Microbes

In situ conservation, conservation of endophytic microbial diversity assured in their natural habitats, i.e., in different geographical locations notable by their exceptional levels of biodiversity and endemism and such locations are defined as hotspots. The copper mine wasteland, agronomic crops and prairie plants, deepwater rice, grass ecosystem, pea cultivars in field condition, tropical grasses, terrestrial and marine geothermal sites, deep ocean trenches, and polar regions are considered as one of them. Conservation of the areas with natural populations of high-valued endophytic microbes is a fundamental condition for the conservation of endophytic microbial diversity. Conservation strategies of endophytic species in agriculturally important plant genus *Zea* were studied as it became domesticated from its wild ancestors (teosinte) to modern maize (corn) and moved from Mexico to Canada. Kernels from populations of four different teosintes and ten different maize varieties were screened for endophytic bacteria by culturing, cloning, and DNA fingerprinting using terminal restriction fragment length polymorphism (TRFLP) of 16S rDNA. Analysis of data showed that there was a core microbiota of endophytes that was conserved in *Zea* seeds across boundaries of evolution, ethnography, and ecology (Johnston-Monje and Raizada 2011). *Silene paradoxa*, a well-adapted plant to extreme environments, were evaluated for their seeds, roots, and aerial plant parts associated with their endophytic bacterial communities. Molecular analysis of high-throughput sequencing of the 16S rRNA genes (microbiota) of bacterial communities isolated from seeds demonstrates that the endophytic bacterial communities were transferred to the next generation of plants as seed endophytes. Their study proves that the plants with special characters and from specific location could provide an in situ conservation of endophytic microbes (Mocali et al. 2017). In situ conservation is seen as the most appropriate means of conserving endophytic biodiversity.

An all-taxa biodiversity inventory (ATBI) has been proposed for systematic investigation and makes a record of the full diversity of living organisms, including species, number, and value of organisms from the selected habitats. The microbial ATBIs have been proposed to search new species utilized for resolving the biotechnological exploration and discovery leading to expose the exact relationship between various environments and different genotypes; to investigate the occurrence and richness of un-culturable microbes, cataloging the infrequent microbes; and to collect the information related to genotypic, phenotypic, and ecology of microbial distribution (Cannon 1995; Bull et al. 2000).

15.8.2 Ex Situ Conservation of Fungal and Bacterial Endophytic Microbes

Ex situ conservation of endophytic microbial diversity is the preservation of endophytes and its components outside their natural habitats. Here, the components refer to gene banks of endophytes, man-made wild field bank of plants with potential endophytes, in vitro plant tissues containing endophytes, artificial propagation of plants containing high-valued endophytes, maintaining botanical gardens, and most importantly endophytic microbial culture collections. Ex situ conservation involves conservation of genetic resources of endophytic species and provides insurance against extinction.

The idea of the microbial culture collection was first set up in the late nineteenth century, when the pure culture techniques for the isolation of pure cultures of microbes such as solid culture media (potato, gelatin, and agar) were introduced to the target collection, maintenance, and distribution of microbial strains among researchers and scientist, and it was considered to be a means to preserve microbial diversity ex situ (Smith 2003). The idea of the establishment of the microbial culture collection is important in the context of availability of microbial resources for fundamental research and studies. The repository of culture collection helps in understanding the mechanisms of metabolic and evolutionary patterns of valuable microorganisms as they play a fundamental role in the development and advancement of agriculture, industry, and health sectors (Sharma et al. 2017). Culture collections have provided repositories for the safe storage of live germplasm of all the microbial species along with the threatened and endangered species. It includes nucleic acid sequences and gene banks too (Sly et al. 1990). Preservation of microbes often requires different preservation methods to ensure optimal viability, storage, purity, and stability of individual microbial strains. However, application of molecular methods such as use of polymerase chain reaction (PCR) and sequence analysis of culturable microbes from the sample and analysis of biodiversity of nonculturable microbes by direct extraction of nucleic acids from the environment is a more robust method. The identity and diversity of a microbial species can be evaluated by analysis of its genetic material (16S rRNA gene).

Several repositories at international level like the World Federation for Culture Collections (WFCC, 2011) (www.wfcc.info) and European Culture Collection Organisation (ECCO, 1981) (www.eccosite.org) and at national level like Korean Agriculture Culture Collection (KACC, 1995) (genebank.rda.go.kr/eng/mic/itr/GeneInfo.do), Korean Collection for Type Cultures (KCTC, 1985) (kctc.kribb.re.kr/English/ekctc.aspx), Biological Resource Centre (BRC, 2016) (www.brc.a-star.edu.sg) at Republic of Korea, Belgium Coordinated Collection of Microorganism (BCCM, 2017) (bccm.belspo.be) at Belgium, Czech Collection of Microorganism (CCM, 2017) (www.sci.muni.cz/ccm) at Republic of Czech, Pasteur Culture Collection of Cyanobacteria (PCC, 2015) (cyanobacteria.web.pasteur.fr) at France, Microbial Type Culture Collection (MTCC, 2017) and Gene Bank (www.mtccindia.res.in) at IMTECH, Chandigarh, Indian Type Culture Collection (ITCC, 2010) (www.iari.res.in) at IARI, New Delhi, and National Agriculturally Important Microbial Culture Collection (NAIMCC, 2017) (nbaim.org.in) at ICAR-NBAIM Mau India were recognized by National Biodiversity Authority (NBA), Government of India, being functional for microbial culture collection. In spite of these recognized repositories, few repositories are being also reported which exclusively dealt with endophytic microbes. The North Carolina Arboretum Germplasm Repository (TNCAGR, 2017) founded in 2008 makes an effort to conserve the endophytic microbes from the native plants. It is noted that they were collected and conserved, more than 2000 germplasm samples of seeds and endophytes from the diverse region with the purpose of boosting their use in collaborative projects related with food and health in native region. By the use of advance technology and regional natural resources, TNCAGR identifies and produces endophytic microbial samples and extracts for researchers to discover, develop, and validate natural medicines for human health and wellness. Collected endophytic strains are found to have antibiotic, antiviral, anticancer, antioxidant, antidiabetic, immunosuppressive, and anti-insecticidal activity (ncarboretum.org). Similarly, the personnel collection repository by GRBio, 2017 (Global Registry of Biodiversity Repositories) group collected and maintained the fungal endophytes isolated from different crop and pasture species (grbio.org).

15.8.2.1 Preservation of Fungal and Bacterial Endophytic Microbes

The preservation and maintenance of endophytic microbial cultures are very essential for the study of biodiversity and systematics. Endophytic microbes have huge diversity; therefore various methods of culturing and preservation are needed to secure their morphology, physiology, viability, and genetic constituents over the time. The laborious work, cost, and time for each method must be effective. Preservation of endophytes is highly required for further studies involving their interaction with plants, their mode of action, and their utilization. However, there is scarcity of literature on the endophytic preservation methods, and there is no well-accepted specific preservation strategy for endophytes. Thus, the traditional methods or the preservation strategies generally used for common microbes can be used for endophytes preservation.

The fundamental methods of culturing for preservation are continuous growth, drying, and freezing. In continuous growth methods, the endophytic cultures are inoculated on agar for growth, and this method is used for short-term storage. These cultures either can be stored from 5 to 20 °C or can be frozen to enhance the time interval between the subculturing. The other most useful method of endophytic culture preservation is drying that is helpful to produce spores and resting structures. The materials frequently required for the drying method are silica gel, soil, and glass beads. Certain fungi have been stored up to 11 years on silica gel by drying method (Smith and Onions 1983). The drying method is technically very simple and cost-effective with less labor work. Freezing methods containing cryopreservation are also really useful in broad ranges. Certain less costly methods such as storage in distilled water are also used for preservation. These low-cost techniques are very good, but not considered as permanent preservation (Nakasone et al. 2004).

Microorganisms have the property of genetic instability. The development of new techniques for the preservation of genotype should be the highest priority. However, none of the methods ensure the complete preservation of genotype; long-term storage can be possible by lyophilization, freezing, and L-drying (Kidby 1977). The spores of AM endophytes are viable after L-drying, and the capacity of spores to survive was affected by several factors (Tommerup and Kidby 1979). Glycerol is used for freezing endophytic microbes, very hardly passes into the cell, and acts as a cryoprotectant to reduce the formation of ice crystals both intracellularly and extracellularly (Gherna 1981). The glycerol concentrations between 5 and 50% at freezing from −10 to −80 °C have been tested for various microbes, including endophytes (Kirsop and Snell 1984; Fontaine et al. 1986).

The short-term preservation by serial transfer technique is used for the maintenance of microbial and endophytic culture up to 1 year. The method is very simple, cost-effective, and widely applicable. However, it is very time taking and laborious, but the periodic transfer of culture is a good option for small collections with continuous use of short time (less than 1 year). The microbial culture must be checked time to time for contamination either by mites or by other microbes and for moisture content. The replacement of nutrient-rich media with nutrient-poor one during the transfer helps to preserve healthy cultures. Certain endophytic fungi require specific media for their preservation (Bacon 1990). The cultures are stored either at room temperature or at 4 °C after the establishment. The preserved culture must be assured that no contamination and dehydration over the time (Nakasone et al. 2004). The oil overlay over the agar slants growing endophytic cultures is a low-cost and less maintenance method. The preservation of endophytic cultures from this method is up to several years at room temperature. This method is very effective for non-sporulating endophytic microbes that are not liable to freeze-drying. One advantage of this method is that the overlaid oil reduces the mite infestations (Burdsall and Dorworth 1994). The microbial cultures grown on agar slants are covered up to 10 mm layer of autoclaved high quality of mineral oil or liquid paraffin. The test tubes are kept at room temperature in an upright position. The level of oil in culture tubes should be checked over the time (Nakasone et al. 2004). Most endophytic microbial cultures which are frozen at −20° to −80 °C in freezer stay viable. The microbial endophytic cultures grown on their respective agar slants or test tubes with screw cap bottles can be directly kept in freezer. The repeated freezing and thawing is not recommended because it reduces the viability of cultures significantly. In freezing technique the storage in liquid nitrogen is a very powerful technique of preservation for all those microbes that are not able to lyophilize. The main benefits of liquid nitrogen storage are reduction in the genetic variability of stocks, time-saving, less labor intensive to handle, reduction in repeated pathogenicity tests, prevention of contamination, and enhanced long-term viability in cultures.

Lyophilization, or freeze-drying, is a less expensive and permanent preservation method, but it does not cover all microbes like some fungi. This technique is very effective for the microbes that form numerous and small propagules. This process is effective for many spore-forming fungi that produce more numbers of spores about 10 μm or less in diameter. The dry skimmed milk powder (sterile 5 or 10% solution) and filter sterilized bovine serum are the two most common materials used during the lyophilization; however the other proteinaceous materials also can be used in place of these two. Lyophilization and freezing in glycerol are the most acceptable technique for long-term storage of endophytic microbes. Certain species of bacteria that are preserved by this method are remaining viable up to 30 years (Gherna 1981; Fontaine et al. 1986).

15.9 Conclusion

Endophytes are naturally occurring microorganisms and are an excellent source of bioactive novel compounds. They are widely exploited in pharmaceutical, medical, and agricultural industries. The use of endophytes makes higher and quicker yields of bioactive compounds in contrast to the large number of mature plants processed to achieve low yields of final products. Endophytes minimized threat of extinction of plants as compounds from endophytes are produced in vitro and in a controlled manner. They are also helpful in minimizing the labor cost and time in order to achieve valuable and large quantity of biological compounds. Fungal and bacterial endophytes have been basically ignored for their conservation. Despite of their ignorance for conservation, their role in production of natural novel compounds, secondary metabolites and industrially important natural products such as insecticides, antimicrobials, therapeutics, etc., is well studied. Secondary metabolites are as well responsible for adaptation, immune induction, and interaction with the environment. Plants associated with endophytes resist the intrusion of pathogens in their part or tissues by producing antimicrobial compounds. Fungal endophytes are the well-understood and well-investigated group of microorganisms as compared to bacterial endophytes. The discovery of novel bioactive compounds from the fungal and bacterial endophytes provides possibilities to overcome the problems associated with several medical problems such as cancer, tuberculosis, etc. Environmental problems like pollution control, biodiversity conservation, etc. are being dealt with endophytic microbes which are helpful in detoxification of industrial effluents, treatment of oil spills, and sewage treatments. With all the above concern contents, there is an increasing demand of research to study ecology, to identify and develop new competent inocula, and to investigate and explore new industrial products from endophytes with its commercial potential in life sciences.

Acknowledgments Authors are thankful to the honorable Vice-Chancellor of the Central Agricultural University for encouraging and providing the necessary facilities. The financial grant (BT/04/NE/2009) from the Department of Biotechnology (DBT) Govt. of India for the Project "Establishment of Institutional Biotechnology Hub" is gratefully acknowledged.

References

Ahl P, Voisarad C, Defago G (1986) Iron siderophores, cyanic acid and antibiotics involved in suppression of *Thielaviopsis basicola* by *Pseudomonas fluorescens* strain. J Phytopathol 116:121–134

Akbari DL, Akbari LF (2017) Detection of GFP expression and colonization of wheat by two endophytic bacteria tagged with GFP gene. Int J Pure App Biosci 5:844–848

Algam SA, Guan-lin X, Coosemans J (2005) Delivery methods for introducing endophytic *Bacillus* into tomato and their effect on growth promotion and suppression of tomato wilt. Plant Pathol J 4:69–74

Ambrose KV, Tian Z, Wang Y, Smith J, Zylstra G, Huang B, Belanger FC (2015) Functional characterization of salicylate hydroxylase from the fungal endophyte *Epichloë festucae*. Sci Rep 5:10939. https://doi.org/10.1038/srep10939

Anu Ranjan S (2012) Microbial endophytes of crop plants and their role in plant growth promotion. Ph.D. Thesis, University of Agricultural Sciences, Bangalore, India

Armando CFD, Francisco ECC, Fernando DA, Paulo TL, Manoel AT, Laura CA, Welington LA, Joao LA, Itamar SM (2009) Isolation of micro propagated strawberry endophytic bacteria and assessment of their potential for plant growth promotion. World J Microbiol Biotechnol 25:189–195

Bacilio-JimeÂnez M, Aguilar-Flores S, del Valle MV, Péreze A, Zepedae A, Zenteno E (2001) Endophytic bacteria in rice seeds inhibit early colonization of roots by *Azospirillum brasilense*. Soil Biol Biochem 33:167–172

Backman PA, Sikora RA (2008) Endophytes: an emerging tool for biological control. Biol Control 46:1–3

Bacon CW (1990) Isolation, culture and maintenance of endophytic fungi of grasses. In: Labeda DP (ed) Isolation of biotechnological organisms from nature. McGraw-Hill, New York, pp 259–282

Balbi V, Devoto A (2008) Jasmonate signaling network in *Arabidopsis thaliana*: crucial regulatory nodes and new physiological scenarios. New Phytol 177:301–318

Balsanelli E, Serrato RV, de Baura V Sassaki G, Yates MG, Rigo LU, Pedrosa FO, de Souza EM, Monteiro RA (2010) *Herbaspirillum seropedicae* rfbB and rfbC genes are required for maize colonization. Environ Microbiol 12:2233–2244

Barakat RM, Al-Masri MI (2005) Biological control of gray mold disease (*Botrytis cinerea*) on tomato and bean plants by using local isolates of *Trichoderma harzianum*. Dirasat Agric Sci 32:145–156

BCCM (2017) Belgian Coordinated Collections of Microorganisms. http://bccm.belspo.be/. Accessed 27 May 2017

Berger RG (2009) Biotechnology of flavours-the next generation. Biotechnol Lett 31:1651–1659

Biswas JC, Ladha JK, Dazzo FB Yanni YG, Rolfe BG (2000) Rhizobial inoculation influences seedlings vigor and yield of rice. Agron J 92:880–886

Bodhankar S, Grover M, Hemanth S, Reddy G, Rasul S, Yadav SK, Desai S, Mallappa M, Mandapaka M, Srinivasarao C (2017) Maize seed endophytic bacteria: dominance of antagonistic, lytic enzyme-producing *Bacillus* spp. 3 Biotech 7:1–13

Bomke C, Rojas MC, Gong F, Hedden P, Tudzynski B (2008) Isolation and characterization of the gibberellin biosynthetic gene cluster in *Sphaceloma manihoticola*. Appl Environ Microbiol 74:5325–5339

Boonphong S, Kittakoop P, Isaka M, Pittayakhajonwut D, Tanticharoen M, Thebtaranonth Y (2001) Multiplolides A and B, new antifungal 10-membered lactones from *Xylaria* multiplex. J Nat Prod 64:965–967

BRC (2016) Biological Resource Centre. https://www.brc.a-star.edu.sg/. Accessed 27 May 2017

Brodersen P, Petersen M, Nielsen HB, Zhu S, Newman MA, Shokat KM, Rietz S, Parker J, Mundy J (2006) *Arabidopsis* MAP kinase 4 regulates salicylic acid and jasmonic acid/ethylene-dependent responses via EDS1 and PAD4. Plant J 47:532–546

Bull AT, Ward AC, Goodfellow M (2000) Search and discovery strategies for biotechnology: the Paradigm shift. Microbiol Mol Biol Rev 64:573–606

Burdsall HH Jr, Dorworth EB (1994) Preserving cultures of wood-decaying Basidiomycotina using sterile distilled water in cryovials. Mycologia 86:275–280

Cannon PF (1995) An ATBI-how to find one and what to do with it. Inoculum 46:1–4

Castillo U, Harper JK, Strobel GA, Sears J, Alesi K, Ford E, Lin J, Hunter M, Maranta M, Ge H, Yaver D, Jensen JB, Porter H, Robinson R, Millar D, Hess WM, Condron M, Teplow D (2003) Kakadumycins, novel antibiotics from *Streptomyces* sp. NRRL 30566, an endophyte of *Grevillea pteridifolia*. FEMS Microbiol Lett 224:183–190

Castillo U, Strobel GA, Ford EJ, Hess WM, Porter H, Jensen JB, Albert H, Robison R, Condron MA, Teplow DB, Stevens D, Yaver D (2002) Munumbicins, wide spectrum antibiotics produced by *Streptomyces munumbi*, endophytic on *Kennedia nigriscans*. Microbiology 148:2675–2685

Cavalcante VA, Döbereiner J (1988) A new acid tolerant nitrogen-fixing bacterium associated with sugar cane. Plant Soil 108:23–31

CCM (2017) Czech Collection of Microorganism. http://www.sci.muni.cz/ccm/. Accessed 27 May 2017

Chen T, Chen Z, Ma GH, Du BH, Shen B, Ding YQ, Xu K (2014) Diversity and potential application of endophytic bacteria in ginger. Genet Mol Res 13:4918–4931

Chincholkar SB, Choudhari BL, Talegaenkar SK, Kothari RM (2000) Microbial chelators, a sustainable tool for the biocontrol of plant diseases. In: Upadhyay RK, Mukerji KG, Chamola BP (eds) Biocontrol potential and its exploitation in sustainable agriculture, Crop disease, weeds and nematodes, vol I. Kluwen Academic/Plenum, New York

Compant S, Clément C, Sessitsch A (2010) Plant growth-promoting bacteria in the rhizo- and endosphere of plants: their role, colonization, mechanisms involved and prospects for utilization. Soil Biol Biochem 42:669–678

Compant S, Duffy B, Nowak J, Clement C, Barka EA (2005) Use of plant growth promoting bacteria for biocontrol of plant diseases: principles, mechanisms of action, and future prospects. Appl Environ Microbiol 71:4951–4959

de Abreu CS, Figueiredo JEF, Oliveira CA, dos Santos VL, Gomes EA, Ribeiro VP, Barros BA, Lana UGP, Marriel IE (2017) Maize endophytic bacteria as mineral phosphate solubilizers. Genet Mol Res 16:gmr16019294

De Bary A (1866) Morphologie und Physiologie der Pilze, Flechten, und Myxomyceten, Hofmeister's handbook of physiological botany, vol II. W. Engelmann, Leipzig, Germany

De Melo FMP, Fiore MF, de Moraes LAB, Silva-Stenico ME, Scramin S, de Araújo Teixeira M, de Melo IS (2009) Antifungal compound produced by the cassava endophyte *Bacillus pumilus* MAIIIM4A. Sci Agric 66:583–592

De Weger LA, Bakker PAHM, Schippers B, van Loosdrecht MCM, Lugtenberg B (1989) *Pseudomonas* spp. with mutational changes in the O-antigenic side chain of their lipopolysaccharides are affected in their ability to colonize potato roots. In: Lugtenberg BJJ (ed) Signal molecules in plant-microbe interactions. Springer, Berlin, Germany, pp 197–202

Dekkers LC, Phoelich CC, Fits LV, Lugtenberg BJJ (1998) A site-specific recombinase is required for competitive root colonization by *Pseudomonas fluorescens* WCS365. Proc Natl Acad Sci U S A 95:7051–7056

Ding L, Maier A, Fiebig H-H, Lin W-H, Hertweck C (2011) A family of multicyclic indolosesquiterpenes from a bacterial endophyte. Org Biomol Chem 9:4029–4031

Dombrowski JE, Hollenbeck VG, Martin RC (2017) Isolation and identification of bacterial endophytes from grasses along the Oregon coast. Am J Plant Sci 8:574–601

Dorr J, Hurek T, Reinhold- Hurek B (1998) Type IV pili are involved in plant-microbe and fungus-microbe interactions. Mol Microbiol 30:7–17

ECCO (1981) European Culture Collections' Organisation. https://www.eccosite.org/. Accessed 27 May 2017

Edwards CL, Maguire RO, Alley MM, Thomason WE, Whitehurst GB (2016) Plant-available phosphorus after application of synthetic chelating agents. Commun Soil Sci Plant Anal 47:433–446

Egamberdieva D, Wirth SJ, Shurigin VV, Hashem A, Abd Allah EF (2017) Endophytic bacteria improve plant growth, symbiotic performance of chickpea (*Cicer arietinum* L.) and induce suppression of root rot caused by *Fusarium solani* under salt stress. Front Microbiol 8:1887. https://doi.org/10.3389/fmicb.2017.01887

Eyberger AL, Dondapati R, Porter JR (2006) Endophyte fungal isolates from *Podophyllum peltatum* produce podophyllotoxin. J Nat Prod 69:1121–1124

Ezra D, Castillo UF, Strobel GA, Hess W, Porter H, Jensen J, Condron M (2004) Coronamycins, peptide antibiotics produced by a verticillate *Streptomyces* sp. (MSU-2110) endophytic on *Monstera* sp. Microbiology 150:785–793

Firáková S, Šturdíková M, Múčková M (2007) Bioactive secondary metabolites produced by microorganisms associated with plants. Biologia 62:251–257

Fontaine MS, Young PH, Torrey JG (1986) Effects of long-term preservation of *Frankia* strains on infectivity, effectivity, and in vitro nitrogenase activity. Appl Environ Microbiol 51:694–698

Forchetti G, Masciarelli O, Alemano S, Alvarez D, Abdala G (2007) Endophytic bacteria in sunflower (*Helianthus annuus* L): isolation, characterization, and production of jasmonates and abscisic acid in culture medium. Appl Microbiol Biotechnol 76:1145–1152

Freeman EM (1904) The seed fungus of *Lolium temulentum* L. Phil Trans R Soc Lond (Biol) 196:1–27

Gaiero JR, Mccall CA, Thompson KA, Day NJ, Best AS, Dunfield KE (2013) Inside the root microbiome: bacterial root endophytes and plant growth promotion. Am J Bot 100:1738–1750

Gangadevi V, Muthumary J (2008) Taxol, an anticancer drug produced by an endophytic fungus *Bartalinia robillardoides* Tassi, isolated from a medicinal plant, *Aegle marmelos* Correa ex Roxb. World J Microbiol Biotechnol 24:717–724

Gangadevi V, Muthumary J (2009) Taxol production by *Pestalotiopsis terminaliae*, an endophytic fungus of *Terminalia arjuna* (arjun tree). Biotechnol Appl Biochem 52:9–15

Gautam AK (2013) Diversity of fungal endophytes in some medicinal plants of Himachal Pradesh, India. Arch Phytopathol Plant Protect 47:537–544. https://doi.org/10.1080/03235408.2013.813678

Gazis R, Chaverri P (2010) Diversity of fungal endophytes in leaves and stems of wild rubber trees (*Hevea brasiliensis*) in Peru. Fungal Ecol 3:240–254

Germaine K, Keogh E, Garcia-cabellos G (2004) Colonisation of poplar trees by gfp expressing bacterial endophytes. FEMS Microbiol Ecol 48:109–118

Gherna RL (1981) Preservation. In: Gerhardt P, Murray RGE, Costilow RN et al (eds) Manual of methods for general bacteriology. American Society for Microbiology, Washington DC, pp 208–217

Giller KE, Merckx R (2003) Exploring the boundaries of N2-fixation in cereals and grasses: a hypothetical and experimental framework. Symbiosis 35:3–17

Gonzalez F, Tkaczuk C, Dinu MM, Fiedler Z, Vidal S, Zchori-Fein E, Messelink GJ (2016) New opportunities for the integration of microorganisms into biological pest control systems in greenhouse crops. J Pest Sci 89:295–311

GRBio (2017) The global registry of biodiversity repositories. http://grbio.org/non-institutional-collection/fungal-endophytes-collection-research-group-agronomy. Accessed 27 May 2017

Gurikar C, Naik MK, Sreenivasa MY (2016) Azotobacter: PGPR activities with special reference to effect of pesticides and biodegradation. In: Singh DP, Singh HB, Prabha R (eds) Microbial inoculants in sustainable agricultural productivity: research perspectives. Springer, New Delhi

Hallman J, Quadt-Hallman A, Mahafee WF, Kloepper JW (1997) Bacterial endophytes in agricultural crops. Can J Microbiol 43:895–914

Hallmann A, Quandt-Hallmann A, Miller WG, Sikora RA, Lindow SE (2001) Endophytic colonization of plants by the biocontrol agent *Rhizobium etli* G12 in relation to *Meloidogyne incognita* infection. Phytopathology 91:415–421

Hamayun M, Hussain A, Khan SA, Kim HY, Khan AL, Waqas M, Irshad M, Iqbal A, Rehman G, Jan S, Lee IJ (2017) Gibberellins producing endophytic fungus *Porostereum spadiceum* AGH786 rescues growth of salt affected soybean. Front Microbiol 8:686. https://doi.org/10.3389/fmicb.2017.00686

Hamayun M, Khan SA, Iqbal I, Na CI, Khan AL, Hwang YH, Lee BH, Lee IJ (2009) *Chrysosporium pseudomerdarium* produces gibberellins and promotes plant growth. J Microbiol 47:425–430

Hardoim PR, Overbeek LSV, Elsas JDV (2008) Properties of bacterial endophytes and their proposed role in plant growth. Trends Microbiol 16:463–471

Hardoim PR, van Overbeek LS, Berg G, Pirttilä AM, Compant S, Campisano A, Döring M, Sessitsch A (2015) The hidden world with in plants: ecological and evolutionary considerations for defining functioning of microbial endophytes. Microbiol Mol Biol Rev 79:293–320

Harish S, Kavino M, Kumar N, Saravanakumar D, Soorianathasundaram K, Samiyappan R (2008) Biohardening with plant growth promoting rhizosphere and endophytic bacteria induces systemic resistance against banana bunchy top virus. Appl Soil Ecol 39:187–200

Harper JK, Arif AM, Ford EJ, Strobel GA, Tomer DP, Grant DM, Porco J, Oneill K (2003) Pestacin: a 1,3- dihydro isobenzofuran from *Pestalotiopsis microspora* possessing antioxidant and antimycotic activities. Tetrahedron 59:2471–2476

Hastuti RD, Lestari Y, Suwanto A, Saraswati R (2012) Endophytic *Streptomyces* spp. as biocontrol agents of rice bacterial leaf blight pathogen (*Xanthomonas oryzae* pv. *oryzae*). HAYATI J Biosci 19:155–162

He R-L, Wang G-P, Liu X-H, Zhang C-L, Lin F (2009) Antagonistic bioactivity of an endophytic bacterium isolated from *Epimedium brevicornu* Maxim. Afr J Biotechnol 8:191–195

Herrera SD, Grossi C, Zawoznik M, Groppa MD (2016) Wheat seeds harbour bacterial endophytes with potential as plant growth promoters and biocontrol agents of *Fusarium graminearum*. Microbiol Res 186–187:37–43

Hong CE, Jo SH, Moon JY, Lee J-S, Kwon S-Y, Park JM (2015) Isolation of novel leaf-inhabiting endophytic bacteria in *Arabidopsis thaliana* and their antagonistic effects on phytopathogens. Plant Biotechnol Rep 9:451–458

Hu H-J, Chen Y-L, Wang Y-F, Tang Y-Y, Chen S-L, Yan S-Z (2017) Endophytic *Bacillus cereus* effectively controls *Meloidogyne incognita* on tomato plants through rapid rhizosphere occupation and repellent action. Plant Dis 101:448–455

ITCC (2010) Indian Type Culture Collection. http://www.iari.res.in/?option=com_content&view=article&id=1251&Itemid=1809. Accessed 27 May 2017

James EK (2000) Nitrogen fixation in endophytic and associative symbiosis. Field Crop Res 65:197–209

James EK, Gyaneshwar P, Mathan N, Barraquio WL, Reddy PM, Iannetta PP, Olivares FL, Ladha JK (2002) Infection and colonization of rice seedlings by the plant growth-promoting bacterium *Herbaspirillum seropedicae* Z67. Mol Plant-Microbe Interact 15:894–906

Ji SH, Gururani MA, Chun S-C (2014) Isolation and characterization of plant growth promoting endophytic diazotrophic bacteria from Korean rice cultivars? Microbiol Res 169:83–98

Johnston-Monje D, Raizada MN (2011) Conservation and diversity of seed associated endophytes in *Zea* across boundaries of evolution, ethnography and ecology. PLoS One 6:1–22

Johri BN (2006) Endophytic to the rescue of plant. Curr Sci 90:1315–1316

KACC (1995) Korean Agricultural Culture Collection – Rural Development Administration. http://genebank.rda.go.kr/eng/mic/itr/GeneInfo.do. Accessed 27 May 2017

Karthik M, Pushpakanth P, Krishnamoorthy R, Senthilkumar M (2017) Endophytic bacteria associated with banana cultivars and their inoculation effect on plant growth. J Hortic Sci Biotechnol 92:1–9. https://doi.org/10.1080/14620316.2017.1310600

Katoch M, Pull S (2017) Endophytic fungi associated with *Monarda citriodora*, an aromatic and medicinal plant and their biocontrol potential. Pharm Biol 55:1528–1535

KCTC (1985) Korean Collection for Type Cultures. http://kctc.kribb.re.kr/English/ekctc.aspx. Accessed 27 May 2017

Kelemu S, Fory P, Zuleta C, Ricaurte J, Rao I, Lascano C (2011) Detecting bacterial endophytes in tropical grasses of the *Brachiaria* genus and determining their role in improving plant growth. Afr J Biotechnol 10:965–976

Khan AL, Hamayun M, Kang S-M, Kim YH, Jung HY, Lee JH, Lee IJ (2012) Endophytic fungal association via gibberellins and indole acetic acid can improve plant growth under abiotic stress: an example of *Paecilomyces formosus* LHL10. BMC Microbiol 12:3. https://doi.org/10.1186/1471-2180-12-3

Khan AL, Hamayun M, Kim YH, Kang SM, Lee JH, Lee IJ (2011) Gibberellins producing endophytic *Aspergillus fumigatus* sp. LH02 influenced endogenous phytohormonal levels, plant growth and isoflavone biosynthesis in soybean under salt stress. Process Biochem 46:440–447

Khan MS, Zaidi A, Wani PA, Ahemad M, Oves M (2009) Functional diversity among plant growth-promoting rhizobacteria. In: Khan MS, Zaidi A, Musarrat J (eds) Microbial strategies for crop improvement. Springer, Berlin, Heidelberg, pp 105–132

Khan R, Shahzad S, Choudhary MI, Khan SA, Ahmad A (2010) Communities of endophytic fungi in medicinal plant *Withania somnifera*. Pak J Bot 42:1281–1287

Khan SA, Hamayun M, Yoon HJ, Kim HY, Suh SJ, Hwang SK, Kim JM, Lee IJ, Choo YS, Yoon UH, Kong WS, Lee BM, Kim JG (2008) Plant growth promotion and *Penicillium citrinum*. BMC Microbiol 8:231–239

Kidby DK (1977) Culture maintenance and productivity. In: Sandford PA, Laskin A (eds) Extracellular microbial polysaccharides. American Chemical Society, Washington DC, pp 1–13

Kirsop BE, Snell JJS (eds) (1984) Maintenance of microorganisms. Academic, London

Kittiwongwattana C (2015) Biodiversity of endophytic bacteria isolated from duckweed (*Landoltia punctata*) and their IAA production. Thammasat Int J Sci Tech 20:1–11

Kogel K-H, Franken P, Huckelhoven R (2006) Endophyte or parasite- what decides? Curr Opin Plant Biol 9:358–363

Kolbas A, Kidd P, Guinberteau J, Jaunatre R, Herzig R, Mench M (2015) Endophytic bacteria take the challenge to improve Cu phytoextraction by sunflower. Environ Sci Pollut Res 22:5370–5382

Kour A, Shawl AS, Rehman S, Sultan P, Qazi PH, Suden P, Khajuria RK, Verma V (2008) Isolation and identification of an endophytic strain of *Fusarium oxysporum* producing podophyllotoxin from *Juniperus recurva*. World J Microbiol Biotechnol 24:1115–1121

Kumar A, Singh R, Yadav A, Giri DD, Singh PK, Pandey KD (2016) Isolation and characterization of bacterial endophytes of *Curcuma longa* L. 3 Biotech 6:60. https://doi.org/10.1007/s13205-016-0393-y

Kumar PMP, Ramesh S, Thipeswamy T, Sivaprasad V (2015) Antagonistic and growth promotion potential of endophytic bacteria of mulberry (*Morus* spp.). Int J Ind Entomol 31:107–114

Kumaran RS, Muthumary J, Hur BK (2008) Production of taxol from *Phyllosticta spinarum*, an endophytic fungus of *Cupressus* sp. Eng Life Sci 8:438–446

Kusari S, Lamshoft M, Spiteller M (2009) *Aspergillus fumigates* Fresenius, an endophytic fungus from *Juniperus communis* L. Horstmann as a novel source of the anticancer pro-drug deoxypodophyllotoxin. J Appl Microbiol 107:1019–1030

Kusari S, Lamshoft M, Zuhlke S, Spiteller M (2008) An endophytic fungus from *Hypericum perforatum* that produces hypericin. J Nat Prod 71:159–162

Lacava PT, Silva-Stenico ME, Araújo WL, Simionato AVC, Carrilho E, Tsai SM, Azevedo JL (2008) Detection of siderophores in endophytic bacteria Methylobacterium spp. associated with *Xylella fastidiosa* sub sp. Pauca. Pesq agropec bras, Brasília 43:521–528

Lahlali R, Hijri M (2010) Screening, identification and evaluation of potential biocontrol fungal endophytes against *Rhizoctonia solani* AG3 on potato plants. FEMS Microbiol Lett 311:152–159

Lanna-Filho R, Souza RM, Magalhães MM, Villela L, Zanotto E, Ribeiro-Júnior PM, Resende MLV (2013) Induced defense responses in tomato against bacterial spot by proteins synthesized by endophytic bacteria. Trop Plant Pathol 38:295–302

Lee J, Lobkovsky E, Pliam NB, Strobel G, Clardy J (1995) Subglutinols A & B: immunosuppressive compounds from the endophytic fungus- *Fusarium subglutinans*. J Org Chem 60:7076–7077

Li E, Jiang L, Guo L, Zhang H, Che Y (2008) Pestalachlorides A-C, antifungal metabolites from the plant endophytic fungus *Pestalotiopsis adusta*. Bioorg Med Chem 16:7894–7899

Li L, Sinkko H, Montonen L, Wei G, Lindström K, Räsänen LA (2012) Biogeography of symbiotic and other endophytic bacteria isolated from medicinal Glycyrrhiza species in China. FEMS Microbiol Ecol 79:46–68

Li Y, Wang Q, Wang L, He LY, Sheng XF (2016) Increased growth and root Cu accumulation of *Sorghum sudanense* by endophytic *Enterobacter* sp. K3-2: implications for *Sorghum sudanense* biomass production and phytostabilization. Ecotoxicol Environ Saf 124:163–168

Li Y, Cheng C, An D (2017) Characterisation of endophytic bacteria from a desert plant *Lepidium perfoliatum* L. Plant Prot Sci 53:32–43. https://doi.org/10.17221/14/2016-PPS

Liu B, Huang L, Buchenauer H, Kang Z (2010a) Isolation and partial characterization of an antifungal protein from the endophytic *Bacillus subtilis* strain EDR4. Pestic Biochem Physiol 98:305–311

Liu JY, Song YC, Zhang Z, Wang L, Guo ZJ, Zou WX, Tan RX (2004) *Aspergillus fumigatus* CY018, an endophytic fungus in *Cynodon dactylon* as a versatile producer of new and bioactive metabolites. J Biotechnol 114:279–287

Liu K, Ding X, Deng B, Chen W (2010b) 10-Hydroxycamptothecin produced by a new endophytic *Xylaria* sp., M20, from *Camptotheca acuminate*. Biotechnol Lett 32:689–693

Liu X, Dong M, Chen X, Jiang M, Lv X, Zhou J (2008) Antimicrobial activity of an endophytic *Xylaria* sp.YX-28 and identification of its antimicrobial compound 7-amino- 4-methylcoumarin. Appl Microbiol Biotechnol 78:241–247

Loaces I, Ferrando L, Scavino AF (2011) Dynamics, diversity and function of endophytic siderophore-producing bacteria in rice. Microb Ecol 61:606–618

Lodewyckx C, Vangronsveld J, Porteous F (2002) Endophytic bacteria and their potential applications. Crit Rev. Plant Sci 21:583–606

Logeshwaran P, Thangaraju M, Rajasundari K (2009) Hydroxamate siderophores of endophytic bacteria *Gluconacetobacter Diazotrophicus* isolated from sugarcane roots. Aust J Basic App Sci 3:3564–3567

Lorenzo O, Chico JM, Sánchez-Serrano JJ, Solano R (2004) Jasmonate-insensitive encodes a MYC transcription factor essential to discriminate between different jasmonate regulated defence responses in *Arabidopsis*. Plant Cell 16:1938–1950

Lu H, Zou WX, Meng JC, Hu J, Tan RX (2000) New bioactive metabolites produced by *Colletotrichum* sp., an endophytic fungus in *Artemisia annua*. Plant Sci 151:67–73

Lugtenberg B, Kamilova F (2009) Plant-growth-promoting-rhizobacteria. Annu Rev. Microbiol 63:541–556

Mastretta C, Taghavi S, van der Lelie D, Mengoni A, Galardi F, Gonnelli C, Barac T, Boulet J, Weyens N, Vangronsveld J (2009) Endophytic bacteria from seeds of *Nicotiana tabacum* can reduce cadmium phytotoxicity. Int J Phytoremediation 11:251–267

Meneses CHSG, Rouws LFM, Simoes-Araujo JL, Vidal MS, Baldani JI (2011) Exopolysaccharide production is required for biofilm formation and plant colonization by the nitrogen-fixing endophyte *Gluconacetobacter diazotrophicus*. Mol Plant-Microbe Interact 24:1448–1458

Mercado-Blanco J, Bakker PAHM (2007) Interactions between plants and beneficial Pseudomonas spp.: exploiting bacterial traits for crop protection. Antonie Van Leeuwenhoek 92:367–389. https://doi.org/10.1007/s10482-007-9167-1

Miche L, Battistoni F, Gemmer S (2006) Up regulation of jasmonate-inducible defense proteins and differential colonization of roots of *Oryza sativa* cultivars with the endophyte *Azoarcus* sp. Mol Plant-Microbe Interact 19:502–511

Miliūtė I, Buzaitė O (2011) IAA production and other plant growth promoting traits of endophytic bacteria from apple tree. Biologija 57:98–102

Miliute I, Buzaite O, Gelvonauskiene D, Sasnauskas A, Stanys V, Baniulis D (2016) Plant growth promoting and antagonistic properties of endophytic bacteria isolated from domestic apple. Zemdirbyste-Agriculture 103:77–82

Miller SH, Browne P, Prigent-Cambaret C, Combes-Meynet E, Morrissey JP, O'Gara F (2010) Biochemical and genomic comparison of inorganic phosphate solubilisation in *Pseudomonas* species. Environ Microbiol Rep 2:403–411

Mocali S, Fabiani A, Chiellini C, Gori G, Gonnelli C (2017) Endophytic and rhizospheric bacteria associated with *Silene paradoxa* grown on metal-contaminated soils are selected and transferred to the next generation of plants as seed endophytes. GRA 19:EGU2017–EGU9137

Montalbán B, Thijs S, Lobo MC, Weyens N, Ameloot M, Vangronsveld J, Pérez-Sanz A (2017) Cultivar and metal-specific effects of endophytic bacteria in Helianthus tuberosus exposed to Cd and Zn. Int J Mol Sci 18:2026. https://doi.org/10.3390/ijms18102026

Monteiro MCP, Alves NM, de Queiroz MV, Pinho DB, Pereira OL, de Souza SMC, Cardoso PG (2017) Antimicrobial activity of endophytic fungi from coffee plants. Biosci J 33:381–389

Mousa WK, Shearer CR, Limay-Rios V, Zhou T, Raizada MN (2015) Bacterial endophytes from wild maize suppress *Fusarium graminearum* in modern maize and inhibit mycotoxin accumulation. Front Plant Sci 6:805. https://doi.org/10.3389/fpls.2015.00805

Moyes AB, Kueppers LM, Pett-Ridge J, Carper DL, Vandehey N, O'Neil J, Frank AC (2016) Evidence for foliar endophytic nitrogen fixation in a widely distributed subalpine conifer. New Phytol 210:657–668

MTCC (2017) The Microbial Type Culture Collection and Gene Bank. https://www.mtccindia.res.in/. Accessed 27 May 2017

Muangthong A, Youpensuk S, Rerkasem B (2015) Isolation and characterisation of endophytic nitrogen fixing bacteria in sugarcane. Trop Life Sci Res 26:41–51

Mukherjee A, Bhattacharjee P, Das R, Pal A, Paul AK (2017) Endophytic bacteria with plant growth promoting abilities from *Ophioglossum reticulatum* L. AIMS Microbiol 3:596–612. https://doi.org/10.3934/microbiol.2017.3.596

NAIMCC (2017) National Agriculturally Important Microbial Culture Collection. http://nbaim.org.in/pages/services-culture-collectionnaimccculture-collectionnaimcc. Accessed 27 May 2017

Nakasone KK, Peterson SW, Jong SC (2004) Preservation and distribution of fungal cultures. Biodiversity of fungi: inventory and monitoring methods. Elsevier, Amsterdam, pp 37–47

Nascimento MA, Nascimento CV, Antunes JE, Figueiredo MV, Tabosa JN, Martínez CR (2014) Selection of plant growth promoting bacteria in sweet sorghum (*Sorghum bicolor* (L.) Moench) under the effects of salinity. BMC Proc 8:P113

Ngamau CN, Matiru VN, Tani A, Muthuri CW (2012) Isolation and identification of endophytic bacteria of bananas (*Musa spp.*) in Kenya and their potential as biofertilizers for sustainable banana production. Afr J Microbiol Res 6:6414–6422

Nielson TH, Christopheresen C, Anthoni U, Sorensen J (1999) Viscosinamide, a new cyclic depsipeptide with surfactant and antifungal properties produced by *Pseudomonas fluorescens* DR54. J Appl Microbiol 87:80–90

Nielson TH, Tharne C, Christophersen C, Anthoni U, Sorensen J (2000) Structure, production characteristics and fungal antagonism of tensin-a new antifungal cyclic lipopeptide from *Pseudomonas fluorescens* strain 96.578. J Appl Microbiol 89:992–1001

Oliveira CA, Alves VM, Marriel IE, Gomes EA, Scotti MR, Carneiro NP, Guimaraes CT, Schaffert RE, Sa NMH (2009) Phosphate solubilizing microorganisms isolated from rhizosphere of maize cultivated in an oxisol of the Brazilian Cerrado biome. Soil Biol Biochem 41:1782–1787

Oteino N, Lally RD, Kiwanuka S, Lloyd A, Ryan D, Germaine KJ, Dowling DN (2015) Plant growth promotion induced by phosphate solubilizing endophytic *Pseudomonas* isolates. Front Microbiol 6:745. https://doi.org/10.3389/fmicb.2015.00745

Pan D, Mionetto A, Tiscornia S, Bettucci L (2015) Endophytic bacteria from wheat grain as biocontrol agents of *Fusarium graminearum* and deoxynivalenol production in wheat. Mycotoxin Res 31:137–143

Pandey PK, Yadav SK, Singh A, Sarma BK, Mishra A, Singh HB (2012) Cross-species alleviation of biotic and abiotic stresses by the endophyte *Pseudomonas aeruginosa* PW09. J Phytopathol 160:532–539

Pandey PK, Singh AK, Singh S, Chandrakumar Singh M (2013) Bacterial Endophytes: current status and future prospects. In: Tiwari SP, Sharma R, Singh RK (eds) Recent advances in microbiology, vol 1. Nova Science, New York, pp 495–518

Pandey PK, Singh S, RNS Y, Singh AK, MCk S (2014) Fungal endophytes: promising tools for pharmaceutical science. Inter J Pharm Sci Rev Res 25:128–138

Pandey PK, MCk S, Singh AK, Singh S, Pandey AK, Pathak M, Kumar M, Shakywar RC, Patidar RK, Devi MB (2016) Arsenal of endophytic actinobacterial microbes. Int J Curr Microbiol App Sci 5:62–66

Pandey PK, MCk S, Singh S, Singh AK, Kumar M, Pathak M, Shakywar RC, Pandey AK (2017) Inside the plants: endophytic bacteria and their functional attributes for plant growth promotion. Int J Curr Microbiol App Sci 6:11–21

Pandi M, Manikandan R, Muthumary J (2010) Anticancer activity of fungal taxol derived from *Botryodiplodia theobromae* Pat., An endophytic fungus, against 7, 12 dimethyl benz(a)anthracene (DMBA)-induced mammary gland carcinogenesis in Sprague dawley rats. Biomed Pharmacother 64:48–53

Park J-H, Choi GJ, Lee HB, Kim KM, Jung HS, Lee SW, Jang KS, Cho KY, Kim J-C (2005) Griseofulvin from *Xylaria* sp. strain F0010, an endophytic fungus of *Abies holophylla* and its antifungal activity against plant pathogenic fungi. J Microbiol Biotechnol 15:112–117

Park M, Lee H, Hong SG, Kim O-S (2013) Endophytic bacterial diversity of an Antarctic moss, *Sanionia uncinata*. Antarct Sci 25:51–54

PCC (2015) Pastuer cutlure collection of cyanobacteria. Web site – Institut Pasteur. http://cyanobacteria.web.pasteur.fr/. Accessed 27 May 2017

Pathak KV (2011) Purification and characterization of antifungal compounds produced by banyan endophytic *Bacilli.* PhD Thesis, Sardar Patel University, Gujarat

Paul NC, Ji SH, Deng JX, Yu SH (2013) Assemblages of endophytic bacteria in chili pepper (*Capsicum annuum* L.) and their antifungal activity against phytopathogens in vitro. Plant Omics J 6:441–448

Pavithra N, Sathish L, Ananda K (2012) Antimicrobial and enzyme activity of endophytic fungi isolated from Tulsi. J Pharm Biomed Sci 16:1–6

Perez-Rosales E, Alcaraz-Meléndez L, Puente ME, Vázquez-Juárez R, Quiroz-Guzmán E, Zenteno-Savín T, Morales-Bojórquez E (2017) Isolation and characterization of endophytic bacteria associated with roots of jojoba (*Simmondsia chinensis* (Link) Schneid). Curr Sci 112:396–401

Petrini O (1991) Fungal endophytes in the tree leaves. In: Andrews JH, Hirano SS (eds) Microbial ecology of leaves. Springer, New York, pp 179–197

Pongcharoen W, Rukachaisirikul V, Phongpaichit S, Kühn T, Pelzing M, Sakayaro J, Taylor WC (2008) Metabolites from the endophytic fungus *Xylaria* sp. PSU-D14. Phytochemistry 69:1900–1902

Priyadharsini P, Muthukumar T (2017) The root endophytic fungus *Curvularia geniculata* from *Parthenium hysterophorus* roots improves plant growth through phosphate solubilization and phytohormone production. Fungal Ecol 27(A):69–77

Puente ME, Li CY, Bashan Y (2009) Rock-degrading endophytic bacteria in cacti. Environ Exp Bot 66:389–401

Qadri M, Johri S, Shah BA, Khajuria A, Sidiq T, Lattoo SK, Abdin MZ, Riyaz-Ul-Hassan S (2013) Identification and bioactive potential of endophytic fungi isolated from selected plants of the Western Himalayas. Springerplus 2:8. https://doi.org/10.1186/2193-1801-2-8

Rabe F, Ajami-Rashidi Z, Doehlemann G, Kahmann R, Djamei A (2013) Degradation of the plant defence hormone salicylic acid by the biotrophic fungus *Ustilago maydis*. Mol Microbiol 89:179–188

Rai R, Dash PK, Prasanna BM, Singh A (2007) Endophytic bacterial flora in the stem tissue of a tropical maize (*Zea mays* L.) genotype: isolation, identification and enumeration. World J Microbiol Biotechnol 23:853–858

Ratul N, Sharma GD, Barooah M (2013) Screening of endophytic bacterial isolates of tea (*Camellia sinensis* L.) roots for their multiple plant growth promoting activities. IJAEB 6:211–215

Reinhold-Hurek B, Maes T, Gemmer S (2006) An endoglucanase is involved in infection of rice roots by the not-cellulose-metabolizing endophyte *Azoarcus* sp. strain BH72. Mol Plant-Microbe Interact 19:181–188

Reva ON, Smirnov VV, Pattersson B, Priest FG (2002) *Bacillus* endophyticus sp. Nov isolated from the inner tissues of cotton plants (*Gossypium sp.*). Int J Syst Bacteriol 52:101–107

Rivas-San Vicente M, Plasencia J (2011) Salicylic acid beyond defence: its role in plant growth and development. J Exp Bot 62:3321–3338

Robinson RJ, Fraaije BA, Clark IM, Jackson RW, Hirsch PR, Mauchline TH (2015) Endophytic bacterial community composition in wheat (*Triticum aestivum*) is determined by plant tissue type, developmental stage and soil nutrient availability. Plant Soil 4:1–6. https://doi.org/10.1007/s11104-015-2495-4

Rodriguez RJ, White JF Jr, Arnold AE, Redman RS (2009) Tansley review; fungal endophytes: diversity and functional roles. New Phytol 182:314–330. https://doi.org/10.1111/j.1469-8137.2009.02773.x

Rogers A, Mcdonald K, Muehlbauer MF (2012) Inoculation of hybrid poplar with the endophytic bacterium Enterobacter sp. 638 increases biomass but does not impact leaf level physiology. Glob Change Biol Bioenergy 4:364–370

Romero FM, Marina M, Pieckenstain FL (2014) The communities of tomato (*Solanum lycopersicum* L.) leaf endophytic bacteria, analyzed by 16S-ribosomal RNA gene pyrosequencing. FEMS Microbiol Lett 351:187–194

Rosenblueth M, Martinez Romero E (2004) *Rhizobium etli* maize populations and their competitiveness for root colonization. Arch Microbiol 181:337–344

Rungin S, Indananda C, Suttiviriya P, Kruasuwan W, Jaemsaeng R (2012) Plant growth enhancing effects by a siderophore producing endophytic streptomycete isolated from a Thai jasmine rice plant (*Oryza sativa* L. cv. KDML105). Antonie Van Leeuwenhoek 102:463–472

Ryan RP, Germaine K, Franks A, Ryan DJ, Dowling DN (2008) Bacterial endophytes: recent developments and applications. FEMS Microbiol Lett 278:1–9

Sauvêtre A, Schröder P (2015) Uptake of carbamazepine by rhizomes and endophytic bacteria of *Phragmites australis*. Front Plant Sci 6:83. https://doi.org/10.3389/fpls.2015.00083

Schloter M, Kirchhof G, Heinzmann U, Döbereiner J, Hartmann A (1994) Immunological studies of the wheat root colonization by the *Azospirillum brasilense* strains Sp7 and Sp 245 using strain specific monoclonal antibodies. In: Hegazi N, Fayez M, Monib M (eds) Nitrogen fixation with non-legumes. The American University in Cairo Press, Cairo, Egypt, pp 290–295

Senthilkumar M, Swarnalakshmi K, Govindasamy V (2009) Biocontrol potential of soybean bacterial endophytes against charcoal rot fungus, *Rhizoctonia bataticola*. Curr Microbiol 58:288–293. https://doi.org/10.1007/s00284008-9329-z

Sessitsch A, Reiter B, Berg G (2004) Endophytic bacterial communities of field grown potato plants and their plant–growth–promoting and antagonistic abilities. Can J Microbiol 50:239–249

Sev TM, Khai AA, Aung A, Yu SS (2016) Evaluation of endophytic bacteria from some rice varieties for plant growth promoting activities. JSIR 5:144–148

Shaanker RU, Ramesha BT, Ravikanth G, Gunaga R, Vasudeva R, Ganeshaiah KN (2008) Chemical profiling of *N. nimmoniana* for camptothecin, an important anticancer alkaloid: towards the development of a sustainable production system. In: Ramawat KG, Merillion J (eds) Bioactive molecules and medicinal plants. Springer, Berlin, Germany, pp 198–210

Shankar NB, Shashikala J (2010) Diversity and structure of fungal endophytes in some climbers and grass species of Malnad region, Western Ghats, Southern India. Mycosphere 1:265–274

Sharma PK, Sarita S, Prell J (2005) Isolation and characterization of an endophytic bacterium related to *Rhizobium*/*Agrobacterium* from wheat (*Triticum aestivum* L) roots. Curr Sci 89:608–610

Sharma S, Roy S (2015) Isolation and identification of a novel endophyte from a plant *Amaranthus spinosus*. Int J Curr Microbiol App Sci 4:785–798

Sharma SK, Saini S, Verma A, Sharma PK, Lal R, Roy M, Singh UB, Saxena AK, Sharma AK (2017) National Agriculturally Important Microbial Culture Collection in the global context of microbial culture collection centres. Proc Natl Acad Sci India, Sect B Biol Sci. https://doi.org/10.1007/s40011-017-0882-8

Shcherbakov AV, Bragina AV, Kuzmina EY, Berg C, Muntyan AN, Makarova NM, Malfanova NV, Cardinale M, Berg G, Chebota VK, Tikhonovich IA (2013) Endophytic bacteria of Sphagnum mosses as promising objects of agricultural microbiology. Microbiology 82:306–315

Sherameti I, Shahollari B, Venus Y, Altschmied L, Varma A, Oelmüller R (2005) The endophytic fungus *Piriformospora indica* stimulates the expression of nitrate reductase and the starch-degrading enzyme glucan-water Dikinase in tobacco and *Arabidopsis* roots through a homeodomain transcription factor that binds to a conserved motif in their promoters. J Biol Chem 280:26241–26247

Shi Y, Lou K, Li C (2009) Promotion of plant growth by phytohormone-producing endophytic microbes of sugar beet. Biol Fertil Soils 45:645–653

Shinozaki K, Yamaguchi-Shinozaki K (2007) Gene networks involved in drought stress response and tolerance. J Exp Bot 58:221–227

Sikora RA, Schäfer K, Dababat AA (2007) Modes of action associated with microbially induced in planta suppression of plant-parasitic nematodes. Australas Plant Pathol 36:124–134

Silva-Froufe LG, Boddey RM, Reis VM (2009) Quantification of natural populations of *Gluconaceto bacterdiazo trophicus* and *Herbaspirillum* spp. in sugar cane (*Saccharum* spp.) using different polyclonal antibodies. Braz J Microbiol 40:866–878

Simons M, van der Bij AJ, de Weger LA, Wijffelman CA, Lugtenberg BJ (1996) Gnotobiotic system for studying rhizosphere colonization by plant growth-promoting *Pseudomonas* bacteria. Mol Plant-Microbe Interact 9:600–607

Singh MJ, Padmavathy S (2015) Hydrocarbon biodegradation by endophytic bacteria from neem leaves. LS: An Int J Life Sci 4:33–36

Sly LI, Lijima T, Kirsop B (eds) (1990) 100 years of culture collections proceedings of the Kral Symposium to celebrate the centenary of the first recorded service culture collection. International House, Osaka

Smith D (2003) Culture collections over the world. Int Microbiol 6:95–100

Smith D, Onions AH (1983) A comparison of some preservation techniques for fungi. Trans Br Mycol Soc 81:535–540

Souza A, Cruz JC, Sousa NR, Procópio AR, Silva GF (2014) Endophytic bacteria from banana cultivars and their antifungal activity. Genet Mol Res 12:8661–8670

Stintzi AC, Barnes JX, Raymond KN (2000) Microbial iron transport via a siderophore shuttle: a membrane ion transport paradigm. PNAS 97:10691–10696

Strobel G, Daisy B (2003) Bioprospecting for microbial endophytes and their natural products. Microbiol Mol Biol Rev 67:491–502

Strobel G, Daisy B, Castillo U, Harper J (2004) Natural products from endophytic microorganisms. J Nat Prod 67:257–268

Strobel G, Ford E, Worapong J, Harper JK, Arif AM, Grant DM, Fung PC, Ming Wah Chau R (2002) Isopestacin, an isobenzofuranone from *Pestalotiopsis microspora*, possessing antifungal and antioxidant activities. Phytochemistry 60:179–183

Strobel GA, Hess WM, Ford E, Sidhu RS, Yang X (1996) Taxol from fungal endophytes and issue of biodiversity. J Ind Microbiol 17:417–423

Strobel GA, Hess WM, Li JY, Ford E, Sears J, Sidhu RS, Summerell B (1997) *Pestalotiopsis guepinii*, a taxol producing endophyte of the Wollemi Pine, *Wollemia nobilis*. Aust J Bot 45:1073–1082

Sturz AV, Christie BR, Nowak J (2000) Bacterial endophytes: potential role in developing sustainable systems of crop production. Crit Rev. Plant Sci 19:1–30

Sutjaritvorakul T, Whalley AJS, Sihanonth P, Roengsumran S (2011) Antimicrobial activity from endophytic fungi isolated from plant leaves in Dipterocarpous forest at Viengsa district Nan province, Thailand. J Agric Tech 7:115–121

Suto M, Takebayashi M, Saito K, Tanaka M, Yokota A, Tomita F (2002) Endophytes as producers of xylanase. J Biosci Bioeng 93:88–90

Swarnalatha Y, Saha B, Lokeswara Choudary Y (2015) Bioactive compound analysis and antioxidant activity of endophytic bacterial extract from *Adhatoda beddomei*. Asian J Pharm Clin Res 8:70–72

Tenguria RK, Khan FN, Quereshi S (2011) Endophytes- mines of pharmacological therapeutics. World J Sci Tech 1:127–149

Tervet IW, Holli JP (1948) Bacteria in the storage organs of healthy plants. J Phytopathol 38:960–967

Thamizhvendan R, Yu YJ, Lee SH, Rhee YH (2010) Diversity of endophytic bacteria in ginseng and their potential for plant growth promotion. J Microbiol 48:559–565

TNCAGR (2017) The North Carolina Arboretum Germplasm Repository. http://www.ncarboretum.org/impact/germplasm-repository/. Accessed 27 May 2017

Tommerup IC, Kidby DK (1979) Preservation of spores of vesicular-arbuscular endophytes by L-drying. Appl Environ Microbiol 37:831–835

Upreti R, Thomas P (2015) Root-associated bacterial endophytes from *Ralstonia solanacearum* resistant and susceptible tomato cultivars and their pathogen antagonistic effects. Front Microbiol 6:255. https://doi.org/10.3389/fmicb.2015.00255

Van Loon LC, Bakker PA, Pieterse CMJ (1998) Systemic resistance induced by rhizosphere bacteria. Annu Rev. Phytopathol 36:453–483

Vega FE, Posada F, Aime MC, Pava-Ripoll M, Infante F, Rehner SA (2008) Entomopathogenic fungal endophytes. Biol Control 46:72–82

Verma SC, Ladha JK, Tripathi AK (2001) Evaluation of plant growth promoting and colonization ability of endophytic diazotrophs from deep water rice. J Biotechnol 91:127–141

Verma VC, Singh SK, Prakash S (2011) Bio-control and plant growth promotion potential of siderophore producing endophytic *Streptomyces* from *Azadirachta indica* A. Juss. J Basic Microbiol 51:550–556

Vetrivelkalai P, Sivakumar M, Jonathan EI (2010) Biocontrol potential of endophytic bacteria on *Meloidogyne incognita* and its effect on plant growth in bhendi. J Biopest 3:452–457

Villacieros M, Power B, S'anchez-Contreras M, Lloret J, Oraezabal RI, Martin M, Fernandez-pinas F, Bonilla I, Whelan C, Dowling DN, Rivilla R (2003) Colonization behaviour of *Pseudomonas fluorescens* and *Sinorhizobium meliloti* in the alfalfa (*Medicago sativa*) rhizosphere. Plant Soil 251:47–54

Vinale FK, Sivasithamparam EL, Ghisalberti R, Marra R, Woo SL, Lorito M (2008) *Trichoderma* plant pathogens interactions. Soil Biol Biochem 40:1–10

Vitorino LC, Silva FG, Soares MA, Souchie EL, Costa AC, Lima WC (2012) Solubilization of calcium and iron phosphate and in vitro production of Indole acetic acid by endophytic isolates of *Hyptismar rubioides* Epling (Lamiaceae). Inter Res J Biotechnol 3:47–54

Vogl AE (1898) Mehl und die anderen mehlprodukte der cerealien und leguminosen. Nahrungsm Unters Hyg Warenk 12:25–29

Wang FW, Jiao RH, Cheng AB, Tan SH, Song YC (2007) Antimicrobial potentials of endophytic fungi residing in *Quercus variabilis* and brefeldin A obtained from *Cladosporium* sp. W J Microbiol Biotech 23:79–83

Wang H, Wen K, Zhao X, Wang X, Li A, Hong H (2009) The inhibitory activity of endophytic *Bacillus* sp. strain CHM1 against plant Pathogenic fungi and its plant growth-promoting effect. Crop Prot 28:634–639

Wang X, Liang G (2014) Control efficacy of an endophytic *Bacillus amyloliquefaciens* strain BZ6-1 against peanut bacterial wilt, *Ralstonia solanacearum*. Biomed Res Int 465435:1–11

Wani ZA, Mirza DN, Arora P, Riyaz-Ul-Hassan S (2016) Molecular phylogeny, diversity, community structure, and plant growth promoting properties of fungal endophytes associated with the corms of saffron plant: an insight into the microbiome of *Crocus sativus* Linn. Fungal Biol 120:1509–1524

Wasilewska A, Vlad F, Sirichandra C, Redko Y, Jammes F, Valon C, Frei dit Frey N, Leung J (2008) An update on abscisic acid signaling in plants and more. Mol Plant 1:198–217

Weber D, Sterner O, Anke T, Gorzalczancy S, Martino V, Acevedo CJ (2004) Phomol, a new anti-inflammatory metabolite from an endophyte of the medicinal plant *Erythrina crista* galli. J Antibiot 57:559–563

Weyens N, van'derLelie D, Taghavi S, Newman L, Vangronsveld J (2009) Exploiting plant-microbe partnerships to improve biomass production and remediation. Trends Biotechnol 27:591–598

WFCC (2011) World Federation for Culture Collection. http://www.wfcc.info/. Accessed 27 May 2017

Whipps J (2001) Microbial interactions and biocontrol in the rhizosphere. J Exp Bot 52:487–511

Wildermuth MC, Dewdney J, Wu G, Ausubel FM (2001) Isochorismate synthase is required to synthesize salicylic acid for plant defence. Nature 414:562–565

Yadav A, Yadav K (2017) Exploring the potential of endophytes in agriculture: a minireview. Adv Plants Agric Res 6(4):00221. https://doi.org/10.15406/apar.2017.06.00221

Yazdani M, Bahmanyar MA, Pirdashti H, Esmaili MA (2009) Effect of phosphate solubilization microorganisms (PSM) and plant growth promoting rhizobacteria (PGPR) on yield and yield components of corn (*Zea mays* L.). World Acad Sci Eng Technol 49:90–92

Yonebayashi K, Katsumi N, Nishi T, Okazaki M (2014) Activation of nitrogen-fixing endophytes is associated with the tuber growth of sweet potato. Mass Spectrom (Tokyo) 3(1):A0032. https://doi.org/10.5702/massspectrometry.A0032

Yonghong L, Jianwei G, Li L, Asem MD, Yongguang Z, Mohamad OA, Salam N, Wenjun L (2017) Endophytic bacteria associated with endangered plant *Ferula sinkiangensis* K. M. Shen in an arid land: diversity and plant growth-promoting traits. J Arid Land 9:432–445

Zachow C, Fatehi J, Cardinale M, Tilcher R, Berg G (2010) Strain-specific colonization pattern of *Rhizoctonia* antagonists in the root system of sugar beet. FEMS Microbiol Ecol 74:124–135

Zhang XX, George A, Bailey MJ, Rainey PB (2006) The histidine utilization (hut) genes of *Pseudomonas fluorescens* SBW25 are active on plant surfaces, but are not required for competitive colonization of sugar beet seedlings. Microbiology 152:1867–1875

Zhao J, Zhou L, Wang J, Shan T, Zhong L, Liu X, Gao X (2010) Endophytic fungi for producing bioactive compounds originally from their host plants. In: Mendez-Vilas A (ed) Current research, technology and education topics in applied microbiology and microbial biotechnology. Formatex Research Center, Badajoz, Spain, pp 567–576

Zhao LF, Xu YJ, Zheng A, Lai XH (2017) Antagonistic endophytic bacteria associated with nodules of soybean (*Glycine max* L.) and plant growth-promoting properties. Braz J Microbiol 49:269–278. https://doi.org/10.1016/j.bjm.2017.06.007

Zhenhua X, Dongmei G, Xiuli S, Ying X (2012) A review of endophyte and its use and function. In: Proceedings of the International Conference on Environmental Engineering and Technology, Advances in Biomedical Engineering (ICEET'12), vol 8. Environmental Engineering and Technology, Zhangjiaji, China, pp 124–130

Zinniel DK, Lambrecht P, Harris NB, Feng Z, Kuczmarski D, Higley P, Ishimaru CA, Arunakumari A, Barletta RG, Vidaver AK (2002) Isolation and characterization of endophytic colonizing bacteria from agronomic crops and prairie plants. Appl Environ Microbiol 68:2198–2208

Chapter 16
Archaea: Ecology, Application, and Conservation

Dipak T. Nagrale and Shailesh P. Gawande

Abstract Archaea are the inhabitants of extreme environments on the earth. They commonly live at extreme acidity, temperature, and alkalinity or in hypersaline water, hot springs, hydrothermal vents, and glaciers and at extreme pressure and radiation. Some of the members live in deep oceans at extreme pressure and temperature above 100 °C. With the advancement of archaeal taxonomy, diversity, and identification of new strains, their functional role has increased in industrial and biotechnological applications in recent years. The extremophilic archaea are well-known sources of extracellular enzymes and biocatalyst and accelerate fermentation process. Some novel antimicrobial compounds and biomolecules have been discovered in certain archaea. Many of archaeal strains have applications in eco-friendly wastewater treatment plants, biodegradation of marshy lands contaminated with organic solvents, and hydrocarbons. In mineralization process, ammonia-oxidizing archaea (AOA) has key role in nitrogen cycle. The long-term preservation of extremely halophilic and thermoacidophilic archaea has been reported successful by L-drying method but it is labile to freeze and freeze-drying. Viability of thermoacidophilic archaea like *Thermoplasma* sustained at 5 °C for more than 15 years. The halophilic archaea may be preserved in the Petri dishes or in the refrigerator at 4 °C for quite longer periods with proper sealing and in deep freezing at −80 °C with specific media at proper salt concentration in 20% supplemented glycerol. In the case of hyperthermophilic archaea like *Pyrococcus furiosus*, the glass capillary tube kept over liquid nitrogen with dimethyl sulfoxide is preferred. The lyophilization method of preservation generally results in loss of viability in most of archaea cultures. Likewise, in situ methods of conservation of archaea in their natural habitats become noteworthy since most of archaea are extremophiles in those particular habitats with unique characteristics and specific traits with several applications. Hence, preservation of archaea requires specific preservation techniques for certain groups, and therefore, it is important to be focused on their maintenance, preservation, and conservation. Hence, it is very important for the

D. T. Nagrale (✉) · S. P. Gawande
Division of Crop Protection, ICAR-Central Institute for Cotton Research, Nagpur, Maharashtra, India
e-mail: Dipak.Nagrale@icar.gov.in

S. K. Sharma, A. Varma (eds.), *Microbial Resource Conservation*, Soil Biology 54,
https://doi.org/10.1007/978-3-319-96971-8_16

development of reliable, simple, and durable preservation technique for particular groups of archaea for long-term preservation with stable viability for over the years.

16.1 Introduction

The Archaea as a major domain of life was not considered for a long time. However, in the late 1970s, a team of Dr. Carl Woese at the University of Illinois was studying the relationships among the prokaryotes with the help of DNA sequences that suggested a new domain of organisms known as the Archaea. These researchers studied bacteria (prokaryotes) that are known to live at higher temperatures andproduce methane were clustered together as a separate group and placed as entirely in a new group different from those of bacteria and the eukaryotes. As there was wide genetic difference among the other domains, Dr. Carl Woese proposed that life form be divided into three domains as Eukaryota, Eubacteria, and Archaebacteria. Further, he suggested that the term Archaebacteria was a misnomer and should be restricted only to Archaea, as the organisms under this group have wide difference from others. The taxonomic classification of these three initial groups was based on basesequence studies of 16S and 18S ribosomal RNA (rRNA) molecules (Woese and Fox 1977). The word Archaea comes from the Ancient Greek thus meaning "ancient things" (http://www.merriam-webster.com/dictionary/archaea). It is considered that the Archaea originated from the common ancestor at the time of evolution and are therefore regarded as the most primitive group of organisms in the life form. The group of methanogens was in separate domain, and domain Archaea was placed in extremophiles found only in extreme habitats, i.e., hot springs, cold deserts, extreme pH, salt ponds, and hypersaline lakes. During the early twenty-first century, researchers and the microbiologists accepted that the Archaea are a large, new, and diverse group of organisms, widely distributed in nature, and are also common in non-extreme habitats, such as soils and oceans (DeLong 1998). It is seen that most of the Archaea under this domain group are highly adapted to extreme conditions and the group can be easily divided into hyperthermophiles, halophiles, and methanogens. Despite the morphological studies, the Archaea are biochemically more closely related to the Eukarya than to the Eubacteria (Bullock 2000).

16.2 Evolution of the Archaea

The Archaea are prokaryotes like bacteria and are members of the third domain of life which depicting many unique genotypic as well as phenotypic properties, testifying for their peculiar evolutionary status. It is a general perception that the archaeal ancestor was probably a hyperthermophilic anaerobe under evolutionary process (Forterre et al. 2000). Therefore, the evolution of organisms and Archaea as distinct domain is an important field in the study of evolutionary biology. Thus, new domain Archaea has a vast range of phenotypic and genotypic characters; it would be an interesting field to study the historical background of the archaea domain, so as to understand the ancestral origin and evolution of these characters. With the extensive study of data from comparative genomics, we can now sum up with more traditional or conventional phylogenetic and taxonomic approaches. It is known that more than 12 genomes of Archaea have been completely sequenced and are now available in public databases (http://www.ncbi.nlm.nih.gov/PMGifs/Genomes/). Similarly, detailed description of identified archaeal species recently has been published in new Volume I of Bergey's Manual edition (Boone and Castenholz 2001).

Furthermore, archaeal rRNA probes have been developed which are widely been used by molecular biologist and ecologists to study and investigate the worldwide distribution of the organisms of this domain, as well as its phylogenetic relatedness (Massana et al. 2000; Lopez-Garcia et al. 2001). The phylogenetic relatedness of Archaea at molecular level is well-advanced, well-documented, and well-studied through comparative genomics (Olsen and Woese 1997; Forterre 1997; Fitz-Gibbon and House 1999; Forterre and Philippe 1999; Snel et al. 1999; Tekaia et al. 1999; Makarova et al. 1999; Wolf et al. 2001). In comparative genomics and evolutionary studies, among the archaeal genomes sequenced until now, it is seen that most encoded proteins match and give first hits with other archaeal proteins when their homologues are searched in public databases. This is very much true for informational proteins (those involved in DNA replication, transcription, and translation) which are usually present in most archaeal genomes and are very close between one archaeon and another compared to those between one archaeon and any other prokaryote or eukaryote organism.

These archaeal distinctive informational proteins are usually showing more similarity to those of eukaryote than to those of prokaryote (bacteria). It is generally believed that protein information comprise the "core" of any organism's genome, since they have less probability of lateral gene transfer (LGT) and therefore are considered as more representative of the ancestral or evolutionary closeness of the organisms (Jain et al. 1999). In contrast to this, there is a well-recorded LGT between archaea and bacteria in aminoacyl-tRNA synthetases where few LGTs of informational proteins have been identified between the two prokaryotic domains (Wolf et al. 1999; Woese et al. 2000).

16.3 Taxonomy of the Archaea

On the basis of rRNA analysis, there are two major groups within Archaea: the kingdom Crenarchaeota and Euryarchaeota. The third kingdom Korarchaeota branches off close to the root (Grant and Larsen 1989; Dawson et al. 2006). A fourth kingdom Nanoarchaeota has been recently discovered in 2002 (Huber et al.

2002). More recently the new kingdom Thaumarchaeota has been proposed in 2008 and 2011 (Tourna et al. 2011; Brochier-Armanet et al. 2008).

A. Crenarchaeota

The kingdom Crenarchaeota contains organisms that live in extreme temperatures like very hot and very cold environments. Most of the members under culturable crenarchaeotes are hyperthermophiles. The members of hyperthermophilic archaea have been isolated from geothermal soils, water, or wastes containing elemental sulfur, sulfides, heavy metals, and solvents. On contrary to the hyperthermophiles, crenarchaeotes under extreme cold have been identified by the analysis of community sampling of ribosomal RNA genes from many nonthermal environments and/or habitats. The developments of fluorescent phylogenetic probes have enabled to find crenarchaeotes in marine waters worldwide. However, these marine crenarchaeotes thrive even in frigid waters, such as those of the Arctic and Antarctic. These organisms are planktonic in nature and occur in significant numbers ($\sim 10^4$/ml) in waters that are nutritionally very poor and even under very cold condition (Madigan et al. 2009). Despite the members are found in sulfur-rich hot springs, the environmental rRNA indicated that they are most abundant in marine habitats (Madigan and Martinko 2005).

B. Euryarchaeota

The kingdom Euryarchaeota consists of a wide range of ecological archaeal diversity which includes variety of characteristics group like hyperthermophiles, methanogens, halophiles, and thermophilic methanogens. Also, a large group of uncultured marine Euryarchaeotes is included in this kingdom. The members of this group are mainly separated from other archaea on the basis of rRNA gene sequences. They may be either gram positive or gram negative and differentiate on the basis of presence of pseudomurein in the cell wall. The diverse archaeal groups like methanogens are obligate anaerobes. The members under these archaeal groups are known to thrive under anaerobic environments and habitats including seawater and freshwater bodies, deep soils, intestinal tracts of animals, industrial processing plants, and sewage treatment facilities. Extremely halophilic archaea or haloarchaea are among the diverse group of prokaryotes that inhabit under hypersaline niches or environments such as crystallizer ponds, saltern pans, solar salt evaporation ponds and natural salt lakes, or artificial saline habitats such as the surfaces of heavily salted foods like certain fish, marine food products, and meats. Such habitats are often called hypersaline. Extreme halophilic archaea are mostly aerobic. These organisms require high salt concentrations for growth and development, however, in some cases near saturation point (Madigan et al. 2009). Currently widely accepted taxonomy is based on *List of Prokaryotic names with Standing in Nomenclature (LPSN), NCBI* database, and 16S rRNA-based LTP release 121 (full tree) by "The All Species LTP."

C. Korarchaeota

The 16S rRNA gene analysis revealed that phylogenetic lineage is not closely related to common archaeal groups, i.e., Crenarchaeota and Euryarchaeota, therefore, suggesting deep branching lineage (Elkins et al. 2008). The members of

Korarchaeota are only found in hot springs and hydrothermal vents, but low in numbers. They are found in habitats like iron- and sulfur-rich Yellowstone hot spring in Obsidian Pool, USA, and hot springs of Kamchatka, Russia, etc.

The Korarchaeota Kingdom of hyperthermophilic archaea is located very close to the archaeal root. It is a part of archaeal TACK superphylum comprising major archaeal groups (Guy and Ettema 2011). Therefore, the biological properties of archaea under this category reveal interesting feature of ancient organisms. The representative culturable archaea under this group have now been studied, but little knowledge is available about them except that they are obvious as hyperthermophiles growing optimally at 85 °C (Madigan et al. 2009).

D. Nanoarchaeota

The kingdom Nanoarchaeota has been discovered recently as a group of Archaea and currently having only one representative, *Nanoarchaeum equitans*. *Nanoarchaeum equitans* is a species of very tiny microbe which was discovered in 2002 in a hydrothermal vent off the coast of Iceland. It is a hyperthermophile growing in temperatures near to boiling. Further study showed that *Nanoarchaeum* appears to be an obligatory symbiont on the archaeon genus *Ignicoccus* (Huber et al. 2002). The morphological studies revealed that the cells are only 400 nm diameters in size which made it to place next to the smallest known living organism except possibly nanobacteria and nanomicrobes. Primarily the examination of single-stranded ribosomal RNA (ssrRNA) indicated a considerable difference between this group and the existing well-known kingdoms—Crenarchaeota and Euryarchaeota. On the other hand, the detailed studies related to open reading frames have suggested that the initial sample of ribosomal RNA was biased and *Nanoarchaeum* actually belongs to Euryarchaeota (Brochier 2005). The superphylum DPANN (Diapherotrites, Parvarchaeota, Aenigmarchaeota, Nanoarchaeota, Nanohaloarchaea) was proposed by Rinke et al. (2013) for extremophile archaea.

E. Thaumarchaeota

The recently established phylum of Archaea Thaumarchaeota (derived from the Greek word "*thaumas*" meaning wonder) was proposed in 2008 after whole-genome sequencing was done and has been noticed to have significant difference from other members of hyperthermophilic phylum Crenarchaeota (Brochier-Armanet et al. 2008; Tourna et al. 2011). There are also three known species in addition to *Cenarchaeum symbiosum*: *Nitrososphaera viennensis*, *Nitrososphaera gargensis*, and *Nitrosopumilus maritimus* (Brochier-Armanet et al. 2008). The organisms which closely belong to this phylum are recognized as chemolithoautotrophic ammonia oxidizers and may play significant role in biogeochemical cycles like carbon cycle and nitrogen cycle and other mineralization processes.

This new phylum was proposed in 2008 on the basis of phylogenetic data obtained from the study of the sequences of these organisms such as ribosomal RNA genes and also the presence of a form of type I topoisomerase that was previously thought to be unique to the eukaryotes only (Brochier-Armanet et al. 2008). This research result was later established by further study in 2010 through

Nitrosopumilus maritimus and *Nitrososphaera gargensis*, the genomes analysis of ammonia-oxidizing archaea (AOA), which summarized that these spp. form a distinct relatedness that includes *Cenarchaeum symbiosum*, which was the first member of the new phylum Thaumarchaeota (Spang et al. 2010).

16.4 Ecology of Archaea

Archaea are known to have existed in a broad range of habitats consisting of large part of earth's ecosystem contributing up to 20% of earth's biomass (DeLong 1998; DeLong and Pace 2001). The first discovered archaeon was grouped under extremophiles (Valentine 2007). Interestingly, members of some archaea survive in high temperatures more commonly above 100 °C, as found in extreme conditions like industrial furnace, processing plants, geysers, black smokers, and oil plants. In addition to these, other common habitats include very cold habitats such as cold deserts like Arctic, Antarctica, etc. and highly saline, acidic, or alkaline sites. Whereas, some members of archaea include mesophiles which can grow in moderate conditions like in marshy land, sewage water, the oceans, sea, and soils. Extremophile archaea are members of four major physiologicalgroups. These are the halophiles, thermophiles, alkaliphiles, and acidophiles (Pikuta et al. 2007). These groups are neither strictly restricted to specific groups only nor comprehensive or phylum-specific as some archaea belong to several groups. Despite this, they are very useful from classification point of view.

In the hypersaline habitats, archaea like halophiles which include the genus *Halobacterium* are known to live in extremely saline environments such as salt lakes and crystallizer ponds and have higher microbial population as compared to bacteria at salinities higher than 20–25% (Valentine 2007). Similarly, thermophiles grow best at temperatures above 45 °C in places such as hot springs; the members of hyperthermophilic archaea grow optimum at temperatures greater than 80 °C. Recently a specific archaeal member *Methanopyrus kandleri* strain 116 is noticed to grow at 122 °C, which is the highest registered temperature for any organism (Takai et al. 2008).

The other archaeal group exists in very acidic or alkaline conditions are acidophiles or alkaliphiles (Pikuta et al. 2007). For example, one of the most extreme archaeon acidophiles is *Picrophilus torridus*, which grows at pH = 0 and is equivalent to thriving in 1.2 M sulfuric acid (Ciaramella et al. 2005).

These properties of archaea as resistance to extreme environments have made the possibility to have extraterrestrial life (Javaux 2006). Some of these extremophile habitats are similar to those found on Mars (Nealson 1999) supporting strongly those viable microbes could have been transferred onto planets through meteorites (Davies 1996).

Despite the extreme habitats of archaea, there are several studies which have shown that archaea exist in moderate conditions like mesophilic and thermophilic environments. They are present there sometimes in high numbers at low temperatures. For example, archaea are common in cold oceanic environments such as polar seas which are known as psychrophiles (Lopez-Garcia et al. 2001). Large archaeal groups of archaea have been reported worldwide in oceans under normal habitats of plankton community of picoplankton (Karner et al. 2001). Although, these archaea may be present in extraordinarily high in numbers (approximately 40% of the microbial biomass), no single species have been isolated and studied in culturable form since they are unculturable (Giovannoni and Stingl 2005).

Similarly, our understanding of archaea as their role in ocean ecology is very limited, so their importance and role on global biogeochemical cycles remain largely unexplored (DeLong and Karl 2005). Some members of marine Crenarchaeota have potential of nitrification, thus affecting nitrogen cycle in oceans (Konneke et al. 2005). However, they may also use other energy sources (Agogue et al. 2008). Significant numbers of archaea are also observed in seafloor sediments, 1 m below ocean bottom, and resistant to extreme pressure known as piezophiles (Teske and Sorensen 2008 and Lipp et al. 2008).

16.5 Biotechnological and Industrial Applications

Extremophilic archaea, especially under thermophiles, acidophiles, or alkaliphiles, are important source of enzymes, proteins, and various metabolites that function under these extreme conditions (Breithaupt 2001; Egorova and Antranikian 2005). The enzymes from psychrophilic archaea and other psychrophiles generally are cold active and heat sensitive, which have major significance in biotechnological applications with particular activity and works at ambient temperature (Vester et al. 2015). Extremozymes from the halophiles have a great economic potential in many industrial processes, including agricultural, chemical, and pharmaceutical applications. These enzymes have multiple uses in human life as well as in the industry. Important enzymes such as DNA polymerases have been obtained from *Thermococcus littoral*, *Pyrococcus woesei*, and *P. furiosus* for their application in polymerase chain reaction (PCR) which has significant role in molecular biology (Satyanarayana et al. 2005). In the same way, Kim and Dordick (1997) reported that an extracellular protease produced by *Halobacterium halobium* has been employed for effective peptide synthesis in water/N0-N0-dimethylformamide.

Recently, a p-nitrophenylphosphate phosphatase (p-NPPase) from *Halobacterium salinarum* was used in an organic medium at very low salt concentrations after entrapping the enzyme in reversed micelles (Marhuenda-Egea et al. 2002). The archaeon *Pyrococcus furiosus* produces thermostable DNA polymerases like Pfu DNA polymerase which has transformed molecular biology through polymerase chain reaction technique, which is a simple and rapid method for DNA cloning.

In the industries, the enzymes like amylases, galactosidases, and pullulanases in some other species of *Pyrococcus* function at over 100 °C allow the food processing at very high temperatures, such as the production of low-lactose milk and whey (Synowiecki et al. 2006).

The thermophilic archaea produce many enzymes and proteins that have been recorded to thermostable in organic solvents; hence, it may be very useful in green chemistry as eco-friendly processes synthesizing organic compounds (Egorova and Antranikian 2005). Therefore, this stability makes them easier to be use in the structural biology. Also, the counterparts of prokaryotic (bacteria) or eukaryotic enzymes from extremophile archaea are frequently used in structural studies (Jenney and Adams 2008).

On contrary, the wide applications of archaea enzymes and the use of the archaea as organisms in biotechnology are not well developed. A very significant role of methanogenic archaea is in sewage treatment plant, as they are major part of the community of microorganisms that carry out anaerobic digestion of biomass and produce biogas (Schiraldi et al. 2002). In biomining or mineral processing, the acidophilic archaea showed great potential for the extraction of metals from ores including gold, cobalt, and copper (Norris et al. 2000). Most of the members of Halobacteriaceae like *Halobacterium* spp., *Haloferax mediterranei*, and *Haloferax volcanii* are known for producing extracellular protease, poly(β-hydroxybutyric acid) (PHB), bacteriorhodopsin, exopolysaccharides (EPS), etc.

One of the most important features of archaea is that they are the major host of new class of potentially useful antibiotics known as archaeocins. Many archaea have been reported to produce antimicrobials known as archaeocins, i.e., halocins and sulfolobicins, inhibiting closely related species (Aravalli et al. 1998; Prangishvili et al. 2000). A few of these archaeocins have been characterized, but many are believed to be unexplored especially within genus *Sulfolobus* (O'Connor and Shand 2002). These antibiotic compounds differ in structure from bacterial antibiotics so they may have novel modes of action, thus can be used in bacterial disease management. In addition to this, archaeal studies may allow the creation of new selectable markers for their use in archaeal molecular biology (Shand and Leyva 2008).

16.6 Methods and Approaches in Archaea Conservation

A diverse number of microbial strains and species existed, but only 1–10% were characterized, preserved, and used for several applications. Among these, very few archaea species are fully characterized, and the taxonomic position of many is newly described. The genetic resources and application in members of kingdom Archaea have not been fully utilized, and preservation methods in archaea are challenging and sophisticated as compared to other microbial conservation methods. Hence, it becomes important to conserve archaea in their natural habitats. Generally, microorganisms are conserved as "in situ," "ex situ," and "in-factory" form.

"In situ" conservation may be highly effective for halophile, acidophile, thermophile, and alkaliphile group of archaebacteria in their natural ecosystem. Whereas "ex situ" (in laboratory) conservation practices maintain and preserve isolated genetic stocks and strains/species on synthetic media and are detailed characterized by polyphasic methods. On the other hand, in industrial or commercial application, "in-factory" method of conservation for archaea is used for mass utilization of genetic resources, metabolites, and their useful traits.

16.6.1 In Situ Conservation

Archaea thrive in extreme habitats of hot springs, hydrothermal vent, hypersaline niches, cold deserts like Antarctica and Arctic, stratosphere, extreme pressure and radiations, etc. Likewise, in situ methods of conservation of archaea in their natural habitats are very important since most of the archaea are extremophiles in those particular habitats that are known for their unique characteristics and specific traits with several industrial, biotechnological, and environmental applications. Any disturbance in their natural habitat by physical, chemical, biological, and/or environmental factors will lead to loss of community abundance, genetic diversity, and any particular trait(s). Hence, it is very important to conserve archaeal microflora in their natural habitats. In absence of their specific habitats, certain group of archaea may lose those trait(s) permanently. The Archaea members are widely distributed under extremophilic environments in particular habitats on earth. Most of the hyperthermophilic archaea are radiation resistant (e.g., *Thermococcus gammatolerans* in deep-sea hydrothermal vent). The hyperthermophilic member *Methanopyrus kandleri* strain 116 can grow at 122 °C, whereas *Picrophilus torridus* is reported as extreme acidophilic microbe known to grow at a pH = 0.06. Haloarchaea are extremely halophilic aerobic with pink- to red-pigmented colonies (e.g., *Halobacterium*, *Haloarcula*, *Halorubrum*, *Haloferax*, etc.). These haloarchaea consist of bacterioruberins, carotenoids (C_{50}), bacteriorhodopsin, extracellular hydrolytic enzymes, polyhydroxyalkanoates (PHAs), etc. which have industrial application. They are found in hypersaline habitats like the Dead Sea, salt-soda lakes, salterns, subterranean-solar salts, salted foods, and coastal marshy areas (Grant et al. 2001). The haloarchaea from coastal marshy sediments can grow at lower salinities (Purdy et al. 2004). In Slovenia salterns, two groups of *Halorubrum* were highly dominated in the crystallizers. Burns et al. (2004) observed abundant square haloarchaea from brine samples collected from a crystallizer saltern pond in Geelong, Victoria, Australia.

Two methanogenic archaea have been isolated from permanently frozen Lake Fryxell, Antarctica. Out of the two clusters of methanogens detected, one was predicted to be methanotrophic.

Euryarchaeota was found in the anoxic water level above the sediment, whereas another crenarchaeote was detected just below the oxycline. They may have major role in native biogeochemical cycle, nitrification, and sulfur cycling (Howes and Smith 1990; Karr et al. 2006; Pouliot et al. 2009). Singh et al. (2005) isolated methylotrophic methanogens, *Methanococcoides alaskense* sp. nov. and

Methanosarcina baltica, from anoxic marine sediments in Skan Bay, Alaska. In polar region like an arctic ecosystem of riverine and coastal area, Euryarchaeota community was commonly associated with specific particle-rich waters; however, Crenarchaeota members are particularly reported as free-living natives of marine waters (Galand et al. 2008). Ammonia-oxidizing archaea under Crenarchaeota group in deep sea, which possibly are chemoautotrophs, have been noticed in samples at the depth of 2000–3000 m and ocean sediments (Francis et al. 2005; Nakagawa et al. 2007). Likewise, ammonia-oxidizing archaea were reported in high-altitude soils (4000–6500 m), such as Mount Everest (Zhang et al. 2009). The Antarctic soils are less dominant in archaeal community: mostly belong to Crenarchaeota (Aislabie and Bowman 2010). The polyextremophile archaea, *Sulfolobus acidocaldarius*, thrive at pH = 3 and temperature of 80 °C, which were isolated from Congress Pool, Norris Geyser Basin, Yellowstone National Park, USA. A nitrate-reducing chemoautolithotroph, *Pyrolobus fumarii* (Crenarchaeota), can grow as high as 113 °C (Blochl et al. 1997). In Japanese soils permeated with solfataric gases, the isolated extreme acidophile aerobic heterotrophs, i.e., *Picrophilus oshimae* and *Picrophilus torridus*, grow at pH 0.7 and 60 °C (Schleper et al. 1995). Similarly, acid mine drainage inhabitant of iron in California, *Ferroplasma acidarmanus* found capable to grow at pH = 0, in the presence of sulfuric acid and high concentration of heavy metals like copper, arsenic, cadmium, and zinc (Edwards et al. 2000). Cold-loving archaea in the world's oceans has significant contribution to the biomass (10^{28} cells) in the hypoxic and/or anoxic condition (Horn et al. 2003). The Archaea may act as an effective model organism for astrobiology. The haloarchaea members under Euryarchaeota are found abundantly in oil-contaminated soils and have role as in "in situ" biodegradation in particular geochemical conditions (Al-Mailem et al. 2010; Bonfa et al. 2011; Wang et al. 2011). Archaea are heterogeneous with diverse physiology. They are heterotrophic on several compounds and in combination with chemiosmosis can utilize substrate level phosphorylation (SLP) to synthesize ATP. The energy is conserved by various means. In anaerobic conditions, energy is conserved by anaerobic photorespiration, fermentation, and anaerobic respiration with nitrate by utilizing bacteriorhodopsin through sodium ion-pumping methyltransferase and proton-pumping hydrogenases (Schäfer et al. 1999; Mayer and Müller 2013). Thus, archaea community has great potential to conserve for long-term preservation in their natural habitats for their particular species or strains through adaptations. It may be targeted at particular species or entire ecosystems (Heywood and Dulloo 2005).

16.6.2 Maintenance of Archaea

The maintenance is the process of preserving a particular condition, techniques, method, or situation that is being preserved. The axenic or pure culture is maintained for all future research and references. The cultures may get contaminated from other microbes or strains. Hence, it becomes necessary to have enough stock of cultures in storage. The sufficient stock may be prepared and subcultured on specialized media with multiple replicas. It is then incubated in BOD at proper temperature to obtained proper growth. The pure culture plates are sealed with Parafilm and kept in refrigerator at 4 °C for suitable time period as per requirement and specification. The culture replicas should be kept in different preservation storage unit to ensure the safety of cultures under any adverse conditions. The stock archaea cultures may be maintained on specialized agar plates and slants with proper condition considering their taxonomic description. The maintenance of haloarchaea on agar plates and slants has limitation as crystals may be formed in medium with shrinkage. Therefore, the cultures are stored at low temperature for proper maintenance and to avoid the practice of frequent subculturing. Also, it may cause mutation in the strain or genetic instability. Therefore, stored cultures must be monitored regularly with periodic observation and if necessary subcultured them. Some culturable archaea may lose viability when kept at longer period of time especially for aerobic archaea. Thus, media should be changed at specific time interval with cultivation specification. The halophilic archaea require subculturing after 5–6 months when stored at 5 °C; however, some strains may show genetic instability by frequent subculturing at variable temperature and medium composition. Nagrale et al. (2015) stated that haloarchaea isolates were highly prone to low-temperature fluctuations with formation of salt crystals in haloarchaea agar. However, strains of *Haloarcula* spp. and *Halorubrum* spp. can be maintained at 4 °C for 2 months without crystallization of salts in haloarchaea agar (Fig. 16.1). But, strain *Haloarcula quadrata* M4 (2) showed genetic instability when preserved in haloarchaea agar slants at room temperature for short period.

Several media have been recommended for the cultivation of various genera and species of family *Halobacteriaceae* (Larsen 1981; Oren 2001a; Tindall 1992; Das Sarma et al. 1995; Rodriguez-Valera 1995). The website of the Deutsche Sammlung von Mikroorganismen und Zellkulturen GmbH (DSMZ, http://www.dmsz.de, http://www.microbiol.unimelb.edu.au/micro/staff/mds/HaloHandbook/index.html), and American Type Culture Collection (http://www.atcc.org) provides detailed information on the growth and cultivation of both halophilic Archaea.

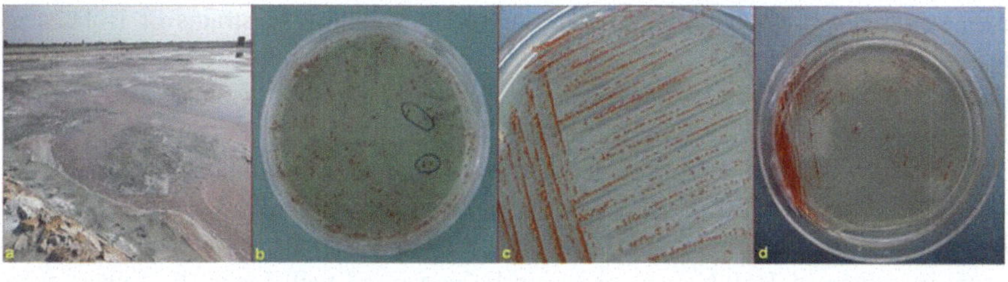

Fig. 16.1 Extremely halophilic archaea: Sampling site, colonies development and strains (**a**) Pink colouration in Sambhar salt lake, Rajasthan (India) by haloarchaea bloom (**b**) Development of pin point colonies on haloarchaea agar at 37 °C after 7 days (**c**) Colonies of strain *Haloarcula marismortui* M3(1) [GenBank accession: KJ526223] (**d**) Colonies of strain *Haloarcula argentinensis* M4(1) [GenBank accession: KJ526221]

Media used for haloarchaea differ significantly both in their salt concentration and ionic composition. The haloalkaliphilic strains are cultivated in medium (pH 9.5) with very low concentrations of divalent cations ($Mg^{2+}$and Ca^{2+}). Some bacteriological peptone (Difco) may cause disintegration of many haloarchaea. However, starch and sugars neutralize toxic effects and enhance growth of some species (Kamekura et al. 1988; Oren 1990). The members of family *Halobacteriaceae* are well suited for growth in dark condition. The haloarchaea media should be amended with antibiotics such as penicillin or ampicillin inhibiting halophilic bacteria. For the isolation and maintenance of halophilic archaea, a higher agar concentration is recommended as high salinity generally interferes with solidification of media.

Haloarchaea cultures may be maintained on agar slants at 4 °C, to be subcultured 3–6 months by specifying their taxa. Loss of character or mutation may occur due to frequent subculturing hence stored by freezing or drying method. Vacuum drying method is quite satisfactory for members of *Halobacteriaceae* and preferred by most of the microbial culture collection centers. Sakane et al. (1992) reported that L-drying has been successfully used for the preservation of certain members of *Halobacteriaceae*, especially aerobic haloarchaea. However, for anaerobic haloarchaea like *Halorhodospira*, *Ectothiorhodospira*, and some other members require special techniques.

Haloarchaea are also stored in liquid nitrogen in specialized media supplemented with DMSO 5% (w/v). It can also be stored at −60 to −80 °C by supplementing media with 10–20% glycerol (Tindall 1992; Hochstein 1988; Jones et al. 1984). *Halanaerobiales* require anaerobic techniques for growth and cultivation. Oren (2001b) suggested boiled anaerobic media amended with nitrogen (80:20) and reducing agents, i.e., cysteine, dithionite, or ascorbate. In addition to this, methanogenic archaea preserved aerobically by freeze and heat drying techniques to store at short and long periods (Bhattad 2012).

16.6.3 Method of Preservation

Different standard techniques are available for the preservation of archaea and other extremophiles. It is also suggested to submit the culturable strains or isolates at recognized culture collection center for publishing research article with proper passport data providing detailed information of the strain or isolate (Table 16.1). Most of the halophilic archaea strains can be preserved in the Petri plates or in the refrigerator at 4 °C for longer periods with suitable sealing. Some strains may lyse rapidly if stored at −20 °C in 20–50% glycerol; however, many strains survive at room temperatures up to 6 months, but this needs to be specified. It is also ported that some members of *Halobacteriaceae* can be preserved well for up to 2 years with cryoprotectants like glycerol and sucrose at −70 °C. The most successful method of haloarchaea storage is in liquid nitrogen storage tank with 15% glycerol mixed in culture media. The cultures can be stored for minimum of 15 years in quality storage tank.

Table 16.1 Archaea cultures available worldwide at different culture collection centers

Sr. no.	Name of the organization	Archaea available	Source
1	Leibniz Institute DSMZ-German Collection of Microorganisms and Cell Cultures, Inhoffenstraβe 7B 38,124 Braunschweig, Germany	500 strains	https://www.dsmz.de/contact.html
2	American Type Culture Collection (ATCC), 12,301 Parklawn Drive, Rockville, MD-20852-1776, USA, PO Box 1549, Manassas, Virginia, 20,108 1549, USA	14 (Type strains)	https://www.atcc.org
3	Japan Collection of Microorganisms (JCM), RIKEN Bio Resource Center, 3-1-1 Koyadai, Tsukuba, Ibaraki 305-0074 Japan	520 strains	http://jcm.brc.riken.jp/en/
4	National Agriculturally Important Microbial Culture Collection (NAIMCC), ICAR-National Bureau of Agriculturally Important Microorganisms, Kushmaur, Mau Nath Bhanjan-275103 (U.P.), India	17 strains	http://www.mgrportal.org.in
5	Microbial Type Culture Collection and Gene Bank, CSIR—Institute of Microbial Technology Sector 39-A, Chandigarh 160036, India	6 strains	https://www.mtccindia.res.in
6	National Centre for Microbial Resources (NCMR), National Center for Cell Science, First floor, Central Tower, Sai Trinity Building Garware Circle, Sutarwadi, Pashan Pune, Maharashtra 411021, India	16 strains	http://www.nccs.res.in

Connaris et al. (1991) evaluated method for the preservation of the hyperthermophile archaeon *Pyrococcus furiosus.* The application of glass capillary tubes kept over liquid nitrogen with dimethyl sulfoxide (DMSO) is preferred for preservation. Lyophilization techniques result in loss of viability of most of the archaea cultures. *Pyrococcus furiosus* lost viability when preserved by lyophilization. However, this technique was quite successful against hyperthermophile archaea like *Desulfurococcus* and *Thermococcus.* Jannasch et al. (1992) stated that hyperthermophiles like *Pyrococcus* spp. may be isolated from refrigerated as well as oxygenated samples in storage after 5 years.

On the basis of duration, the preservation techniques can be classified as:

- Short-term preservation
- Medium-term preservation
- Long-term preservation

16.6.3.1 Short-Term Preservation

Refrigeration

This is preferred method of storage of haloarchaea at 4 °C for only short period. The cultures can be maintained on haloarchaea agar slants or in Petri dishes for routine study. Petri dishes should be sealed properly with Parafilm to avoid any contamination and maintenance of moisture in agar media. For aerobic archaea, cotton plugs are preferred over screw-capped tubes in sterilized slants with agar media and then inoculate the culture in aseptic condition. The cultures may be transferred in new agar media with proper time period to maintain viability. Erauso et al. (1993) demonstrated that *Pyrococcus abyssi* sp. nov., a hyperthermophilic archaeon culture, can be stored at 4 °C when gas phase flushed with N_2 to remove H_2S emitted during cultivation. The cultures may be stored for a minimum of 1 year. Arab et al. (2000) stored two novel hyperthermophilic archaea (*Thermococcus aegaeicus* sp. nov. and *Staphylothermus hellenicus* sp. nov.) at 4 °C for a short period. Godfroy et al. (1996) stated that purified culture of *Thermococcus fumicolan*, a novel hyperthermophilic archaeon, was stored at 4 °C for short period of 1 year.

16.6.3.2 Medium-Term Preservation

L-Drying

L-drying or liquid state drying is a method where culture is protected from freezing. The method of drying is practiced in vacuum below 4 °C. Sakane et al. (1992) successfully used L-drying method for long-term preservation of extremely haloarchaea and thermoacidophilic archaea, labile to freeze and freeze-drying. Accelerated storage test is thus effective for estimating the stability of the dried specimens of archaea during preservation. Even the most sensitive archaea, *Thermoplasma*, survived for more than 15 years at 5 °C. Preservation of haloarchaea cultures requires skim milk "sponges" or "plugs" followed by freeze-drier.

Storing of Cultures (Dyall-Smith 2009)

1. Cells are harvested from 20 ml fresh grown culture and resuspended in a prepared solution (per 100 ml):

   Monosodium glutamate: 10 g
   Adonitol (adonite): 1.5 g
   D-Sorbitol: 2.0 g
   Sodium thioglycolate: 0.05 g
   Sodium chloride: 20 g
   Prepare the suspension in 0.1 M phosphate buffer and sterilized by filtration (0.45 μm). Maintain the pH at 7.0

2. Place 1–2 drops of resuspended culture onto a skim milk "sponge," and put in a freeze dryer for drying for half to 1 h.
3. Then, small tube should be kept in larger test tube and sealed by heating.

Opening and Revival Cultures

1. The outer test tube is opened by heating the top over burner.
2. Remove the inner glass tube, and add fresh small quantity of sterilized growth medium onto skim milk sponge or plug, and mix by a Pasteur pipette.
3. Then transfer the contents in sterilized flask with growth medium, and incubate at 37 °C in shaking incubator, and observe the growth.

16.6.3.3 Long-Term Preservation

Ultralow Freezing

Archaea cultures can be stored for several years by ultralow freezing. Ultralow temperature minimizes chemical reaction within culture. The American Type Culture Collection (ATCC) utilizes freeze-drying method for preservation of archaea including several other microorganisms. Higher methanogenic activity has been reported in freeze-dried cultures than heat-dried cultures. In limited oxygen level, higher methanogenic activity of archaea is noticed than in complete anaerobic conditions. In freeze-drying, Bhattad (2012) reported glucose as cryoprotectant, more effective for methanogenic activity compared to heat drying. The *Halohandbook* (Dyall-Smith 2009) demonstrated that haloarchaea strains can be preserved at −80 °C with 80% glycerol and 6% SW solution. Rieger et al. (1997) stated that cryofixation of hyperthermophilic archaea with very high cell densities in cellulose capillary tubes results in improved preservation of their fine structures, whereas Rengpipat et al. (1988) stated that halophilic archaea may be preserved through lyophilization or at −80 °C in anaerobic suspensions with specified media at proper salt concentration and 20% glycerol. Erauso et al. (1993) also stated that pure culture of *Pyrococcus abyssi*, a novel species of hyperthermophilic archaeon, was stored at −80 °C in anaerobic condition with growth medium supplemented with 20% (w/v) glycerol. Huber et al. (2000) demonstrated that two novel hyperthermophilic and chemolithoautotrophic *Ignicoccus* spp. stock cultures can be stored at −140 °C with 5% (v/v) DMSO over liquid nitrogen, and cultures were found viable for a minimum of 3 years. Arab et al. (2000) reported two novel species of hyperthermophilic archaea (*Thermococcus aegaeicus* and *Staphylothermus hellenicus*) which can be stored for long term in pure form at −70 °C in anaerobic condition when supplemented with cryoprotectant 5% (w/v) DMSO (Sigma). Birrien et al. (2011) stated that *Pyrococcus yayanosii*, a novel obligate hyperthermophilic piezophile archaeon, can be kept for long-term storage under anaerobic condition in 1.8 ml cryotubes at −80 °C containing cryoprotectant as DMSO 5% (v/v) (Sigma).

16.7 Conclusion and Perspectives

Several groups of archaea develop in diverse ecosystem and environments. The general classification of archaea is based on rRNA gene sequences to reveal molecular phylogenetics. Most of the culturable archaea members have two main phyla, i.e., Euryarchaeota and Crenarchaeota. The phylum Nanoarchaeota has only one member as *Nanoarchaeum equitans*. The members in other phylum Korarchaeota have generally thermophilic species, sharing characteristics of major phylum, but are close to Crenarchaeota. The culturable halophilic archaea are generally red-pigmented species that belong to Halobacteriaceae which are aerobic or facultative anaerobe. In 2006, new some smallest-sized group of archaea was detected, designated as archaeal Richmond Mine acidophilic nanoorganisms (ARMAN) consisting of Micrarchaeota and Parvarchaeota. Similarly, a superphylum TACK has been proposed including the members from major phyla. Archaea have significant role in biotechnological and biogeochemical transformation with several applications in industry, pharmaceutical, biotechnology, food, chemical industries, environmental sciences, bioremediation, and ecosystem management. Archaea preservation requires very specific preservation techniques, since they are highly specific in their cultivation parameters and, thus, need more focused research for maintenance, preservation, conservation, and cultivation. Hence, it necessitates for the development of reliable, simple, and durable preservation technique for certain groups of archaea for long-term preservation with stable viability over the years. For effective in situ conservation, the particular habitats and niche must be protected from disturbance by means of any anthropogenic and environmental factors through biodiversity regulatory agency. The specific biomolecules, secondary metabolites, genes, and biopolymers produced by archaea may play important role in industrial growth. Many species of archaea are being utilized as biological model to study extraterrestrial life. The unique feature of domain Archaea exhibits characteristic features from the other domains that continue to stimulate discussions among evolutionary biologists. Thus, culture collections or microbial repository has major role and challenges to preserve this treasure archaeal group by maintaining specific traits for utilization in research areas and industrial application. Likewise, most of the members of domain Archaea are very sensitive for isolation and maintenance in laboratories; hence, there are challenges before scientists and researchers develop modified, novel, and simple technique(s) which will be effective for long-term preservation with distinct characteristics and their maintenance for future use.

References

Agogue H, Brink M, Dinasquet J, Herndl GJ (2008) Major gradients in putatively nitrifying and non-nitrifying Archaea in the deep North Atlantic. Nature 456:788–791

Aislabie J, Bowman JP (2010) Archaeal diversity in Antarctic ecosystems. In: Bej AK, Aislabie J, Atlas RM (eds) Polar microbiology: the ecology, biodiversity and bioremediation potential of microorganisms in extremely cold environments. CRC Press, Boca Raton, FL, pp 31–61

Al-Mailem DM, Sorkhoh NA, Al-Awadhi H, Eliyas M, Radwan SS (2010) Biodegradation of crude oil and pure hydrocarbons by extreme halophilic archaea from hypersaline coasts of the Arabian Gulf. Extremophiles 14:321–328

Arab H, Volker H, Thomm M (2000) *Thermococcus aegaeicus* sp. nov. and *Staphylothermus hellenicus* sp. nov., two novel hyperthermophilic archaea isolated from geothermally heated vents off Palaeochori Bay, Milos, Greece. Int J Syst Evol Microbiol 50:2101–2108

Aravalli RN, She Q, Garrett RA (1998) Archaea and the new age of microorganisms. Trends Ecol Evol 13:190–194

Bhattad UH (2012) Preservation of methanogenic cultures to enhance anaerobic digestion. Dissertations (2009). Paper 209. http://epublications.marquette.edu/dissertations_mu/209

Birrien JL, Zeng X, Jebbar M, Cambon-Bonavita MA, Quérellou J, Oger P, Bienvenu N, Xiao X, Prieur D (2011) *Pyrococcus yayanosii* sp. nov., an obligate piezophilic hyperthermophilic archaeon isolated from a deep-sea hydrothermal vent. Int J Syst Evol Microbiol 61:2827–2831

Blochl E, Rachel R, Burggraf S, Hafendradl D, Jannasch HW, Stetter KO (1997) *Pyrolobus fumarii*, gen. and sp. nov., represents a novel group of archaea, extending the upper temperature limit for life to 113 °C. Extremophiles 1:14–21

Bonfa MRL, Grossman MJ, Mellado E, Durrant LR (2011) Biodegradation of aromatic hydrocarbons by Haloarchaea and their use for the reduction of the chemical oxygen demand of hypersaline petroleum produced water. Chemosphere 84:1671–1676

Boone DR, Castenholz RW (2001) Bergey's manual of systematic bacteriology, vol 1, 2nd edn. Springer, New York, Berlin, Heidelberg

Breithaupt H (2001) The hunt for living gold. The search for organisms in extreme environments yields useful enzymes for industry. EMBO Rep 2:968–971

Brochier C (2005) Nanoarchaea: representatives of a novel archaeal phylum or a fast evolving euryarchaeal lineage related to *Thermococcales*. Genome Biol 6:R42

Brochier-Armanet C, Boussau B, Gribaldo S, Forterre P (2008) Mesophilic crenarchaeota: proposal for a third archaeal phylum, the Thaumarchaeota. Nat Rev Microbiol 6:245–252

Bullock C (2000) The Archaea- a biochemical perspective. Biochem Mol Biol Edu 28:186–191

Burns DG, Camakaris HM, Janssen PH, Dyall-Smith ML (2004) Cultivation of Walsby's square haloarchaeon. FEMS Microbiol Lett 238:469–473

Ciaramella M, Napoli A, Rossi M (2005) Another extreme genome: how to live at pH0. Trends Microbiol 13:49–51

Connaris H, Cowan D, Ruffett M, Sharp RJ (1991) Preservation of the hyperthermophile *Pyrococcus furiosus*. Lett Appl Microbiol 13:25–27

DasSarma S, Fleischmann EM, Rodriguez-Valera F (1995) Appendix 2: media for halophiles. In: DasSarma S, Fleischmann EM (eds) Archaea. A laboratory manual. Halophiles. Cold Spring Harbor Laboratory Press, Cold Spring Harbor, pp 225–230

Davies PC (1996) The transfer of viable microorganisms between planets. Ciba Found Symp 202:304–314

Dawson S, DeLong EF, Pace NR (2006) Phylogenetic and ecological perspectives on uncultured Crenarchaeota and Korarchaeota. In: Dworkin M, Falkow S, Rosenberg E, Schleifer K-H, Stackebrandt E (eds) The prokaryotes, vol 3, 3rd edn. Springer, New York

DeLong EF (1998) Everything in moderation: Archaea as 'non-extremophiles. Curr Opin Genet Dev 8:649–654

DeLong EF, Pace NR (2001) Environmental diversity of bacteria and Archaea. Syst Biol 50:470–478

DeLong EF, Karl DM (2005) Genomic perspectives in microbial oceanography. Nature 437:33642

Dyall-Smith M (2009) The halohandbook – protocols for haloarchaeal genetics. Available online at: http://www.haloarchaea.com/resources/halohandbook/index.html.(Accessioned on 02.11.2017)

Edwards KJ, Bond PL, Gihring TM, Banfield JF (2000) An archaeal iron-oxidizing extreme acidophile important in acid mine drainage. Science 287:1796–1799

Egorova K, Antranikian G (2005) Industrial relevance of thermophilic Archaea. Curr Opin Microbiol 8:649–655

Elkins JG, Podar M, Graham DE, Makarova KS, Wolf Y, Randau L, Hedlundg BP, Brochier-Armanet C, Kunin V, Anderson I, Lapidus A, Goltsman E, Barry K, Koonin EV, Hugenholtz P, Kyrpides N, Wanner G, Richardson P, Keller M, Stetter KO (2008) A korarchaeal genome reveals insights into the evolution of the Archaea. Proc Nat Acad Sci USA 105:8102–8107

Erauso G, Reysenbach AL, Godfroy A, Meunier JR, Crump B, Partensky F, Baross JA, Marteinsson V, Barbier G, Pace NR, Prieur D (1993) *Pyrococcus abyssi* sp. nov., a new hyperthermophilic archaeon isolated from a deep-sea hydrothermal vent. Arch Microbiol 160:338–349

Fitz-Gibbon ST, House CH (1999) Whole genome-based phylogenetic analysis of free-living microorganisms. Nucleic Acids Res 27:4218–4222

Forterre P (1997) Archaea: what can we learn from their sequences? Curr Opin Genet Dev 7:764–770

Forterre P, Philippe H (1999) Where is the root of the universal tree of life? Bioassay 21:871–879

Forterre P, Bouthier De La Tour C, Philippe H, Duguet M (2000) Reverse gyrase from hyperthermophiles: probable transfer of a thermoadaptation trait from Archaea to bacteria. Trends Genet 16:152–154

Francis CA, Roberts KJ, Beman JM, Santoro AE, Oakley BB (2005) Ubiquity and diversity of ammonia-oxidizing Archaea in water columns and sediments of the ocean. Proc Natl Acad Sci U S A 102:14683–14688

Galand PE, Lovejoy C, Pouliot J, Vincent WF (2008) Heterogeneous archaeal communities in the particle-rich environment of an arctic shelf ecosystem. J Marine Syst 74:774–782

Giovannoni SJ, Stingl U (2005) Molecular diversity and ecology of microbial plankton. Nature 427:343–348

Godfroy A, Meunier JR, Guezennec J, Lesongeur F, Raguenes G, Rimbault A, Barbier G (1996) *Thermococcus fumicolans* sp. nov., a new hyperthermophilic archaeon isolated from a deep-sea hydrothermal vent in the North Fiji basin. Int J Syst Bacteriol 46:1113–1119

Grant WD, Larsen H (1989) Group 111. Extremely halophilic archaebacteria, Order Halobacteriales ord. nov. In Bergey's Manual of Systematic Bacteriology, vol. 3, pp. 2216-2233. J. T. Staley, M. P. Bryant, N. Pfennig and J. G. Holt. Baltimore, MD: Williams & Wilkins

Grant WD, Kamekura M, McGenity TJ, Ventosa A (2001) Class III. *Halobacteria* class. nov. In Bergey's Manual of Systematic Bacteriology, 2nd edn, vol. 1, pp. 294–301. D. R. Boone, R. W. Castenholz and G. M. Garrity. New York: Springer

Guy L, Ettema TJG (2011) The archaeal 'TACK' superphylum and the origin of eukaryotes. Trends Microbiol 19:580–587

Heywood VH, Dulloo ME (2005) In situ conservation of wild plant species: a critical global review of good practices. IPGRI Technical Bulletin No. 11. International Plant genetic Resources Institute, Rome, Italy

Hochstein LI (1988) The physiology and metabolism of the extremely halophilic bacteria. In: Rodriguez-Valera F (ed) Halophilic bacteria, vol II. CRC Press, Boca Raton, pp 67–83

Horn MA, Matthies C, Kusel K, Schramm A, Drake HL (2003) Hydrogenotrophic methanogenesis by moderately acid-tolerant methanogens of a methane-emitting acidic peat. Appl Environ Microbiol 69:74–83

Howes BL, Smith RL (1990) Sulfur cycling in a permanently ice covered amictic Antarctic lake, Lake Fryxell. Ant J US 25:230–233

Huber R, Huber H, Stetter KO (2000) Towards the ecology of hyperthermophiles: biotopes, new isolation strategies and novel metabolic properties. FEMS Microbiol Rev 24:615–623

Huber H, Hohn MJ, Rachel R, Fuchs T, Wimmer VC, Stetter KO (2002) A new phylum of Archaea represented by a nanosized hyperthermophilic symbiont. Nature 417:63–67

Jain R, Rivera MC, Lake JA (1999) Horizontal gene transfer among genomes: the complexity hypothesis. Proc Natl Acad Sci USA 96:3801–3806

Jannasch HW, Wirsen CO, Molyneaux SJ, Langworthy TA (1992) Comparative physiological studies on hyperthermophilic Archaea isolated from deep-sea hot vents with emphasis on *Pyrococcus* strain GB-D. Appl Environ Microbiol 58:3472–3481

Javaux EJ (2006) Extreme life on Earth—past, present and possibly beyond. Res Microbiol 157:37–48

Jenney FE, Adams MW (2008) The impact of extremophiles on structural genomics and vice versa. Extremophiles 12:39–50

Jones D, Pell PA, Sneath PHA (1984) Maintenance of bacteria on glass beads at −60 °C to −70 °C. In: Kirsop BE, Snell JSS (eds) Maintenance of microorganisms. A manual of laboratory methods. Academic, London, pp 35–40

Kamekura M, Oesterhelt D, Wallace R, Anderson P, Kushner DJ (1988) Lysis of halobacteria in bactopeptone by bile acids. Appl Environ Microbiol 54:990–995

Karner MB, DeLong EF, Karl DM (2001) Archaeal dominance in the mesopelagic zone of the Pacific Ocean. Nature 409:507–510

Karr EA, Ng JM, Belchik SM, Sattley WM, Madigan MT, Achenbach LA (2006) Biodiversity of methanogenic and other archaea in the permanently frozen Lake Fryxell, Antarctica. Appl Environ Microbiol 72:1663–1666

Kim J, Dordick JS (1997) Unusual salt and solvent dependence of a protease from an extreme halophile. Biotechnol Bioeng 55:471–479

Konneke M, Bernhard AE, de la Torre JR, Walker CB, Waterbury JB, Stahl DA (2005) Isolation of an autotrophic ammonia-oxidizing marine archaeon. Nature 437:543–546

Larsen H (1981) The family Halobacteriaceae. In: Starr MP, Stolp H, Truper HG, Balows A, Schlegel HG (eds) The prokaryotes. A handbook on habitats, isolation and identification of bacteria, vol I. Springer, Berlin, pp 985–994

Lipp JS, Morono Y, Inagaki F, Hinrichs KU (2008) Significant contribution of Archaea to extant biomass in marine subsurface sediments. Nature 454:991–994

Lopez-Garcia P, Lopez-Lopez A, Moreira D, Rodriguez-Valera F (2001) Diversity of free-living prokaryotes from a deep-sea site at the Antarctic Polar Front. FEMS Microbiol Ecol 3:193–202

Madigan MT, Martinko JM (2005) Brock biology of microorganisms (11 edn). Pearson, San Francisco, CA, p. 136. isbn:0-13-196893-9

Madigan MT, Martinko JM, Dunlap PV, Clark DP (eds) (2009) Brock biology of microorganisms, 12th edn. Pearson Benjamin Cummings, San Francisco

Makarova KS, Aravind L, Galperin MY, Grishin NV, Tatusov RL, Wolf YI, Koonin EV (1999) Comparative genomics of the Archaea (Euryarchaeota): evolution of conserved protein families, the stable core, and the variable shell. Genome Res 9:608–628

Marhuenda-Egea FC, Piere-Velazquez S, Cadenas C, Cadenas E (2002) An extreme halophilic enzyme active at low salt in reversed micelles. J Biotechnol 93:159–164

Massana R, DeLong EF, Pedros-Alio C (2000) A few cosmopolitan phylotypes dominate planktonic archaeal assemblages in widely different oceanic provinces. Appl Environ Microbiol 66:1777–1787

Mayer F, Müller V (2013) Adaptations of anaerobic archaea to life under extreme energy limitation. FEMS Microbiol Rev 38:449–472. https://doi.org/10.1111/1574-6976.12043

Nagrale DT, Renu and Sharma AK (2015) Preservation and maintenance of extremely halophilic archaea at low temperature. In: International Conference on "Low Temperature Science and Biotechnological Advances" during 27–30th April 2015), NASC Complex, Pusa Campus, New Delhi-110 012, India (Poster in theme session: Microbial storage for biotechnological application, pp.111

Nakagawa T, Mori K, Kato C, Takahashi R, Tokuyama T (2007) Distribution of cold-adapted ammonia-oxidizing microorganisms in the deep ocean of the Northeastern Japan Sea. Microbes Environ 22:365–372

Nealson KH (1999) Post-Viking microbiology: new approaches, new data, new insights. Orig Life Evol Biosph 29:73–93

Norris PR, Burton NP, Foulis NA (2000) Acidophiles in bioreactor mineral processing. Extremophiles 4:71–76

O'Connor EM, Shand RF (2002) Halocins and sulfolobicins: the emerging story of archaeal protein and peptide antibiotics. J Ind Microbiol Biotechnol 28:23–31

Olsen GJ, Woese CR (1997) Archaeal genomics: an overview. Cell 89:991–994

Oren A (1990) Starch counteracts the inhibitory action of bacto-peptone and bile salts in media for the growth of halobacteria. Can J Microbiol 36:299–301

Oren A (2001a) The order Haloanaerobiales. In: Dworkin M, Falkow S, Rosenberg E, Schleifer KH, Stackebrandt E (eds) The prokaryotes. A handbook on the biology of bacteria: ecophysiology, isolation, identification, applications, third edn. Springer, New York (electronic publication)

Oren A (2001b) The order Halobacteriales. In: Dworkin M, Falkow S, Rosenberg E, Schleifer KH, Stackebrandt E (eds) The prokaryotes. A handbook on the biology of bacteria: ecophysiology, isolation, identification, applications, 3rd edn. Springer, New York (electronic publication)

Pikuta EV, Hoover RB, Tang J (2007) Microbial extremophiles at the limits of life. Crit Rev Microbiol 33:183–209

Pouliot J, Ganland PE, Lovejoy C, Vincent WF (2009) Vertical structure of archaeal communities and the distribution of ammonia monooxygenase A gene variants in two meromictic High Arctic lakes. Environ Microbiol 11:687–699

Prangishvili D, Holz I, Stieger E, Nickell S, Kristjansson JK, Zillig W (2000) Sulfolobicins, specific proteinaceous toxins produced by strains of the extremophilic archaeal genus *Sulfolobus*. J Bacteriol 182:2985–2988

Purdy KJ, Cresswell-Maynard TD, Nedwell DB, Mc Genity TJ, Grant WD, Timmis KN, Embley TM (2004) Isolation of haloarchaea that grow at low salinities. Environ Microbiol 6:591–595

Rengpipat S, Langworthy TA, Zeikus JG (1988) *Halobacteroides acetoethylicus* sp. nov., a new obligately anaerobic halophile isolated from deep subsurface hypersaline environments. Syst Appl Microbiol 11:28–35

Rieger G, Müller K, Hermann R, Stetter KO, Rachel R (1997) Cultivation of hyperthermophilic archaea in capillary tubes resulting in improved preservation of fine structures. Arch Microbiol 168:373–379

Rinke C, Schwientek P, Sczyrba A, Ivanova NN, Anderson IJ, Cheng JF, Dodsworth JA (2013) Insights into the phylogeny and coding potential of microbial dark matter. Nature 499:431–437

Rodriguez-Valera F (1995) Cultivation of halophilic Archaea. In: DasSarma S, Fleischmann EM (eds) Archaea. A laboratory manual. Halophiles. Cold Spring Harbor Laboratory Press, Cold Spring Harbor, pp 13–16

Sakane T, Fukuda I, Itoh T, Yokota A (1992) Long-term preservation of halophilic archaebacteria and thermoacidophilic archaebacteria by liquid drying. J Microbiol Methods 16:281–287

Satyanarayana T, Raghukumar C, Shivaji S (2005) Extremophilic microbes: diversity and perspectives. Curr Sci 89:78–90

Schäfer G, Engelhard M, Müller V (1999) Bioenergetics of the archaea. Microbiol Mol Biol Rev 63:570–620

Schiraldi C, Giuliano M, De Rosa M (2002) Perspectives on biotechnological applications of archaea. Archaea 1:75–86

Schleper C, Puehler G, Holz I, Gambacorta A, Janekovic D, Santarius U, Klenk HP, Zillig W (1995) *Picrophilus* gen. nov., fam. nov.: a novel aerobic, heterotrophic, thermoacidophilic genus and family comprising archaea capable of growth around pH0. J Bacteriol 177:7050–7059

Shand RF, Leyva KJ (2008) Archaeal antimicrobials: an undiscovered country. In: Blum P (ed) *Archaea:* new models for prokaryotic biology. Caister Academic Press, Norfolk. isbn:978-1-904455-27-1

Singh N, Kendall MM, Liu Y, Boone DR (2005) Isolation and characterization of methylotrophic methanogens from anoxic marine sediments in Skan Bay, Alaska: description of Methanococcoides alaskense sp. nov., and emended description of Methanosarcina baltica. Int J Syst Evol Microbiol 55:2531–2538

Snel B, Bork P, Huynen MA (1999) Genome phylogeny based on gene content. Nat Genet 21:108–110

Spang A, Hatzenpichler R, Brochier-Armanet C, Rattei T, Tischler P, Spieck E, Streit W, Stahl DA, David A, Wagner M, Schleper C (2010) Distinct gene set in two different lineages of ammonia-oxidizing archaea supports the phylum Thaumarchaeota. Trends Microbiol 18:331–340

Synowiecki J, Grzybowska B, Zdziebło A (2006) Sources, properties and suitability of new thermostable enzymes in food processing. Crit Rev Food Sci Nutr 46:197–205

Takai K, Nakamura K, Toki T, Tsunogai U, Miyazaki M, Miyazaki J, Hirayama H, Nakagawa S, Nunoura T, Horikoshi K (2008) Cell proliferation at 122°C and isotopically heavy CH_4 production by a hyperthermophilic methanogen under high-pressure cultivation. Proc Natl Acad Sci USA 105:10949–10954

Tekaia F, Lazcano A, Dujon B (1999) The genomic tree as revealed from whole proteome comparisons. Genome Res 9:550–557

Teske A, Sorensen KB (2008) Uncultured archaea in deep marine subsurface sediments: have we caught them all? ISME J 2:3–18

Tindall BJ (1992) The family Halobacteriaceae. In: Balows A, Truper HG, Dworkin M, Harder W, Schleifer KH (eds) The prokaryotes. A handbook on the biology of bacteria: ecophysiology, isolation, identification, applications, vol I. Springer, New York, pp 768–808

Tourna M, Stieglmeier M, Spang A, Konneke M, Schintlmeister A, Urich T, Engel M, Schloter M, Wagner M, Richter A, Schleper C (2011) *Nitrososphaera viennensis*, an ammonia oxidizing archaeon from soil. Proc Natl Acad Sci U S A 108:8420–8425

Valentine DL (2007) Adaptations to energy stress dictate the ecology and evolution of the Archaea. Nat Rev. Microbiol 5:316–323

Vester JK, Glaring MA, Stougaard P (2015) Improved cultivation and metagenomics as new tools for bioprospecting in cold environments. Extremophiles 19:17–29

Wang X, Han Z, Bai Z, Tang J, Ma A, He J, Zhuang G (2011) Archaeal community structure along a gradient of petroleum contamination in saline-alkali soil. J Enviro Sci 23:1858–1864

Woese CR, Fox GE (1977) Phylogenetic structure of the prokaryotic domain: the primary kingdoms. Proc Natl Acad Sci USA 74:5088–5090

Woese CR, Olsen GJ, Ibba M, Soll D (2000) Aminoacylt RNA synthetases, the genetic code and the evolutionary process. Microbiol Mol Biol Rev 64:202–236

Wolf YI, Aravind L, Grishin NV, Koonin EV (1999) Evolution of amino-acyl tRNA synthetases—analysis of unique domain architectures and phylogenetic trees reveals a complex history of horizontal gene transfer events. Genome Res 9:689–710

Wolf YI, Rogozin IB, Grishin NV, Tatusov RL, Koonin EV (2001) Genome trees constructed using five different approaches suggest new major bacterial clades. BMC Evol Biol 1:8. https://doi.org/10.1186/1471-2148-1-8

Zhang LM, Wang M, Prosser JI, Zheng YM, He JZ (2009) Altitude ammonia-oxidizing bacteria and archaea in soils of Mount Everest. FEMS Microbiol Ecol 70:208–217

The manufacturer's authorised representative in the EU is Springer Nature Customer Service Centre GmbH, Europaplatz 3, 69115 Heidelberg, Germany. If you have any concerns regarding our products, please contact ProductSafety@springernature.com

Printed and bound by CPI Group (UK) Ltd, Croydon, CR0 4YY
05/10/2026
02228147-0005